INSIDE ARC/INFO®

Revised Edition

Michael Zeiler

INSIDE ARC/INFO®

Revised Edition
By Michael Zeiler

Published by
OnWord Press
2530 Camino Entrada
Santa Fe, NM 87505-4835 USA

SAN 694-0269
First Edition, 1994
Revised Edition, 1997

10 9 8 7 6 5 4 3 2 1

Printed in the United States of America

Cataloging-in-Publication Data

Michael Zeiler
INSIDE ARC/INFO
Includes index.

1. ARC/INFO (computer software) 2. Geographic Information Systems I. Title

93-85027
ISBN 1-56690-111-1

Trademarks

Warning and Disclaimer

About the Author

Michael Zeiler is a MapObjects product specialist with Environmental Systems Research Institute (ESRI) in Redlands, California. Prior to joining ESRI, Michael was an ARC/INFO programmer with seven years of application development experience in Santa Fe, New Mexico.

Michael earned a B.A. in Physics from the University of California at Berkeley in 1982. Although quantum mechanics has no obvious relation to GIS, this education taught Michael to think abstractly, an indispensable skill for software development.

Michael began as a CAD programmer for Diginetics, Inc. in Albuquerque, New Mexico, in 1984 and wrote mapping software integrated with civil engineering analysis. There, he learned the fundamentals of cartography and the discipline of application development.

Next, Michael began Envision's GIS applications in 1987. By that time, he had learned about the advantages of GIS for combining analysis with maps. He evaluated the available GIS software and selected ARC/INFO. Since then, Michael and his colleagues have developed a suite of GIS applications tailored for the electric utility industry, which are the Electric GIS, Distribution Analysis/GIS, System Inventory/GIS, and the GPS/GIS Bridge.

In his spare time, Michael enjoys classical music, playing piano, silk-screening, and time outdoors with his family.

Acknowledgments

Writing this book was an unexpectedly large venture and would have been impossible but for the assistance of many people.

This book is inspired by a remarkable company, Environmental Systems Research Institute. What is most striking to an outside observer about the people at ESRI is their collective passion for making a difference in the world. With the leadership of Jack Dangermond, they are implementing GIS as a technology to solve many real problems in the world.

In my work and while writing this book, I have received insight and assistance from many friends within ESRI. I'd like to acknowledge the developers of ArcTools and ArcView, from whose work I borrowed many concepts for user interface and application design. I'd like to specifically thank Bill Miller, Gary Kabot, Randy Worch, and Jack Horton for their careful technical review of this book and Ron Schroeder for expediting an early release of version 7 of ARC/INFO to me.

Next, I'd like to acknowledge the special support I received from my employer, Envision Utility Software Corporation. I was encouraged by my managers, Gary Thompson and John Kernaghan, and was allowed access to Envision's computer hardware and ARC/INFO software licenses. I'd like to also acknowledge the users at Envision's customer utilities who over the years have set me straight on just what makes a user interface usable.

The GIScity applications contained on the companion CD-ROM are original and were written expressly for this book, but they borrow much of their design from Envision's GIS applications. ARC/INFO users at electric utilities will find Envision GIS applications to be a rich superset of the functions in the GIScity applications, tailored for utility needs.

The GIS data on the companion CD-ROM and illustrated throughout the book is courtesy of the Planning Division of the City of Santa Fe, an ARC/INFO user. I'd like to express gratitude to Gar Clarke for his enthusiastic support of this project and exceeding my expectations in providing a superior sample GIS dataset for this book.

For graciously reviewing this book, I'd like to acknowledge Irvin Goldblatt, W. Lynn Shirley, and Gar Clarke. Their effort positively influenced this book and led to a chapter reorganization that improved the flow of the book.

I'd like to thank Laura Sanchez and Margaret Burns, editors at High Mountain Press, for patiently guiding me through this process and ensuring the quality of the book. Also, Dan Raker and David Talbott for giving me the crazy idea that I could write a book. Perhaps I'll do it again (but not right away!).

Michael Zeiler
October 1994

Book Production

This book was produced in Ventura for Windows 4.1.1 by Michelle Mann. The cover design is by Lynne Egensteiner, using QuarkXpress and Aldus FreeHand. Additional software used during the creation of this book includes ARC/INFO and Hijaak Pro for Windows.

OnWord Press

OnWord Press is dedicated to the fine art of professional documentation.

In addition to the author, members of the OnWord Press team contributed to making this book.

First Edition
Dan Raker, President
Kate Hayward, Publisher
Gary Lange, Associate Publisher
David Talbott, Acquisitions Editor
Frank Conforti, Managing Editor
Margaret Burns, Project Editor
Laura Sanchez, Project Editor
Carol Leyba, Production Manager
Michelle Mann, Production Editor
Janet Dick, Marketing Manager
John Tollett, Illustrator
Lynne Egensteiner, Cover designer, Illustrator
Kate Bemis, Indexer
Bob Leyba and Barbara Kohl, Production Assistance

Revised Edition
Dan Raker, Publisher and President
Gary Lange, Vice-President
Janet Leigh Dick, Associate Publisher and Director of Channel Marketing
David Talbott, Director of Acquisitions
Daniel Clavio, Director of Market Development and Strategic Relations
Rena Rully, Senior Manager, Editorial and Production
Daril Bentley, Project Editor
Carol Leyba, Sr. Production Manager
Cynthia Welch, Production Editor
Lynne Egensteiner, Cover designer, Illustrator

To Elizabeth, my wife. Thank you for your love, support and strength. This is your achievement as well as mine. To Petra, my daughter. May you have many dreams and attain one of them.

Table of Contents

Introduction

Who Should Read This Book

Congratulations on starting with ARC/INFO and for choosing this book. ARC/INFO is the premier Geographic Information Systems (GIS) software package available from Environmental Systems Research Institute, Inc. (ESRI). This book will help validate your organization's good judgment in choosing ARC/INFO.

INSIDE ARC/INFO is for the professional who needs to become productive quickly with ARC/INFO. You will find that ARC/INFO is an exceptionally powerful analytic tool, but with that power comes a degree of complexity.

When your company or agency first selected ARC/INFO, you probably received an impressive demonstration of the potential of GIS. You may have seen applications such as the following:

- ❑ Locating the optimal location of a new store based on consumer demographics, the proximity of competitors, and driving time and distance.
- ❑ Mapping the progression of an oil spill and analyzing the effects on marine animals and coastline ecology.
- ❑ Generating a mailing list of all property owners living within a 500-foot radius of a parcel requesting a zoning variance.
- ❑ Natural resource management using gorgeous visual displays that integrate aerial or satellite images with map data.

Let's face it: A well-presented ARC/INFO demonstration is like a magic show. More than most other types of computer software, GIS is visually exciting. Also, GIS is being used to integrate diverse information throughout organizations for the simple reason that most information ties in some way to a place on the earth. Because of these factors, and the skill of the ESRI marketing staff, it is very easy for expectations to be set very high within an organization that is just adopting ARC/INFO. Your management has gotten a briefing on the potential of GIS, but may not fully realize that creating that magic is going to take some hard work.

You might be experiencing a scenario like this:

After an ARC/INFO demonstration, your mind is reeling with the possibilities. You imagine your data integrated within the GIS, and you extrapolate what you saw in demonstration into cutting-edge methods for your field. Your management makes a substantial investment in computer hardware and GIS software and assigns you the task of realizing this amazing technology for your organization.

However, when you open the boxes containing your ARC/INFO distribution, you find a number of manuals and several CD-ROMs with your software release. As you read the installation guide and load the software, you realize that the flashy demonstration you saw is not there. Instead, you discover over 2000 ARC/INFO commands that you must somehow piece together into an application for your organization. Your stomach begins to turn as you wonder how you will ever satisfy your organization's expectations; you just know that wizardry is expected of you.

Relax. You won't become a "GIS ninja" or build your ARC/INFO applications overnight, but you can get to that level in a reasonable time.

Learning ARC/INFO doesn't have to be Rocket Science or GIS 101. *INSIDE ARC/INFO* will help you in these ways:

- ❑ You will immediately get started with an ARC/INFO application called GIScity with real GIS data from the city of Santa Fe, New Mexico.
- ❑ The most important commands in ARC/INFO will be identified, explained, and graphically depicted for you.
- ❑ The inner logic of ARC/INFO, which might take a year or more of experience to solidify in your mind, will be revealed and explained in plain English.
- ❑ The key to productivity with ARC/INFO is to develop applications with the Arc Macro Language. This book gives an introduction to AML, and GIScity contains many valuable examples of code that you can use as-is, copy, or learn from.

This book is not a substitute for ARC/INFO training. ARC/INFO is not like some popular software on PCs or Macintoshes that you can casually pick up in your spare time. It requires a time commitment and your careful attention. It is emphatically recommended that you receive qualified training in ARC/INFO.

With a competitive economy and limited budgets, training is sometimes overlooked or short-changed. Please try to impress on your management the benefits of ARC/INFO training. The cost of training is a small portion of your organization's GIS investment and will directly affect your productivity. This book can be used as a prelude to your ARC/INFO training, to reinforce what you learn through examples, and as a springboard for creating your own GIS applications.

Also highly recommended is participation in local, national, and international ARC/INFO Users Conference meetings. Every May, there is an excellent conference in Palm Springs, California, where you have many opportunities to talk to other users with the same applications and visit with the programming and support staff of ESRI.

GIS and CAD

There are two basic types of software tools you can choose to create maps with: Computer Aided Design (CAD) software and GIS software. One is not better than the other, but each suits certain requirements better. In the simplest terms, the difference between the two is that CAD software creates maps as drawings and GIS interacts with maps as geographic databases.

CAD software is used for general design, drafting, and engineering needs. You can use CAD for diverse purposes such as designing an integrated circuit board, an engine part, or a profile view of a road. Some qualities of CAD software are ease of use, good graphic display, and usually lower cost than GIS software. Some successful CAD software systems are AutoCAD, designPro, and Intergraph's MicroStation.

GIS software is designed for the express purpose of managing geographical information. GIS software includes or is directly layered on top of a relational database management system. Also, GIS software manages topology, which means that the information about how geographic features are adjacent to each other is stored in the database. Some examples of topology are the information about how road segments are connected to each other, which water wells are within certain resource management areas, and which parcel lots are adjacent to each other. ARC/INFO is the predominant GIS software on the market world-wide. ARC/INFO's GIS competitors include gds, GeoVision, and Intergraph's MGE.

Because of the marketing success of GIS, sometimes CAD software is touted as GIS. The following are the basic guidelines for distinguishing true GIS software from CAD:

- ❑ Is the software expressly designed for geographic analysis, and not general drafting functions?
- ❑ Is the software based upon, and not merely linked to, a database management system?
- ❑ Does the data model contain, and can the software system manage, the information of how geographic features are related to one another?

Sometimes, people get caught in a GIS versus CAD discussion. The truth is that GIS and CAD can happily coexist in most organizations. ARC/INFO has the ability to display CAD drawings in a GIS context. For example, a CAD drawing of a house floor plan can be retrieved and displayed on your workstation screen by selecting the outline or symbol for a house within ARC/INFO. Also, CAD can be useful for low-cost digitizing of graphical features and then that data can be later imported into ARC/INFO. (However, if the data to be digitized have significant attributes to be entered or coded, then it is more cost-effective to digitize directly with ARC/INFO.)

Four Secrets to Success with ARC/INFO

ARC/INFO is a big, big software package. Very few individuals, and perhaps no one, is completely familiar with every aspect of ARC/INFO. For the author, and for many seasoned users, the first year of using ARC/INFO was a time of both excitement and confusion. The excitement comes from a realization of the possibilities you discover but is tempered by the overwhelming volume of commands and the sometimes inconsistent terminology. You can become very productive with ARC/INFO with the mastery of only a portion of all the commands. **The first secret to success with ARC/INFO is to understand its underlying principles, internal organization, and most important commands.**

In ARC/INFO, you may find several or many ways to accomplish any particular geographic task. In complex analysis, database design, map design, or application development, beginning users are at first grateful to find one solution, any solution, and are reluctant to exchange that way of solving a problem for another, better and simpler way. **The second secret to success with ARC/INFO is to develop the ability to identify and implement the simplest and most elegant solution from all the alternatives.**

You will fully tap the power of ARC/INFO by exploiting its intrinsic description of how objects connect in the real world (topology), its relational database foundation, and, ability to relate to other databases used by your firm or agency. When you digitize and edit a map, you are actually updating a geographic database, and when you plot a map, you are producing a report from the geographic database and related attribute databases. When you are an advanced user, you will naturally think of database operations such as lookup tables or relates in conjunction with the common commands. **The third secret to success is to learn how to use ARC/INFO as an integrated geographic database manager.**

While this book necessarily covers ARC/INFO at the level of its most commonly used basic commands, in productive use you will perform most of your work through menus you or other people design. Just as you can't purchase the Oracle or Sybase database management system and expect completed applications such as Consumer Billing or General Ledger, ARC/INFO is not an application, but rather a geographic toolkit. Because of the "toolbox" design of ARC/INFO, what you can do in one CAD command may take a half dozen or more commands in ARC/INFO. For this reason, few users use ARC/INFO primarily at the command level. **The fourth secret to success is to automate your geographic update and query functions by developing ARC Macro Language (AML) menu-based applications.**

You can do this by purchasing applications from third-party vendors, hiring experienced ARC/INFO programmers, or progressively gaining AML experience by starting with simple AML scripts. This book provides an introduction to AML, which will assist you in writing scripts and evaluating consultant services and third-party application software.

While gaining mastery of ARC/INFO will take time, the rewards are substantial: the knowledge to solve almost any geographic and database query, the satisfaction of building useful applications, and an entry to an exciting field in which qualified professionals are in great demand.

ARC/INFO Systems Documented in This Book

This book discusses the three key programs in ARC/INFO: ARC, ARCPLOT, and ARCEDIT. Optional programs, such as GRID or NETWORK, are not documented.

When you begin an ARC/INFO session, you first enter the ARC program, which is a mostly nongraphic system for maintaining and managing your GIS databases and performing some geographic processing commands. ARC is the entry point for the two other programs, ARCPLOT and ARCEDIT.

ARCPLOT is a graphic subsystem for display and visualization of your GIS databases and for producing maps for output to printers and plotters. Database values can be updated in ARCPLOT, but geographic features cannot be edited.

ARCEDIT is for editing your geographic features. Input can be accepted from the mouse, keyboard, digitizing table, and other GIS databases.

ARC/INFO Systems Not Documented in This Book

The following optional ARC/INFO modules and programs are beyond the scope of this book:

- ❑ **ArcView** is an inexpensive and easy-to-use GIS visualization and query program available on workstations, IBM-compatible PCs, and Macintoshes. ArcView makes GIS databases accessible to everyone.
- ❑ **ArcStorm** is a new geographic data manager expressly designed for customers who have large amounts of data and require multiple users to simultaneously edit GIS data.
- ❑ **ArcScan** extends the capabilities of ARCEDIT to include semiautomated conversion from raster to vector data sets.
- ❑ **ArcExpress** is an optional module that optimizes graphics display performance in ARCPLOT and ARCEDIT.
- ❑ **COGO** (Coordinate Geometry) is an extension of ARCPLOT and ARCEDIT for surveyors and civil engineers. It comprises commands for inputting data from survey instruments, designing parcel subdivisions, and the preparation of plats.

- **TIN** (Triangulated Irregular Network) is an extension of ARCPLOT and ARCEDIT designed to model, visualize, and analyze surfaces. While not a full three-dimensional GIS package, TIN offers powerful surface modeling and perspective views.
- **NETWORK** is an extension of ARCPLOT and ARCEDIT for traversing networks and allocating resources along networks, such as needed to optimize delivery routes and business locations. **Dynamic Segmentation** is sometimes confused with NETWORK, but the two are mostly independent and Dynamic Segmentation can also satisfy many network modeling needs.
- **GRID** is a separate subsystem used to model and analyze geographic information that is organized by cells, such as the pixels in satellite images of natural resources. GRID contains commands to convert cell-based information to and from the vector-based information common to the rest of ARC/INFO.
- **ArcCAD** is a personal computer product that bridges the CAD and GIS worlds. ArcCAD is a result of an alliance between AutoDesk, the developers of AutoCAD, and ESRI. While not offering the wealth of capabilities of ARC/INFO, it can be valuable in streamlining the automation of geographic data from AutoCAD to ARC/INFO.
- **PC ARC/INFO** is another personal computer product that offers a subset of the workstation ARC/INFO functionality.

How This Book Is Organized

This book is organized in four sections:

Section I, Introduction, introduces you to ARC/INFO and its working environment through a GIS application provided on the companion CD-ROM called GIScity. Next, what a GIS is and what it can do is discussed, followed by a chapter on how ARC/INFO stores geographic information. This section is completed with a general discussion on managing attribute information in ARC/INFO.

Section II, ARCPLOT, explores the techniques for displaying information through maps, both interactively on a workstation's graphics display and through finished maps that can be plotted or printed. The basic commands in ARCPLOT are covered both at the command level and through a part of the GIScity application, called Display GIScity.

Section III, ARCEDIT, documents the steps for editing ARC/INFO's primary geographic data set, which is called a coverage. We will work through examples in the ARCEDIT program, both through the command level and a GIScity application called Edit GIScity.

Section IV, AML, covers some general methods for creating, removing, copying, and managing coverages. ARC is a nongraphic program for maintaining and troubleshooting coverages. Also, we touch on automating GIS applications through the Arc Macro Language (AML).

Paths Through the Book

New ARC/INFO User

You are a professional assigned to becoming productive quickly in editing, analyzing, and plotting geographic information with ARC/INFO. You already deal with maps in some way in your work. You might have experience with a CAD system such as AutoCAD; you've finally figured that out, and now you are wondering why your management has just made your life more difficult!

Start with Section I: Introduction. This will give you some bright ideas about what GIS is all about, and you'll see that this is something quite different, interesting, and worth your effort. The chapter on the internals of ARC/INFO by necessity provides you with quite a bit of detail. Don't worry if you don't grasp all this information at first. Concentrate on understanding label points, arcs, and polygons and return to this chapter for reference on the other types of features. Likewise, the chapter on managing attributes gives you quite a bit of detail on selecting and updating attributes. Focus on understanding selected sets, feature attribute tables, and items. Return later to this chapter for reference on topics such as database cursors and the INFO and TABLES programs.

If your main job will be to produce maps from the GIS database, proceed to Section II, ARCPLOT. This will be the key section in the book for you. At first, the commands in ARCPLOT are presented at the command level. You will see that the commands are flexible and powerful, but the toolkit design leads to a lot of typing. Don't worry; most of the ARCPLOT functions you learn at the command level are revisited in Section II through the Display GIScity application.

If you are more interested in digitizing and editing maps, then go to Section III, ARCEDIT. This is a good overview of all the steps to enter and update GIS data. At first, you can get confused with all the commands in ARCEDIT, but this section breaks down the main steps to editing GIS databases into six clear and understandable steps. You can visit Sections II and III in either order.

Section IV, AML, shows you how to create, copy, process, troubleshoot, and optimize coverages, and the rudiments of customizing ARC/INFO through the ARC Macro Language (AML).

If programming is not your cup of tea, then pick and choose the information in the chapter on AML that's useful for you. When you find yourself doing the same sequence of commands over and over in ARCPLOT or ARCEDIT, be brave and try writing some simple AML scripts. It's not as hard as it first seems and you will be rewarded for your valor.

Better yet, try this: show the GIScity application to your organization's programmer. Tell her you like it and that you don't think she can do as good a job. Remember, programmers have fragile egos, and the surest way to get them to do your bidding is to challenge their skill. After a year or two, they might have something as good, but don't worry, your gainful employment is secure.

Users Familiar with ARC/INFO

You have some ARC/INFO experience under your belt and understand the basics of ARCPLOT and ARCEDIT, but many details still elude you and certain concepts are confusing. Mutterings about pseudo-nodes, dangling arcs, and fuzzy tolerances qualify you as a social outcast.

You have some basic ideas about which commands precede which, but wish there was an easier way than a Vulcan mind meld with the documentation. This book will help you to clearly understand the important commands in ARCPLOT and ARCEDIT.

Go through Section I. You'll realize through the GIScity application that ARC/INFO can actually be easy to use. If you don't have a well-crafted AML interface like this in your organization, you'll become an agitator within your organization to write or acquire such an AML application.

Some of Sections II and III on ARCPLOT and ARCEDIT will be old hat, but other parts will reveal commands you never knew existed or understood. Some key concepts for you to solidify are feature attribute tables, selected sets, and relates. You're part way there, but some study and overview will help you become swift and sure footed in ARC/INFO.

Once you have a basic command of ARC/INFO, the last chapter will give you some guidance toward exploring the ARC Macro Language. You can get started automating ARC/INFO tasks with simple AML scripts, if you have not done so already. If you are so inclined, you can tinker with the GIScity application; it's not too difficult to adapt it to your data. Concentrate on modifying the theme and view definitions to point to your coverages and attributes tables, and with the help of your system administrator, you could have a tidy ARC/INFO application and fool your peers. (You'll still be a social outcast.)

Application Programmers Starting with ARC/INFO

You are an application programmer with some common-sense knowledge of how mapping relates to your organization's needs, but no ARC/INFO and other specific mapping experience. In fact, you are probably the least qualified person in your organization to tackle GIS, because everyone knows programmers don't read and can't run other people's software, but let's at least try.

You are experienced in programming procedural languages such as C or FORTRAN, perhaps experienced in object-oriented languages such as C++ or Smalltalk, and probably experienced in 4GL or macro languages. Good software design standards and techniques apply to

AML just as they do to other languages. Don't jump in and start hacking AML code; do the proper system design and review process. Although AML is an interpreted language with polymorphic variables, the GIScity application incorporates some elements of object-oriented technology, especially in the graphical user interface.

Quickly go through the GIScity applications in Section I, II, and III. The user interface closely follows the command syntax at ARC/INFO's command level. Two commands in GIScity, Echo On and Watch On, are especially important for you because you can follow and study the machinations of the AML. This is like a debugger with a trace mode in AML. Acquire ESRI's book on AML programming and study the code provided on the CD-ROM as programming examples. You are free to use the AML code in any way you wish.

You should quickly become an AML programming wizard; it's not particularly difficult. But unless you understand the command level of ARC, ARCPLOT, and ARCEDIT, you can't be productive as an application developer. After you have AML all figured out, study the chapters on ARC/INFO's command level, look up ARC/INFO commands you find in the GIScity application, and get familiar with ARC/INFO's on-line documentation system called ArcDoc.

Watch that programmer's ego! Put aside the "Not Invented Here" syndrome and reuse any menu objects from the GIScity application you deem worthy. GIScity can be applied to diverse applications; first try modifying the view and theme definitions to match your applications, symbol tables, and GIS database design. Then extend or simplify the menu objects for your needs.

Track down and study the Arc Tools AML applications that came with your ARC/INFO distribution. You'll find that Arc Tools is complex and unwieldy for the average end user, but it is a gold mine for programmers. Many ideas and techniques in the GIScity application are borrowed from Arc Tools and simplified.

The essence of a successful GIS application is a consistent and well designed interface that provides the needed analytical capabilities without encumbering the user with unnecessary complexity. Your success in this endeavor is crucial to the success of ARC/INFO in your organization. Good luck (and don't read the final paragraph in the part for the New ARC/INFO users)!

Typographical Conventions

In command usage, the command prompt appears in `mono` type. What the user types in after the prompt appears in **`mono bold`** type.

```
Arc: WORKSPACE {workspace}
```

Anything in angle brackets (< and >) is a mandatory command argument. Command arguments in curly brackets ({ and }) are optional. If two or more choices are separated by a vertical line (| , called a pipe), they are mutually exclusive choices. The first choice is the default value, if you don't type anything in for that value.

```
Arcedit: SELECT <BOX | POLYGON | CIRCLE | SCREEN>
         {WITHIN | PASSTHRU}
```

For anything in lowercase letters, you must substitute a value, such as a coverage name, numeric value, item name, or other value. For anything in uppercase, you should type exactly that word or one of the displayed words.

```
Arcedit: UNSELECT {FOR} <logical_expression>
```

If you want to skip one or more optional arguments in a list of command arguments, you must type a pound sign (#) in its place. Whenever you use the pound sign, ARC/INFO applies a default value, which you can learn about in the ArcDoc message for that command.

Ellipsis (...) between two arguments means that you can give one or more names or values for that argument.

The asterisk (*) indicates an interactive coordinate entry.

This book documents the most common usages for selected commands. Refer to the ARC/INFO documentation for all usages and all commands.

In the Arc Macro Language, there are two basic types of commands: directives and functions. Directives execute the assignment of variables, and functions return information to you or ask for information. Directives are preceded by an ampersand (&). Functions are specified by typing the keyword, without brackets.

```
&SHOW <parameter> {argument...argument}
```

Special Icons

The following are special icons that designate certain kinds of text or chapters.

TIP: *The Tip icon shows shortcuts and hints that can help you be more productive.*

NOTE: *The Note icon presents important information or concepts that might otherwise be overlooked.*

WARNING: *Warning points out functions and procedures that could get you in trouble if you are not careful.*

Display GIScity: This icon, located at the bottom outside margin of certain chapters, indicates that the chapter is about the Display GIScity application enclosed on the bonus CD-ROM.

Edit GIScity: The Edit GIScity icon, located at the bottom outside margin of some chapters, indicates that the chapter is about the Edit GIScity application enclosed on the bonus CD-ROM.

Introducing the GIScity Applications

On the companion CD-ROM, you'll find an AML application written for this book that demonstrates common and practical GIS operations for a medium sized city. These applications show that GIS can be simple, fun, and powerful.

Display GIScity is an application for easy display of GIS data by a diverse set of users, including city planners, engineers, and yes, tourists. This application takes place in ARCPLOT and demonstrates access to information from other databases, panning and zooming, selecting features and displaying information through symbols and text, and producing maps.

Edit GIScity is an application for modifying GIS data. It can be used to add new streets, utility lines, or buildings. This application gets the new user familiar with the basic concepts in ARCEDIT.

Both applications use real city GIS data courtesy of the planning division of the city of Santa Fe, New Mexico, an ARC/INFO user. Some of the GIS themes present on the diskette include roads, streams, contours, buildings, sewer system, survey control, census statistics, and a digital orthophoto.

Installing the Companion CD-ROM

The companion CD-ROM contains AMLs and data. The AMLs comprise the Display GIScity and Edit GIScity applications documented throughout this book. The data are the coverages and an image referenced by the AML application.

Below are instructions for Windows NT, UNIX, and VMS workstations. Because some commands are very specific to the type of workstation you are using, please pull out and reference the ARC/INFO CD-ROM installation notes. These are the notes in the box that your ARC/INFO CD-ROMs came in and contain the precise commands for mounting the CD-ROM disk on your system.

Windows NT Workstations

Mount the CD-ROM and open it in Windows Explorer or My Computer. Double-click on the readme.rtf file in the top level of the CD-ROM for further instructions on installation.

UNIX Workstations

First, enter the root user mode on your workstation by typing

```
% su
```

You may not have privileges to enter the root user mode. If so, please ask your system administrator to execute these steps.

Second, check if a /cdrom directory exists. If it doesn't, create one like this:

```
# cd /
# mkdir /cdrom
```

Next, mount the CD-ROM. Each brand of UNIX workstations has different syntax.

Reference the ARC/INFO CD-ROM notes. Some examples are:

```
Alpha AXP/OSF:    mount /t cdfs -r -o nodefperm /<dev> /cdrom
HP 700/8x7:        mount -rt cdfs /<dev> /cdrom
Sun SPARCstation (Solaris): mount -F hsfs -o ro /<dev> /cdrom
```

Next, run the installation command like this:

```
% /cdrom/install -load
```

Many of the examples in the book reference an environment variable to specify where you have installed GIScity. In your .cshrc file, add the following line:

```
setenv GCHOME <path to the GIScity directory>/giscity
```

After you have edited the .cshrc file, run it by typing at your UNIX prompt

```
% source .cshrc
```

VMS Workstations

Place the CD-ROM into your CD drive and use this command

```
$ MOUNT/OVERRIDE=ID <cd-device>
```

Next, install the contents of the CD-ROM to your hard drive by typing

```
$ @<cd-device>:[INSTALL]INSTALL_VMS load
```

The GIScity application uses a logical to reference its location on your disk. To add this logical, edit your LOGIN.COM file in your personal directory and add this line:

```
$ define $GCHOME <disk>:[<install_directory>.GISCITY.]
```

If you encounter problems in the installation, you may send a message to the author care of ipgorders@hmp.com.

ARC/INFO Updates

Throughout its history, ARC/INFO has evolved with the architectural changes in computer hardware and operating systems. The early versions of ARC/INFO operated on mainframe and mini computers. Later, when computer workstations became the platform of choice, ARC/INFO was ported to the UNIX operating system. Now that IBM-compatible personal computers have become capable of hosting complex software and large data sets, version 7.1 of ARC/INFO is being released to support Windows NT, the emerging operating system of choice for desktop computing.

Version 7.1 provides the full functionality of ARC/INFO for Windows NT on the Intel platform. Version 7.1.1 supports Windows NT on the DEC platform and all UNIX platforms previously supported. To protect the investment of ARC/INFO users who wish to migrate from UNIX to Windows NT, the ESRI development staff defined and met these goals:

- ❑ Applications written with the ARC Macro Language (AML) can be moved from UNIX to Windows NT with very minor changes, which are summarized in material that follows.
- ❑ ARC/INFO coverages and other data sources can be moved from UNIX to Windows NT without any modification. You can also directly copy ARC/INFO workspaces (directories or folders that contain ARC/INFO data sets) across a computer network from a UNIX system to a Windows NT system. From your personal computer, you can access ARC/INFO coverages and other data sources on a UNIX computer, provided you have network software that supports remote file access. ARC/INFO now has the look and feel of Windows NT's user interface. Although you will see changes such as new buttons in the title bar that close, minimize, maximize, and restore windows, the application area in which you see graphical display or menu interfaces will have only minor cosmetic changes from UNIX.

The ARC/INFO development staff also took the opportunity of the port to Windows NT to perform a major re-architecture of the low-level functions within ARC/INFO. Although these changes are not directly apparent to ARC/INFO users, you are the beneficiary of improved performance, reliability, and a software base for future development. While support for Windows NT is the major story of ARC/INFO 7.1, this version also contains a number of specific enhancements. These are summarized in sections that follow.

These notes are not a complete listing of the changes within ARC/INFO 7.1, but highlight those changes that are likely to be encountered by the majority of ARC/INFO users. Please refer to ESRI's ArcDoc online help, especially the What's New topics, for an authoritative reference of most-recent changes.

Changes in ARC

These are the key enhancements to the ARC module at version 7.1:

- ❑ There is a new option for the GENERALIZE command called **BENDSIMPLIFY**, which improves the cartographic representation of lines in which you filter out points. Rather than using the previous default method, now called POINTREMOVE, you can use the BENDSIMPLIFY option to improve line generalization at sharp corners and bends.
- ❑ **AREAQUERY** is a new ARC command that streamlines polygon overlay functions. AREAQUERY lets you create a coverage by combining polygon overlay functions such as those performed by the INTERSECT, UNION, and IDENTIFY commands upon many polygon and region subclasses. Within an interactive dialog, you can specify up to 32 coverages with overlay and selection operators.
- ❑ The **RENODE** command has been modified to items for elevation values as parameters. This is useful when you are performing path tracing on transportation networks, because you can more naturally model features such as underpasses and overpasses. RENODE is designed to take advantage of elevation values provided by some vendors of navigational data. It can also be used to better model networks such as electric distribution, where power lines may cross on a map but do not necessarily form a physical junction.

Changes in ARCPLOT

The following are the major changes to ARCPLOT under version 7.1:

- ❑ At version 7.1, a set of commands has been added to client-side support for ESRI's Spatial Database Engine (SDE), a high-performance server of geographic data. Data managed by SDE resides in a commercial DBMS, such as Oracle, as relation tables, with geographic information represented as a column. These new ARCPLOT commands serve the functions of connecting SDE data sets, managing SDE layers, importing and exporting SDE layers and coverages, and display and query support for SDE layers. Most of the new SDE support commands begin with LAYER. Some of these commands include **LAYERANNO**, **LAYERCALCULATE**, **LAYERCOLUMNS**, **LAYERDELETE**, **LAYERDRAW**, and **LAYEREXPORT**. Consult ArcDoc for a full listing.

- Other changes in ARCPLOT include the following:
  - A BENDSIMPLIFY operator for the WEEDDRAW and WEEDTOLERANCE commands
  - New options for rotating and offsetting marker symbols with POINTMARKERS, NODEMARKERS, and LABELMARKERS
  - Improvements to the user interface of LINEEDIT, MARKEREDIT, SHADEEDIT, and TEXTEDIT
  - A new tool called COLOREDIT, which lets you specify colors
  - Support for Windows Bitmap format (*.BMP) and Windows Metafile format (*.WMF)
  - A new SUM option for SPOT commands such as ARCSPOT and to let you set spot symbol sizes based on the sum of spot slices and current SPOTSIZE
- In addition, three new image formats are accessible in the Image Integrator in ARCPLOT: **BMP**, **GeoTIFF**, and **JFIF**. BMP denotes Windows Bitmap, which is the native bitmap format for Windows. GeoTIFF is an extension of TIFF (tagged image file format), which includes translation and scaling factors (like those in world files) and JFIF, which is an image format that incorporates the JPEG compression format. Version 7.1 also delivers six TrueType fonts containing cartographic symbols, as well as a number of new markersets that provide support for applications in areas such as AM/FM, municipal government, environmental, oil and gas, real estate, and transportation.

Changes in ARCEDIT

The following are changes under version 7.1 to ARCEDIT:

- The GENERALIZE command now supports two new options for generalizing linear features: **POINTREMOVE** and **BENDSIMPLIFY**. These options have the same effect as the GENERALIZE command in the ARC module.
- Most personal computers have a two-button mouse, instead of the three-button mouse widely used on UNIX workstations. The UNIX version of ARCEDIT makes wide use of all three buttons, so a new command, **2BUTTON** (implemented as an ATOOL AML), is provided to translate the functions of a three-button mouse to a two-button mouse for commands such as ADD, MOVE, SELECT, and SPLIT. Consult ArcDoc for a detailed listing of the new mouse key settings. When you use ARCEDIT on a personal computer, you can use a two-button mouse with the 2BUTTON command at the beginning of your edit session, or you can purchase a three-button mouse for use with your PC. You can employ either option according to your preference.

Changes in AML

These are some of the general enhancements to AML in version 7.1:

- New AML math functions are provided for logarithmic and exponential functions and random number generation. These new AML functions are **[EXP]**, **[LOG]**, **[LOG10]**, and **[RANDOM]**.
- The **FORMEDIT** program for generating AML form menus has a new user interface. Forms created with the new FORMEDIT program work in both Windows NT and UNIX AML applications.
- New functions have been added to support the Spatial Database Engine (SDE). These new functions are **[GETDATALAYER]**, **[GETDEFLAYERS]**, and **[GETLAYERCOLS]**. The **[EXISTS]** function has been extended to detect SDE layers.
- A new AML directive, **&CODEPAGE**, sets the native code page for localizing your AML application for international use.

- A new function, **[QUOTEEXISTS]**, is provided to simplify the detection of quoted arguments, quoted strings, or quote characters.
- The **[SORT]** function now supports numerical sorting.

Changes in Other ARC/INFO Modules

This book focuses on the core ARC/INFO modules of ARC, ARCPLOT, and ARCEDIT. The extension modules such as TIN, NETWORK, and COGO are beyond the scope of this book, but in brief, here are a few of the changes in the extension modules:

- The TIN module includes a new command called **SURFACE-SCENE**, which rapidly generates realistic perspective scenes in a fraction of the generation time of previous commands. Tools include the generation of effects such as sky, haze, and fog.
- Tools are provided to generate animations for MPEG viewers.
- The TIN module provides a new conversion to VRML (virtual reality modeling language) so that 3D models can be transmitted across the Internet to Web browsers.
- The **GRID** module contains enhancements for working directly from point and line data sets to generate grids with density, proximity, or other statistics. New functions are available to generate contours and to convert between grids and shapefiles. Also, GRID includes support for the BMP, GeoTIFF, and JFIF image formats.

General Changes in ARC/INFO

The following are general changes to ARC/INFO under version 7.1:

- The ArcDoc online help system has been converted to the **Win-Help viewer**, which is native to Windows NT. You have an easy-to-use interface that is consistent with other Windows applications to quickly locate information about ARC/INFO commands and concepts. ArcDoc also offers enhancements for printing documentation for selected topics, sections, or entire books.

- Improvements have been made so that international users of ARC/INFO can display and access data using local character sets and conventions. These enhancements for **National Language Support** include TrueType fonts as well as Intellifonts for single-byte and multi-byte characters and code-page support between UNIX and Windows NT. This lets you display attribute information in any local language from INFO tables, feature attribute tables, or external databases such as Oracle. You can also generate graphics files in formats such as GRA, PostScript, and Adobe Illustrator in the local language.

Changes for This Edition

This book was first printed for ARC/INFO version 7.0. Very little of the core functionality within ARC/INFO as documented in the first printing has changed. The areas in which changes in version 7.1 affect the documentation within this book are:

- Discussion of the ArcDoc system (pages 6 and 7)
- Windowing systems, operating systems, and mouse input and windows (pages 27 to 32)
- The availability of SDE as a geographic data server (page 95 and ARCPLOT chapters)
- Support for TrueType marker symbols (Chapter 6)
- Improvements to the user interface for MARKEREDIT, LINEEDIT, and SHADEEDIT (Chapter 6)
- File system structure of ARC/INFO workspaces (pages 520 to 522)
- The improved user interface for the FORMEDIT program (pages 574 to 582)

The illustration of sample screen output in this book reflects the Motif windowing system. The user interface of ARC/INFO 7.1 on Windows NT is very similar except for Windows user interface elements such as window title bars.

The Companion CD-ROM to this book has been modified so that you can install the sample AML application and sample data for Windows NT as well as UNIX. As of version 7.1, ARC/INFO is no longer supported on the VMS operating system, but the AML application and sample data for VMS remain on the CD-ROM for those users who are using version 7.0 on VMS. For instructions on installing and using the Windows NT version of the sample AML application from the CD-ROM, read the README.TXT file in the top-level folder.

On a personal note, writing *INSIDE ARC/INFO* and updating it for the Revised Edition has led me to an unexpected opportunity to join ESRI. I now enjoy the rare privilege of being a long-time user of ESRI software products with the chance to influence the development of new GIS products. I am continually impressed by the ingenuity of ARC/INFO users in applying GIS technology in ways unanticipated by its developers. Innovation is indeed driven by you, the reader and user, who are making a difference by applying GIS technology to improve our world.

Michael Zeiler, 1996

Part I

Introduction

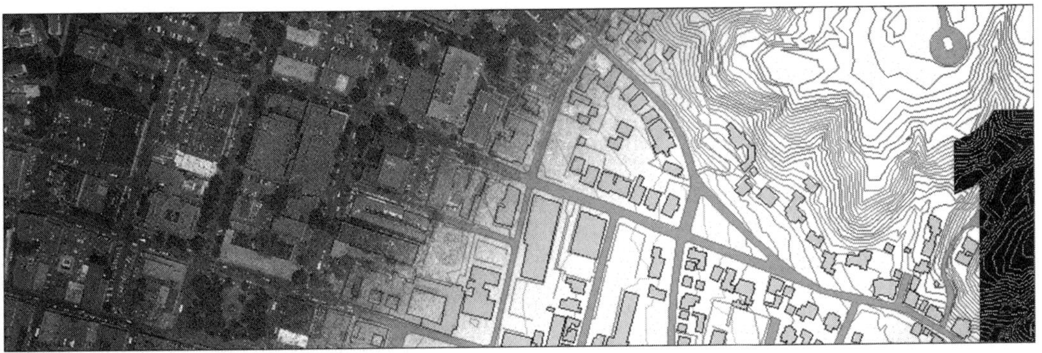

Getting Started with ARC/INFO

Navigating Windows in Display GIScity

ARC/INFO is the world's leading geographic information systems (GIS) software. It delivers geographic information to you through graphical display and printed maps. With ARC/INFO, you have the power to explore relationships between your data that is not possible with any other technology. ARC/INFO's strengths include an internal database management system, geographic data management tools, and a rich application development environment.

Yet becoming productive and knowledgeable with ARC/INFO can be a daunting task. This is because the power accessible to you inevitably results in complexity. ARC/INFO's options and possibilities in expressing and managing geographic information comprise over 2000 commands, some with many usages.

The truth is that few people use the command level of ARC/INFO for the majority of their productive work. Rather, success is most often achieved with ARC/INFO when you use it through applications built with the Arc Macro Language (AML).

This first section of the book introduces you to the principles and capabilities of ARC/INFO. To run the GIScity application enclosed with this book, no understanding of AML programming is necessary, nor is any prior experience with ARC/INFO, other GIS, or even CAD software assumed. This application, GIScity, introduces you to the basic methods and concepts in accessing geographic information in ARC/INFO. You will also find that ARC/INFO can be easy to learn and use through a well-designed application.

This chapter starts you with a part of that application, called Display GIScity. This application accesses and visualizes GIS data within the ARC/INFO program called ARCPLOT. You will learn how to start a session, what to do if you get stuck, how to control your windows, and how to set the display and extent of the geographic information you will see.

Because of the tool kit design of ARC/INFO, it is not practical or desirable to begin with the command level of ARC/INFO. We will get started with GIScity in this chapter. The next few chapters will cover basic concepts in ARC/INFO, and later we will return to the GIScity application.

Getting Started with ARC

ARC, ARCPLOT, and ARCEDIT are the central trilogy of programs within ARC/INFO. ARC/INFO also comprises other programs such as GRID and LIBRARIAN, but they are not as central to understanding the fundamentals of ARC/INFO. When you understand ARC, ARCPLOT, and ARCEDIT, you will be poised to learn and use the other ARC/INFO programs and modules.

Most optional ARC/INFO modules—TIN, COGO, NETWORK, ArcScan, and ArcStorm—are implemented as additional commands in ARC, ARCEDIT, and ARCPLOT. The exception is GRID. Because GRID works exclusively on raster data sets, it requires its own program.

LIBRARIAN is a program separate from ARC, ARCEDIT, and ARCPLOT which is provided with the core ARC/INFO system. Because its functions

are being replaced by a newer module called ArcStorm, LIBRARIAN is not documented in this book.

Starting ARC

When ARC/INFO is properly installed on your workstation, and the appropriate definitions have been made in your system initialization files (the .cshrc file in UNIX or the LOGIN.COM file in Open VMS), you can start the ARC program by simply typing `ARC` at the operating system command prompt.

When you begin ARC, you will first see a copyright notice by Environmental System Research Institute, Inc., followed by the version number and a short description of the terms and limitations for using the ARC/INFO software.

Next, you will see this command prompt, which tells you that the program is waiting for you to issue a command:

```
Arc:
```

Quitting ARC

When you are ready to leave ARC and return to your operating system, type `QUIT` at the ARC command prompt and press [Enter]. In your dialog window, the command will look like this:

```
Arc: quit
```

On the next line, you will see your operating system command prompt. You have completely left the ARC/INFO environment. The operating system command prompt varies from one type of workstation to another.

Starting Other ARC/INFO Programs

You must enter ARC before you can enter the other ARC/INFO programs. To enter another programs, just type that program's name at the Arc: prompt. You can initiate the following programs from ARC:

```
Arc: ARCPLOT
Arc: ARCEDIT
Arc: TABLES
Arc: LIBRARIAN
Arc: INFO
Arc: GRID
```

To leave any of these programs, type QUIT at that program's prompt. You will be returned to the ARC program.

NOTE: *There is no ARC command to enter most of the ARC/INFO optional modules: TIN, NETWORK, COGO, ArcStorm, and ArcScan. They are implemented as additional commands in the ARC, ARCPLOT, and ARCEDIT environment.*

Getting Help in ARC

Most of the documentation provided with ARC/INFO is in an on-line documentation system called ArcDoc. This is a hypertext application on your workstation that is based on FrameViewer, an on-line help system. FrameViewer is provided by the Frame Technology Corporation, the makers of a popular desktop publishing system, FrameMaker.

To begin ArcDoc from ARC, simply type `HELP` and press [Enter], and you will be returned to the `Arc:` prompt. At first, nothing seems to be happening. In a couple of seconds, ArcDoc will appear as a separate help window like that in the preceding picture. You can also start ArcDoc this way from ARCPLOT and ARCEDIT.

At the bottom of the help window, there are a number of commands. The commands vary depending on where you are within the system. They include Master Contents, Index, Local Contents, Functional List, Alphabetical List, Go Back, First Page, Previous Page, Next Page, Print, ViewerHelp, and Quit. Click the Viewer Help command for an explanation of these commands.

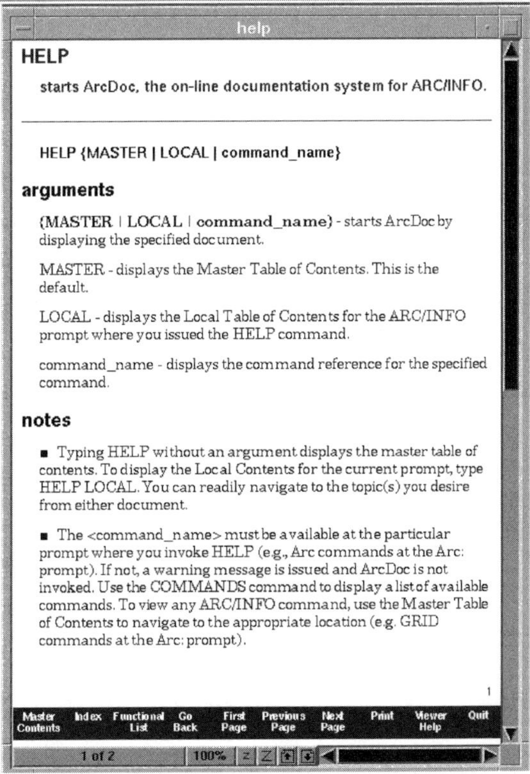

help
HELP
starts ArcDoc, the on-line documentation system for ARC/INFO.
HELP {MASTER | LOCAL | command_name}
arguments
{MASTER | LOCAL | command_name} - starts ArcDoc by displaying the specified document.
MASTER - displays the Master Table of Contents. This is the default.
LOCAL - displays the Local Table of Contents for the ARC/INFO prompt where you issued the HELP command.
command_name - displays the command reference for the specified command.
notes
■ Typing HELP without an argument displays the master table of contents. To display the Local Contents for the current prompt, type HELP LOCAL. You can readily navigate to the topic(s) you desire from either document.
■ The <command_name> must be available at the particular prompt where you invoke HELP (e.g., Arc commands at the Arc: prompt). If not, a warning message is issued and ArcDoc is not invoked. Use the COMMANDS command to display a list of available commands. To view any ARC/INFO command, use the Master Table of Contents to navigate to the appropriate location (e.g. GRID commands at the Arc: prompt).
1
Master Contents
Index
Functional List
Go Back
First Page
Previous Page
Next Page
Print
Viewer Help
Quit
1 of 2
100%

ArcDoc, the on-line ARC/INFO help system. This window shows the help message for the ARC HELP command.

You will see that some of the text in your help window is colored blue and underlined. These words are hyperlinks topics. To get more information on one of these topics, move your screen cursor over that phase and click mouse button 1. Immediately, the help window will jump to that topic.

TIP: *Sometimes, you'd like to see several help screens at the same time. To do so, press the [Shift] key when you click on a help command or hyper link. You'll get a new help window and the existing help window will remain.*

Listing Commands and Usages

No person can memorize all the commands available in ARC/INFO. To get a listing of all the available commands in the ARC/INFO program you are in, just type `COMMANDS` at the `Arc:` prompt. To narrow your search, you can also type in a letter or the first few letters of a command you are searching for. Here's what you will see if you type in `COMMANDS A` at the ARC command prompt:

```
        Arc: COMMANDS A

ABBREVIATIONS        ADDIMAGE           ADDITEM          ADDressBUILD
ADDressCREATE        ADDressERRORS      ADDressMATCH     ADDressPARSE
ADDressTEST          AddRouteMeasure    ADDTEXT          ADDXY
ADJUST               ADRGGRID           ADS              ADSARC
ANNOCLIP             APPEND             ARCADS           ARCDFAD
ARCDIME              ARCDLG             ARCDXF           ArcEdit
ARCFONT              ARCIGDS            ARCIGES          ARCLABEL
ARCMOSS              ArcPlot            ARCPOINT         ARCSCITEX
ARCROUTE             ARCSECTION         ARCSLF           ARCTIGER
ASCIIGRID
ATOOL directory      abbreviate
/arcexe70/atool/arc
                     addindexatt        annocopy         arcshell
arctools
```

These are all the ARC commands that begin with the letter A. Most are native commands in ARC. The commands at the end under the phrase "ATOOL directory" are some additional commands that are written in the Arc Macro Language and are supplied for your convenience with ARC/INFO.

Command Abbreviations

Some of the most frequently used commands in ARC/INFO have abbreviations. These commands will be displayed as both upper- and lowercase letters. To use one of these command abbreviations, type in just the uppercase letters. In the preceding listing, the command ARCPLOT is written as "ArcPlot." This tells you that the valid abbreviation for that command is "AP." When you use ARC, ARCEDIT, and ARCPLOT at the command level, abbreviations will speed up your work, especially if you are one of the "typing challenged."

The following list includes all the commands in ARC that can be abbreviated.

Command Name	**Abbreviation**
ADDressBUILD	*ADDBUILD*
ADDressCREATE	*ADDCREATE*
ADDressERRORS	*ADDERRORS*
ADDressMATCH	*ADDMATCH*
ADDressPARSE	*ADDPARSE*
ADDressTEST	*ADDTEST*
AddRouteMeasure	*ARM*
ArcEdit	*AE*
ArcPlot	*AP*
CalibrateRoutes	*CR*
Commands	*C*
CoPyWorkspace	*CPW*
CouNTVeRTices	*CNTVRT*
CreateWorkspace	*CW*
DeleteWorkspace	*DW*
DIRectory	*DIR*
DissolveEVents	*DEV*
EventTransform	*ET*

Command Name	**Abbreviation**
ListCoverages	*LC*
ListGrids	*LG*
ListImages	*LI*
ListSTacKs	*LSTK*
ListWorkspaces	*LW*
MeasureRoute	*MR*
OverlayEvents	*OE*
Quit	*Q*
RenameWorkspace	*RW*
Workspace	*W*

Command Usages

Most commands in ARC/INFO have usages. This means that to execute that command, you type the name of the command followed by some values such as coverage names, tolerance values, and other options. These values are called *command arguments.*

After you've located a command, you need to know which arguments you must enter to use it. If you are using a command for the first time, use ArcDoc to learn about the command. If you're using a familiar command, but need to refresh your memory, type USAGE followed by the command name. Here's an example:

```
Arc: usage clean

Usage: CLEAN <in_cover> {out_cover} {dangle_length}
       {fuzzy_tolerance} {POLY | LINE}
```

The usage information tells you how to use that command in a compact notation. Anything in angle brackets (< and >) is a mandatory command argument. Command arguments in curly brackets ({ and }) are optional. If you see two or more choices separated by a vertical line (|) (called a pipe), those are mutually exclusive choices. The first choice is the default value, if you don't type anything in for that value.

For anything in lowercase letters, you must substitute a value, such as a coverage name, numeric value, item name, or other value. For anything in uppercase letters, you should type exactly that word or one of the displayed words.

If you want to skip one or more optional arguments in a list of command arguments, you must type a pound sign (#) in its place. Whenever you use #, ARC/INFO applies a default value, which you can learn about in the ArcDoc message for that command.

The example above shows the important CLEAN command, which we'll cover in detail later. Here's how to read the usage for the CLEAN command: After you type in `CLEAN`, you must enter a coverage name—the input coverage. Next, you can optionally type in the name of an output coverage. Two more optional choices are the dangle length and fuzzy tolerance. Lastly, you can optionally type in either `POLY` or `LINE`.

For example, if you want to perform the CLEAN command on a coverage named SFROAD, send the output to a new coverage named SFROADCL, specify a dangle length of 5 feet, accept the default fuzzy tolerance, and specify the LINE option, then you would type in:

```
Arc: CLEAN SFROAD SFROADCL 5 # LINE
```

TIP: *Unlike the UNIX operating system, ARC/INFO is almost completely indifferent to whether you use upper- or lowercase characters when typing any command and usage. The only notable exception is when you enter the INFO program, which we'll cover later.*

Feature Classes

Some ARC commands are sensitive to which feature attribute tables are present. For your reference, below are some terms used as command arguments to denote the combination of feature attribute tables in a coverage, also known as *feature classes* in ARC. Regrettably, these terms, especially LINK, are confusing because they have a meaning distinct from how the same word is used elsewhere. You'll only encounter these terms

for command arguments for ARC commands.

POINT. A coverage that has a Point Attribute Table
LINK. A coverage that has a Point Attribute Table and an Arc Attribute Table.
LINE. A coverage that has an Arc Attribute Table.
NET. A coverage that has an Arc Attribute Table and a Polygon Attribute Table.
POLY. A coverage that has a Polygon Attribute Table.

These feature classes do not take into account the other feature attribute tables, such as TAT, RAT, SEC, and so on.

Command Interpretation

When you type a command in ARC, it can be interpreted in several ways. First, ARC will look up whether that command exists as a native command. If your entry is not a valid command, ARC will search for a macro with that name in an ARC macro program directory called "atool." If that doesn't work, ARC will look for a macro with that name in the directory you are in, and attempt to execute it. Failing that, ARC will issue the command to your operating system.

If all these steps fail, you will get an error message. Command interpretation works this way in all the ARC/INFO programs. If you type in a misspelled command, you will see a message like this:

```
Arc: CLEAF
Submitting command to Operating System ...
CLEAF: Command not found.
```

Exercise 1: Exploring ARC

1. Let's get started with the ARC program. At your operating system prompt, type `arc`. You should see several lines in your window with copyright and version information about ARC/INFO.

2. If you do not get the `Arc:` prompt, you can try the following steps: For UNIX systems only: Check that the "flexlm" license manager has been started. This requires superuser status, changing directory to $ARCHOME/sysgen, and running the "lmgrd" program. Check your platform's ARC/INFO system guide for the details of this command.

   Verify that ARC/INFO was installed on your system by doing a directory listing on the main ARC/INFO directory. In UNIX, type `ls $ARCHOME`. In VMS, type `DIRECTORY $ARCHOME`. You should see a listing with about 30 directories. If you do not, either ARC/INFO has not been installed, or your account does not recognize the directory alias or logical called $ARCHOME.

   Check your .cshrc or LOGIN.COM file. Several aliases (logicals) should be set in these files which are run every time you log on your system.

3. At the `Arc:` prompt, type `COMMANDS` (or just `C`). You will see several screens of every command in ARC. Next, type `COMMANDS` followed by a letter. You'll see only the commands beginning with that letter.

4. Try specifying a wildcard. Type `COMMANDS` followed by letters with an asterisk, which is the wildcard character. For example, to see all commands that contain "ARC," you can type `COMMANDS *ARC*`. Also try `COMMANDS *ARC`, and `COMMANDS ARC*` to get a good idea about how wildcard characters work.

5. Now that you've seen several commands listed, try listing the syntax for some of the commands. Pick one or several of the command names you've seen and type in `USAGE` followed by the command name. (Unfortunately, wildcard characters don't work for the USAGE command.)

6. Let's enter the ArcDoc on-line help system. Type `HELP`, and a separate window will appear after a few moments. From the window, click on the line that says "ARC/INFO Data Model, Concepts & Key Terms." Next click on the line that says "The Coverage." You will get a screen with an index to all of the

components of a coverage: arcs, nodes, labels, and so on. Click on any of these. Try moving through the on-line document by clicking the Previous Page and Next Page commands on the bottom.

7. Next, click on Go Back on the bottom of the menu. This will take you back to the previous window. Do this successively until you have returned to the original window.

8. When you are back at the original window, click on Command Lists. This will bring up a window with all the ARC/INFO programs and modules, with two columns of buttons under the headings of Alphabetical and Functional. Click the button for the ARC program and under the Alphabetical heading.

9. Keep this window active as you work through this chapter. For any commands that you are curious about, press the letter on the top that the command begins with. Then click on one of the command names in BLUE. You'll be able to access the full on-line documentation for each command. At first, you'll be overwhelmed by the volume of explanation for many commands. As you gain experience with ARC/INFO, you'll come to value the detail of the on-line help system. The explanations in this book present the most basic information you need to get started with ARC/INFO. ArcDoc is your resource for comprehensive and detailed explanations of all of ARC/INFO.

Viewing Maps in Display GIScity

Display GIScity operates within the ARC/INFO program called ARCPLOT. It is written in the Arc Macro Language (AML) to demonstrate the most important ARCPLOT commands. Keep in mind that this application is not meant to represent a comprehensive GIS application, but it is a good introduction to the basic capabilities of ARC/INFO.

Getting Started

Let's get started right away with ARC/INFO and Display GIScity.

At this point, you should verify that the programs and data in the companion CD-ROM have been successfully installed. If so, proceed. If not, review the steps in the Companion Disk Installation section. If it is inconvenient or not possible to load the CD-ROM, you will see every GIScity command fully illustrated in the following chapters.

The data contained on the CD-ROM is from the city of Santa Fe, New Mexico. Santa Fe is a city of about 55,000 people, located at the base of the Sangre de Cristo range of the Rocky Mountains. Santa Fe is noted for its varied cultural activities, outdoor recreation opportunities, and scenic beauty. The economic base of Santa Fe is tourism and state government. Santa Fe is also the home of OnWord Press and the author.

Roughly one square mile of downtown Santa Fe is included on the CD-ROM. Compiled mainly through *aerial photogrammetry*, a technique to convert aerial photographs to a map with proper scale relations, this is very high quality data and perhaps more detailed than the GIS data you will usually work with. This data, courtesy of the Planning Division of the city of Santa Fe, includes roads, streams, contours, buildings, census statistical data, and other information.

To start Display GIScity on a UNIX system, type this command at the system prompt:

```
arc &run $GCHOME/display/start
```

On a Open VMS system, type this command at the system prompt:

```
arc &run $GCHOME:[DISPLAY]START
```

On a Windows NT system, start the Arc program from Windows Explorer or My Computer. At the Arc: prompt in the command window, type:

```
&run $GCHOME\display\start
```

NOTE: *On some workstations, you may need to break this command on two lines. If you get an error message, type:*

```
arc
&run $GCHOME/display/start (UNIX)
```

or,

```
arc
&run $GCHOME:[DISPLAY]START
```

Shortly after entering either of these commands, you will see a graphics window appear on your workstation screen. Display GIScity will next draw a map of downtown Santa Fe on the graphics window. After the map is complete, the main menu will appear to begin the application.

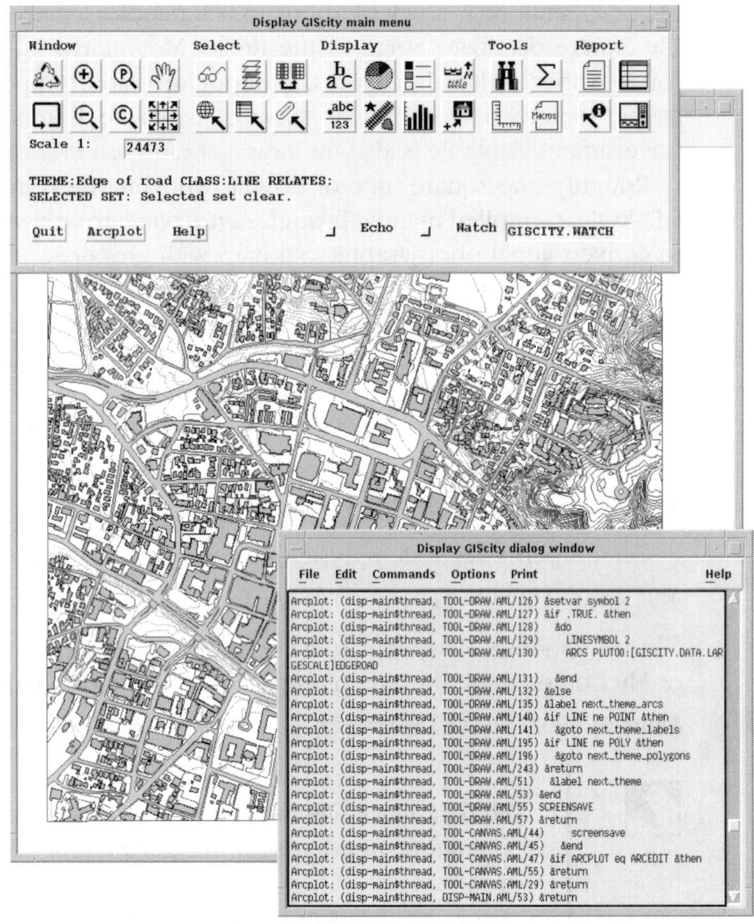

This is the initial appearance of Display GIScity. The large window is your graphics window. On the bottom is a dialog window through which you initiate ARC/INFO. On top is the Display GIScity main menu, which will be present throughout your session.

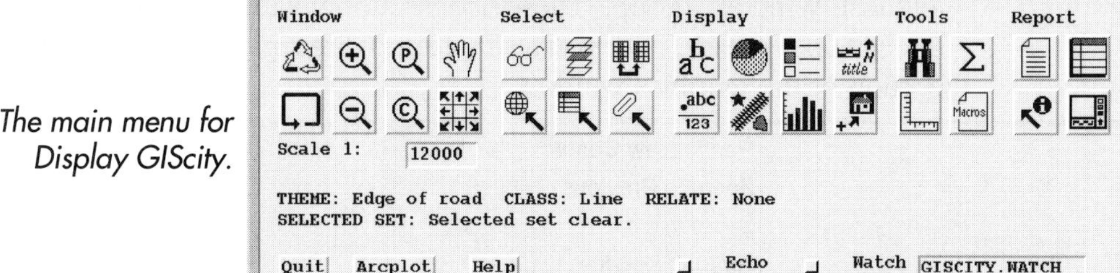

The main menu for Display GIScity.

NOTE: *If Display GIScity does not appear as pictured, verify the following with your system administrator: (1) That all steps in installing the programs and data from the companion disk were successfully completed; (2) that your user account has the system definitions to run ARC/INFO and the necessary privileges to access the installed data; (3) that version 7 of ARC/INFO is properly installed on your workstation.*

TIP: *Normally, you will start ARC/INFO by typing "arc" at the system prompt in a dialog window. However, it is possible to tailor many windowing environments so that ARC/INFO automatically begins when you log in to your workstation, by clicking an icon on your screen or by selecting a menu choice. If you want to simplify the starting of an ARC/INFO session, consult with your system administrator to enable one of these techniques on your workstation.*

Window Commands

Now that we have started Display GIScity, let's take a closer look at the GIS data.

The GIScity main menu contains groups of icons labeled Window, Select, Display, Tools, and Report. Eight icons are included in the leftmost Window group. These icons, along with a map scale box directly underneath, offer you a simple and direct means to zoom to any portion of the graphics window.

Pan to New Center
Zoom to Previous
Zoom In
Refresh Window

Redraw Window
Zoom Out
Zoom to City
Page to Direction

Scale Box
Scale 1: 21000

The Window commands within the main menu for Display GIScity.

NOTE: *At any time, you can get the menu name for any icon by moving your cursor over the icon and pressing the rightmost mouse button.*

Refresh Window is a handy command to use whenever you want to quickly refresh your geographic display after using the commands in the Display portion of the main menu. The Display commands draw special symbols or text at your command, and Refresh Window instantly restores your graphics window to its prior appearance. This command acts immediately and requires no cursor input.

Redraw Window will completely redraw your graphics window at the current scale. You can use this command after using a command in the Select portion of the main menu. This command acts immediately and requires no cursor input.

Zoom In lets you zoom to any visible portion of your graphics window. You can move the cursor within the graphics window, press the left mouse button to mark one corner, press the left mouse button again to mark another corner, and the graphics window will redraw to span those two points. You will use this command frequently.

NOTE: *If you specify two points that define a box wider or thinner than the aspect of the graphics window, the Zoom In command will fill in the remainder of the graphics window.*

Zoom Out will redraw the graphics window while zooming out by a factor of two. This command acts immediately and requires no cursor input.

Pan to New Center will redraw your graphics window at the same scale, centered at a new point. Move your cursor within the graphics window, and press the left-most mouse button to mark a new center. This command will then redraw the graphics window about the new center.

NOTE: *The two Window commands Zoom In and Pan to New Center require cursor input. To provide cursor input, move the cursor within the graphics window and press the left mouse button. Without cursor input after these two commands, Display GIScity will seem to freeze on you.*

Zoom to Previous will redraw the graphics window at the previous extent and scale. This command is useful when you are doing detail work in a small portion of the map and then want to return to the previous scale and extent. This command acts immediately and requires no cursor input.

Zoom to City will redraw the graphics window to the original scale and extent of the city. This will recreate the graphics window as it was when you first started Display GIScity. This command acts immediately and requires no cursor input.

Page to Direction is different from the other commands. Instead of acting directly, or through cursor input in the graphics window, this command invokes another menu with arrows pointing to eight directions. When you select an arrow, your graphics window shifts over nearly a full page in that direction. You may keep the Page to Direction menu visible as long as you want, and you can successively page across the graphics window. At any time, to reduce visual clutter, you can remove the page menu by pressing the Dismiss button.

Scale Box is a command that is run by typing in a scale value. You will notice that whenever you zoom in or out, the value in the scale box changes. The value is your present scale in a unitless scale. Cartographers prefer unitless scales, because they imply no units of measurement systems. This scale works equally well whether you are familiar with Metric or English systems of measurement.

To relate English or Metric scales to unitless scales, simply multiply the scale divisor by the value of unit multipliers. For example, to convert a Metric scale of 1 centimeter = 1000 meters, multiply 1000 (the scale divisor) by 100 (centimeters in a meter) to derive a unitless scale of 1:100 000. To convert an English scale of 1 inch = 200 feet, multiply 200 (the scale divisor) by 12 (inches in a foot) to derive a unitless scale of 1:2400.

You can use this command to zoom to an exact scale, which is often useful when you are duplicating the appearance of a finished map plot. This command acts immediately after you type in a value and press [Enter] on your keyboard.

NOTE: *As you use the Window commands to increase your scale, you will see additional map detail appear. Display GIScity has the intelligence to determine the scale you are presently using and to selectively display map detail appropriate for that scale. This is done through the use of* **scale thresholds**, *which will be discussed later.*

Exercise 1: Using Window Commands

Let's go through all of these commands in Display GIScity and see how to use them.

1. Click on the Zoom In command.
2. Using the left mouse button, select two diagonal points on your graphics window separated by about two inches. Look at the scale box and observe the scale value before and after this command.
3. Now try the Zoom Out command. It will redraw your graphic window zoomed out by a factor of two. Click on the Zoom Out icon; the command will execute immediately without requiring cursor input. The value in the scale box will reappear multiplied by two.
4. Now we'll try the Pan command. Click on the Pan icon, move your cursor to a new center point inside the graphics window, and press the left mouse button. Immediately after you select a new center, the graphics window will redraw centered around that point. Note that the value in the scale box has not changed.
5. Now, try using the Scale box to change the graphics window scale. Click on the Scale box using the left mouse button.
6. Type in a numeric value between 100 and 50,000, and press [Enter] on your keyboard. Try this several times using different values. You'll see some types of map features turning off and on at different scales.

This graphics window shows a sample map extent with two corners marked at the corners of a box. (The box and corners do not normally appear; they are shown for illustrative purposes). The next diagram shows the results of selecting the Zoom In icon, and then selecting these two points with the left mouse button.

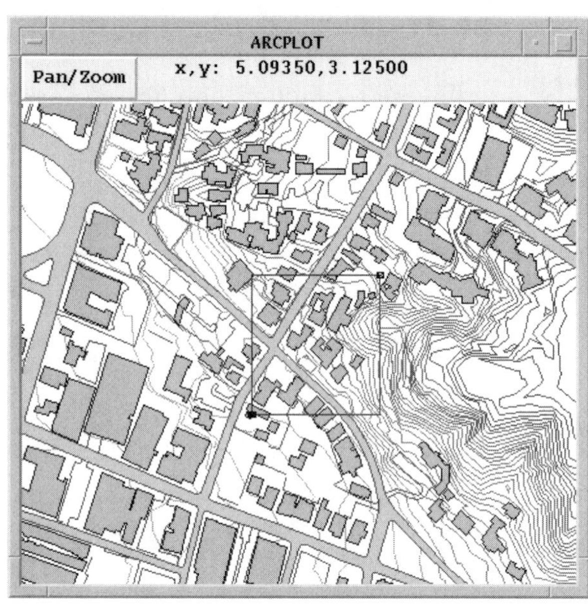

The result of the completed Zoom In command. Note that some street names now appear. Later, we'll find out how we can set scale thresholds for drawing features.

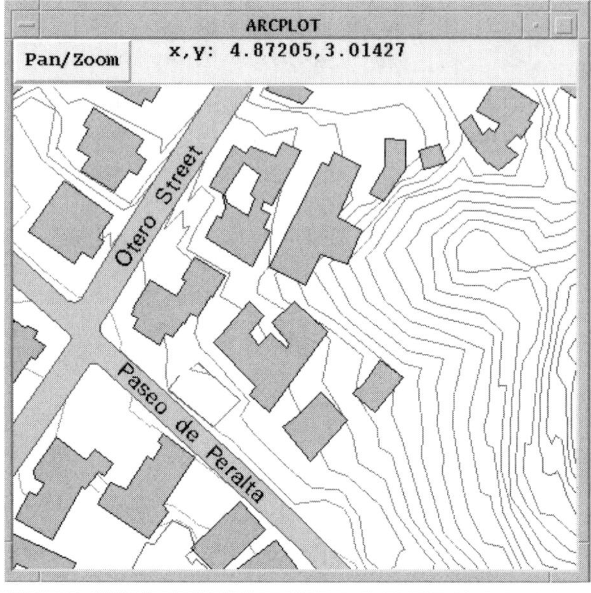

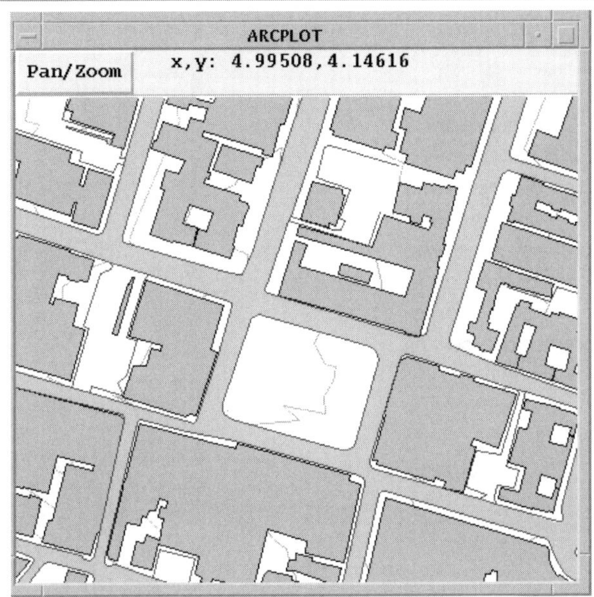

The graphics window shows a sample map extent before using the Zoom Out command.

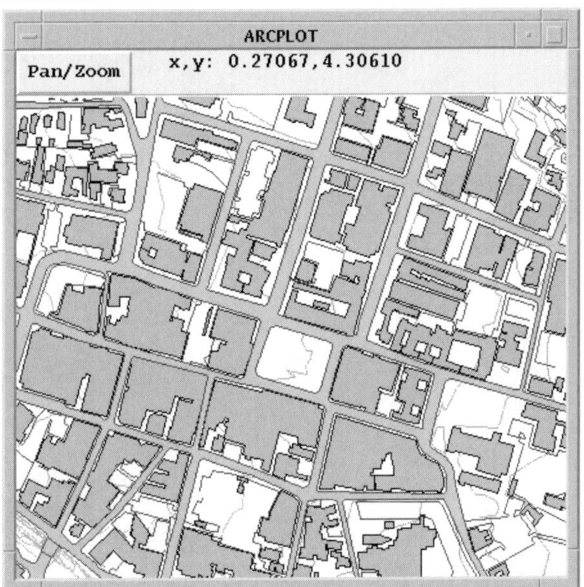

The map extent zoomed out by a factor of two, immediately after issuing the Zoom Out command.

This graphics window shows a sample map extent with a center marked. After you select the Pan to New Center command, you will see a vertical and horizontal line moving along with the cursor.

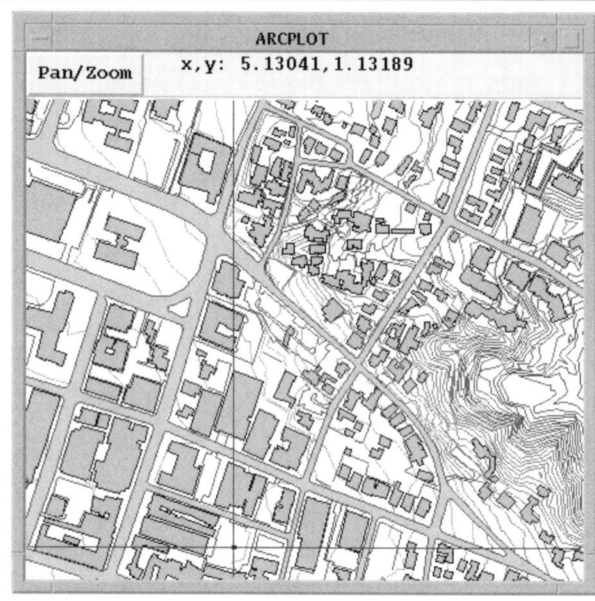

The map extent after using the Pan command. The scale is the same, but the map is repositioned within the window.

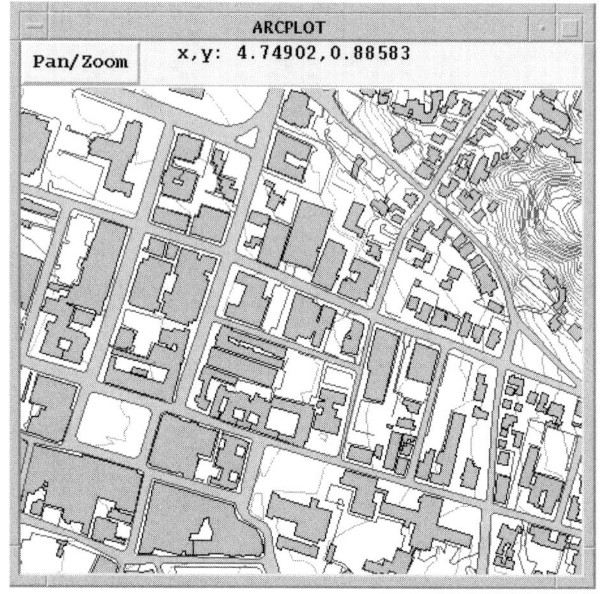

7. Click on Zoom to Previous to get back to the previous map extent and scale. This command is very handy when you want to inspect an area at a large scale and then quickly return to your previous extent.
8. Click on Zoom to City to return to your original view, displayed when you started Display GIScity.
9. Click on Refresh Window to instantly redraw your window. This command acts very quickly and restores the contents of the graphics window from your workstation's memory. Right now, you might not notice the effects of this command because there will be no visible difference in your graphics window. However, this command's effect will become evident to you later as we explore the Display commands.
10. Click on Redraw Window to redraw your window from scratch. Later, when we use the Select commands, you will see how using Redraw Window updates the content of your graphics window after you modify settings for your display. This command is not as fast as Refresh Window because it completely redraws your graphics window.

Getting Help

There are several ways of getting help within GIScity. Here are two simple ones.

First, every menu has a Help button. Simply clicking on Help will bring up a pop-up window that displays a detailed summary of all the commands in that menu.

There is another way of getting help in GIScity. While some menu icons may be clear to you immediately, others may be confusing at first. Move your cursor over the icon in question, and press the right mouse button. On the bottom of your menu, the name of that icon command will appear.

The pop-up window shows a sample Help message for the Display GIScity menu you've just been using. The pop-up window contains scroll bars for extended help messages. When you've finished, you can dismiss the popup window with the Quit button.

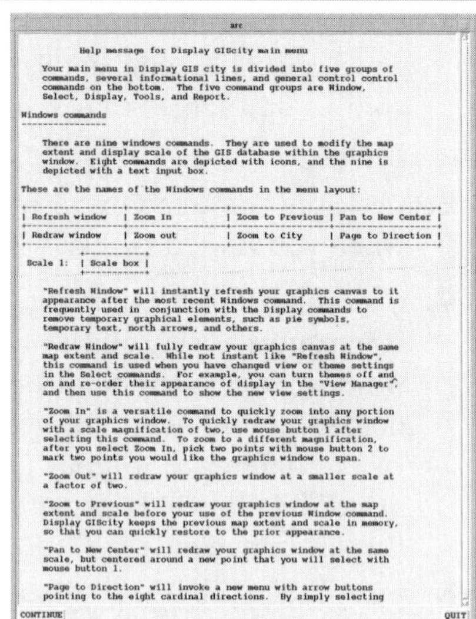

arc

Help message for Display GIScity main menu

Your main menu in Display GIS city is divided into five groups of commands, several informational lines, and general control control commands on the bottom. The five command groups are Window, Select, Display, Tools, and Report.

Windows commands

There are nine windows commands. They are used to modify the map extent and display scale of the GIS database within the graphics window. Eight commands are depicted with icons, and the nine is depicted with a text input box.

These are the names of the Windows commands in the menu layout:

Refresh window	Zoom In	Zoom to Previous	Pan to New Center
Redraw window	Zoom out	Zoom to City	Page to Direction

Scale 1: | Scale box |

"Refresh Window" will instantly refresh your graphics canvas to it appearance after the most recent Windows command. This command is frequently used in conjunction with the Display commands to remove temporary graphical elements, such as pie symbols, temporary text, north arrows, and others.

"Redraw Window" will fully redraw your graphics canvas at the same map extent and scale. While not instant like "Refresh Window", this command is used when you have changed view or theme settings in the Select commands. For example, you can turn themes off and on and re-order their appearance of display in the "View Manager", and then use this command to show the new view settings.

"Zoom In" is a versatile command to quickly zoom into any portion of your graphics window. To quickly redraw your graphics window with a scale magnification of two, use mouse button 1 after selecting this command. To zoom to a different magnification, after you select Zoom In, pick two points with mouse button 2 to mark two points you would like the graphics window to span.

"Zoom Out" will redraw your graphics window at a smaller scale at a factor of two.

"Zoom to Previous" will redraw your graphics window at the map extent and scale before your use of the previous Window command. Display GIScity keeps the previous map extent and scale in memory, so that you can quickly restore to the prior appearance.

"Pan to New Center" will redraw your graphics window at the same scale, but centered around a new point that you will select with mouse button 1.

"Page to Direction" will invoke a new menu with arrow buttons pointing to the eight cardinal directions. By simply selecting

CONTINUE QUIT

Besides giving you the name of menu commands, the bottom line of every menu will provide information about that menu, actions you have just issued, and whether GIScity requires further input from you. Watch this line after you issue menu commands.

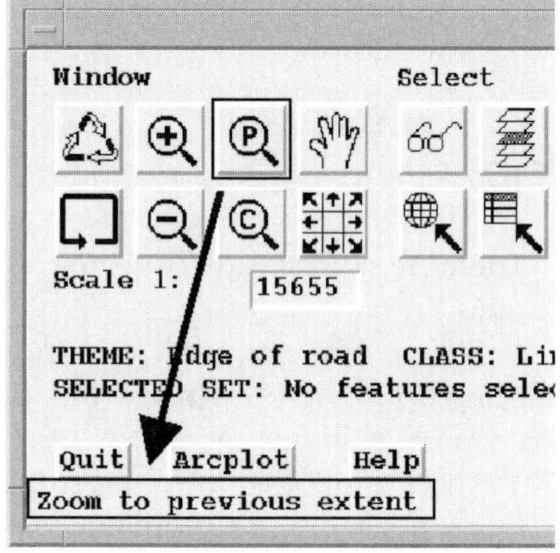

ARC/INFO's Working Environment

Now that we have begun Display GIScity, let's take time out and review some facts about ARC/INFO's working environment.

ARC/INFO runs on all major workstations and window terminals. Examples of compatible workstations are those manufactured by Hewlett Packard, IBM, Digital Equipment Corporation, Data General, and Sun MicroSystems.

A *workstation* is a self-contained computer system that includes a central processing unit (CPU), random access memory (RAM), disk drives, and a graphics window monitor. Workstations are commonly networked with other computers so that tape drives, disk drives, printers, plotters, and other devices can be shared and centrally managed.

A *window terminal* contains specialized graphics chips, RAM, and a display monitor, but no CPU. Window terminals are always connected to another workstation or computer and are popular because they can be a cost-effective solution for providing multiple users with workstation software. (Window terminals are often called *X-terminals* because they employ the industry standard graphics protocol called X-windows.)

ARC/INFO is more demanding of RAM than most workstation software. ESRI recommends a minimum of 16 megabytes, and you will notice significant improvements in performance with 32, 64, or more megabytes of RAM. Additional memory is often the most economical way to improve the performance of ARC/INFO on your system, especially if your workstations or computers are serving window terminals.

Windowing Systems

All workstations and window terminals operate through a graphical user interface that is similar to Microsoft Windows or the Macintosh system. All but one vendor use a standard windowing system called Motif. Motif is a windowing standard defined by a consortium of computer manufacturers called the Open Systems Foundation. However, the standout among workstation vendors is Sun Microsystems, which uses a proprietary windowing system called OpenLook. Sun Microsystems is moving

to Motif, and if you are running ARC/INFO on a Sun SPARCstation, it may be running under either OpenLook or Motif.

ARC/INFO operates and appears identical on all workstations that use Motif. Although there are visual differences between the Motif and OpenLook systems, fortunately there are very few functional differences in the use of ARC/INFO between the two windowing systems.

Operating Systems

Another key difference between workstations is the operating system. You can think of the operating system as a continuously running low-level computer program that other programs use to access the computer's memory and devices. There are two main operating systems supported by ARC/INFO: UNIX (and its many variants) and Open VMS. Digital Equipment Corporation is the standout with its proprietary operating system, Open VMS. All other workstation vendors use a version of the UNIX operating system.

The differences between using ARC/INFO under the UNIX or Open VMS operating systems are minor and infrequent. One notable difference occurs when you reference the location of your data on the disk drive, a process that involves specifying paths and directories. Wherever this operation appears in this book, you will see sample commands in both UNIX and Open VMS. The GIScity application runs identically under both UNIX and Open VMS, and you won't need to be aware of the differences until you use ARC/INFO at the command level.

TIP: *You can customize many aspects of ARC/INFO's user interface, such as menu colors, fonts, mouse pointer symbols, and other aspects of the ARC/INFO interface by copying the file named Arcinfo in UNIX or ARCINFO.DAT file in VMS from the ARC/INFO system file directory into your home directory. By editing the Arcinfo or ARCINFO.DAT file in a text editor, you can modify any of the listed settings. For details, consult the ARC/INFO System Dependencies manual for your platform.*

Mouse, Input Keys, and Cursor

ARC/INFO is programmed to accept menu input through nine input keys. These keys can be input through the digitizer, keyboard, or most commonly, through the mouse. For the Display GIScity application, we will use input keys executed through the mouse.

Your mouse has (at most) three buttons, so where do we get the rest of the input keys? From the combination of pressing keyboard keys and mouse buttons simultaneously. On the lower left corner of your keyboard, you will find [Shift] and [Ctrl] keys. Input keys 1, 2, and 3 are executed by pressing the left, middle, and right buttons on your mouse. Input keys 4, 5, and 6 are executed by pressing the left, middle, and right mouse buttons while simultaneously pressing the [Shift] key on the keyboard. Input keys 7, 8, and 9 are executed by pressing the left, middle, and right mouse buttons while simultaneously pressing the [Ctrl] key on the keyboard.

With three mouse buttons and the [Shift] and [Ctrl] keys, you can specify all nine input keys that ARC/INFO uses. For input keys 4 through 9, first press [Shift] or [Ctrl], and then press the mouse button while holding down the key.

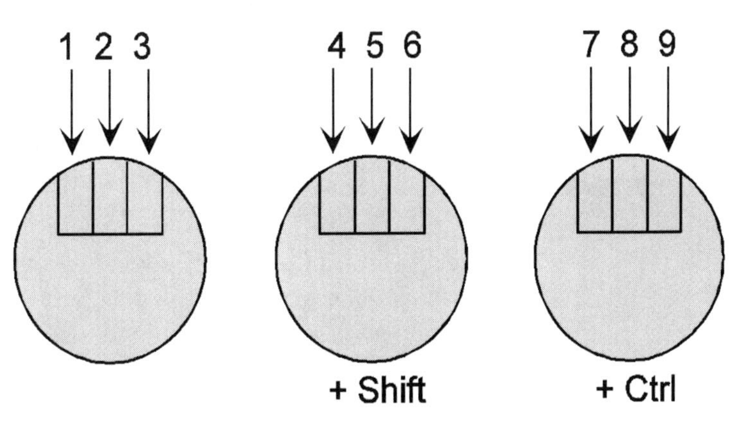

The great majority of commands in GIScity are executed with input key 1, by using the cursor to move the mouse over the icon or button desired and pressing the left mouse button. You will use input key 3 for getting context-sensitive help or the names of icons in menus. Input keys 2, and 4 through 8, are not used in Display GIScity, and only

occasionally in the other GIScity application, Edit GIScity. Input key 9 is frequently used for ARC/INFO commands that accept input until terminated.

TIP: *You will frequently use input key 9, but you may find the combination of pressing the [Ctrl] key simultaneously with the right mouse button to be cumbersome. A quick shortcut in executing input key 9 is to simply press the [9] key on the top row of your keyboard. The numeral keys on your keyboard will work for any of the input keys. Unlike most keyboard input, the effect will be immediate; you will not need to press the [Enter] key.*

When you move your mouse on the mouse pad, you move a cursor on the workstation monitor. You will notice that the appearance of the cursor will sometimes change as you move it from one window to another within the workstation screen. The cursor may appear as an arrow over the dialog window and as a cross-hair within the graphics window.

Dialog Windows, Graphics Windows, and Pop-up Windows

You begin ARC/INFO through a window, which displays scrolling text and is usually 80 columns wide by 24 lines. This window is referred to as the *dialog window*.

Windows that display graphics in ARC/INFO are called *graphics windows*. Later you will learn how to control graphics window size and placement yourself. When you resize the graphics window, the graphics window is immediately blanked. When you are in the Display GIScity application, you can quickly restore the graphics window with the Redraw Window command.

GIScity utilizes dialog windows, graphics windows, pop-up windows, and form menus.

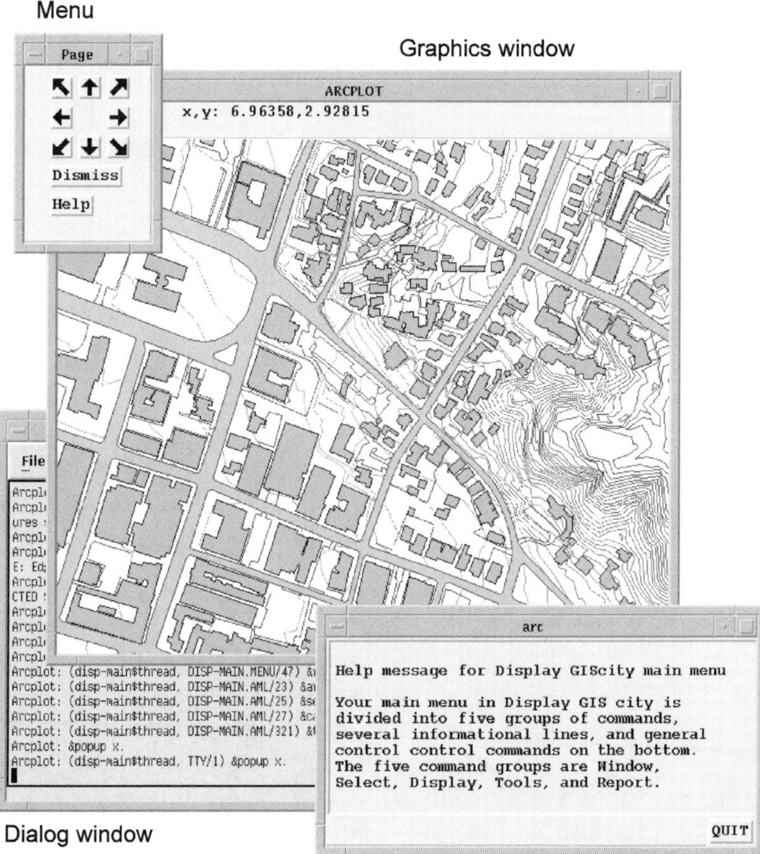

The Anatomy of a Window

All windows on your workstation, whether they are dialog, graphic, pop-up, or other windows, share the same basic components.

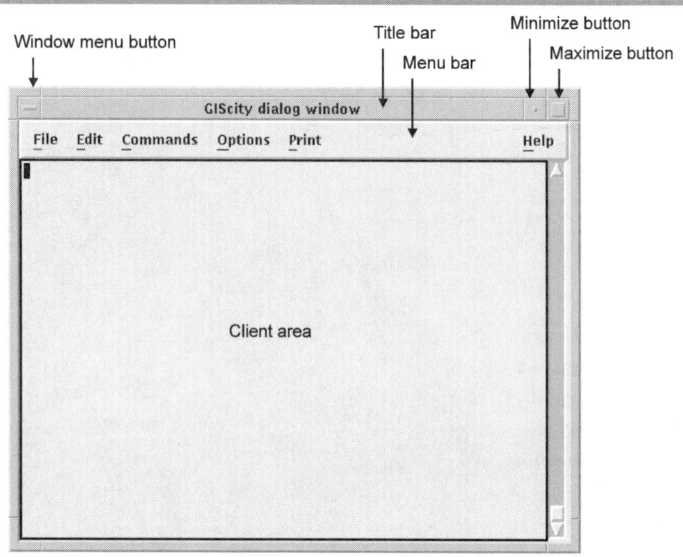

Pressing the *Window menu button* will display a menu for that window's general control operations, such as closing, resizing, and bringing a window to the foreground.

The *title bar* states the type of window or program. If the window is an ARC/INFO menu, it is usually a descriptive title for that menu. In the Arc Macro Language, the title bar is referred to as a *menu stripe*.

The *Menu bar* contains the commands that can be issued for that type of window. The choices under the menu bar vary by which windowing environment and operating system you are using. The commands in the menu bar are independent of ARC/INFO.

The *Minimize button* will shrink the window to an icon. If GIScity seems to freeze on you, you may have accidentally pressed this button and turned a dialog window, pop-up window, graphics windows, or menu into an icon. This icon will be in a corner or side of the menu. Double-clicking that icon will restore it to its previous position and size.

The *Maximize button* will increase the size of the window to encompass the entire workstation screen. You will rarely use this button,

but if you inadvertently press it, you can restore the normal size and position of the window with the Resize border.

Along the edges of each window is a thin border, called the *Resize border.* If you position the cursor in that border, you can change the size of the window by depressing the left input key and dragging the cursor to a new position delineating the new window border. Notice that the corners of the Resize borders are set apart from the side borders. Within the corner area, you can resize the window in the horizontal and vertical directions at the same time.

You will rarely need to resize any menus or pop-up windows. However, it is often useful to resize the dialog window so that it has minimal overlap with the graphics window. You may also want to resize the graphics window to your personal preferences. Whenever you resize the graphics window in Display GIScity, you will need to issue the Redraw Window command to restore graphics. (In Edit GIScity, when you resize the graphics window, it will automatically be redrawn for you.)

The center of each window is called the *Client area.* Within this area, you will find text, buttons, icons, or graphical display.

Form Menus, Widgets, and Icons

GIScity uses a type of ARC/INFO menu called a *form menu.* GIScity uses form menus exclusively because they are the most powerful and efficient menu that ARC/INFO offers.

You interact with form menus through the use of *widgets.* A widget is a computer industry term for an interactive element within a window. GIScity makes extensive use of all available widget types.

Although GIScity operates in both the Motif and OpenLook windowing systems, the pictures in this book illustrate only Motif windows because Motif is predominant on workstations and is gradually replacing OpenLook on Sun SPARCstations.

There are nine possible widgets in a form menu:

- ❑ Buttons
- ❑ Check boxes
- ❑ Choices

- ❑ Data lists
- ❑ Display fields
- ❑ Sliders
- ❑ Symbol lists
- ❑ Text
- ❑ Text input fields.

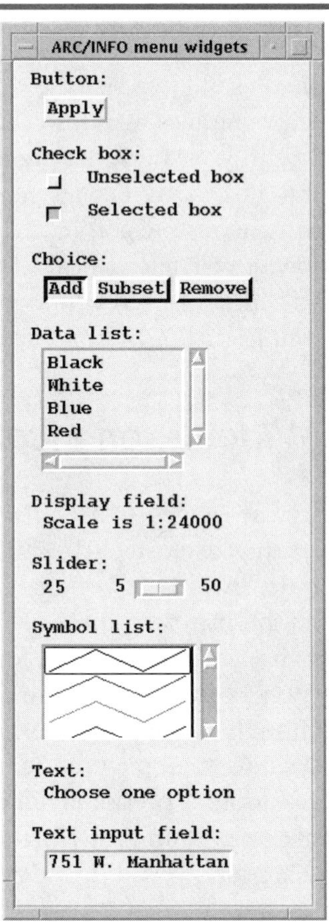

A sample form menu with every available widget.

You can click on a *button* to quickly execute a particular action. Buttons can either appear as words surrounded by rounded boxes, or they can appear as small graphic pictures called *icons*. Many ARC/INFO

applications, including GIScity, use icons as an important part of an intuitive user interface.

An application will use *check boxes* for specifying on or off status, such as which map themes are active for display or selection. Usually, setting a check box status will not change that attribute until you select another menu command or dismiss the menu.

Choices serve a purpose similar to check boxes but are used for short lists of mutually exclusive choices. Setting a choice usually does not perform an action until a command is executed.

Data lists are useful for making a choice from a long list. That list may comprise almost anything GIScity will offer, including themes, symbols, data sets, and files. Data lists can be *dynamic*. That is, they may change depending on other actions you take. Sometimes a choice from a data list results in an immediate action; at other times, a follow-up command is required to effect the action.

Display fields dynamically display changes within ARC/INFO showing, for example, information about which data your application is interacting with. Display fields are not used for input; they are for information purposes only.

Sliders are best suited for displaying or entering numeric data along a predefined range of values, such as temperature, scale, or size. Sliders have minimum and maximum values defined within the application, and an increment value. Some sliders have an input field for you to type in a value, but you cannot exceed the minimum or maximum values.

Symbol lists display a range of symbols available for selection. They are used to quickly choose a point, line, shade, or text symbol for interactive drawing.

Text widgets can be placed anywhere in a form menu for description of the menu or an adjacent menu element.

Text input fields are used to type any kind of data into a menu. They can be used for either numeric or character input. Sometimes an application will control what type of data is acceptable and will inform you if another type of input is required. The Scale box you used earlier is an example of a text input field.

Getting Unstuck

Here is a summary of steps to follow if you get stuck anywhere within the GIScity application:

1. Depending on the computer you are using, some commands may take a while to complete. Wait for the command to complete.
2. You may have a menu hidden behind the graphics window. Using the menu control button, or other means on your windowing system, move the graphics window to the background so that the menu is in the foreground and can be accessed.
3. You may have an active cross-hair cursor hidden by a form menu. Move the cursor to within the graphics window.
4. An active menu or graphics window may have been turned into an icon in your windowing system. If this is the case, double-click on the icon to restore the graphics window or menu.
5. Move to the dialog window. If you see an ARC, ARCEDIT, or ARCPLOT prompt, type `&RETURN` to get back to Display GIScity.
6. If you see messages that seem to be error messages, have your system administrator check the proper system settings for ARC/INFO, proper installation of ARC/INFO, and privilege to access the GIScity data.
7. If all else fails and you simply cannot regain control of GIScity, move your cursor to the dialog window, and press the [Ctrl] and [C] keys simultaneously. This will interrupt your GIScity session. You may need to do this more than once. ARC/INFO will then ask if you really want to terminate your ARC/INFO session. If you do, answer by typing [Y].
8. If step 7 doesn't work, select the Motif or OpenLook window menu button and select Close from the menu that appears. This has the same effect as step 7.
9. If you exhaust all of the above steps and still can't terminate ARC/INFO, use whatever method is available in your workstation's graphical user interface to log out.

Basics of Geographic Information Systems

Geographic Analysis and Cartographic Output

A GIS is the combination of skilled persons, analytic methods, spatial and descriptive data, and computer hardware and software—all organized to automate, manage, and deliver information through geography.

When you look at an atlas, the amount of information present is staggering. Much more than an inventory of places of human settlement and the thoroughfares connecting them, an atlas invites you to explore how people interact with the environment, how markets are shaped by demographics interwoven with traffic patterns, and how culture diffuses through patterns of population movement.

A GIS resembles the admixture of an atlas and an encyclopedia. Using a GIS is like placing your finger at a point on a map and invoking the history of a place, or revealing which products are manufactured and which resources are at hand, or what the habits of the denizens are.

With a GIS, you can access encyclopedic information through the paradigm of an atlas.

This, then, is what distinguishes GIS from other information technologies: a GIS has the unique ability to present vast amounts of information through geography, and it contains the means for you to rapidly drill down to the information that is of interest to you. GIS integrates spatial and descriptive information. You can display data from disparate sources and find relations you may never have expected.

For example, imagine how a GIS can assist an epidemiologist during the outbreak of a new disease. A GIS could overlay the locations of where people were stricken with the geographic display of transportation networks, bodies of water, wildlife habitats, vegetative zones, industrial sites, patterns of prevailing winds, and every suspect man-made or environmental factor. You could postulate correlations and test them. Once you have isolated the infectious agent, you could further use the GIS to study modes of transport.

A related discipline to geography is cartography. In a GIS, cartography becomes the output from geography. A GIS does not remove the artistry required to produce elegant maps; instead, it offers a new palette of symbolization for cartographers. A GIS also empowers cartographers to create maps that are dynamic representations of data. For example, a cartographer can define a map layout that shows voting precincts and the demographics of the most recent vote. Using the same map layout and the results of a new election, a cartographer can produce a new map with little effort.

Despite the dramatic advances we've seen in computer graphics display, in multimedia integration of graphics, video, and voice, and in updated user interfaces, this basic fact remains for most GIS applications: a published map is still the most important product from a GIS. Accordingly, an important emphasis for this book is what cartographic methods are available in GIS.

Data Elements in a GIS

There are two ways to represent information in a GIS:

- ❑ Spatial information stores the positions and connectivity of geographic features.
- ❑ Descriptive information relates the attributes of those features.

Spatial information spans two broad types of information: *vector* and *raster data sets.*

Raster Data Sets in ARC/INFO

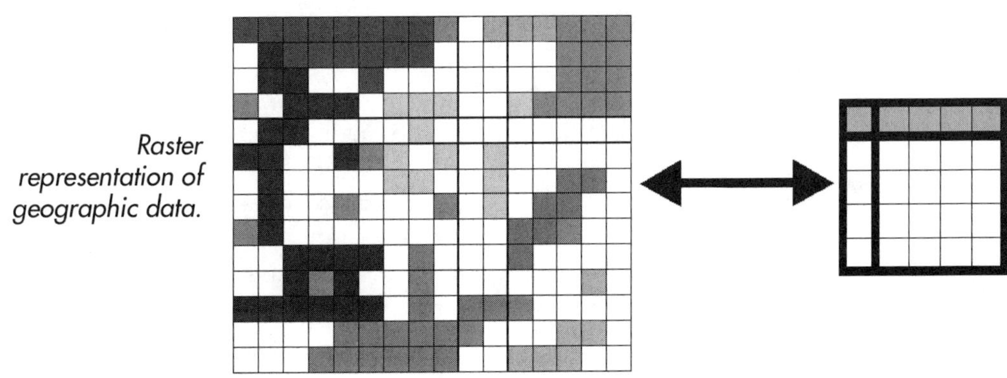

Raster data sets store spatial data as cells within a two-dimensional matrix. Raster data sets are established when the means of data capture is through a grid. Examples are satellite imagery and scanned aerial photographs. Raster data sets are also used to simplify certain types of complex resource analysis, such as diffusion of wildfires.

Each cell in a raster dataset can have descriptive information, such as vegetation type, associated with it. In ARC/INFO, the basic type of raster data set is called a *grid*. An ARC/INFO module, called *GRID*, lets you perform display and analysis upon grids.

Portion of the orthophoto image in GIScity, the application provided with this book.

Another raster data set in ARC/INFO is the *image*. An image is used for strictly graphical display, and no analysis is performed on images. Examples of images are scanned aerial photographs, scanned photographs of features such as buildings, and scanned map manuscripts.

The main difference between images and grids is that you can define any descriptive information you desire to be associated with each cell in a grid, and an image does not have any descriptive information associated with it.

NOTE: *The raster processing capability in GRID is not documented within this book. GRID is an optional module that is an extension to the core ARC/INFO programs.*

Vector Data Sets in ARC/INFO

Vector data sets contain features with discrete positions. These positions are stored as X and Y coordinate values, using the Cartesian coordinate system you learned in secondary school.

To describe arcs, a GIS stores the positions of each node and vertex, in order, and assigns an internal identification number to that arc.

How vector data sets store the positions of features.

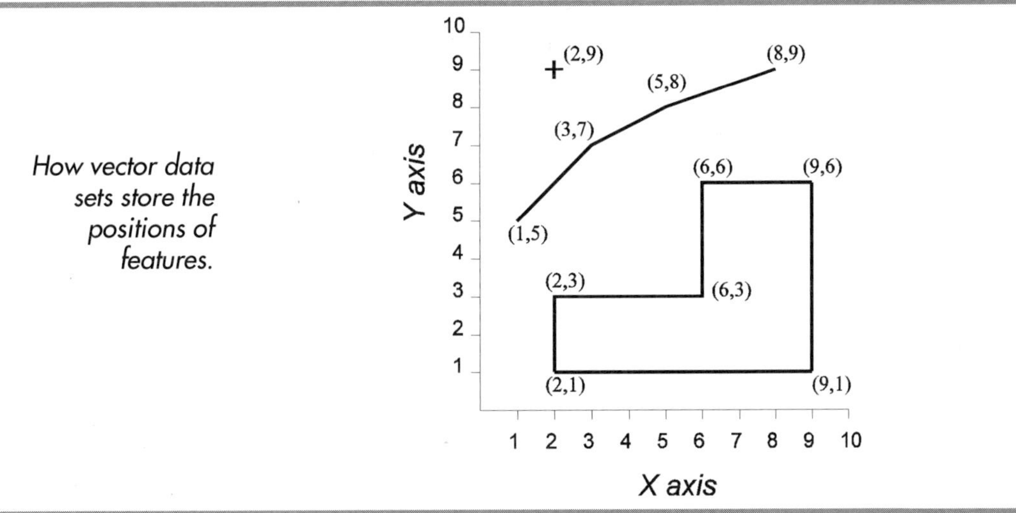

The building blocks of GIS data are points, lines, and polygons.

- ❑ **Points** are used to represent geographic features that are too small to delineate on a map. Examples of point features are springs, utility poles, and survey control points. In ARC/INFO, these are called *label points.*
- ❑ **Lines** depict features that comprise linear networks, such as streams and roads. In ARC/INFO, they are called *arcs.*
- ❑ **Polygons** outline areas with a common attribute, such as a body of water, a parcel of land, or a environmental zone.

Other graphics software, such as popular draw programs or CAD software, share these same building blocks. Yet they are not thought of as GIS. What distinguishes GIS from other graphics software?

In a nutshell, GIS embraces two important characteristics: *topology* and a *relational database.*

Topology means that GIS software maintains an internal description of how features are connected to each other. A GIS stores information about which lines are connected to other lines and which polygons are adjacent.

All commercial GIS software also contains an embedded *relational database manager*. The systems either directly contain, or can be layered on top of, a database management system to manage descriptive information about geographic features.

Coordinate Systems and Map Projections

Positions are stored in a Cartesian coordinate system.

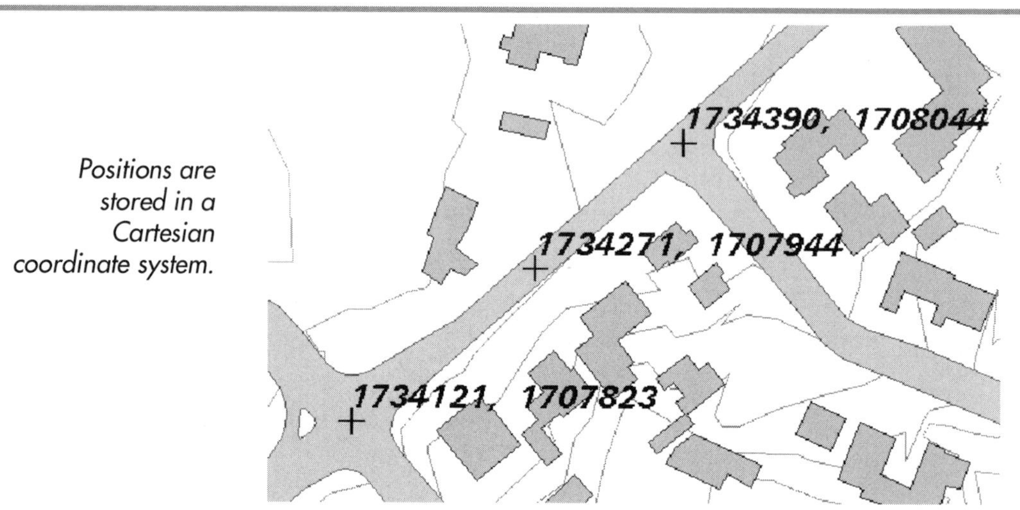

All maps use coordinate systems to mark the placement of geographic features. Each point on a map has two values associated with it: its position along the X axis of a coordinate system and another along the Y axis. Most of the time, you aren't directly concerned with these coordinate values. You are, however, when you are comparing map positions with survey results or positions collected with Global Positioning System (GPS) receivers.

In the preceding example, the numbers are kept small. However, real-world coordinates are quite large. Typically, the values for X and Y positions in a real-world coordinate system are in the hundreds of thousands or millions.

Distortion is inevitable when you map a sphere onto a plane.

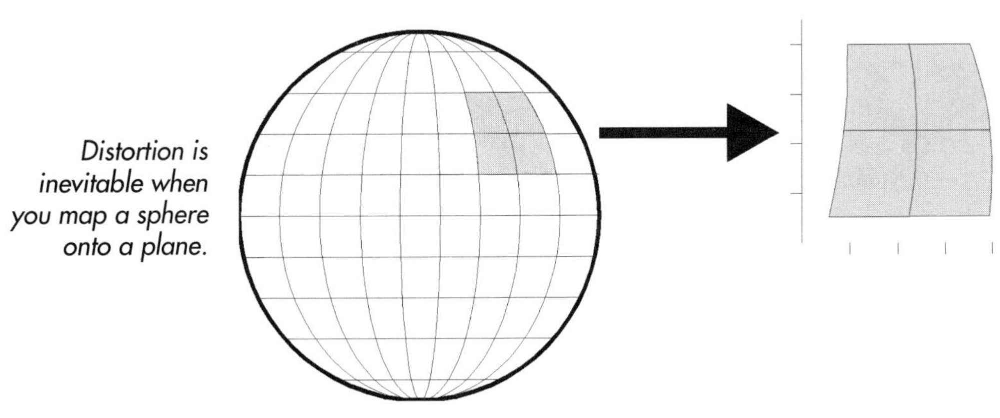

A map, by definition, is a representation of a portion of the Earth's surface upon a flat sheet. It is an approximation called a map projection. All *map projections* introduce distortion. This distortion can be mainly in distances, areas, or angles. So someone designing or applying a map projection chooses whether true distances, areas, or angles are most important.

It is important to realize the distinction between *distortion* and *inaccuracy*. In a mapping system, distortions can be calculated exactly and compensated for. They are not "error" but known deviations from measurements. Positional inaccuracies in a mapping system are errors accumulated from the limits of survey instruments, the methods used in compiling maps, or just plain errors. Inaccuracies can never be improved upon, short of adding newly surveyed and more accurate positions.

The two most common map projections are *cylindrical* and *conical*. Cylindrical projections use an imaginary cylinder touching the surface of the Earth. A start point (or *origin*) is defined, usually along the equator. Then X and Y positions from the cylinder are projected upon the Earth's surface to define the cylindrical position.

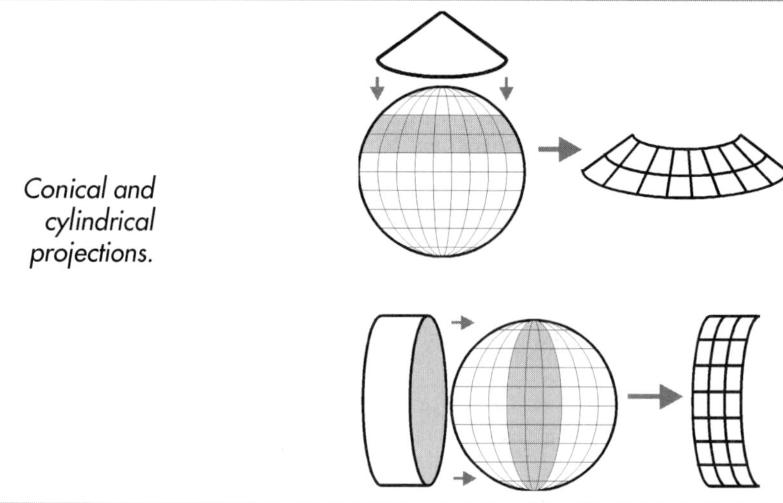

Conical and cylindrical projections.

A cylinder projected along the Earth's equator is called a *Mercator projection.* A cylinder projected along a line of longitude, going from the North Pole to the South Pole, is called a *Transverse Mercator projection.*

A *conical projection* wraps an imaginary cone along the Earth's surface at a line of latitude. Some conical projections are also called *Lambert conical projections.*

Why bother with cylinders, cones, or any of the other possible projection methods? Because you can control the distortion in a map. Projections are especially important in reducing distortion when you make maps of very large areas. Lambert Conical projections are best for areas that are mostly east-west in extent. Transverse Mercator projections are best for areas that are mainly north-south in extent.

For example, nationwide maps of the United States usually use a Lambert Conical projection. (Note how the "straight" boundary between Canada and the United States is curved on most national maps).

If you look at the edges of many maps, you will see reference coordinates. These coordinates are sometimes expressed in latitude and longitude. At other times, they use a grid system, such as the Universal Transverse Mercator (UTM), a world-wide coordinate reference, the State Plane Coordinate System (SPCS), commonly used in the United States, or other grid systems defined for countries or regions.

All coordinate systems (UTM, SPCS, and others) are based on projection systems (or several projection systems) and have single defined origins.

Sometimes, you may see thin lines at even intervals crossing the length and height of a map. These lines, along with the coordinate intervals annotated on the side of the map, are a reference to help you orient features to the specified coordinate system. This is called a *graticule*. Sometimes you may even see two different graticules on the same map.

NOTE: *Very rarely are geographic coordinates (latitude and longitude) used to directly store coordinates in a GIS dataset. Latitude and longitude are spherical coordinates, rather than planar coordinates. The problem is that a degree of longitude gets smaller in real-world length the farther north (or south) you are from the equator. Planar coordinates are nearly always used in GISs because one unit in the X direction is (very nearly) exactly the same as one unit in the Y directory.*

The basic reason GISs use coordinate systems is to allow you to overlay geographic information from different sources and have the features from those sources precisely aligned on top of each other.

If different map sources have different coordinate systems and you want to combine them in a GIS, you have two choices. You can apply GIS commands to permanently convert all the different map sources to one common coordinate system, or you can keep the different coordinate systems and set up your GIS for on-the-fly map projection. Usually, permanent coordinate conversion is preferable, so that display is not slowed by the intensive calculations of repeated on-the-fly map projections. Sometimes, you can make a case for keeping differing coordinate systems, and accepting the slower display time of "on-the-fly" map projections. The trade-off is between convenience and performance.

Yet another wrinkle in coordinate systems is a *datum*. The datum essentially defines which precise survey network your coordinates are based on. In the United States, most maps are still defined in the North American Datum of 1927 (NAD27). New GISs are adopting its replacement, the North American Datum of 1983 (NAD83). The use of either is usually acceptable, but using NAD83 is smart for brand new GISs. GPS technology is encouraging the transition to NAD83, because GPS coordinates are essentially in NAD83.

It is not important for most users of a GIS to understand the details of map projections, coordinate systems, or datums; but it is important to know several facts about how coordinates in your GIS data are defined. When you receive or send geographic data, you will always need to know the coordinate system, datum, and unit of measurement, so that, if necessary, you can apply GIS commands to convert from one system to another.

Don't worry about the time or complexity to convert from one coordinate system to another. It can be as easy as a one-line command and does not take very long. It is best, though, to select and stick to one coordinate system to prevent creeping errors from mathematical round-offs.

Topology in a GIS

Topology is how a GIS performs many of its functions. Topology simply refers to GIS software's knowledge about how features are connected and which features are adjacent to each other. In ARC/INFO, there are three important forms of topology:

- ❑ Connectivity refers to how arcs are connected to each other.
- ❑ Polygon definition refers to the way a series of connected arcs identify a polygon.
- ❑ Contiguity refers to how polygons are associated with their neighboring polygons.

One of the strengths of ARC/INFO is that it creates and maintains topology automatically for you. If you follow some recommended steps when digitizing maps, topology is easily established.

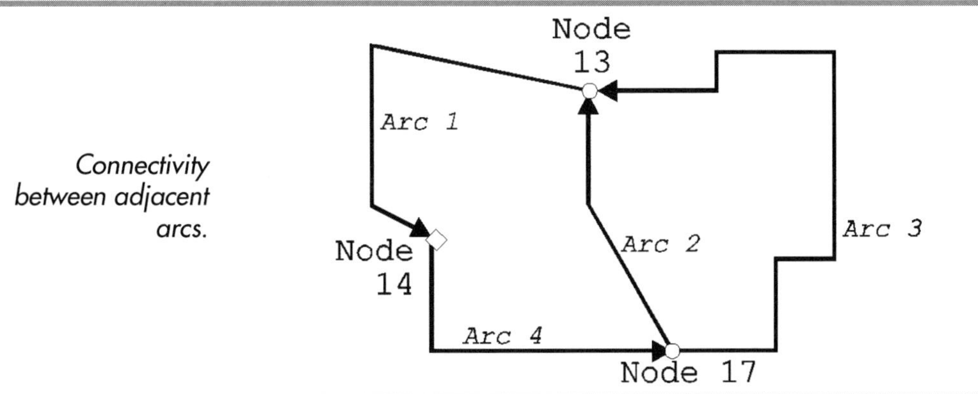

Connectivity between adjacent arcs.

Nodes are the endpoints of arcs and the common points to adjacent arcs. ARC/INFO keeps track of which arcs are connected to other arcs through shared nodes. This is called *Arc-Node topology* in ARC/INFO and is a basic GIS concept.

Listing of connectivity information.

Electric GIS

File Edit Commands Options Print Help

Record	FNODE#	TNODE#	LPOLY#	RPOLY#	LENGTH	EDGEROAD#	EDGEROAD-ID
1017	933	942	4	115	24.303	1017	1183
1028	933	952	115	115	140.405	1028	1182
1054	980	974	148	4	21.567	1054	1155
1063	980	974	85	148	419.558	1063	1205
1070	942	987	115	115	139.488	1070	1175
2718	942	929	0	0	1238.999	0	1183
2725	1114	1114	0	0	126.679	0	1315

Arcedit:

You can see the "from node" and "to node" whenever you list attribute records for a theme containing lines. They are stored in an Arc Attribute Table as the FNODE and TNODE items.

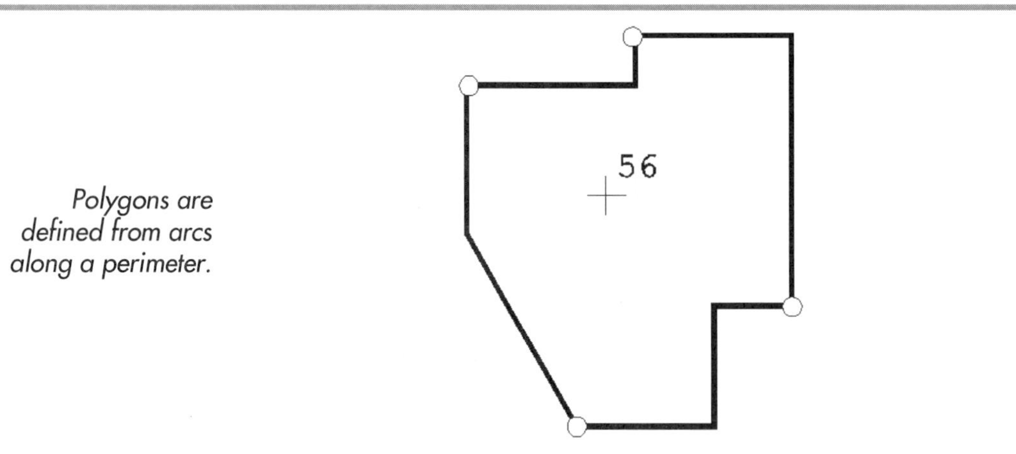

Polygons are defined from arcs along a perimeter.

A polygon is defined as a list of arcs. In ARC/INFO, this is referred to as *polygon-arc topology.*

Polygons are not stored as a closed series of coordinates because that would cause redundancy. For adjacent polygons, all common coordinates would be repeated, and you would end up with twice as many coordinates.

Using a list of arcs neatly solves the problem of redundant coordinates. If you were digitizing polygons such as forest stands, without polygon-arc topology you would have to digitize each forest stand boundary twice.

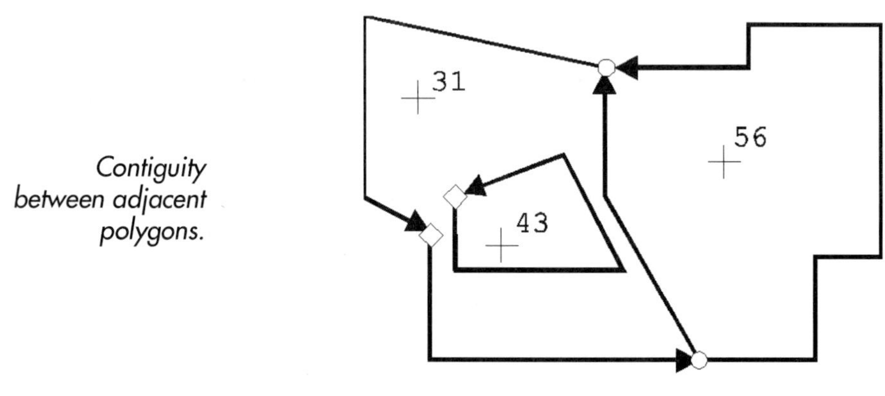

Contiguity between adjacent polygons.

The third basic topology is *contiguity*. In ARC/INFO, this is also known as *left-right topology*. Here's how this topology works. Each arc has a list of which polygons are on the right side and which are on the left side. Commands in ARC/INFO use this information to determine, from one polygon, what the adjacent polygons are.

A GIS Is a Geographic Database Manager

The most important point to grasp about a GIS is that it truly is a geographic database manager. This means that a GIS treats all geographic features as records in a database and not simply as graphics.

Common Concepts

Nearly all the concepts in traditional relational databases apply to GIS, but with the added dimension of geography. Here are some like concepts:

- Relational databases are defined to be well normalized, which means the tables are designed so that there is no duplication of data, just common database keys. A GIS also applies this concept through functions like polygon-arc topology and relates between attribute tables and other tables.
- A set of closely related tables form a relational database. A set of attribute tables closely related to geographic data sets form a theme.
- Although data in a relational database can be spread throughout a number of tables, a user can define a *view*, which is a way of looking at desired information from different tables all at once. GIS applications also present diverse information through a view, which is a list of themes with the context for display and descriptive information defined.
- All relational database managers provide the ability to produce reports, either on single tables or on views (multiple tables). With all the visualization techniques in GIS, a displayed or published map is, in essence, a report from the geographic database manager.

The Georelational Model

The georelational model is the cornerstone of GIS.

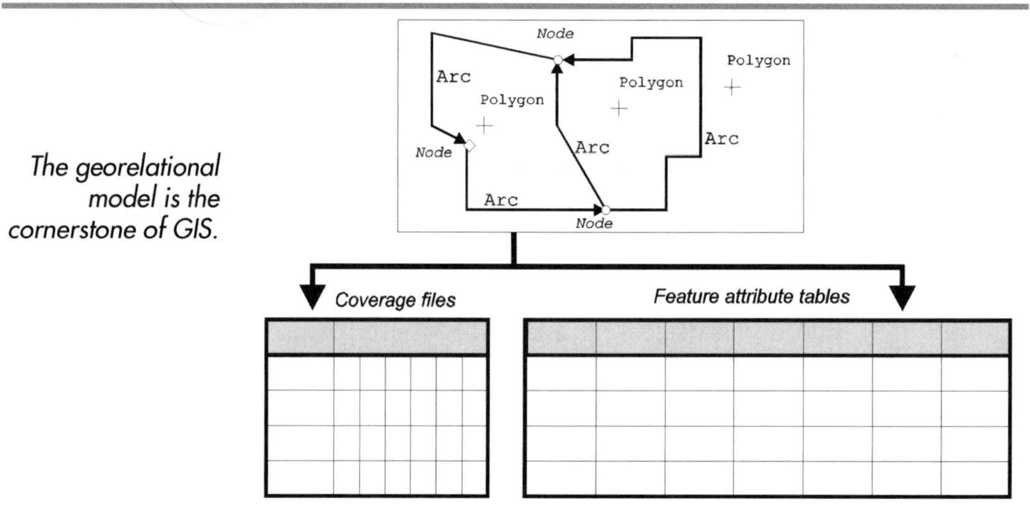

GIS builds the bridge between geography and descriptive information through the *georelational model.*

In brief, the georelational model means that for a given geographic data set, you can define an attribute table, for which there is a one-to-one correspondence between the two. Some items (database fields) are predefined for you in attribute tables, but you can add any other items you desire.

The beauty of the georelational model is that it lets you connect to any tabular databases, not just the ones embedded in the GIS. The attribute table is your gateway to other databases through the use of database keys. The other databases can either be internal or external to the GIS software.

We will discuss the georelational model in later chapters because this is the single most important characteristic that defines GIS.

How a GIS Visualizes and Reports Information

Output Options

These are some ways to present information from your GIS:

- You can use interactive screen display on a *workstation* or *windowing terminal.* A workstation is a self-contained computer with a graphics display. In most respects, it is similar to a personal computer. The chief difference between workstations and PCs has been the operating system, but with Windows NT, that distinction is diminishing.
- Most mapping shops are equipped with pen plotters, also known as vector plotters. The chief advantages of vector plotters have been ease-of-use and economy, but with low cost alternatives now appearing, vector plotters are gradually diminishing in popularity.
- Raster plotters are now taking the mapping industry by storm. There are two main varieties: electrostatic and ink jet plotters. Electrostatic plotters provide the best output but are costly. New, large format ink jet plotters provide near-electrostatic quality for a very reasonable cost. Ink jet plotters are capturing the major part of the GIS output market.
- Laser printers can also be used for presenting maps and geographic reports. ARC/INFO supports several standard formats used by desktop publishing software and printers, such as PostScript and Adobe Illustrator. Today, most laser printers are monochrome and small format, and this is the main limit to their more widespread use. As affordable color laser printers with larger formats appear, laser printers are poised to become the output option of choice.

Visualization Techniques

Within each of the above output options, there are a variety of ways to display information. These apply to point, line, and polygon features. Following are some of the many techniques that a GIS can use to communicate information.

Pie symbols.

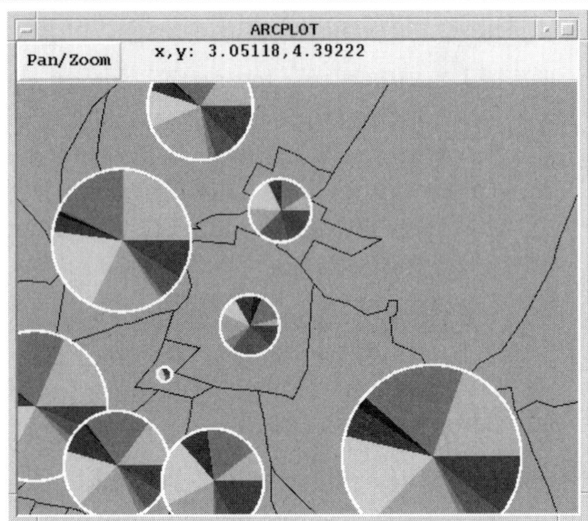

For features that contain information that can be organized into classes, *pie symbols* can effectively display the distribution of complex data at a glance. Besides showing the relative distribution of classes, the size of each pie symbol depicts the sample size of each feature. In Display GIScity, you can invoke pie symbols for U.S. Census attributes.

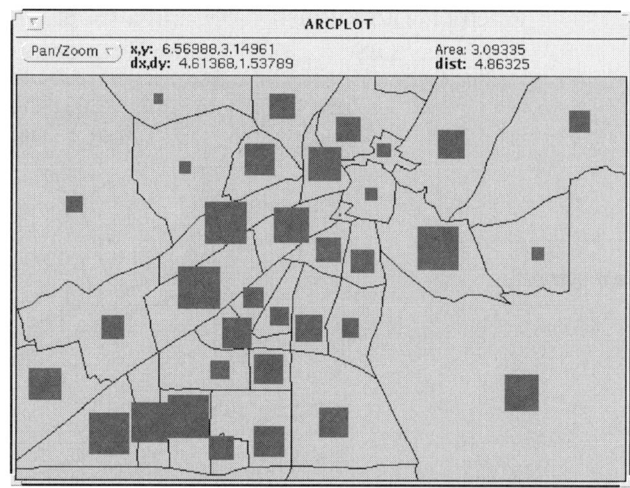

Graduated symbols.

When you want to display a numeric attribute, *graduated symbols* can illustrate that value clearly. When you display features, you can specify the output size of the feature to be portrayed in proportion to a numeric attribute.

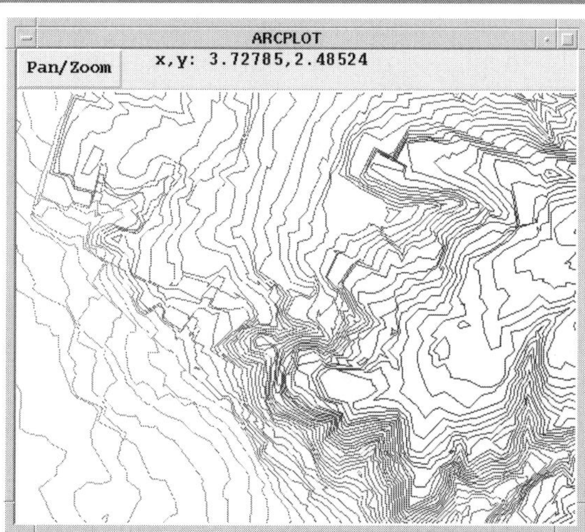

Color ramps.

An alternative to graduated symbols for symbolizing numeric attributes is *color ramps*. You can define a start color and attribute value, together with a finish color and attribute value to define a color ramp. In Display GIScity, a color ramp for contours is predefined for you.

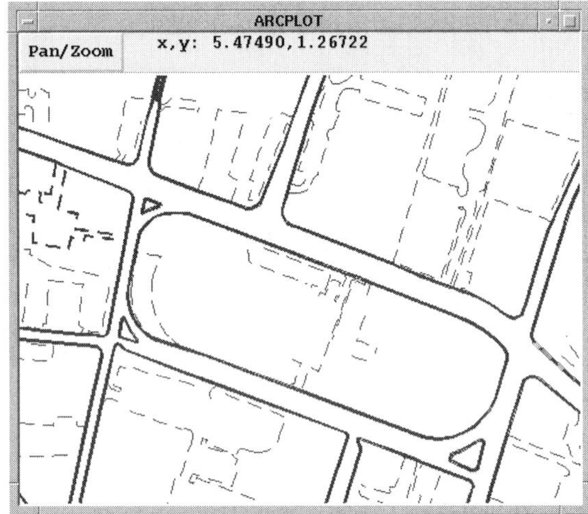

Discrete symbols.

Most maps utilize *discrete symbols* to differentiate feature types. For example, a road map distinguishes interstate highways from secondary roads through line symbols. This illustration shows discrete line symbols depicting road and driveway edges.

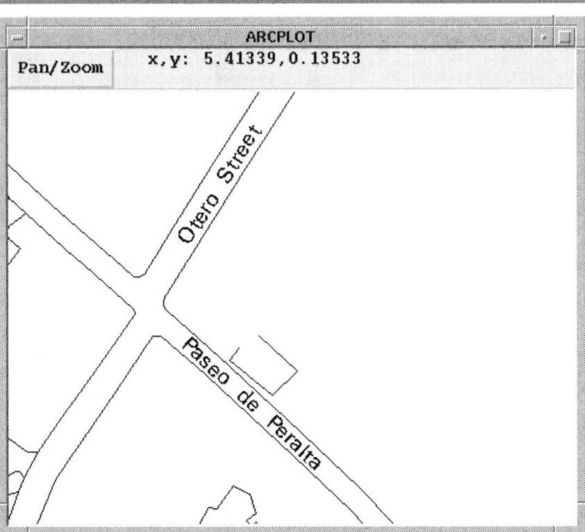

Text on features.

Maps also display descriptive information through *text*. The text can be generated upon demand, or edited and stored.

Graphs.

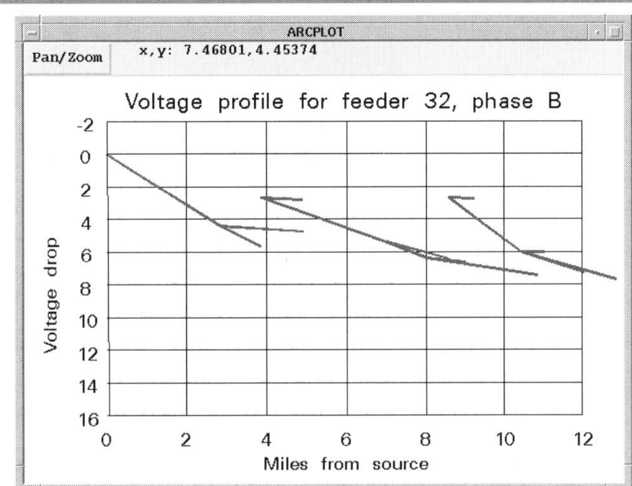

Complex information can be shown as *graphs*. The relationship between two attributes is shown. In a GIS, distance is a frequently used attribute. In the illustration, the graph shows electrical voltages downstream from a substation. The X axis is the distance downstream, and the Y axis is the voltage level.

Reporting Techniques

Descriptive data can be presented through ways similar to relational database management systems. Following are three examples.

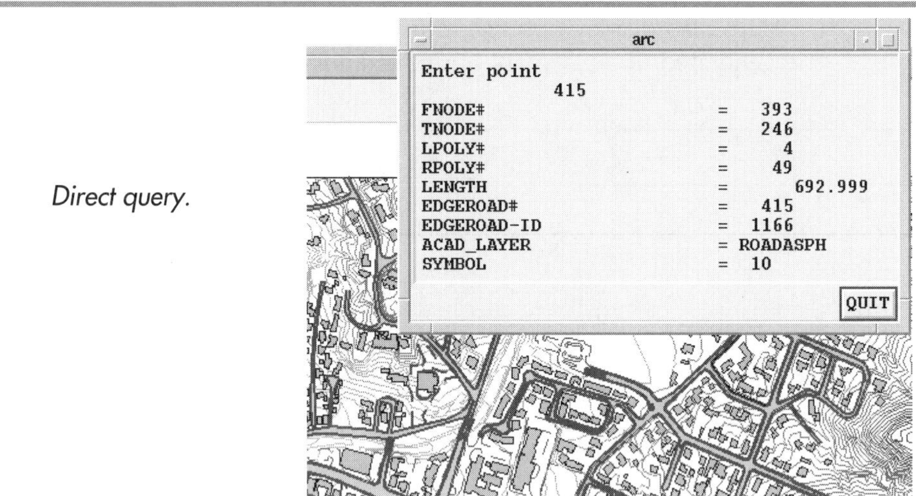

Direct query.

Individual features can be selected, and all attributes for that feature appear. A GIS provides general *query* capability.

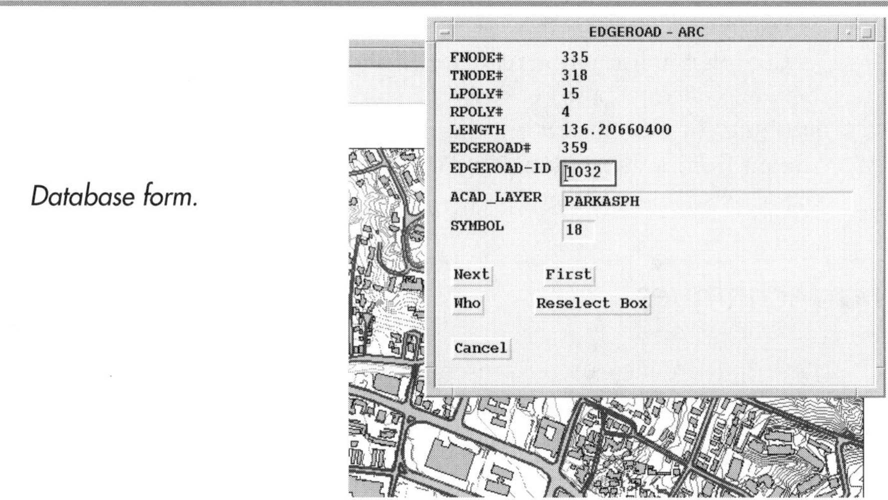

Database form.

When you have selected a group of features, you can invoke a *database form* for combined query and update. With a form, you can step through the selected features one by one.

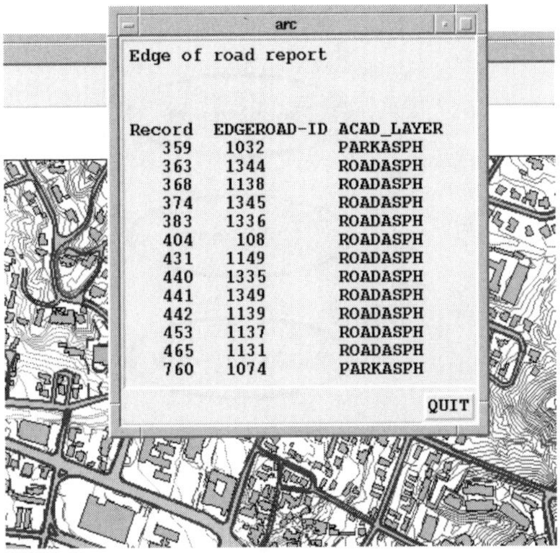

arc

Edge of road report

Record	EDGEROAD-ID	ACAD_LAYER
359	1032	PARKASPH
363	1344	ROADASPH
368	1138	ROADASPH
374	1345	ROADASPH
383	1336	ROADASPH
404	108	ROADASPH
431	1149	ROADASPH
440	1335	ROADASPH
441	1349	ROADASPH
442	1139	ROADASPH
453	1137	ROADASPH
465	1131	ROADASPH
760	1074	PARKASPH

QUIT

Tabular reports.

You can produce summaries of selected information for groups of features. Reports are the most common way database management systems provide information. GIS software typically contains basic internal tabular report generation capabilities. For more sophisticated reports, GIS software provides a gateway to the report generation tools provided with your commercial database management system.

Cartographic Techniques

While a GIS provides powerful database manipulation, cartographic output is an important subject that should not be neglected. We will see most of these techniques in Display GIScity, and we will learn more about cartographic commands in ARCPLOT. Following is a review of some techniques that GIS software employs to communicate information visually.

Sample themes within a view.

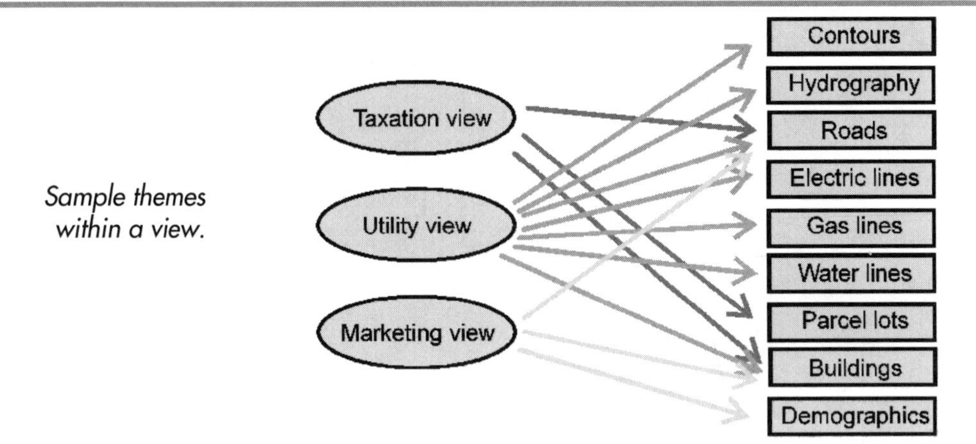

Views and *themes* specify which geographic data sets are to be presented and in which order. Polygons are usually drawn first. Otherwise, they would obscure the other map features. Then, lines are drawn, followed by points, and then text.

Scale thresholds.

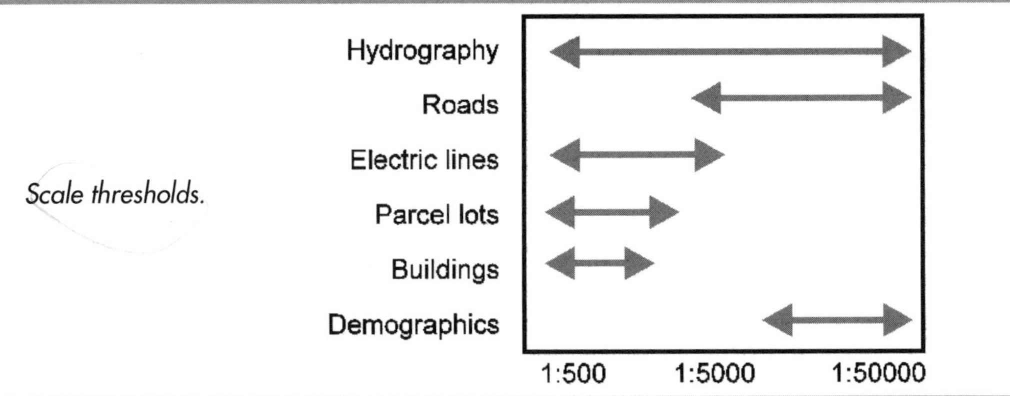

Scale thresholds are important to prevent cluttered map displays, especially for interactive display. Scale thresholds allow certain geographic data to be displayed, suppressed, or simplified, according to the scale used when a map is displayed or published.

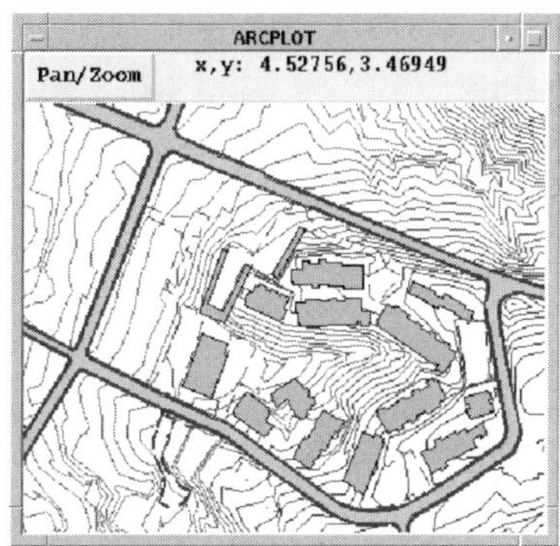

Displaying selected features.

When you publish or display geographic data, you can use *selected sets* to present only selected features. You can also use selected sets to highlight selected features and contrast them with other features. You can make special-purpose maps of objects, such as buildings, meeting any criteria you specify.

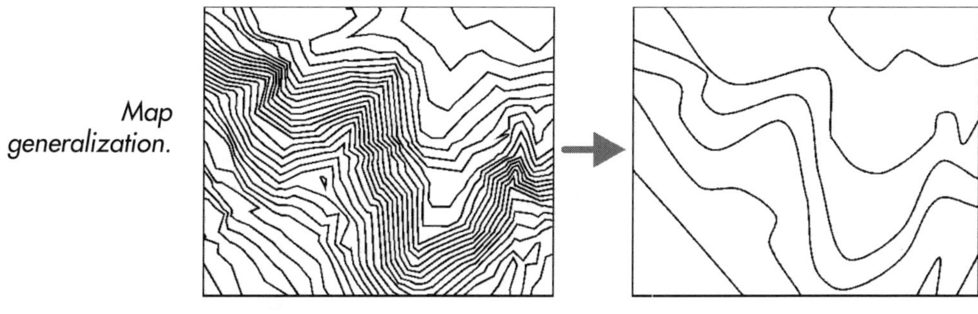

Map generalization.

Map generalization is a set of tools and standards to deliver geographic information in a form and detail appropriate for a given scale.

Map generalization works for taking detailed information and presenting it at a less detailed scale, but it does not work the other way around. Some of the techniques comprising map generalization are filtering out coordinates, changing polygon features into point features (or line features into point features), and resizing text.

This is a very sketchy description of map generalization. Entire books have been written on this subject.

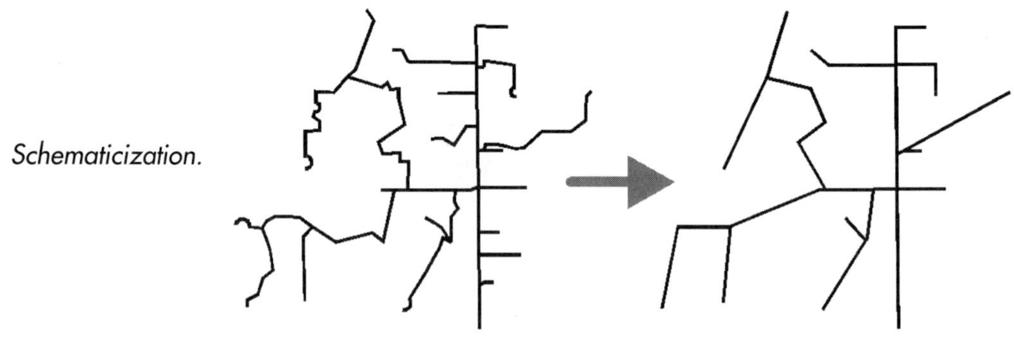

Schematicization.

Related to map generalization, *schematicization* is a mapping process that simplifies geographic features in order to portray only pertinent features and positions. Examples of schematics are bus route maps, subway maps, and utility schematic maps.

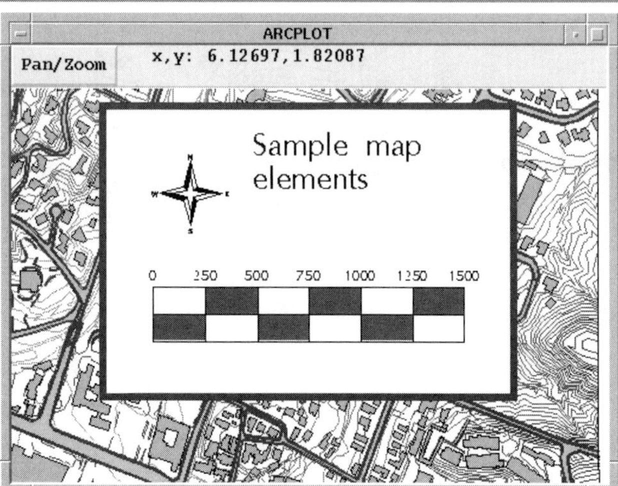

Map elements.

To provide a reference for a map, some *map elements* commonly displayed on published maps are north arrows, scale bars, sheet reference maps, legends, titles, neat lines, trim lines, and references to adjacent map sheets.

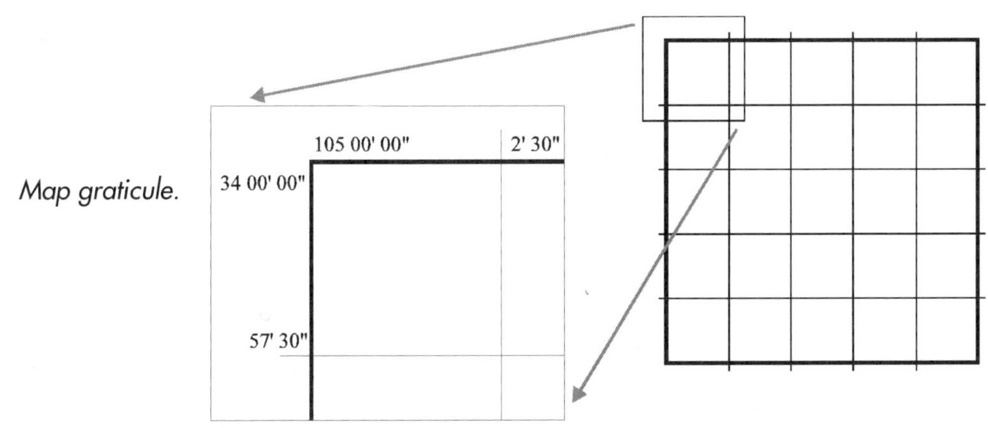

Map graticule.

A positional reference used on maps is the *map graticule.* The graticule comprises reference lines matching even-valued coordinate values, such as 34 degrees, 15 minutes. A map graticule has reference text associated with it.

With some careful work, you can devise a system to automatically publish map pages for a map book or atlas. This requires a significant up-front effort, but once established, you can make publishing maps as easy as selecting a map sheet and executing one command. I have spent a good portion of my work in building GIS applications on map book publication.

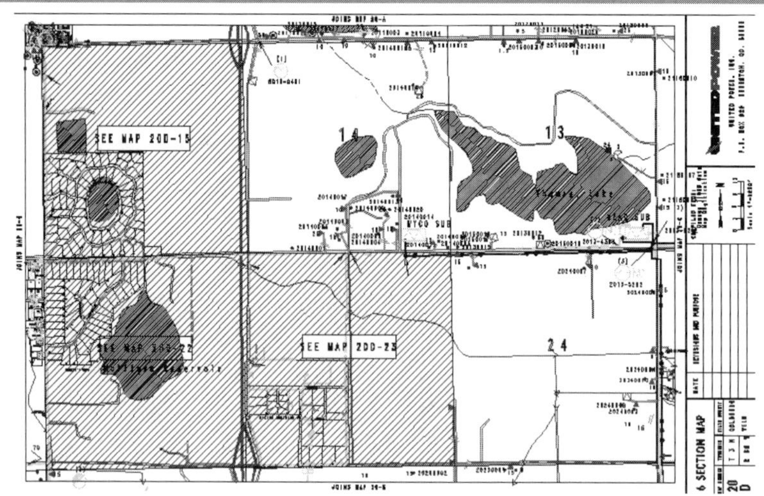

Map page publication.

Defining a GIS

What is GIS? Because GIS is a technology that is now experiencing broad usage, many people have different impressions about it.

A precise definition of GIS is elusive. You can ask many people and get many diverse answers. At the head of the chapter, I put forward a general definition. But the best way to understand GIS is to survey which applications people are using GIS for.

A Sampling of GIS Applications

Here are brief descriptions of just a few GIS applications found in the proceedings of the 1993 Annual ARC/INFO Users' Conference:

Natural Environment

- ❑ A numeric flood model for predicting flood levels and velocities.
- ❑ Mapping of shallow aquifers in karst terrains, which have a high potential for groundwater contamination.

- ❑ A marine surveillance and information system to monitor ocean pollution, temperature, current, salinity, and other characteristics.
- ❑ The adjudication of water rights along rivers.
- ❑ Assessment environmental impact to natural resources and endangered species along pipelines, roadways, and power lines.

Economics and Marketing

- ❑ Evaluation of the market economy's effect on indigenous peoples in the Amazon basin.
- ❑ Evaluation of store location, customer information, and sales boundaries for retail operations.
- ❑ Assessment of whether the placement of additional stores enhances or reduces profitability.

Forestry and Fire Management

- ❑ Fire management studies to plan controlled burns and analyze fire spread rates.
- ❑ Analysis of forest and brush land histories using historic aerial photography coverages.
- ❑ Mapping the risk of wildfires from human sources of ignition.
- ❑ Incorporation of knowledge-based system to provide support for emergency management of forest fires.

Transportation

- ❑ Long-range road and bridge maintenance programs.
- ❑ The management of ship traffic through a port.
- ❑ Modeling of integrated transportation flows—road, air, rail, and water.
- ❑ Mapping of human settlement versus transportation access.

Utilities

- Field inventory collection through GPS receivers for import to GIS.
- Modeling of an underground electric distribution network.
- Development of an interface between distribution analysis and GIS.
- The tracking of an electric plant from construction to retirement.

Data Conversion

- Tools to correct raster-to-vector conversion.
- The automation of road centerline generation from double edge road features.
- Problems in integrating map data from diverse sources.
- The quantification of errors in spatial accuracy.

Hydrographic Modeling

- Solid waste disposal in an urban setting.
- Automated delineation of flood boundaries.
- Calculation of potential recharge rates to an aquifer.
- Well surveys of transport of dissolved chemicals in an agricultural area.

Emergency Response

- Evacuation planning for nuclear reactor accidents.
- Geographic displays to assist fire and police dispatch in a municipality.
- Analysis of crime incidence.
- Coordination of state government resources in response to hurricane disasters.

Hazardous Materials

- Epidemiological study in the vicinity of a toxic waste site.

- ❑ Information system to support rapid deployment in response to oil spills.
- ❑ Dispersal of radio nuclides in the Arctic Ocean.
- ❑ The mapping of radon potential in houses.

Regional Government

- ❑ Site selection for solid waste disposal.
- ❑ Data collection and visualization for sewer management models.
- ❑ Development of address matches between postal codes and digital road maps.
- ❑ Modeling and regulation of urban growth.
- ❑ Identification of comparable property units for real estate appraisal.

New Technologies

- ❑ Integration of neural networks with spatial modeling to predict residence of prehistoric populations.
- ❑ Integration of ARC/INFO with a commercial database to provide environmental data in a heterogeneous client/server environment.
- ❑ Correction of satellite imagery geometry using GIS and GPS.
- ❑ The combination of GIS analysis with video presentations.
- ❑ Integration of pen based computers, GPS, GIS, and digital photography

This listing of GIS applications and topics, while brief and incomplete, illuminates the potential of GIS for analysis. Nearly every area of human endeavor has some application that can be served by GIS.

Internal Organization of ARC/INFO

Coverages and the Features They Contain

A recurring concept in this book is that a GIS is distinct from other graphical programs because it is, in essence, a geographic database manager. Because nearly all commercial GISes are layered on top of a relational database manager, GIS software has a chameleon-like ability to adapt itself to diverse applications.

Through the acquisition or development of macro programs such as GIScity, combined with data management by your GIS database administrator, you can have a GIS tailored to the precise needs of your organization. You can shield your end users from the internals of GIS software. In fact, some users of these applications may be unaware that a GIS tool kit is underneath the application they are running. (When that happens, you have truly succeeded in designing GIS applications!)

An important key to success with ARC/INFO is to become secure in your understanding of how spatial and descriptive information is stored in ARC/INFO. This chapter presents an overview of how ARC/INFO stores, manages, and retrieves geographic information from your computer. We'll look at the chief components of a *coverage*, including the possible feature types in a coverage, coverage files, and their related feature attribute tables. Not every file possible within a coverage is discussed—that would take several chapters—but you will learn about the key files that define a coverage.

The Anatomy of a Coverage

A coverage is the basic unit of storage for vector data sets in ARC/INFO. A coverage contains both descriptive and spatial information. The features you put into individual coverages correspond to a thematic map layer: roads, rivers and streams, buildings, utility lines, and so forth.

General Properties

The illustration shows three sample coverages: CENSUS, STREAMS, and OILWELL. The coverage names are arbitrary and user-definable. These coverages illustrate the most common feature classes within coverages: polygons, arcs, and points. A coverage includes the feature attribute tables as well as other files that store the positions of features.

You can think of a feature attribute table as a standard database table. That's just what it is, except you add database records by adding geographic features. The columns in the feature attribute tables are called *items*.

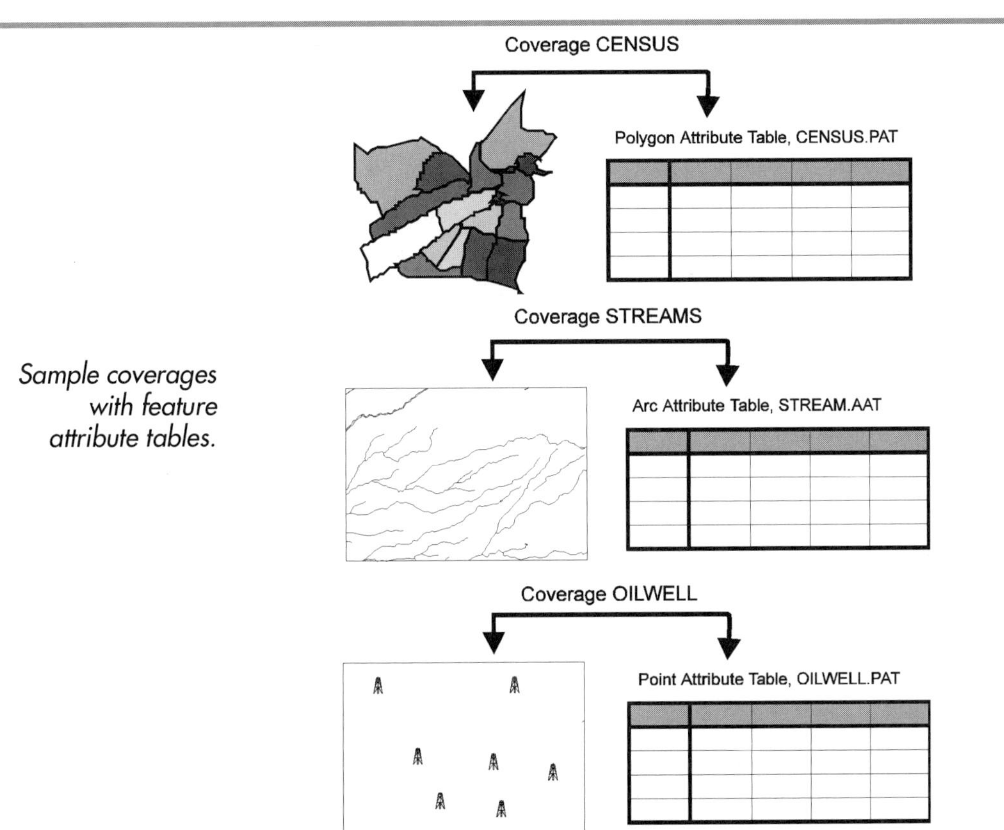

Sample coverages with feature attribute tables.

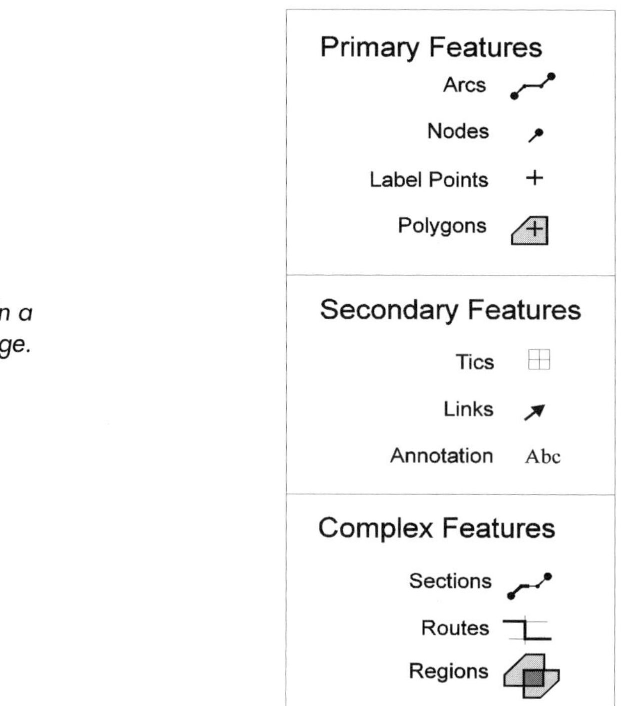

Features in a coverage.

Coverages contain a variety of features: arcs, nodes, label points, polygons, regions, tics, annotation, links, and route systems. These can be logically grouped into three categories: primary, secondary, and complex features.

- The **primary features** are the building blocks in ARC/INFO. They describe points, lines, and polygons.
- **Secondary features** are not fundamental to describing geography, as are the primary features, but they affect registration, correction, and provide description.
- The **complex features** are built upon the primary features. Route systems are aggregates of arcs; regions are aggregates of polygons.

The Georelational Model in ARC/INFO

The georelational model is the key method of data organization in ARC/INFO that distinguishes it from other graphics programs. This model provides the flexibility that we enjoy in tailoring GIS databases, and it is the gateway to external databases. Let's explore exactly how the georelational model is implemented in ARC/INFO.

A coverage includes two types of files: coverage files and feature attribute tables. Throughout this book, we explore feature attribute tables repeatedly, but not coverage files. ARC/INFO does a good job of hiding the details of coverage files from you. Even very experienced ARC/INFO users rarely know the internal details of coverage files. Here's a conceptual overview of coverage files and what they look like if you look at a coverage using your operating system commands:

A unique identifier ties together the coverage file and a feature attribute table.

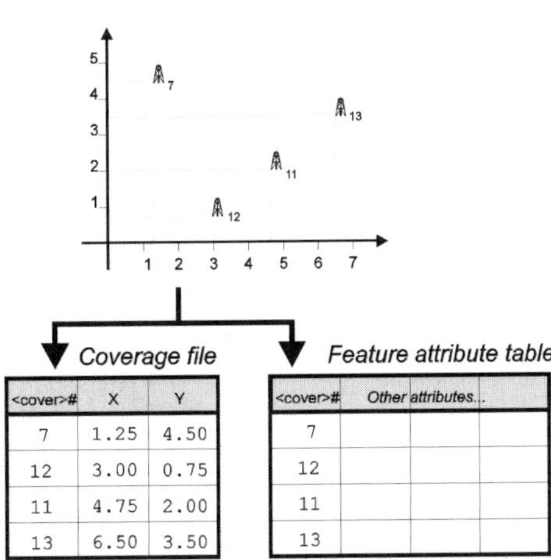

<cover>#	X	Y
7	1.25	4.50
12	3.00	0.75
11	4.75	2.00
13	6.50	3.50

<cover>#	Other attributes...		
7			
12			
11			
13			

Coverage files store all the X and Y coordinates for all features in a coverage. Some coverage files include other information, such as which arcs outline a given polygon.

One important coverage file is the ARC file. This file contains a data record for every arc, and within each record, the X and Y positions for every vertex. Another coverage file is the LAB file. The preceding illustration displays the contents typical of a small LAB file. There are many files that a coverage can contain, and we'll summarize some of them later in this chapter.

Coverages are the first half of the georelational model; feature attribute tables are the second half. There is a one-to-one correspondence between records in coverage files and associated feature attribute tables. This correspondence is managed through a unique identifier, called the *internal sequence number.*

ARC/INFO uses the internal sequence number to update coverage files and feature attribute tables simultaneously. When you digitize a new arc in ARCEDIT, it automatically adds a record with a common internal sequence number to both the ARC coverage file and the Arc Attribute Table (AAT). Likewise, when you delete an arc, ARCEDIT removes a record with the same internal sequence number from both the ARC file and the AAT.

WARNING: *The internal sequence number is the crucial link between coverage files and attribute tables. You should NEVER modify this number in any feature attribute table—a corrupted coverage may be the result.*

Identifiers for Coverage Features

All attribute tables contain two items (or columns) that are identifiers for features. We just looked at the internalsequencenumber, which appears in a listing of items as the coverage name followed by a #, and connects coverage files and feature attribute tables. (In diagrams and listings, we'll use "<cover>#"to denote the internal sequence number). ARC/INFO users commonly refer to this item as the *pound ID.*

A second identifier in a feature attribute table is the *feature ID* (or *user ID*). In a listing of items, you'll see this as the coverage name followed by "-ID". (In diagrams and listings, we'll use "<cover>-ID"to denote the feature ID item.) ARC/INFO users call this the *dash ID* item.

Like the internal sequence number, the feature ID is automatically assigned for you. This item is an optional database "hook" that connects to other databases. Unlike the internal sequence number, you can freely modify the feature ID. (The feature ID is not present within coverage files.)

Feature attribute tables can be connected to external tables two ways; through the feature ID (<cover-ID>) or another user-defined item (INVENTORY in example).

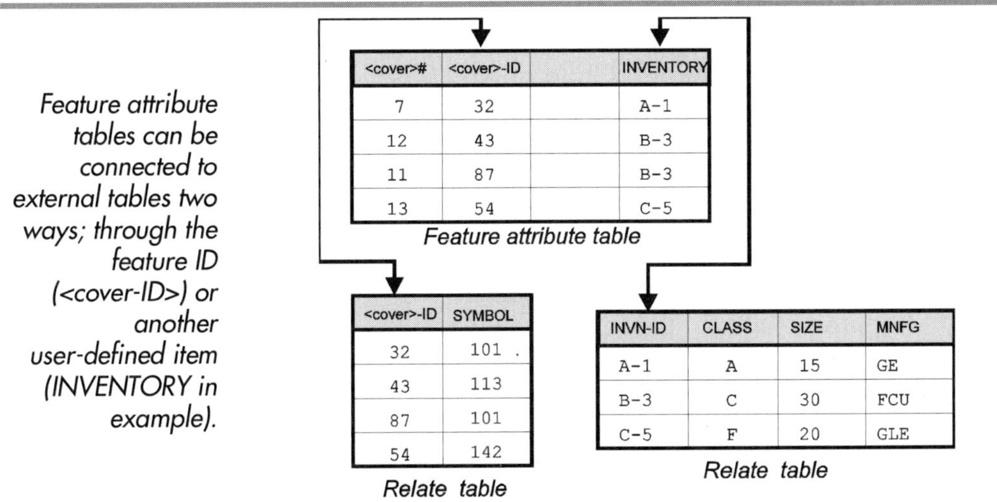

<cover>#
<cover>-ID
INVENTORY
7 32 A-1
12 43 B-3
11 87 B-3
13 54 C-5
<cover>-ID
SYMBOL
32 101
43 113
87 101
54 142
INVN-ID CLASS SIZE MNFG
A-1 A 15 GE
B-3 C 30 FCU
C-5 F 20 GLE

Feature attribute table

Relate table

Relate table

You can connect feature attribute tables to other tables. These can be relate tables, which can be associated with a coverage, or an external database, which can be a preexisting database storing information independent of the GIS.

To connect feature attribute tables to relate tables or external databases, you have two choices: you can use the feature ID, or you can define an additional, custom item for this purpose. You can also use both choices in a coverage, as shown in the preceding illustration.

For example, you can use the feature ID to tie a geographic feature to an external database for inventory purposes. However, there is a limitation to the feature ID—it can only be an integer value. Sometimes a relate table or external database has database keys that contain alphabetic characters, such as "NM28059." In this case, you must add a new item that allows alphabetic values. Many users routinely ignore the feature ID and establish a new item that is the identifier (or foreign key) to relate tables.

Primary Features of a Coverage

The primary features of a coverage are arcs, nodes, label points, and polygons.

Arcs

An arc is a connected string of line segments. Each line segment is delineated by a vertex. The vertices at the endpoints of arcs are called *nodes.*

Positions for arcs are stored in the ARC coverage file; attributes are stored in the Arc Attribute Table.

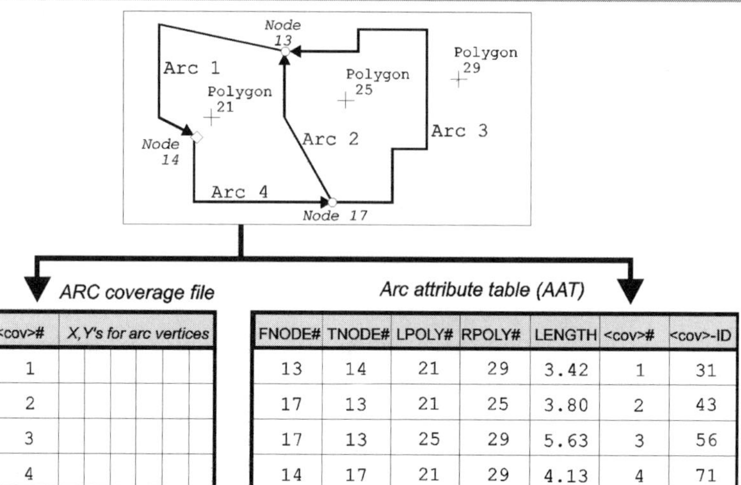

<cov>#	X, Y's for arc vertices
1	
2	
3	
4	

FNODE#	TNODE#	LPOLY#	RPOLY#	LENGTH	<cov>#	<cov>-ID
13	14	21	29	3.42	1	31
17	13	21	25	3.80	2	43
17	13	25	29	5.63	3	56
14	17	21	29	4.13	4	71

The most direct purpose of an arc is to represent linear geographic features such as streams and road centerlines. Other important uses for arcs are to delineate perimeters of polygons and to lay down the positions of route systems and regions.

The positional information for arcs is stored in the ARC coverage file. This file contains the X and Y positions of each vertex for every arc in the coverage, and the internal arc ID (the <coverage># item). The Arc Attribute Table predefines these standard items for you:

- ❑ **FNODE#.** The internal identifier for the "from node" of an arc.
- ❑ **TNODE#.** The internal identifier for the "to node" of an arc.
- ❑ **LPOLY#.** The internal identifier for the polygon on the left side of the arc, in the direction from the "from node" to the "to node." This item will be non-zero only if polygon topology is created for the coverage.

- ❑ **RPOLY#.** The internal identifier for the polygon on the right side of the arc. As with LPOLY#, this item will be non-zero only if polygon topology is created for the coverage.
- ❑ **LENGTH.** The calculated length of an arc. Whenever you modify the arc by changing a vertex position or other edit function, this value will be automatically recalculated for you.
- ❑ **<cov>#.** The internal identifier for the arc. You cannot modify this value; it uniquely ties each arc in the ARC file to a database record in the Arc Attribute Table.
- ❑ **<cov>-ID.** The user-assigned identifier for the arc. ARC/INFO automatically assigns a value for this item, but you may change this value. In most instances, you need not concern yourself with the value of this item.

You can add additional items to the AAT for your application. Any user-defined items must be inserted after the <cov>-ID item.

The FNODE# and TNODE# items are the keystones for arc-node topology. ARC/INFO uses these two items to determine which arcs share common nodes—the basis for network tracing.

The RPOLY# and LPOLY# items enable ARC/INFO to keep track of polygon topology. Because information about the adjacency of polygons is stored in an Arc Attribute Table, there is no need for ARC/INFO to separately store positions in the coverage files for polygons.

Nodes

Nodes are the end vertices of arcs. When two or more arcs join at a common node, there is only one database record in the Node Attribute Table.

Information for nodes is stored in the ARC coverage file and the Node Attribute Table. Nodes are also referenced in the Arc Attribute Table.

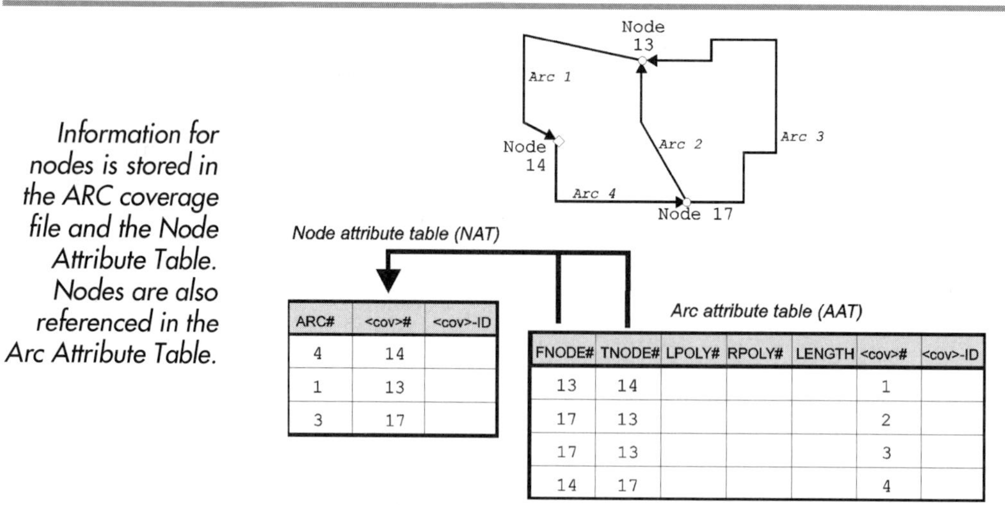

ARC#	<cov>#	<cov>-ID
4	14	
1	13	
3	17	

FNODE#	TNODE#	LPOLY#	RPOLY#	LENGTH	<cov>#	<cov>-ID
13	14				1	
17	13				2	
17	13				3	
14	17				4	

If arcs are the most important building block in a coverage, nodes contain the information about how arcs connect, and whether the arcs correctly define polygons or route systems. Nodes can also be used to store attributes at junctions on a network, such as whether a valve is present at a certain point or junction along a water pipe.

A Node Attribute Table can be optionally defined for nodes. These are the standard items predefined for you in the NAT:

- ❑ **ARC#.** The internal identifier for an arc connected at that node.
- ❑ **<cover>#.** The internal identifier for the node. You cannot modify this value. (This is the identifier pointed to by the FNODE# and TNODE# items in the Arc Attribute Table.)
- ❑ **<cover>-ID.** The user-assigned identifier for the node. ARC/INFO automatically assigns a value for this item, but you may change this value.

You can add additional items to the NAT for your application. Any user-defined items must be inserted after the <cover>-ID item.

The preceding graphic illustrates an interesting feature of the ARC# item. There may be several arcs that connect at any given node. So which one gets assigned as the ARC# for that node? ARC/INFO makes an arbitrary decision for you, and you don't need to care which one it picks.

ARC/INFO then uses that information for functions such as network tracing.

All feature attribute tables are optional. For example, you can have a coverage with arcs and no Arc Attribute Table. In practice, you nearly always define an arc attribute table. Nodes are a bit different. It often is unnecessary to keep attributes for nodes, so you won't define a Node Attribute Table as frequently as you would an Arc Attribute Table.

The three types of nodes.

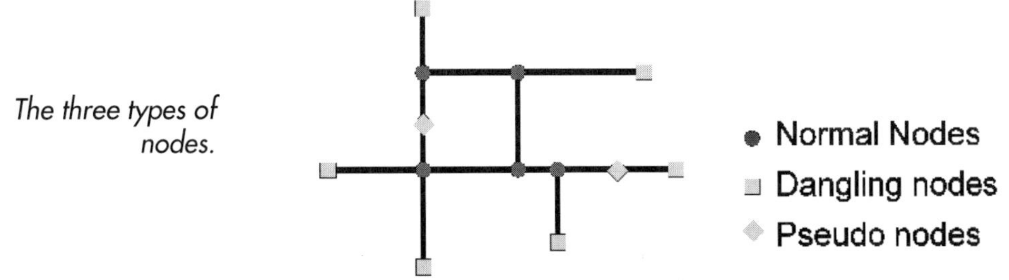

There are three types of nodes: normal, pseudo, and dangling. They are drawn in this illustration as they would be in ARCEDIT. Normal nodes are drawn as circles, pseudo nodes as diamonds, and dangling nodes as squares.

Normal nodes occur at the junctions of three or more arcs. They are called "normal" because, in a coverage with polygon topology, ideally every single node will be of this type. In a coverage like a soil map, dangling nodes are undesirable because they would leave a soil region open. Pseudo nodes may not leave polygons open, but in a strict sense, they are unnecessary to define polygons.

Pseudo nodes are sometimes added so that distinct attributes can be coded to arcs.

12" diameter

8" diameter

16" diameter

Pseudo nodes. If you digitize a free-standing polygon with a single arc joined at the begin and end nodes (an island polygon), there will inevitably be a *pseudo node* at the common begin and end nodes. No matter which ARCEDIT commands you may try, you cannot make that pseudo node disappear. This type of pseudo node is not considered an error.

Another purpose for a pseudo node is to mark where an attribute value changes along a line. For example, if you digitize a water pipe, and its diameter changes midstream, to model that change correctly you would digitize a *pseudo node* at the change point, and code the attribute values for pipe diameter distinctly on both sides. Again, this is not considered an error.

When you add arcs, ARCEDIT will sometimes create a pseudo node automatically for you, particularly when you are digitizing circular arcs or splines. This behavior seems a little peculiar at first. Here's the explanation: ARC/INFO allows a maximum of 500 vertices per arc. When a newly digitized arc exceeds that limit, ARCEDIT will automatically split that arc at the 500th vertex and make a pseudo node at that point.

TIP: *If you find yourself creating pseudo nodes to code different attribute values on arcs along a line, consider implementing dynamic segmentation instead. You can define sections and routes upon arcs to serve the same purpose. The decision whether or not to code attributes by the use of pseudo nodes or by sections and routes depends on whether the effort to model sections and routes justifies the additional benefit.*

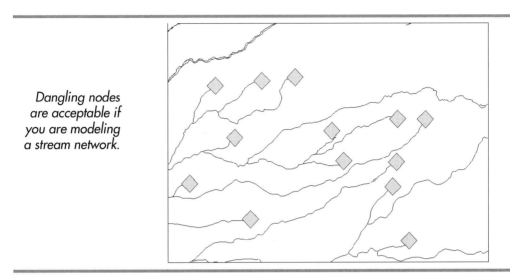

Dangling nodes are acceptable if you are modeling a stream network.

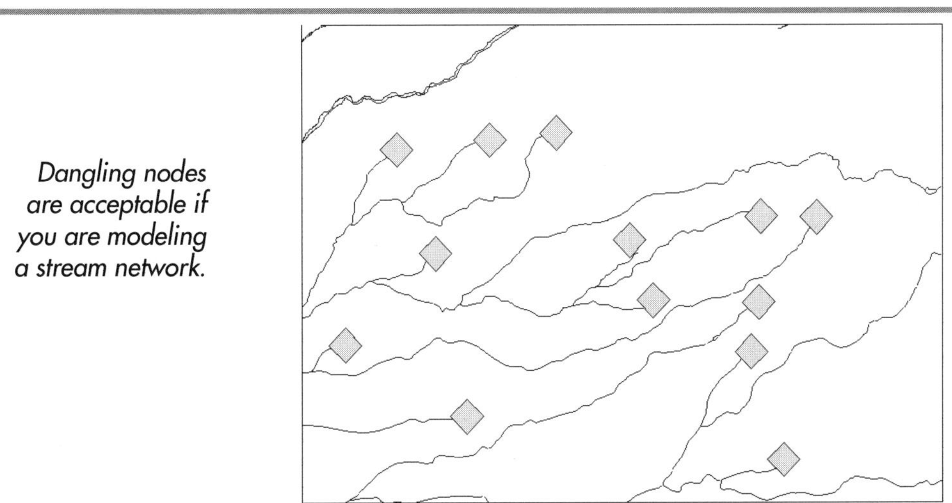

A *dangling node* is simply a node with no other arc connected to it. Dangling nodes are accepted when an arc represents a geographic feature that is unconnected, such as the source of a stream or a dead-end on a road.

In some ARC/INFO commands (such as the DRAWENVIRONMENT command in ARCEDIT), pseudo nodes and dangling nodes are called *error nodes.* The examples just discussed outline several instances where they are not at all errors. Yet, the appellation applies somewhat when you are digitizing coverages oriented toward polygon topology, such as lakes and geological zones. The point is that the presence of pseudo and dangling nodes can signal errors, and that ARC and ARCEDIT provide you with a suite of tools to detect and correct pseudo and dangling nodes.

Label Points

Label points have a dual purpose in ARC/INFO. Label points can represent geographic features that are too small to have a spatial extent when mapped in a coverage. Examples of small features are wells or electrical transformers. Only a single X, Y position is associated with a label point. Label points are also used to mark polygons. All the descriptive information for a polygon is associated with the label point within that polygon.

Two different feature attribute tables can be associated with label points.

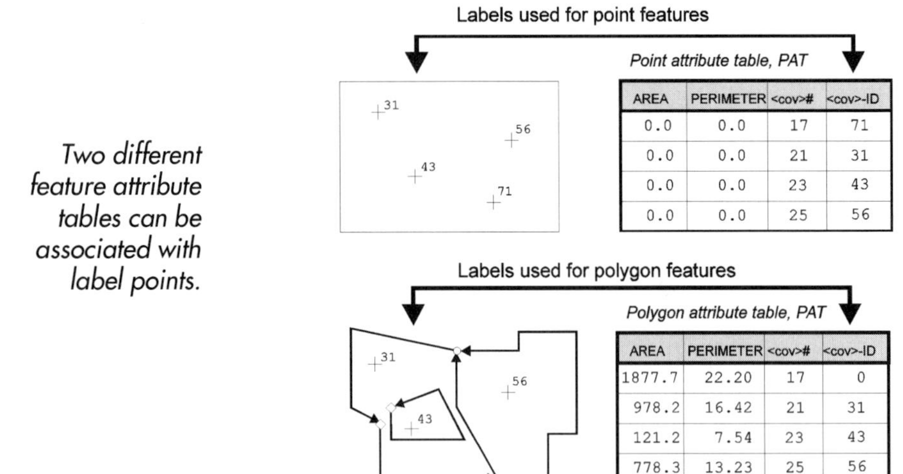

AREA	PERIMETER	<cov>#	<cov>-ID
0.0	0.0	17	71
0.0	0.0	21	31
0.0	0.0	23	43
0.0	0.0	25	56

AREA	PERIMETER	<cov>#	<cov>-ID
1877.7	22.20	17	0
978.2	16.42	21	31
121.2	7.54	23	43
778.3	13.23	25	56

If you set up label points to represent point features in a coverage, they relate to a Point Attribute Table. If you set them up to represent polygons, they relate to a Polygon Attribute Table. You will notice that both have the same items defined. However, the AREA and PERIMETER items are always zero values for Point Attribute Tables.

Whether label points represent very small features or information anchors for very large features, within any given coverage you can use label points for only one of these purposes. This makes sense, because if you allowed a dual identity for label points in a coverage, ambiguities would crop up about whether a label point modeled a point feature or marked an area.

In good GIS database design, this is not a significant restriction. When you determine thematic layers for your application, you usually isolate point, line, and polygon topology to separate coverages anyway. For example, in ARC/INFO it works well to model well heads (points), streams (lines), and lakes (polygons) as separate thematic layers (and consequently separate coverages). Some ARC commands that we will discuss later perform buffering and polygon overlays on these topologically distinct coverages.

If you use a label point for a point feature, then you are very interested in making the X, Y position as accurate as possible. However, if you are using label points to mark polygons, it makes no difference where the label points are placed within the polygon.

Label points can be rotated, which is handy if a symbol needs to align with another feature such as an arc. For example, you could rotate a symbol of a valve so that it is perpendicular to a water pipe.

ARC/INFO refers to both Point Attribute Tables and Polygon Attribute Tables as PATs. They both contain exactly the same predefined items:

- **AREA.** The calculated area of a polygon. For a Point Attribute Table, this item is present but always zero.
- **PERIMETER.** The calculated perimeter length of a polygon. For a Point Attribute Table, this item is present but always zero.
- **<cover>#.** The internal identifier for the point or polygon. You cannot modify this value.
- **<cover>-ID.** The user-assigned identifier for the point or polygon. ARC/INFO automatically assigns a value for this item, but you may change this value. Normally, you do not care about the value of this item.

You can add additional items to the PAT for your application. Any user-defined items must be inserted after the <cover>-ID item.

Polygons

Polygons are geographic features that span a discrete area. Examples are geologic zones, islands, lakes, and property parcels.

A polygon in ARC/INFO is represented by a set of arcs that exactly define its perimeter, and a single label point to associate all descriptive information for that polygon.

Because of the way polygons are pieced together from arcs and a label point, no positional information is stored directly with the coverage files for polygons. Positions are referenced indirectly to the set of arcs that span a polygon.

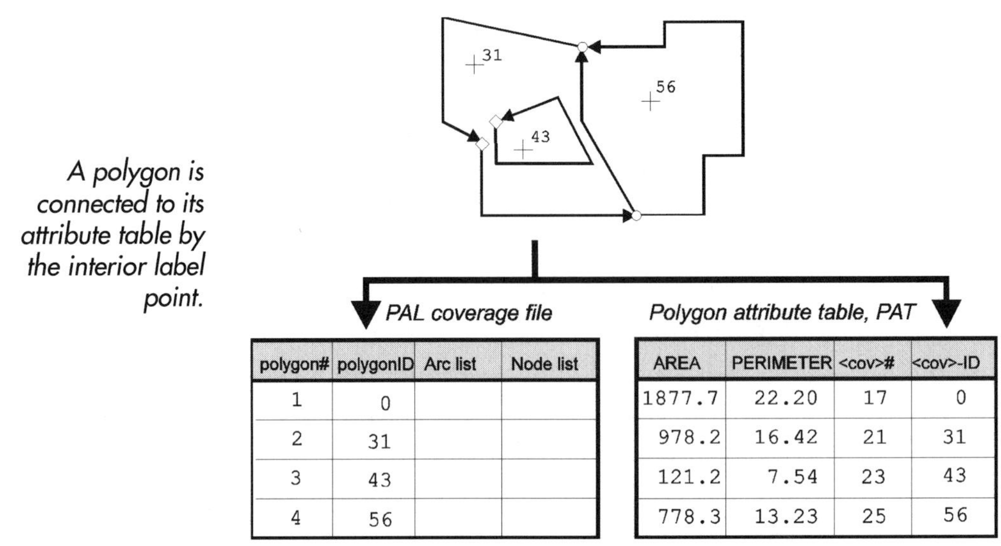

polygon#	polygonID	Arc list	Node list
1	0		
2	31		
3	43		
4	56		

AREA	PERIMETER	<cov>#	<cov>-ID
1877.7	22.20	17	0
978.2	16.42	21	31
121.2	7.54	23	43
778.3	13.23	25	56

A polygon is connected to its attribute table by the interior label point.

This information is contained within a node list and an arc list in the PAL coverage file. These are lists of all the arcs and nodes that delineate each polygon. The PAL coverage file is also designed to handle situations such as island polygons, or polygons that contain dangling arcs. This gets somewhat complicated, so we'll skip the details. All told, four coverage files are necessary to model polygons: PAL, CNT, LAB, and ARC.

In the preceding illustration, you'll notice that the first record in the Polygon Attribute Table has a User ID of 0. This is called the *universe polygon*, and has an area equal to all the other polygons combined. The universe polygon has no explicit label point, and you'll normally ignore this record in the PAT. ARC/INFO creates the universe polygon for its internal program logic.

Polygon Attribute Tables contain these predefined items:

- **AREA.** The calculated area of a polygon.
- **PERIMETER.** The calculated circumference of a polygon.
- **<cover>#.** The internal identifier for the point or polygon. You cannot modify this value. (This is the identifier referred to by the LPOLY# and RPOLY# items in the Arc Attribute Table.)

- **<cover>-ID.** The user assigned identifier for the point or polygon. ARC/INFO automatically assigns a value for this item, but you may change this value.

If a polygon lacks a label point, no descriptive information for the polygon is kept, and this is considered an error. Likewise, two or more label points inside a polygon is considered an error. The information is ambiguous: which label point has the correct descriptive information? For coverages with polygon topology, exactly one label point in the interior of each and every polygon is the goal. ARC and ARCEDIT contain a suite of tools to attain and ensure this.

You can add additional items to the PAT for your application. Any user-defined items must be inserted after the <cover>-ID item.

NOTE: *Later, you will find that some ARC commands will place or shift label points within polygons to a mathematically calculated centroid position. This positioning reduces the chance that a label point may end up outside its polygon during GIS operations such as projecting from one coordinate system to another. Otherwise, positional accuracy of label points marking polygons really is unimportant.*

Secondary Features of a Coverage

The secondary features included in a coverage are annotation, tics, and links.

Annotation

Annotation is used to place descriptions of features in a coverage. Unlike text, which is used for temporary display of attributes, *annotation* is a permanent component of a coverage. You can think of annotation as the intersection between physical geography and human culture. While arcs may depict natural features such as streams and contours, annotation describes features with their historic and cultural context.

Annotation in one coverage can be related to several text attribute tables.

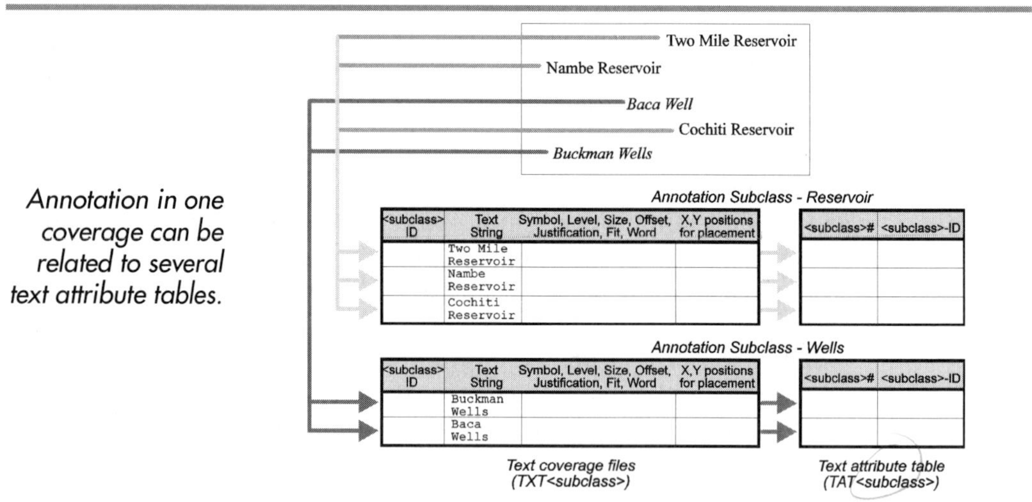

Annotation can be organized into subclasses You can think of subclasses as "layers" that group similar annotations together. Specifying subclasses allows you to easily display and edit annotation selectively. Later, we'll learn how to create subclasses and access annotation in subclasses.

In the illustration, there are only two standard items for the Text Attribute Table, <subclass># and <subclass>-ID. The real usefulness of the TAT arises when you define additional items, such as an identifier to another table.

NOTE: *ARC/INFO maintains an older method of aggregating annotation, called levels. While levels are still supported to provide compatibility with earlier versions of ARC/INFO, there is no reason for new users to use levels—the functions offered by annotation subclasses are a rich superset of levels.*

Several ways to position annotation.

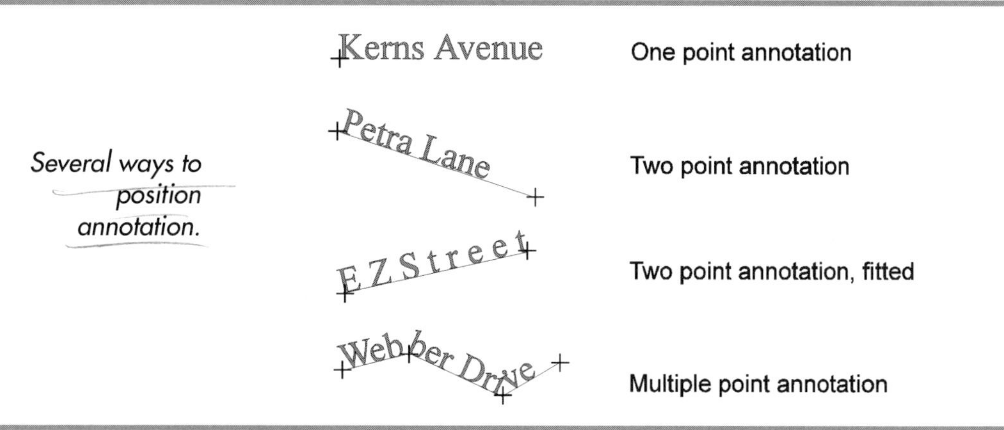

Annotation can be positioned several ways: at a single point for horizontal placement, at two points for placement at an angle, or along several points to place curved annotation.

Offset annotation from features.

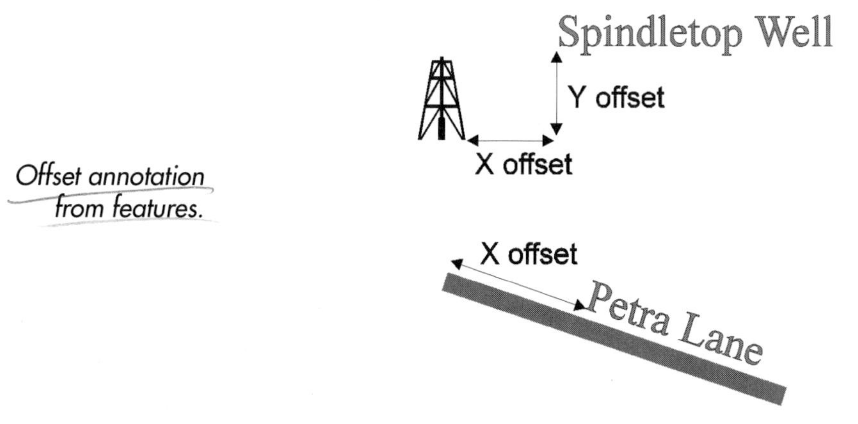

When you place annotation in reference to other features, such as label points or arcs, you can specify an offset distance for annotation to be added.

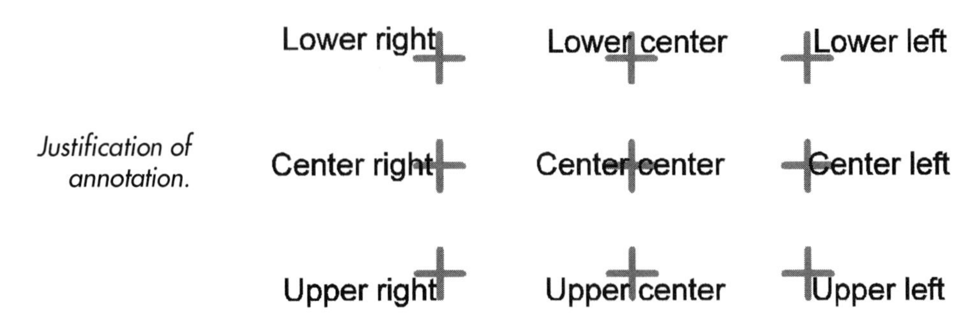

Justification of annotation.

When you add annotation, you can set nine justification positions. The default value is Lower Left. When you place annotation on linear features, the Center Left, Center Center and Center Right positions can be specified for placement.

Tics

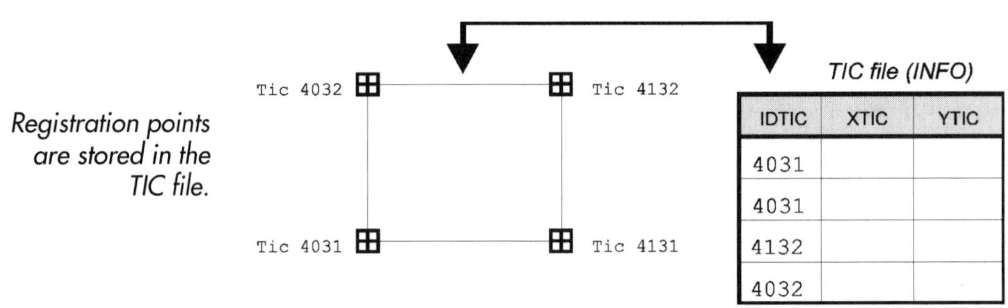

IDTIC	XTIC	YTIC
4031		
4031		
4132		
4032		

Registration points are stored in the TIC file.

Tics are used for map registration. They are entered into coverages by digitizing, conversion, survey input, and increasingly, GPS (Global Positioning System) data collection.

Tics are usually defined as points which you can recognize and precisely locate on the ground, such as a survey control points, or as mathematically defined points, such as the corners of U.S.G.S. quadrangle maps.

Tics are stored in an INFO file. There are three items defined for tics:

- ❑ **IDTIC** is a User ID for tics. You can modify this value.
- ❑ **XTIC** is the precise X coordinate of the tic.
- ❑ **YTIC** is the precise Y coordinate of the tic.

Links

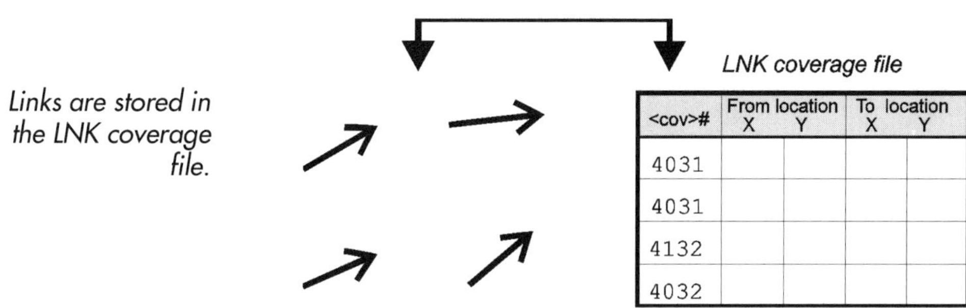
Links are stored in the LNK coverage file.
LNK coverage file

<cov>#	From location X	Y	To location X	Y
4031				
4031				
4132				
4032				

Links are rubbersheeting vectors. They are not normally displayed on maps, but are used to correct distortions in GIS databases. They are most often used to digitize features from old, inaccurate maps and adjust the positions to tie to a more accurate GIS land base positions.

All information about links is stored in a coverage file called LNK. There is no attribute table associated with links.

Complex Features of a Coverage

Routes and Sections

Routes are complex linear features built on arcs. A *route* is a path along connected arcs from one point to another. A *route system* is a collection of routes. Routes are made up of sections A *section* is either a whole arc or a portion of an arc. Many sections are whole arcs.

Routes are primarily used for transportation analysis. A good example of a route system is a bus route system. If you look at a map of city streets, you will see that the major streets are overlaid with city bus routes. Usually these routes are separate, but sometimes bus routes overlap one another. It is possible, but very complex, to model something like a bus route system using just Arc Attribute Tables.

Route systems simplify the modeling of transportation routes by solving certain problems for you. You don't have to digitize multiple arcs for common route segments. You don't have to split arcs if a route begins in the middle of an arc. Instead of representing a location as an X,Y coordinate, you can define a location as a linear distance from the beginning of a route (such as a milepost along a highway).

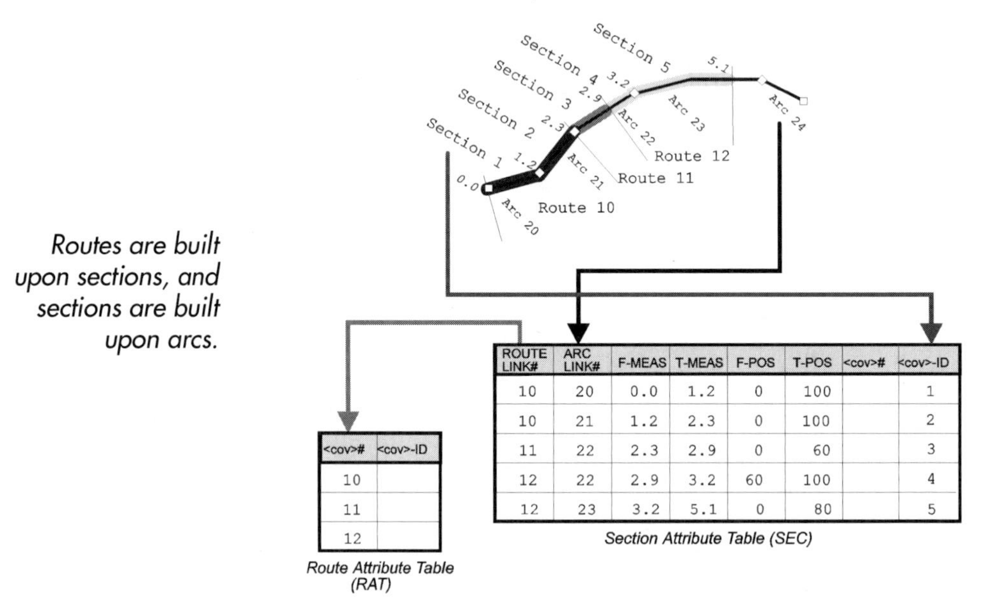

Routes are built upon sections, and sections are built upon arcs.

<cov>#	<cov>-ID
10	
11	
12	

Route Attribute Table (RAT)

ROUTE LINK#	ARC LINK#	F-MEAS	T-MEAS	F-POS	T-POS	<cov>#	<cov>-ID
10	20	0.0	1.2	0	100		1
10	21	1.2	2.3	0	100		2
11	22	2.3	2.9	0	60		3
12	22	2.9	3.2	60	100		4
12	23	3.2	5.1	0	80		5

Section Attribute Table (SEC)

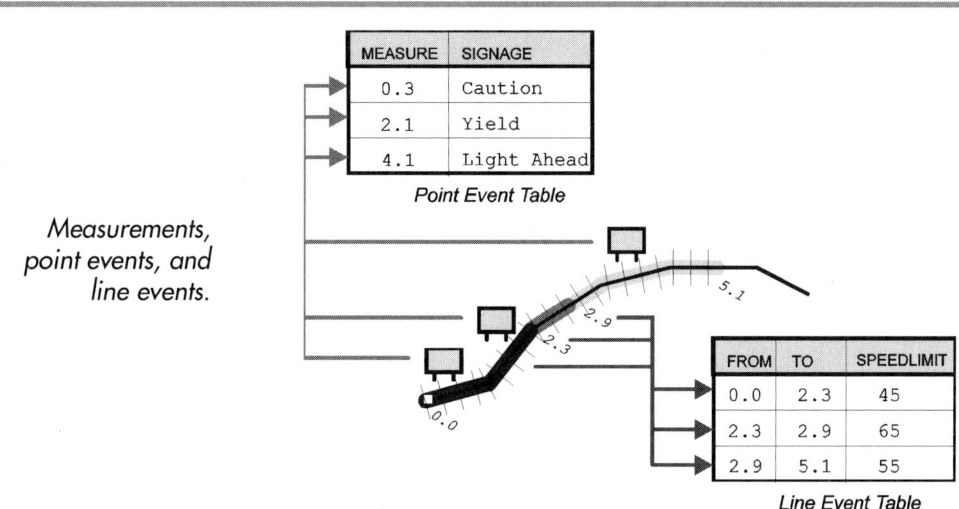

MEASURE	SIGNAGE
0.3	Caution
2.1	Yield
4.1	Light Ahead

Point Event Table

FROM	TO	SPEEDLIMIT
0.0	2.3	45
2.3	2.9	65
2.9	5.1	55

Line Event Table

Measurements, point events, and line events.

Once you have established a route system, ARC/INFO commands can overlay measurements upon routes. From a start point of a route, ARC/INFO calibrates and displays linear measurements downstream from that position.

Upon these measurements, you can define *point events* and *line events.* An example of a point event would be the location of transportation signs. Modeling signs as point events makes sense because transportation departments reference them by a milepost (or station) instead of an X,Y position. Line events could be used for a continuous characteristic of a road, such as the number of lanes, speed limit, or pavement status. Line events have a start milepost and an end milepost.

The capability provided by the route system model and related commands in ARC, ARCPLOT, and ARCEDIT is called *Dynamic Segmentation.* This is sometimes abbreviated as DynSeg. For further information on this complex and specialized function, please see the ARC/INFO documentation.

Regions

Regions are compound polygons. Sometimes you want to associate several polygons into one entity. For example, consider a map of the fifty states making up the United States. Digitizing the majority of them as polygons would not pose any particular difficulty. However, several states would be problematic. Michigan is divided into two parts by Lake Superior. Hawaii comprises several islands. Alaska has many islands in the Aleutian chain.

You would end up with more than 50 polygons in a Polygon Attribute Table, but you want to relate those polygons to a database you have for the fifty states. What to do? By aggregating the discontinuous states into regions, you can avoid the sticky many-to-one relationship between geographic feature and attribute tables.

There are three types of areas that regions are used to model.

- **Nested areas.** Polygons that can be combined to form other polygons. For example, in the U.S., counties can combine to form states.
- **Associated areas.** Polygons that are separate, but model a common type of feature, such as the preceding example of the discontinuous states.
- **Overlapping areas.** Polygons that are distinct, but overlap, such as wildlife habitats. There can be considerable overlap in ranges of different species, but for some types of analysis, you want them in the same coverage.

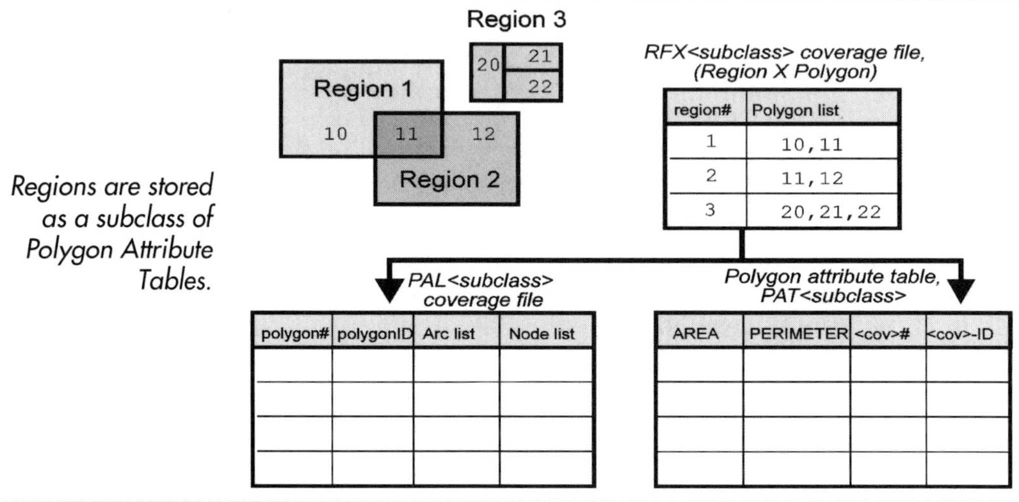

Regions are stored as a subclass of Polygon Attribute Tables.

Like route systems and annotation, regions are subclasses in coverages. From our example for nested areas, you can have one set of regions for counties and another (aggregating counties) to define states.

The relationship between arcs, nodes, and regions is identical to that of polygons. Regions incorporate one additional file called the RXP file, which is essentially a list of polygons that define a region.

NOTE: *Editing and displaying the complex coverage features, sections route systems, and regions, is an advanced topic beyond the scope of this book.*

Contents of Coverages

Summary of the main files present in coverages.

Summary of Coverage Files and Tables	Coverage Files	Attribute Tables
Primary Features		
Arcs	ARC	AAT - Arc Attribute Table
Nodes	ARC	NAT - Node Attribute Table
Label Points	LAB	PAT - Point Attribute Table
Polygons	PAL, CNT, LAB, ARC	PAT - Polygon Attribute Table
Secondary Features		
Tics	–	TIC file
Links	LNK	–
Annotation	TXT<subclass>	TAT<subclass> - Text Attribute Table
Complex Features		
Sections	ARC	SEC<subclass> - Section Attribute Table
Routes	ARC, SEC	RAT<subclass> - Route Attribute Table
Regions	PAL,CNT,ARC	PAT<subclass> - Polygon Attribute Table

These are the main files and tables that make up a coverage. Any coverage contains a subset of these features, depending on what features you add to the coverage.

There are many other files that a coverage can contain which are beyond the scope of this book. Here are three additional files in a coverage:

- **BND** is the boundary file for a coverage. This is an INFO table that just stores the minimum and maximum X and Y coordinate values of all the features in a coverage. These coordinates define a bounding box around the entire coverage.
- **TOL** is the tolerance file for a coverage. Any tolerances, such as weed tolerance and grain tolerance, that you explicitly set for a coverage, are stored here. This is a binary file.
- **PRJ** is the projection file for a coverage. This file contains the information about the coordinate system that goes along with a coverage. For example, a coverage may contain coordinates in the Universal Transverse Mercator system, Zone 13, on the North American Datum of 1983 (NAD83).

Coordinate Precision

One more characteristic of coverages is the *coordinate precision level.*

By default, coordinate precision is set to *single precision,* which means that coordinates are stored to about 7 digits of accuracy. If your coordinate system has X and Y values in the millions, then your expected coordinate accuracy is one or a few units (feet or meters). For many applications, this is sufficient because the intrinsic accuracy of the map positions is less than this. If you digitized from a map manuscript, your geographic accuracy may be 10 or 40 feet, for which single precision coordinates would probably be adequate.

For some coverages, the method of entering coordinates may be quite precise. Examples of these methods are entering survey data through COGO (coordinate geometry), and adding positions entered with GPS (Global Positioning System) receivers. The accuracy from these sources may exceed the precision level of single precision coordinates. For these coverages, you can define coverage coordinates to be stored as *double precision.* This means that the coordinates will remain accurate for up to 16 digits. This is more than adequate for any mapping coordinate system.

If you are in doubt about whether to use single precision or double precision, choose double precision. The only drawback of using double precision coordinates is that your coverages will take more disk space. For most users, the assurance that the coordinates you store are not degraded outweighs the increase in coverage size.

NOTE: *Setting double precision coordinates does not automatically "improve" your coordinate accuracy. It means that the method of storing coordinates does not degrade the accuracy of your collected data to just 7 digits. The procedures and technology that you use for data collection still determine the geographic accuracy of the features you collect into a coverage.*

Design Considerations for Coverages

Two approaches to defining coverages.

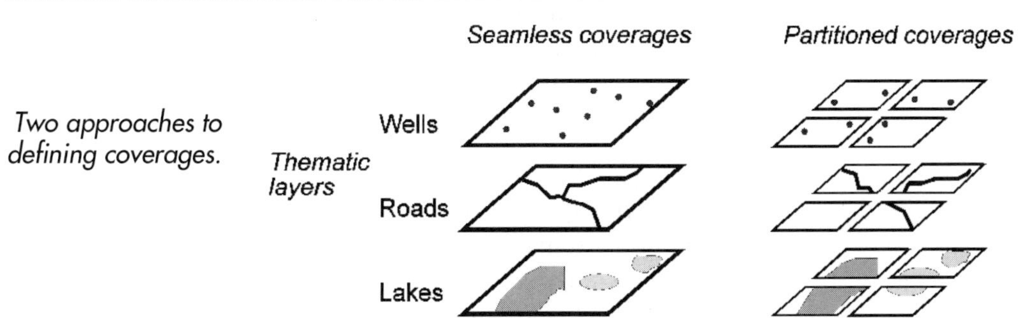

All coverages span thematic layers. All ARC/INFO applications interact with at least several coverages. Within thematic layers, you can set up your coverages in two basic ways:

- **Seamless coverages** span the entire area of interest to your organization.
- **Partitioned coverages** subdivide thematic layers into map sheets or administrative districts.

In the preceding illustration, three thematic layers are shown. The seamless coverage design includes three coverages. The partitioned coverage design contains four map sheets for each thematic layer. Twelve coverages are required in the example for partitioned coverages.

What are the advantages and disadvantages of the two approaches?

The seamless coverage design is usually best. The key advantage is that polygon overlay and network analysis commands are kept simple. You don't have to introduce logic to your AML applications to cross from one map sheet to another, as in a network trace.

Another advantage to the seamless design is that you can establish ties to external databases more naturally. For example, a seamless coverage containing 3,000 label points depicting well heads can cleanly tie to an external database for well head protection that has 3,000 database records. When possible, a one-to-one correspondence between geographic features in a coverage and records in an external database is most desirable.

The seamless design does not preclude map page publication by map sheet boundaries. ARC/INFO AML applications can be constructed to produce plots by map sheets.

The seamless coverage design fails when your coverages become very large (over 100,000 features). In that instance, you have three choices:

- Implement a partitioned coverage design and build an AML application to manage the separate partitions.
- Use the Map Librarian function in ARC/INFO. Map Librarian manages subdivision of thematic layers in tiles, which are basically partitioned coverages.
- Acquire the optional ARC/INFO module called ArcStorm (Arc Storage Manager).

The first option, managing your own partitioned coverages, is possible, but the substantial development effort rarely justifies this approach. The second option, Map Librarian, was formerly the preferred method but had some limitations in linking geographic features to external databases.

The third option, using ArcStorm, is the most desirable. ArcStorm is newly introduced with version 7.0 of ARC/INFO and can transparently partition very large coverages automatically. Therefore, you can enjoy the simplicity of a seamless coverage design and avoid the performance constraints that very large data sets pose.

ArcStorm has other desirable functions. It can create and manage a transactional history on your GIS databases. Additions, changes, and deletions are all logged, and you can restore the GIS database as it was at any time in the past. ArcStorm also has a capability called *feature level locking*, which allows different users to edit a coverage simultaneously, while locking selected geographic features.

ArcStorm is a server of geographic information.

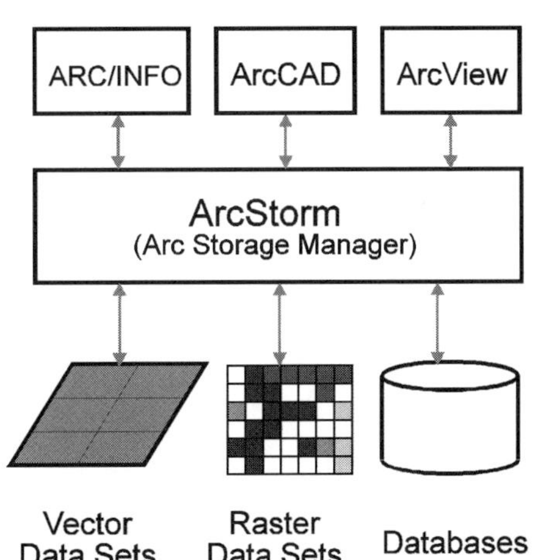

ArcStorm software delivers information from all ARC/INFO geographic data sets, coverages, grids, images, and so forth, and establishes logical connections to external databases such as Oracle, Sybase, Informix, and Ingres. ArcStorm is independent of ARC/INFO, and can deliver data to three ESRI software packages, ARC/INFO, ArcView, and ArcCAD.

While Map Librarian will be maintained for some time within ARC/INFO, there is no good reason for new ARC/INFO users to implement Map Librarian. The functions in ArcStorm are a superset of the functions in Map Librarian.

NOTE: *ArcStorm is an optional product from ESRI. It is not further documented in this book, and interested users should refer to the ESRI documentation set on this product.*

For many users, a good approach will be to begin with a seamless coverage design. When you have accumulated some experience with ARC/INFO and when you require some of the functions of ArcStorm, you can easily migrate your data to the ArcStorm data model.

The themes you explored in Display GIScity are mostly based upon seamless coverages. The exception is the orthophotomap theme, which is a rectified, photographic image of downtown Santa Fe.

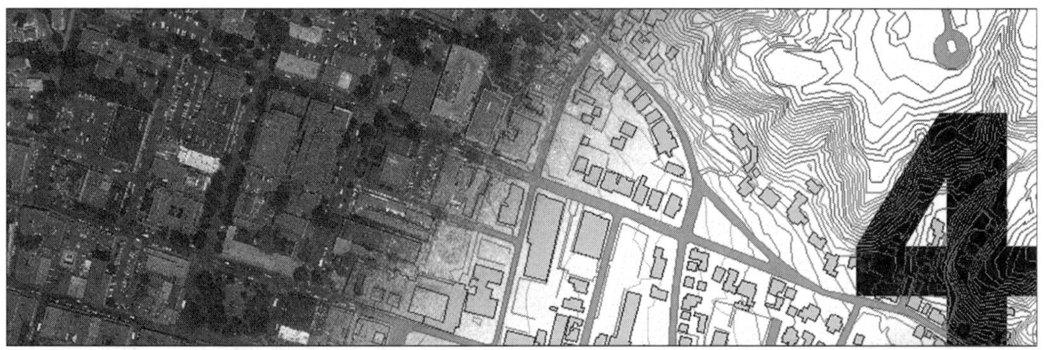

Managing Attribute Data

TABLES, INFO, Selected Sets, and Cursors

The outstanding strength of ARC/INFO is its ability to represent attribute information associated with geographic features and to access and update information in other databases. The GIS design behind this is called the *georelational model.*

The function of creating, querying, and updating attribute information is woven throughout ARC, ARCEDIT, and ARCPLOT. You will find many similarities—and a few differences—in selecting records in these three modules.

Two other programs supplied with ARC/INFO, INFO and TABLES, are specifically designed to manage attributes. The exercises in this chapter walk you through a sample session in INFO and TABLES.

INFO is a basic relational database management system. While INFO is not actually required to use ARC/INFO, it is provided for your convenience to directly manage tables.

TABLES, a program accessed through ARC, also provides table management. The commands in TABLES emulate many commands in INFO. TABLES is a little easier to use than INFO, but it does not span the full functionality of INFO. TABLES and INFO are, for the most part, interchangeable. Which program you use is generally a matter of preference.

In this chapter, we'll cover table management, updating attribute values, selecting table records, and logically scrolling through table records with cursors. We will also briefly cover the DATABASE INTEGRATOR, which is the gateway to commercial database management systems.

An Overview of Tabular Data Storage

First, let's look at some basics: ARC/INFO database terminology, the relationship between coverages and tables, and the layout and components of an INFO table.

Terminology

Some of the database terminology in ARC/INFO is non-standard; it may be confusing at first for users who have experience with another database management system. We'll now review some definitions and terminology.

A standard for accessing information in a database management systems is the Structured Query Language (SQL). With this language, you can pose general questions to any SQL-compliant database and receive consistent answers. We'll compare the definitions in ARC/INFO and SQL.

A *table* is the basic unit of storage in a database management system. It is a two-dimensional matrix of attribute values. Both ARC/INFO and SQL share this terme.

An *item* is a vertical component in a table. In SQL, this is called a *column*. Another term in common use is *field*.

A comparison of basic database terminology as used in ARC/INFO and SQL (in italics.)

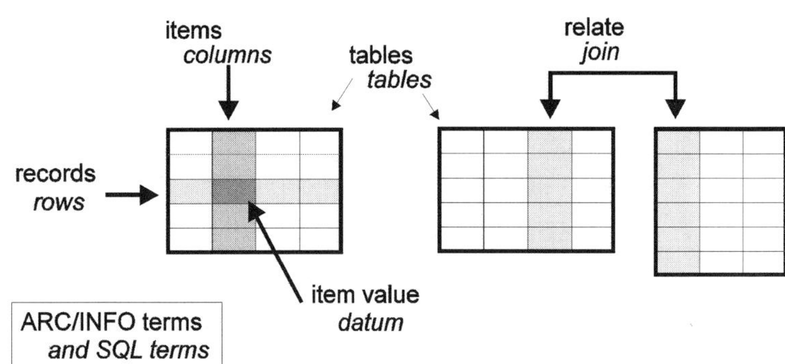

A *record* is the horizontal component in a table. In SQL, this is called a *row.*

The intersection of an item and record is a single, discrete entity called an *item value.* In SQL, this is called a *datum.*

A *relate* is the virtual joining of two tables through a common item. This is called a *join* in SQL.

The relationship between coverages and tables.

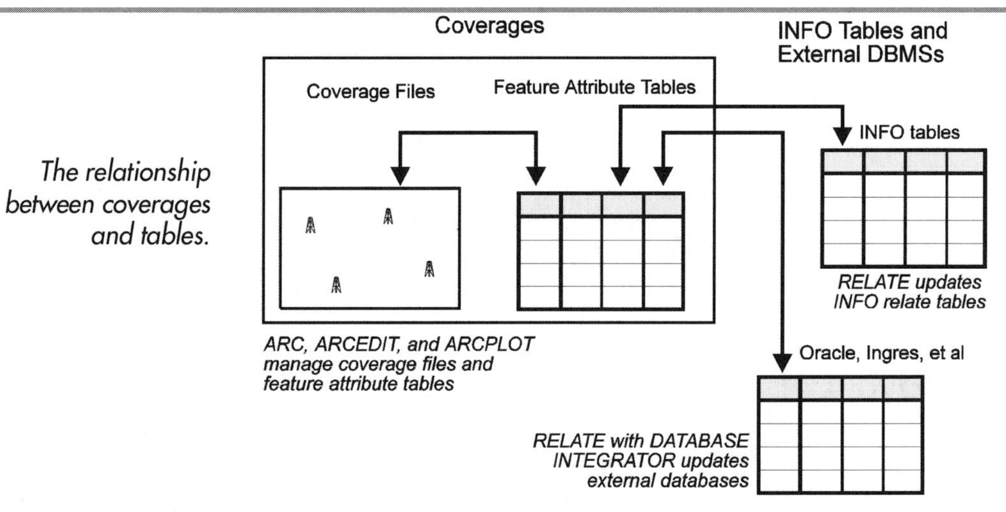

Coverages and Tables

There are three places that attribute data can reside and be accessible from ARC/INFO:

- ❑ In a feature attribute table
- ❑ In an INFO relate table
- ❑ Within a commercial database management system.

A coverage contains coordinate files and feature attribute tables. There can be one or several feature attribute tables in a coverage.

Coverage files contain positional information. As an ARC/INFO user, you never directly see the internal data in a coverage file. Instead, you see the positional information as a graphical display, and interact with coverage files through the commands in ARC, ARCPLOT, and ARCEDIT.

Feature attribute tables are a special type of INFO table that ARC/INFO creates and manages so that they are uniquely associated with the geographic features in a coverage. You can interact with feature attribute tables through commands in ARC, ARCEDIT, ARCPLOT, TABLES, and INFO. In any command that contains an INFO table (or file) as a command argument, you can also specify a feature attribute table.

When you create a feature attribute table in ARC/INFO, a set of standard ARC/INFO items are automatically defined for you. These were summarized in the last chapter. You can add further items as needed for your application.

Feature attribute tables can be logically connected to other tables through the User ID or through any other item that you specify. You can define a relate to make a logical connection between a feature attribute table and INFO tables or external database management systems.

The Layout of INFO Tables

When you create or modify items in an INFO table, you must specify five characteristics for those items:

- ❑ **Item name.** Any name you choose up to 16 characters.

- **Item width.** The width of the item in bytes (characters).
- **Output width.** The width of item values when displayed.
- **Item type.** The type of item, such as integer, character, or floating.
- **Number of decimals.** For floating item types, how many places to the right of the decimal place.

When you add items in ARC/INFO, you have several choices for the item types: Binary, Character, Date, Floating, Integer, and Decimal numbers. These item types, their codes, and definitions apply to both feature attribute tables and any INFO tables you define. Following are all the item types and the codes by which they are referenced:

- **C: Character.** This item type can contain any alphanumeric character, and is used for names, codes, and foreign keys. The maximum item width for this item type is 320 characters.
- **F: Floating.** This item type is for real numbered values, such as 2.71828. The item width must be either 4 or 8. A value of 4 defines single precision accuracy, which means about 7 places of numeric accuracy before mathematical round-off becomes a factor. A value of 8 defines double precision accuracy, or about 14 places of numeric accuracy. When you use the floating item type, you must also specify the number of decimals for the item definition.
- **I: Integer.** This item type is for integer values and can be as large as slightly over 2 billion. The item width can range from 1 to 16.
- **B: Binary.** This item type contains integer values in a binary format. In every attribute table, you will see two binary items—the internal ID (or "pound ID") and the User ID (or "dash ID"). When adding new items for integers, you will more frequently use the I item type.
- **N: Decimal number.** This rarely used item type stores floating values with one character per digit.
- **D: Date.** This item stores dates in a "year-month-day" format. This is useful for time-tagging attribute records.

When you update or query about item definitions in ARC, ARCPLOT, or ARCEDIT, you will enter or see item definitions similar to the following examples:

- ❑ TRANS-VAL 5 5 C denotes a character item called TRANS-VAL, with an input and output width of 5.
- ❑ SYMBOL 3 3 I denotes an integer item called SYMBOL with a width of three. (Items used for referencing symbol tables are nearly always defined just like this.)
- ❑ HDIST-ID 4 5 B denotes a binary item with 4 bytes of storage and 5 places for output. (This is exactly what the User ID for a coverage named HDIST would look like.)
- ❑ AREA 4 12 F 3 is a floating item type and the standard item for polygon areas in Polygon Attribute Tables for single precision coverages. You can see the output width allows up to 12 characters for output display, with a decimal point three places to the right.

Here's an item listing from a sample Polygon Attribute Table.

```
Arc: LIST HDIST.PAT
COL  ITEM NAME    WIDTH  OUTPUT  TYPE  N.DEC  ALTERNATE NAME INDEXED?
 1    AREA            4      12     F      3                    -
 5    PERIMETER       4      12     F      3                    -
 9    HDIST#          4       5     B      -                    -
13    HDIST-ID        4       5     B      -                    -
17    DISTRICT       20      20     C      -                    -
37    ZONE            8       8     C      -                    -
45    SHADESYMBOL     3       3     I      -                    -
```

In this listing, there are seven items. The first four are standard items created with the PAT and updated with ARC/INFO commands. The last three items were defined by the user: DISTRICT, ZONE, and SHADESYMBOL. In Chapter 5, we'll discuss symbology in greater detail, but a common method of implementing shade symbols is through using an item referencing a symbol in a symbol table.

WARNING: *When you create a feature attribute table, there will always be several standard items, as shown in the item listing above. The User ID item (<coverage>-ID) will always be the last standard item. Do not modify the definitions of any items listed before the User ID item. If you do, many ARC/INFO commands will fail because they will not see a required item. Also, do not modify the values of any standard items, with the exception of the User ID item. The User ID item is for your use as an identifier.*

Selection by Query

In this section, we'll cover how to select features logically. In later chapters, we'll cover the graphical selection of features in ARCPLOT and ARCEDIT, and in the Edit GIScity and Display GIScity applications.

Selected Set Basics

Let's review some basic facts about selected sets and then cover the commands to modify selected sets by query.

- A selected set is a list of the features currently selected in ARCPLOT and ARCEDIT.
- Many commands in ARCPLOT and ARCEDIT operate upon selected sets. Generally, the commands that display or modify features use selected sets.
- Selected sets defined within a session remain so until modified. You can combine a spatial selection with a logical selection by performing one type of selection after another.
- A few commands in ARCEDIT are very specific about how many features should be selected. Some commands, like the VERTEX command, require that only one feature be selected. If these criteria

are not met, you will see an error message appear at the `Arcplot:` or `Arcedit:` prompts.

These are some differences regarding selected sets in ARCPLOT and ARCEDIT:

- ARCPLOT can work with any number of selected sets. ARCEDIT can only work with one selected set at a time.
- ARCPLOT allows you to save and restore selected sets to your computer disk. ARCEDIT does not support the storage and retrieval of selected sets.
- ARCPLOT selection commands require you to type in the coverage name and feature class. ARCEDIT assumes that the selected set is for the current coverage and feature class.
- ARCPLOT has a concept of clear *selected set*, which works as if all features are selected. When you begin an ARCPLOT session, all coverages have a clear selected set by default. ARCEDIT does not have clear selected sets. When you begin an ARCEDIT session, no features are selected.
- When you draw features in an ARCPLOT coverage, only the selected features will be drawn. If you have not selected any features, all features will be drawn. When you draw selected features in ARCEDIT with the DRAW command, all features in the coverage will be drawn. There is another command to draw only selected features, which is DRAWSELECT. (More on this later.)

Because of these differences, you will quickly notice that the selection commands in ARCPLOT require more typing, because you need to specify the coverage and class, but selection in ARCPLOT is more flexible. Similar commands in ARCEDIT are shorter to type, but not as flexible.

Selection by Query in ARCPLOT

Diagram of steps in selecting features by query in ARCPLOT.

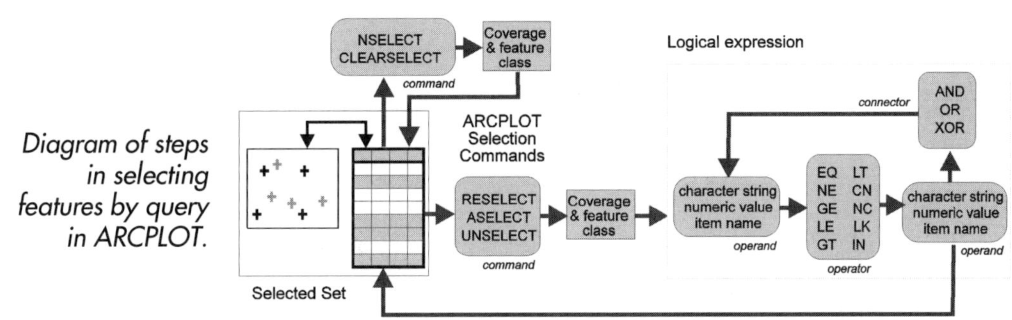

This diagram illustrates the basic methods for logical selection within ARCPLOT. Each command modifies a selected set for the specified coverage and feature class. Remember, in ARCPLOT you can have many selected sets for various coverages and feature classes.

These are the selection commands in ARCPLOT as they are used for logical selection.

```
Arcplot: ASELECT <coverage> <feature_class> {logical_expression}
```

adds the features matching the logical expression to the selected set.

```
Arcplot: RESELECT <coverage> <feature_class> {logical_expression}
```

selects a subset of features already in the selected set that match the logical expression.

```
Arcplot: UNSELECT <coverage> <feature_class> {logical_expression}
```

removes the features matching the logical expression to the selected set.

```
Arcplot: CLEARSELECT <coverage> <feature_class>
```

clears the selected set for the specified coverage and feature class. As you learned in Display GIScity, a clear selected set in ARCPLOT is like having all features selected. If you issue the CLEARSELECT command without the parameters, all selected sets are cleared.

```
Arcplot: NSELECT <coverage> <feature_class>
```

flips the selected set. If you had 4 features selected out of 17, the other 13 features would become selected after this command.

Selection by Query in ARCEDIT

Diagram of steps in selecting features by query in ARCEDIT.

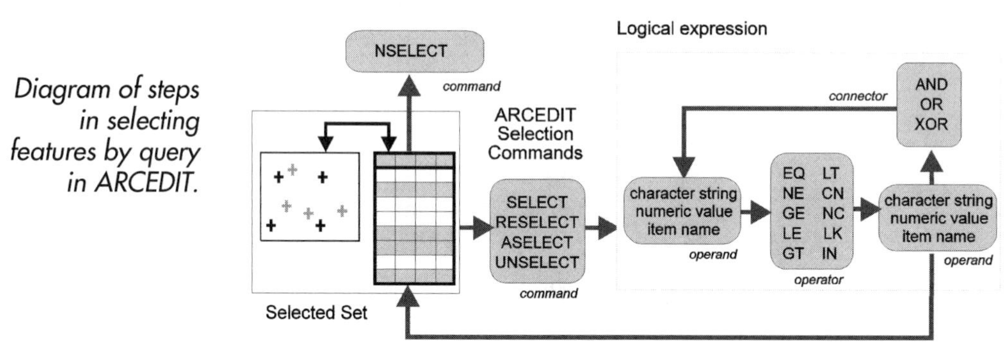

Although the selection commands in ARCEDIT are similar to those in ARCEDIT, there is one important difference: You don't specify a coverage or feature class in the selection command. Instead, the coverage and feature class are implied by the editcoverage and editfeature settings in ARCEDIT. Also, in ARCEDIT, you only have one selected set, which is for the edit coverage and edit feature.

These are the selection commands in ARCEDIT used for logical selection.

```
Arcplot: ASELECT <logical selection>
```

adds the features matching the logical expression to the selected set.

```
Arcplot: RESELECT <logical selection>
```

selects the subset from the current selected set with the features matching the logical expression.

```
Arcplot: SELECT <logical selection>
```

creates a new selected set with the features matching the logical expression.

```
Arcplot: UNSELECT <logical selection>
```

removes features that match the logical expression from the selected set.

```
Arcplot: NSELECT
```

reverses the features in the current selected set.

Logical Expressions

While discussing these ARCPLOT and ARCEDIT selection commands, we've referred to *logical expressions*. What is a logical expression? It is a computer-specific statement that corresponds to English statements such as "Select all buildings that are larger than 4000 square feet and do not have a fire sprinkler system."

The syntax of logical expressions.

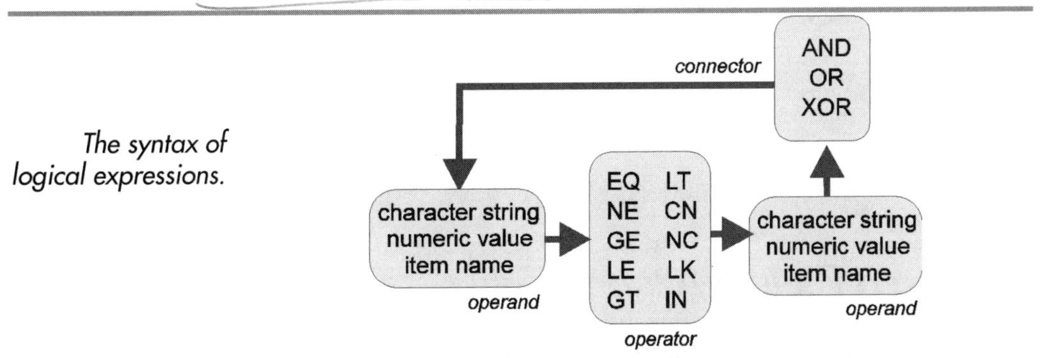

The first component of a logical expression is the *operand*. It is either a fixed character string (such as "West Manhattan"), a numeric value (such as 759), or an item name (such as STREETS-ID).

The second component of a logical expression is always an *operator*, which qualifies the expression. The first six operators in the preceding illustration are "equals," "not equals," "greater than or equal to," "less than or equal to," "greater than," and "less than." The next four operators only apply if one of the surrounding operands is an item name with a character type. These operators are "contains," "not contains," "like," and "inside."

NOTE: *While most operators are identical throughout ARC/INFO, the LK and IN operators are not available in the INFO program.*

After the operator is another operand that can complete the logical expression. Again, the operand can be a character string, numeric value, or item name.

A *connector* can optionally be added to the logical expression to further qualify the selection. The connectors are AND, OR, and XOR. AND or OR are obvious and match their English meaning. XOR stands for "exclusive or" and is rarely used.

Managing Feature Attribute Tables

You will find duplicated commands in the ARC, ARCEDIT, ARCPLOT, INFO, and TABLES programs that perform essentially the same or similar functions. In this section, we'll discuss the most essential commands common to all five programs function by function.

Creating a Feature Attribute Table

The normal method for creating a coverage's new attribute tables is by using the ARC BUILD and CLEAN commands. As discussed in the last chapter, BUILD and CLEAN are similar. CLEAN can perform geometric changes as well as update or create topology, while BUILD will only update topology. One fact to remember about point topology is that only BUILD can make Point Attribute Tables. Otherwise, CLEAN and BUILD update the other types of feature attribute tables.

Alternately, the CREATEATTRIBUTES and CREATEFEATURE commands can be used in ARCEDIT to create feature attribute tables.

```
Arcedit: CREATEATTRIBUTES

Arcedit: CREATEFEATURE <ARC | NODE | POLY | LABEL | TIC |
         LINK | ANNO.subclass | SECTION.subclass |
         ROUTE.subclass | REGION.subclass>
```

The CREATEATTRIBUTES command uses the context of the current edit coverage and edit feature you set during an ARCEDIT session. This is a quick way to create the basic feature attribute tables.

The CREATEFEATURE command requires you to enter some arguments, but it is the only way to create attribute tables for feature types that contain subclasses, such as annotation, sections, routes, and regions. Both these commands are equivalent to running the BUILD command in ARC.

Modifying Feature Attribute Tables

ADD ITEM

Adding an item to a feature attribute table.

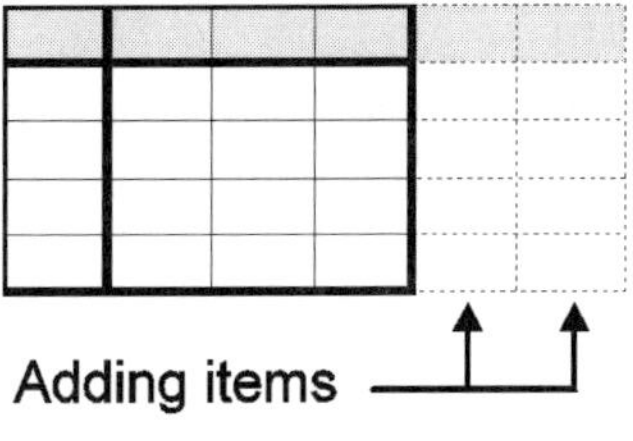

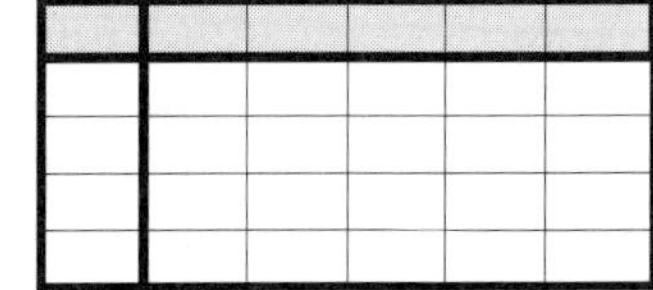

```
Arc: ADDITEM <info_file> <item_name> <item_width>
     <output_width> <item_type> {decimal_places} {start_item}

Arcedit: ADDITEM <add_item> <item_width> <output_width>
         <item_type> {decimal_places}
```

Frequently, you will want to add new items to a feature attribute table. The ADDITEM command is implemented in the ARC, ARCEDIT, and TABLES programs. You'll notice some differences in the arguments among the different programs, but the syntax is similar. The item definitions we covered earlier, such as name, width, and type, are used in this command.

After you add a new item, the item values for every feature in the coverage are blank.

DROP ITEM

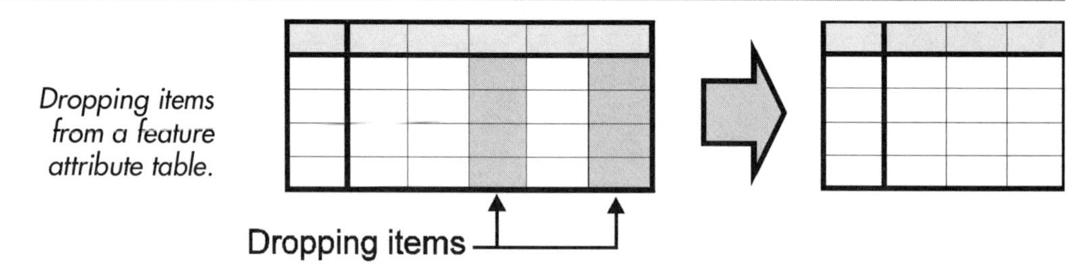

Dropping items from a feature attribute table.

```
Arc: DROPITEM <info_file> <item...item>
```

Dropping an item means that you remove an entire column of item values from a feature attribute table. You often need to do this if you have added items in error, or if another command has created items you are no longer interested in.

This command works on both INFO tables and feature attribute tables. Remember, don't drop any of ARC/INFO's predefined items in feature attribute tables.

PULL ITEM

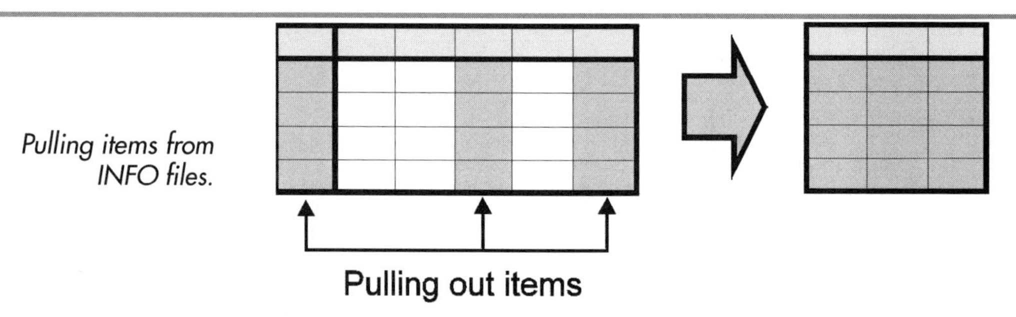

Pulling items from INFO files.

```
Arc: PULLITEMS <in_info_file> <out_info_file>
```

Pulling items means that you create a new INFO file or feature attribute table by pulling out only selected items from another table. This command has a similar result as DROPITEM, but instead of specifying items to be dropped, you specify items you want to save.

WARNING: *If you use this command on a feature attribute table, be sure to specify all the standard predefined coverage items up to the -ID and in the correct order. If you do not, then you will find you've corrupted the feature attribute table and recovery may be tricky. If in doubt, use the DROPITEM command instead.*

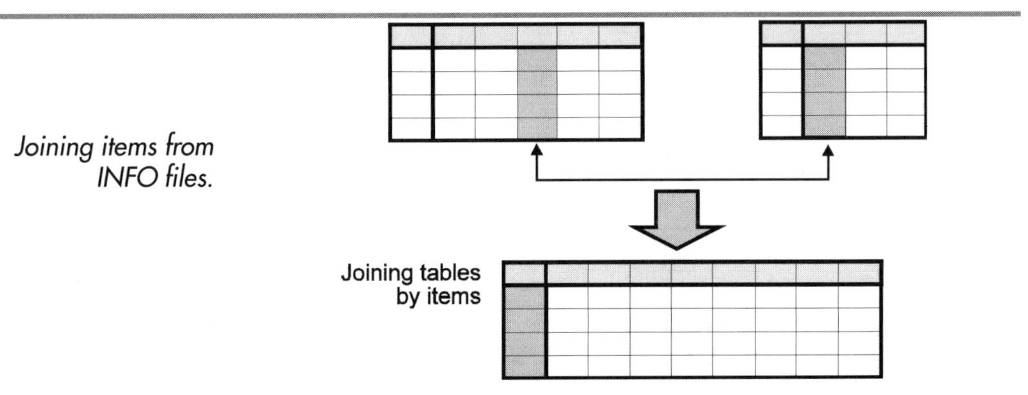

Joining items from INFO files.

```
Arc: JOINITEM <in_info_file> <join_info_file>
     <out_info_file> <relate_item> <start_item>
     {LINEAR | ORDERED | LINK}
```

Joining items resembles creating a relate to another table, but you are permanently merging items from one table to another. Just as with a relate, you specify an item which is an identifier common to both tables.

There are two situations where JOINITEM is justified: when the need for performance is paramount (accessing information directly from a feature attribute table is quicker than through a relate), or when you are permanently merging two separate tables. You should not use JOINITEM when you wish to retain the table you are joining to; using JOINITEM for this purpose will duplicate data entry and management, which is a cardinal sin in relational database management.

Modifying Item Values

There are two basic commands in ARCPLOT and ARCEDIT that modify the values in an item: CALCULATE and MOVEITEM. As with many other commands in ARCPLOT and ARCEDIT, CALCULATE and MOVEITEM only affect the features currently in your selected set.

CALCULATE

CALCULATE assigns numeric values to items. After specifying a target item, you can either place an integer or real value, or an expression.

```
Arcplot: CALCULATE <cover> <feature_class>
         <target_item> = <arithmetic_expression>

Arcedit: CALCULATE <target_item> = <arithmetic_expression>
```

Suppose you are in ARCEDIT and want to assign the integer value "17" to an item called TREE-COUNT. The statement would be simply

```
Arcedit: CALCULATE TREE-COUNT = 17
```

Instead of a numeric value, you can place an arithmetic expression. The expression can contain integer or real values, mathematical operators such as +, -, *, ** and /, and item values. For example, you could set one item value to be equal to the product of two other items.

MOVE ITEM

MOVEITEM assigns a character value to a character item.

```
Arcplot: MOVEITEM <coverage> <feature_class>
         <'character_string' | source_item> <target_item>

Arcedit: MOVEITEM <'character_string' | source_item>
         <target_item>
```

With MOVEITEM, you can transfer either a string you type in, or the contents of a source item. How does this command tell the difference? A character string must have single quotes around it. Otherwise,

ARC/INFO assumes you are specifying a target item. Forgetting the single quotes is a common mistake; if you do so, you'll get an error message stating that an item with the name of the string you entered can't be found.

A curious fact about CALCULATE and MOVEITEM is that the order of arguments is reversed. In CALCULATE, the assigned value is placed after the target item. In MOVEITEM, the assigned value is before the target item.

Listing Item Definitions in a Feature Attribute Table

Because you have considerable freedom in setting up the items in feature attribute tables, you will frequently want to know what the item definitions are. Here are three commands to list the item definitions in a feature attribute table:

```
Arc: ITEMS <info_file>
```

For a feature attribute table, specify the coverage name followed by the extension for that table. For instance, an arc attribute table for a coverage called ROADS would be ROADS.AAT.

TIP: *Often, it's inconvenient to move from workspace to workspace to get information about an INFO table. Here's a tip to reference INFO tables in other workspaces: You must reference the INFO subdirectory and insert this text, '!'ARC!'. If you want to list items on an INFO table called STREET.LUT in the directory $GCDATA/smallscale (or $GCDATA: [SMALLSCALE] in VMS), you would enter $GCDATA/smallscale/info!arc!street.lut (or $GCDATA:[SMALLSCALE.INFO]!ARC!STREET.LUT). (Remember, unlike in UNIX, using upper- or lowercase for directories and files is not critical.) This syntax works in any ARC/INFO command that references INFO tables.*

```
Arcedit: ITEMS {$ALL | info_file}
```

The ARCEDIT ITEMS command works on the coverage and feature class you've presently set. Without any arguments, you will get a listing of the items for the active feature attribute table. The optional argument $ALL will list pseudo-items. You usually don't need to concern yourself with pseudo-items for a feature class, but sometimes they are useful, usually with annotation. You can alternately list the items in an INFO table.

```
Arcplot: ITEMS <cover> <feature_class>
Arcplot: ITEMS <info_file> INFO
```

The ARCPLOT ITEMS command has two different usages, one for feature attribute tables by coverage and feature class, and the other by INFO table explictly.

Indexing Items

When the tables you create become large, you can create an *item index* based on selected items to speed search and retrieval in ARC, ARCPLOT, and ARCEDIT. You can create one or several item indexes for an INFO table. Whenever you do a logical selection in ARC/INFO, if an item index is present, ARC/INFO will use that index. The performance gain may be dramatic in certain circumstances.

```
Arc: INDEXITEM <info_file> <item>
```

This ARC command creates an item index for an INFO table.

```
Arc: DROPINDEX <info_file> <item>
```

With this command, you can remove an item index.

Any ARC command that modifies INFO tables will make an item index obsolete. You (or your AML application) would then have to update the item index by using the INDEXITEM command in the ARC program. ARCEDIT, however, will update item indexes for you.

To find out if you have item indexes defined, use the ITEMS command in ARC, ARCPLOT, or ARCEDIT.

You should only use INDEXITEM on items that you plan to frequently select upon. While it makes selection based on that item faster, selection of other items will be a bit slower. As a rule of thumb, do not exceed three item indexes per INFO table; that's the point of diminishing returns for performance enhancement.

Connecting Tables with RELATE

A *relate* is a logical relation between two tables with an item sharing the same values. It can be a many-to-one relate, a one-to-one relate, or a one-to-many relate between the tables. (Cursor processing is required to access all the elements in a one-to-many relate.)

The following usages for the RELATE command are identical in ARCPLOT and ARCEDIT.

```
RELATE <ADD | DROP>
```

To add or drop a relate. When you specify ADD, you enter a dialog to specify these properties: relation name, table identifier, database name (INFO or commercial DBMS name), INFO item, relate column, relate type, relate access, and relation name. These are all the same properties you set in the relate menu in GIScity.

```
RELATE <RESTORE | SAVE> <info_file>
```

The definitions for a relate (but not the actual relate) can be saved in an INFO table. With the RESTORE and SAVE arguments, you can bring back a relate from a previous session, or save one for a future session.

```
RELATE LIST {relate}
```

This usage lists the properties of either all the relates that are current or just the relate you specify.

Cursor Processing

Two important ways to interact with tables in ARCPLOT and ARCEDIT is through *selected sets* and *cursor processing*.

A selected set is a virtual subset of your current attribute table. When you use an ARC/INFO command that makes a selection, you've made or modified a selected set.

Often, you need to selectively access features within a selected set. Let's say you've selected 17 building features. You want to examine the attributes for each building one by one, starting with the first building and stepping through them until you've reached the last building. Cursors make it possible to do this.

Cursor processing is a technique for "walking through" a selected set. When you initiate a cursor, it points to the first record within the selected set. You can successively step to each following record until you've reached the last record.

Selected sets and cursor processing.

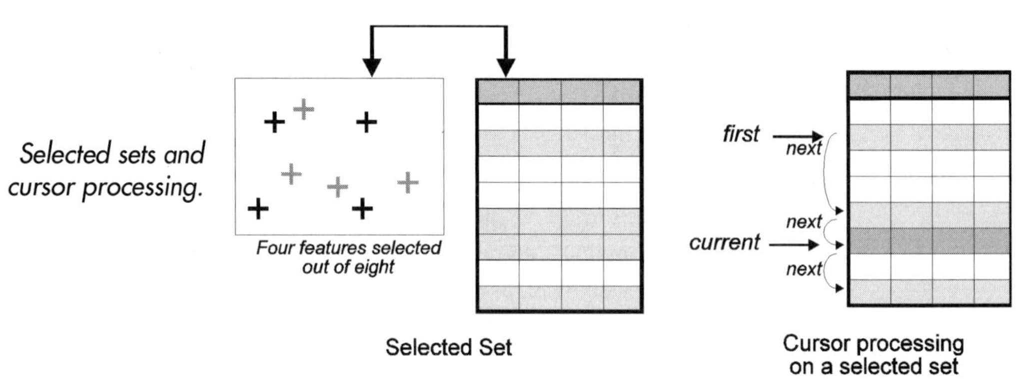

Here are some basic features of cursors:

- ❑ Cursors operate on the current selected set.
- ❑ Cursors can also scroll through records in a relate table tied to the selected set.
- ❑ Cursors in ARCPLOT are named. Cursor names let you keep track of multiple cursors, which you can define against as many as ten selected sets.
- ❑ There is only one cursor available in ARCEDIT.
- ❑ When you open a cursor, it points to the first feature in the selected set.
- ❑ You can move with a cursor in two ways—to the next feature, or back to the first feature in a selected set.
- ❑ Cursors can be used to navigate one-to-many relates.
- ❑ When you have finished using a cursor, your selected set is not modified.

Database cursors are a powerful technique, but they are rarely used interactively at the command line prompt. They are usually programmed into AML applications.

Following is the full syntax for the CURSOR command in ARCPLOT. Before you can use any other CURSOR command, you must first declare a cursor.

```
Arcplot: CURSOR <cursor> DECLARE <coverage> <feature_class> {RO | RW}
Arcplot: CURSOR <cursor> DECLARE <info_file> INFO {RO | RW}
Arcplot: CURSOR <cursor> <OPEN | CLOSE | INSERT | DELETE | REMOVE>
Arcplot: CURSOR <cursor> <NEXT | FIRST | record_number>
Arcplot: CURSOR <cursor> RELATE <relate> {NEXT | FIRST | INSERT | DELETE}
```

The full syntax of the CURSOR command in ARCEDIT is shorter; there is only one cursor available in ARCEDIT, so there is no cursor name to declare.

```
Arcplot: CURSOR <OPEN | CLOSE | NEXT | FIRST | record_number>
Arcplot: CURSOR RELATE <relate> {NEXT | FIRST | INSERT | DELETE}
```

The INFO and TABLES Programs

The ARC, ARCPLOT, and ARCEDIT programs contain many commands for modifying and updating feature attribute tables and INFO tables. Although these programs meet most of your attribute management needs, occasionally you'll enter INFO or TABLES to perform some surgery on your feature attribute tables or INFO tables. For instance, you would enter TABLES or INFO to modify an item definition, or to load or unload item values from an ASCII file. (To modify items, reference the ALTER and REDEFINE commands. To load or unload values to and from ASCII files, reference the ADD and UNLOAD commands.)

To enter either INFO or TABLES, just type in the program name at the `Arc:` prompt.

```
Arc: INFO
Arc: TABLES
```

A full discussion of INFO or TABLES is beyond the scope of this book. Rather, we'll show a simple session in both programs.

There are a few things to remember about INFO:

- After you type INFO at the `Arc:` prompt, you must type `arc` at the `ENTER USER NAME>` prompt.
- You must use the exact upper- or lowercase names of your directory paths and files.
- If you give a command that displays more than one screen on your dialog window, you must specify `Y` to see more or `N` to quit the display.
- To exit INFO, you must type `Q STOP`.

A Sample INFO Session

Below is a sample session in INFO. We will ask INFO for a directory of INFO tables, select one, list some of its contents, change item values, and exit. We'll use the INFO MOVE command, which is similar to the MOVEITEM command in ARCPLOT and ARCEDIT.

```
    Arc: info

ENTER USER NAME> ARC
ENTER COMMAND> DIR
TYPE  NAME              NTERNAL NAME  NO.RECS  LENGTH  EXTERNL
DF   ROOFLINES.TIC      ARC000DAT          4      12      XX
DF   ROOFLINES.BND      ARC001DAT          1      16      XX
DF   ROOFLINES.PAT      ARC002DAT       1565      26      XX
DF   CONTOURS.TIC       ARC003DAT          4      12      XX
DF   CONTOURS.AAT       ARC004DAT        414      49      XX
DF   CONTOURS.BND       ARC005DAT          1      16      XX
DF   EDGEROAD.TIC       ARC006DAT          4      12      XX
DF   EDGEROAD.AAT       ARC007DAT       2716      59      XX
DF   EDGEROAD.BND       ARC008DAT          1      16      XX
DF  EDGEROAD.FRQ       ARC009DAT          8      39
DF   EDGEROAD.PAT       ARC010DAT        418      16      XX
MORE? N
```

```
ENTER COMMAND> SELECT EDGEROAD.AAT
2716 RECORD(S) SELECTED
ENTER COMMAND> LIST
               1
FNODE#          =     1
TNODE#          =     2
LPOLY#          =     1
RPOLY#          =     4
LENGTH          =        26.375
EDGEROAD#       =     1
EDGEROAD-ID     =   281
ACAD_LAYER      = SECT
               2
FNODE#          =     3
TNODE#          =     4
LPOLY#          =     1
RPOLY#          =     6
LENGTH          =        10.500
EDGEROAD#       =     2
EDGEROAD-ID     =   281
ACAD_LAYER      = SECT
               3
FNODE#          =     1
TNODE#          =     5
LPOLY#          =     3
RPOLY#          =     1
MORE? N
ENTER COMMAND> RESELECT ACAD_LAYER EQ 'SECT'
144 RECORD(S) SELECTED
ENTER COMMAND> MOVE 'SECT' TO ACAD_LAYER
ENTER COMMAND> Q STOP
```

A Sample TABLES Session

Now, we'll do exactly the same thing in TABLES. You'll see that TABLES is more forgiving. You don't enter a user name as in INFO because TABLES types in ARC for you. TABLES is not sensitive to upper- and lowercase, but otherwise, it is similar to INFO. TABLES has most, but not all, of the functionality of INFO.

```
Arc: tables
Enter User Name: ARC
Enter Command: dir
TYPE  NAME               INTERNAL NAME  NO.RECS  LENGTH EXTERNL
DF    ROOFLINES.TIC       ARC000DAT        4       12      XX
DF    ROOFLINES.BND       ARC001DAT        1       16      XX
DF    ROOFLINES.PAT      ARC002DAT      1565       26      XX
DF    CONTOURS.TIC       ARC003DAT         4       12      XX
DF    CONTOURS.AAT       ARC004DAT       414       49      XX
DF    CONTOURS.BND       ARC005DAT         1       16      XX
DF    EDGEROAD.TIC       ARC006DAT         4       12      XX
DF    EDGEROAD.AAT       ARC007DAT      2716       59      XX
DF    EDGEROAD.BND       ARC008DAT         1       16      XX
DF   EDGEROAD.FRQ        ARC009DAT         8      39
DF    EDGEROAD.PAT       ARC010DAT       418       16      XX
Continue? N
Enter Command: sel edgeroad.aat
2716 Records Selected.
Enter Command: list
                 1
FNODE#                    =     1
TNODE#                    =     2
LPOLY#                    =     1
RPOLY#                    =     4
LENGTH                    =         26.375
EDGEROAD#                 =     1
EDGEROAD-ID               =   281
ACAD_LAYER                = SECTIONS
                 2
FNODE#                    =     3
TNODE#                    =     4
LPOLY#                    =     1
```

```
RPOLY#                    =    6
LENGTH                    =        10.500
EDGEROAD#                 =    2
EDGEROAD-ID               =   281
ACAD_LAYER                = SECTIONS
              3
FNODE#                    =    1
TNODE#                    =    5
LPOLY#                    =    3
RPOLY#                    =    1
Continue? n
Enter Command: reselect for acad_layer eq 'SECTIONS'
144 Records Selected.
Enter Command: move 'SECT' to acad_layer
Enter Command: q stop
```

The DATABASE INTEGRATOR

The DATABASE INTEGRATOR is the gateway between attribute tables in ARC/INFO and commercial database management systems. Database management systems that are tightly integrated include Oracle, Ingres, Sybase, and Informix. It is possible to connect the DATABASE INTEGRATOR functions to other commercial databases, but this may require additional programming effort.

A detailed discussion of the DATABASE INTEGRATOR is beyond the scope of this book. Also, the details of implementing DATABASE INTEGRATOR depend on the commercial database system you are using.

The four basic commands in ARC/INFO for using the DATABASE INTEGRATOR are CONNECT, DBMSEXECUTE, DBMSCURSOR, and DISCONNECT. They are present in the ARC, ARCPLOT, and ARCEDIT programs, although with different arguments.

The following steps give you a conceptual overview of DATABASE INTEGRATOR.

1. First, establish a logical connection with an external database with the CONNECT command.

2. You can scroll through specific database records that are related to the selected set for a feature attribute table with the DBMSCURSOR command. This is similar to the CURSOR command which operates upon INFO tables.

3. You can issue native mode SQL (Structured Query Language) statements with the DBMSEXECUTE command. This is a way to directly issue commands on database rows (records) that tie to your selected set. The command is directly passed to your commercial database management system.

4. Finally, you can disconnect an external database with the DISCONNECT command.

Part II

ARCPLOT

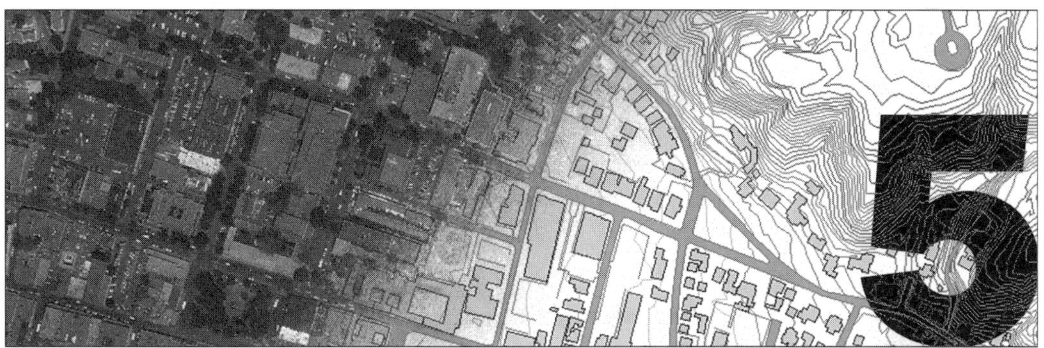

The Basics of Map Production

Displaying and Making Graphics Files in ARCPLOT

Maps are the basic output of a GIS. Although increasing numbers of people are making use of interactive computer displays, most people still receive information from a GIS through published maps.

As we've discussed a number of times, a GIS is fundamentally a database management system, albeit one designed to manage information through geography. A map from a GIS is in actuality a database report.

ARCPLOT is the ARC/INFO program that produces maps. With ARCPLOT, you can display maps on a workstation's graphics display or you can create a graphics file, which is a computer file with plotting instructions for your plotter or printer. Graphics files can be stored on your computer's disk and sent to your plotter or printer whenever you want. After you have established whether the output from ARCPLOT is sent to your workstation's display or sent to a graphics file, virtually all the commands in ARCPLOT apply to both screen display and graphics file output.

This chapter covers the basics of making maps: how to relate map coordinates to the dimensions of the map page, how to set the graphics environment, how to draw coverage features, and how to draw map elements. At the conclusion of this chapter, we'll review what we learned by stepping through a complete ARCPLOT session to make a sample map. In the next chapter, we'll learn more about symbolizing attribute information in a map, and how to create and modify symbol sets.

Starting ARCPLOT

Before we actually start making a map, we'll take a brief look at getting into ARCPLOT, getting help, setting the screen display, making and previewing graphics files, and sending those files to a printer or plotter.

Starting a Session

To start an ARCPLOT session, from the ARC program, type:

```
Arc: ARCPLOT
```

To leave an ARCPLOT session, type:

```
Arcplot: QUIT
```

If you wish to use AML menus during an ARCPLOT session, type:

```
Arcplot: &TERMINAL 9999
```

The &TERMINAL command is required whenever you invoke any AML menus during your ARCPLOT session. &TERMINAL can be abbreviated as &TERM.

The &TERMINAL argument 9999 specifies that you are using an X-windows display, which applies to virtually all workstations.

Getting Help in ARCPLOT

ARCPLOT contains more commands than any other ARC/INFO program. Many of these commands are quite specialized—you will never use a large fraction of them. Even longtime users of ARCPLOT are familiar with only a subset of ARCPLOT commands.

Use the COMMANDS command to get a listing of the available commands in ARCPLOT. Also, you can specify a partial name to get the listing of all commands starting with the partial name. For instance, the following listing shows all the commands which begin with LABEL:

```
Arcplot: COMMANDS LABEL
LABELERRORS   LABELMARKERS    LABELS    LABELSPOT
LABELTEXT
```

You can also specify a wildcard to get a command listing. For example, to find all the commands containing "SPOT," use * as the wildcard specifier. Here's how it works:

```
Arcplot: COMMANDS *SPOT
ARCSPOT          LABELSPOT      NODESPOT           POINTSPOT
POLYGONSPOT      REGIONSPOT     ROUTEEVENTSPOT     ROUTESPOT
SECTIONSPOT      SPOT
```

If you know the command you want, but have forgotten the command arguments, enter the USAGE command:

```
Arcplot: USAGE ARCLINES
Usage: ARCLINES <cover> {item | symbol} {lookup_table}
```

To get more help on ARCPLOT commands, you can invoke the ARC/INFO on-line documentation, ArcDoc, by typing:

```
Arcplot: HELP
```

Setting the Screen Display

The DISPLAY command controls whether graphics are sent to your workstation's display or to a graphics file. You must choose one or the other, but you can alternate between the two within an ARCPLOT session.

If you want to send graphics to your graphics window, use the DISPLAY command with the 9999 argument, followed by a number from 1 to 4.

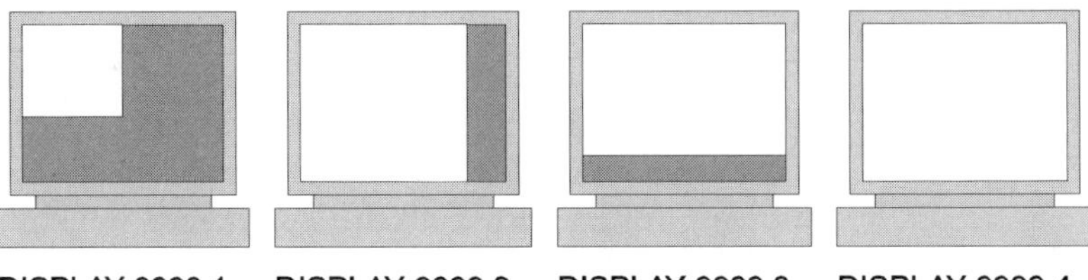

Four usages of the DISPLAY command for workstation display.

In ARCPLOT, you can clear the graphics window at any time. When you do so, the screen will revert to its initial color, defined with the CANVASCOLOR setting. To clear the graphics window, simply type:

```
Arcplot: CLEAR
```

NOTE: *If you prefer white instead of black for a background color, remember to use the CANVASCOLOR definition in your operating system before starting ARC/INFO. In UNIX, type* `setenv CANVASCOLOR WHITE` *at the system prompt. In VMS,* `DEFINE CANVASCOLOR WHITE`. *At the same time, also set your CROSSHAIRCOLOR to black (or another color besides white) so that you can see it against the white background.*

Making Graphics Files

If you want to produce an ARC/INFO graphics file, use the DISPLAY command with an argument of 1040.

NOTE: *There is no special significance to the number 1040, but it supersedes 1039, a previous argument for making an older type of graphics file, the ARC/INFO plot file. You can still produce the older style plot files with DISPLAY 1039, but there is no reason to do so.*

When you use DISPLAY 1040, there are six options that specify which type of graphics file is created.

- ❑ ARC/INFO graphics file. File extension is .gra
- ❑ Encapsulated PostScript. File extension is .eps
- ❑ Adobe Illustrator. File extension is .ai
- ❑ Computer Graphics Metafile, character. File extension is .cgm
- ❑ Computer Graphics Metafile, clear text. File extension is .cgm
- ❑ Computer Graphics Metafile, binary. File extension is .cgm

The following usage sets the output format to the ARC/INFO graphics file, the default option.

```
Arcplot: DISPLAY 1040
```

Immediately after you issue DISPLAY 1040 with any option, you must type in the file name for the graphics file. You can skip the file extension; one will be supplied for you. By default, the graphics file will go into your current workspace. You can insert a directory path, according to the syntax of your operating system, UNIX or VMS.

Here's a sample sequence to open a graphics file called SF_ROADS:

```
Arcplot: DISPLAY 1040
Enter output filename: SF_ROADS
```

A new graphics file called SF_ROADS.GRA has been opened for you. Now you can enter other ARCPLOT commands to add graphics to your graphics file. When you use DISPLAY 1040, any open graphics windows will be removed. You will only see your dialog window.

A common practice is to try a sequence of ARCPLOT commands first with DISPLAY 9999, refine those steps to produce the desired output visually, type those steps to a simple AML, and when done with the refinements, change the DISPLAY 9999 command in the AML to a

DISPLAY 1040 command followed by a graphics file name. For new users of ARC/INFO, this may be the first type of AML you will write.

Previewing Graphics Files

Once you've made some graphics files, you can preview graphics files in both the ARC and ARCPLOT programs. In ARC, use the DRAW command to display graphics files.

```
Arc: DRAW <graphics_file> {device} {option}
```

The device and option arguments are the same as for the DISPLAY command. Using 9999 for the device specifies an X-windows workstation display. The 1, 2, 3, or 4 option sets the initial size of the graphics window. You can skip the device and option arguments if you have already used the DISPLAY command in the ARC, ARCPLOT, or ARCEDIT program in the current ARC/INFO session.

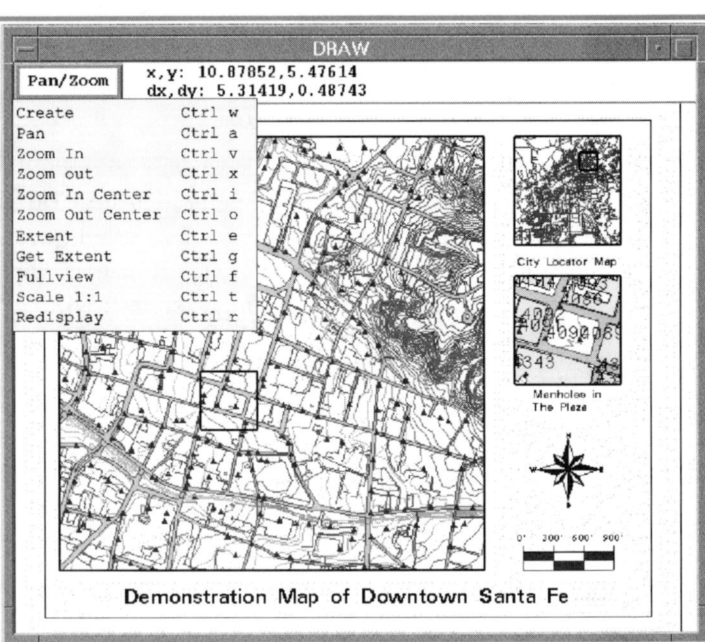

Displaying graphics file with the DRAW command and Pan and Zoom menu.

When you use the DRAW command, you will see a button called Pan/Zoom on the upper left corner of the graphics window. If you click this button, several choices will appear: Create, Pan, Zoom In, Zoom Out, Zoom In Center, Zoom Out Center, Extent, Get Extent, Fullview, Scale 1:1, and Redisplay. Use some of these commands to get a careful look at portions of a graphics file before sending it to a plotter.

In ARCPLOT, use the PLOT command to display graphics files.

```
Arcplot: PLOT <graphics_file> {* | xy}

Arcplot: PLOT <graphics_file> BOX <* | xmin ymin
         xmax ymax>
```

You must have a graphics display open before using PLOT. The handy usage PLOT <graphics_file> BOX * lets you use two opposite corners of your graphics window and scale the graphics file to fit your window.

Sending Graphics Files to a Plotter or Printer

The ARC/INFO graphics file is a device-independent format. It can be sent to any output device supported by ARC/INFO, but it requires translation. Depending on which type of plotter or printer you use, you will need one of these plotter translator commands in the ARC program:

- ❑ For pen plotters, use the CALCOMP, HPGL, or ZETA commands.
- ❑ For electrostatic plotters, use the COLORHCBS, HPGL2, or VCGL commands.

Because the use of these commands depends upon the plotting hardware and your computer's operating system, refer to ArcDoc and the System Dependencies manual for your operating system for details on sending graphics files to output devices.

For printers that accept PostScript, the DISPLAY 1040 2 usage produces an Encapsulated PostScript file. Use an operating system command to spool the EPS file to your printer.

Demystifying Map-to-Page Transformation

For new users of GIS, understanding exactly how to control the placement of coverage features on a page is one of the most confusing topics. Expert ARC/INFO users get confused too! This subject, map-to-page transformation, doesn't have to be so difficult. In this section, we'll explain how to precisely and simply set map extents to page sizes and positions.

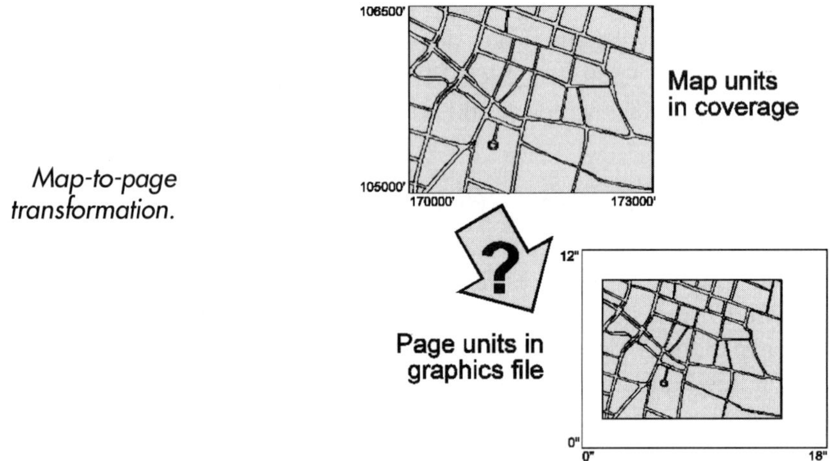

Map-to-page transformation.

Map Units, Page Units, and Map Scale

First, let's review map units and page units.

- ❑ Map units are generally in feet or meters, and the coordinate values in coverages are usually large, typically 6 or 7 digits to the left of the decimal.
- ❑ Page units are usually inches or centimeters. The lower left corner of a page has the value (0,0).

When you are producing a map, there are three basic operations to fit coverages onto pages: rotation, scaling, and positioning. In ARCPLOT, you have a number of ways to do this (which causes some of the confusion).

When you use ARCPLOT with the screen display set (instead of graphics file output), ARCPLOT does a pretty good job of keeping the details of map-to-page transformation hidden from you. However, when you start producing graphics files to be published, you need to understand the relation between map units and page units.

Setting Up the Graphics Page

First, we need to specify the physical dimensions of the media we expect to use. For laser printers in the U.S., this will usually be 8.5″ by 11″ or 8.5″ by 14″. For plotters or large format printers, this may be 11″ by 17″, 22″ by 34″, or larger.

Depending on whether we are in the United States or elsewhere in the world, we need to set the units of measurement for our graphics page.

```
Arcplot: PAGEUNITS <INCHES | CM | pageunits_per_inch>
```

The default value for PAGEUNITS is inches.

Also, we need to express the units of measurement in our coverages:

```
Arcplot: MAPUNITS <INCHES | FEET | CM | METERS |
         mapunits_per_inch>
```

You will nearly always use METERS or FEET for map units.

A number of ARCPLOT commands which accept X and Y coordinates can draw features based on either the currently set page units or map units. Some examples of these commands are BOX, CIRCLE, and LINE.

```
Arcplot: UNITS <PAGE | MAP | PROJECTEDMAP | GRAPH>
```

The UNITS command specifies how succeeding ARCPLOT commands will interpret X and Y coordinates. You can change these units repeatedly in an ARCPLOT session. In the demonstration at the end of this chapter, we'll see all three of these commands, PAGEUNITS, MAPUNITS, and UNITS, used.

The graphics page is the actual medium you are printing or plotting on. The map limits is that portion of the graphics page to which coverage feature will be drawn.

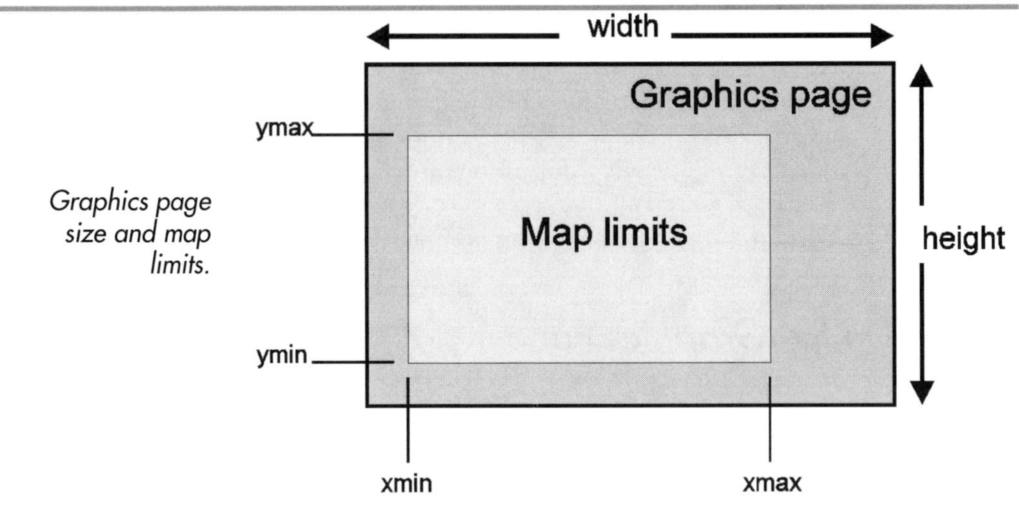

Graphics page size and map limits.

```
Arcplot: PAGESIZE <DEVICE | width height>
```

This command sets the physical size of the medium you are plotting or printing. Normally, you will enter a width and height in the current page units. You can specify DEVICE to automatically set the page size to the size of your graphics window.

```
Arcplot: MAPLIMITS <PAGE | * | xmin ymin xmax ymax>
```

MAPLIMITS sets the range of page units on your graphics page that will accept graphics. With MAPLIMITS set, ARCPLOT will automatically clip coverage features outside the map limits. Using the PAGE argument sets the map limits to be exactly the same as the page size. You can use the cursor, but only if your graphics page is a graphics window. The predominant usage of this command is to specify the X and Y ranges in page units.

NOTE: *You can set MAPLIMITS several times within an ARCPLOT session to create a graphics file with different portions of a coverage. The example at the end of this chapter illustrates making a map with three map limits set.*

NOTE: *Actually, the name MAPLIMITS is a bit confusing. This command would be better called PAGELIMITS, because the X and Y coordinate values are always in page units, never in map units.*

```
Arcplot: MAPSHIFT <x_shift> <y_shift> {PAGE | MAP}
```

MAPSHIFT will shift the map from the lower left corner of the map to the specified position. You can use this command instead of MAPLIMITS if you also use the MAPSCALE and/or MAPEXTENT commands. You can specify whether the shift is in page units (the PAGE option) or map units (the MAP option). You will usually use the PAGE option, because you think of positioning maps on a page through page units (such as inches or centimeters).

Executing Map-to-Page Transformation

Now that we have specified a page size and map limits, let's review some basic ways to place coverage features inside the map limits. These are not the only techniques you can use. For example, you can rotate coverage features with the MAPANGLE command, and you can also project the coverage coordinates into another map projection. But we'll keep things simple for now.

NOTE: *Coordinate systems are a tricky business. The north direction based on a given coordinate system may not be the true north direction based on latitude and longitude. The deviation may be as much as a few degrees. This is not the same as the magnetic deviation you sometimes see on maps, but is a consequence of the type of map projection used. When you make maps, you can either ignore this deviation, or rotate coverage features slightly with the MAPANGLE command, provided you can calculate the angular deviation.*

Let's use some basic mathematics and figure out a simple example.

Map-to-page example.

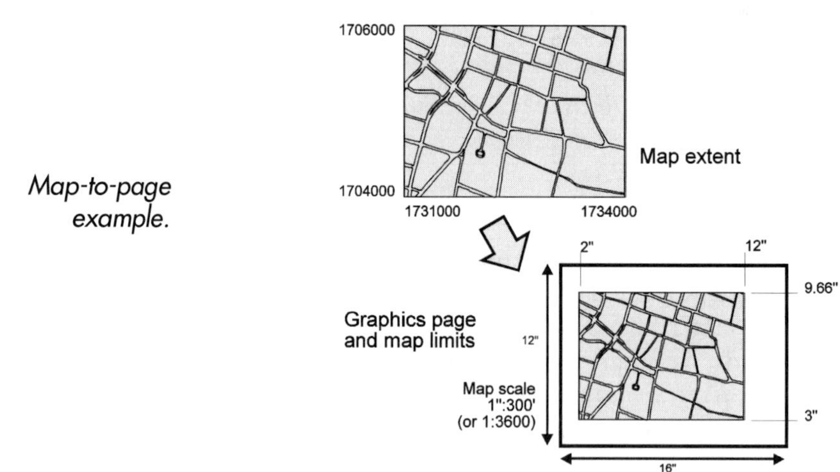

It is always easier to understand subjects like map scales, map extents, and map limits when you use real numbers and actual commands. We'll execute the map-to-page transformation shown in the illustration in several ways.

Now, the features in the coverage range from 1 731 000 to 1 734 000 along the X axis and from 1 704 000 to 1 706 000 along the Y axis. (These are actual coordinate values from the GIScity data.) This is a map extent of part of the Santa Fe Edge of Road theme. The map extent is 3000 feet from east to west and 2000 feet from south to north.

First, let's say you want to make a map at a precise scale. If we pick a map scale of 1″:300′, that means the coverage features on the graphics page will span a distance of (3000 feet) / (300 feet per inch) in the X direction, or 10 inches. In the Y direction, the distance is (2000 feet) / (300 feet per inch), or 6.66 inches.

Now, you also want to set a margin along the graphics page, because maps are rarely plotted right up to the paper's edge. Let's say the margin will be 2 inches across and 3 inches up from the lower left corner of the page. That means the map limits, in inches, will be (2, 3) to (12, 9.66).

Here is a simple ARCPLOT session to make this graphics file from scratch:

```
Arc: ARCPLOT
```

```
Arcplot: &WORKSPACE $GCHOME/data (UNIX or Windows NT) or
         &WORKSPACE $GCHOME:[DATA] (VMS)
Arcplot: DISPLAY 1040
Enter output filename: ROADS
Arcplot: MAPUNITS FEET
Arcplot: MAPEXTENT 1731000 1704000 1734000 1706000
Arcplot: PAGEUNITS INCHES
Arcplot: PAGESIZE 16 12
Arcplot: MAPLIMITS 2 3 12 9.66
Arcplot: ARCS EDGEROAD
Arcplot: QUIT
```

That's it! This is a complete ARCPLOT session which will make a graphics file like that in the illustration. You can try this yourself on your workstation.

You can view this graphics file with this command:

```
Arc: DRAW ROADS.GRA 9999 3
```

The PAGEUNITS INCHES command is not actually necessary because that is the default unit in ARCPLOT, but these commands were included to show how to set other units (the default value for MAPUNITS is inches). Also, we just drew part of the coverage, not the entire coverage. The MAPEXTENT command restricts the part of the coverage that will be drawn.

NOTE: *You can also make a graphics file using the MAPEXTENT <coverage> argument instead of MAPEXTENT <xmin ymin xmax ymax>, but you may be puzzled to see a blank margin. ARCPLOT automatically adds a 5% blank margin inside the map limits when the <coverage> argument is used. ARCPLOT doesn't do so with actual coordinate arguments. So for precise control over your maps, set map extents by actual coordinates instead of by the <coverage> argument.*

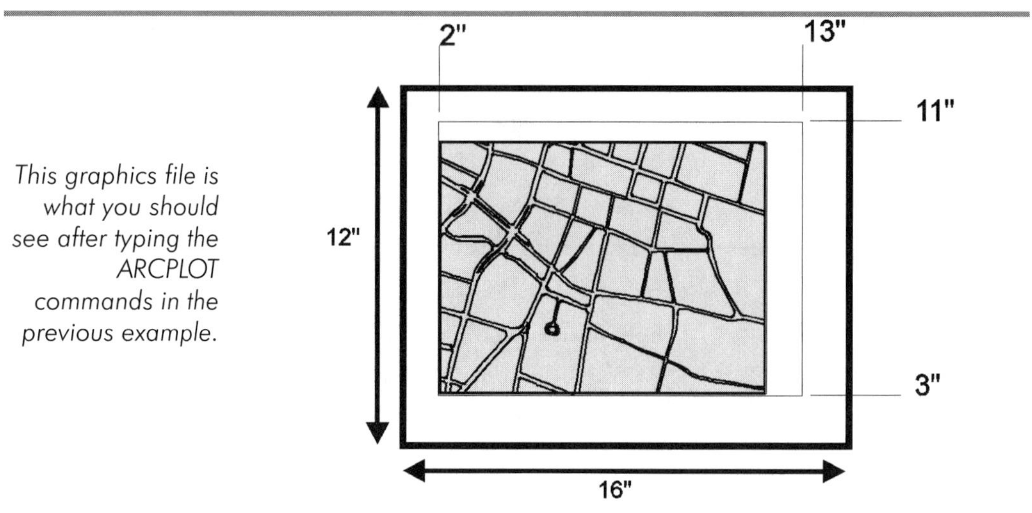

This graphics file is what you should see after typing the ARCPLOT commands in the previous example.

In the previous example, we made a graphics file by knowing a map extent, selecting a desired map scale, and calculating the map limits. Here's another way to create this same graphics file:

```
Arc: ARCPLOT
Arcplot: &WORKSPACE $GCHOME/data (UNIX or Windows NT) or
         &WORKSPACE &GCHOME:[DATA] (in VMS)
Arcplot: DISPLAY 1040
Enter output filename: ROADS
Arcplot: MAPEXTENT 1731000 1704000 1734000 1706000
Arcplot: MAPUNITS FEET
Arcplot: MAPSCALE 300
Arcplot: PAGESIZE 16 12
Arcplot: MAPSHIFT 2 3
Arcplot: ARCS EDGEROAD
Arcplot: QUIT
Arc:
```

In this example, we used the MAPSCALE and MAPSHIFT commands instead of MAPLIMITS. If you typed in the commands above and made a graphics file, you'd find the end result is identical to the previous example.

What happens when we change the map scale? Here's an example where we use a map scale of 1":200' (or 1:2400):

```
Arc: ARCPLOT
Arcplot: &WORKSPACE $GCHOME/data (UNIX or Windows NT) or
Arcplot: &WORKSPACE $GCHOME:[DATA] (VMS)
Arcplot: DISPLAY 1040
Enter output filename: ROADS
Arcplot: MAPUNITS FEET
Arcplot: MAPSCALE 200
Arcplot: MAPEXTENT 1731000 1704000 1734000 1706000
Arcplot: PAGESIZE 16 12
Arcplot: MAPSHIFT 2 3 13 11
Arcplot: ARCS EDGEROAD
Arcplot: QUIT
```

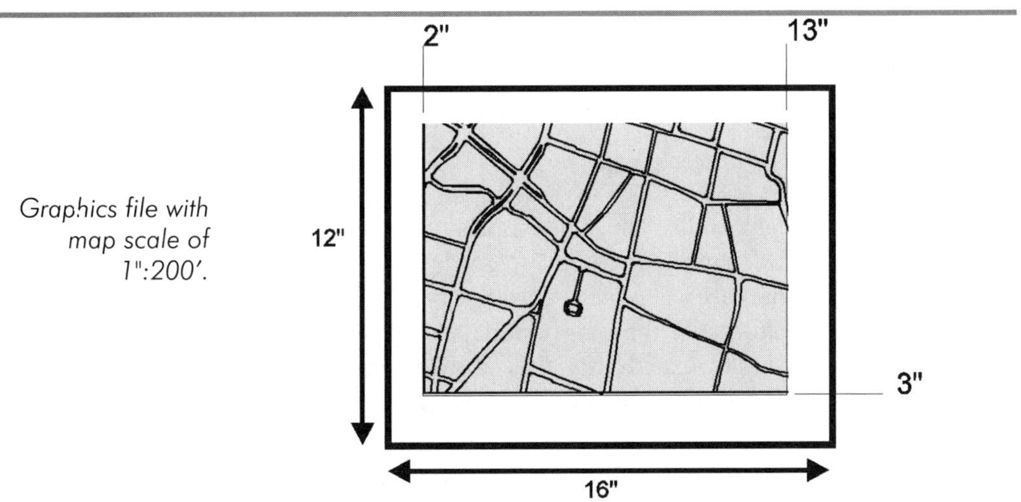

Graphics file with map scale of 1":200'.

The arcs are drawn on the page at a higher magnification (or larger scale). The page size is the same, the part of the page drawn with coverage features (the map limits) is the same, but we are drawing coverage features at a larger magnification.

What if the graphics determined by the map scale and map extent is smaller than the map limit? By default, the graphics would appear at the lower left corner of the map limits, but you can change this positioning with the MAPPOSITION command.

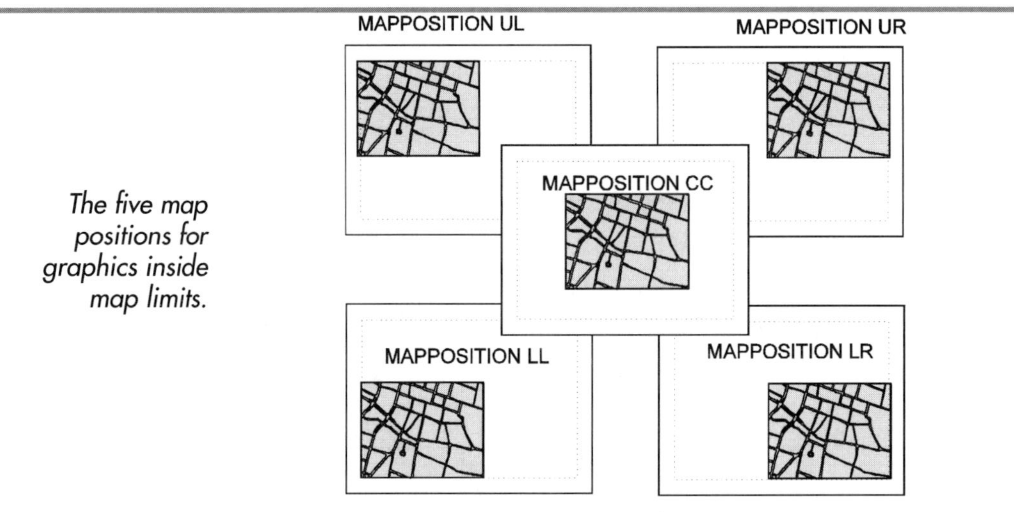

The five map positions for graphics inside map limits.

```
Arcplot: MAPPOSITION <* | LL | LR | UL | UR | CEN | x>
                     <* | LL | LR | UL | UR | CEN | y>
```

The MAPPOSITION command can be used to shift graphics inside the map limits. The illustration shows the effect if the graphics area is smaller than the map limits. MAPPOSITION will also shift features if the graphics area is greater than the map limits.

Note that if the map extent and map scale combined to match the map limits exactly, as they did in our very first example, then the MAPPOSITION command would have no effect.

Summary of Map-to-Page Transformation

These are the key ARCPLOT commands that control map-to-page transformation for making graphics files: PAGESIZE, MAPEXTENT, MAPLIMITS, MAPSCALE, MAPSHIFT, and MAPPOSITION. To make a map, you need use only some of these commands.

If you want to make a graphics file at an exact map scale, use PAGESIZE, MAPEXTENT, MAPSCALE, and MAPSHIFT.

If you want to make a graphics file where the map extent of a coverage fits exactly within a specified area of the page, use PAGESIZE, MAPEXTENT, and MAPLIMITS.

If you use MAPEXTENT together with MAPSCALE and MAPLIMITS, you might have the map area either bigger or smaller than the map limits. Also, the shape of the map extent can be wider or narrower than the shape of the map limits. If either is the case, you can use MAPPOSITION to justify the map to the lower left, lower right, upper left, upper right, or center of the map limits.

Don't forget to set your MAPUNITS (usually meters or feet) and PAGEUNITS (usually centimeters or inches) to the values you use in your GIS.

Disabling Map Extent Clipping

There is one more useful command, especially for when you use ARCPLOT interactively with a graphics window:

```
Arcplot: CLIPMAPEXTENT <ON | OFF>
```

When CLIPMAPEXTENT is on, all coverage features will be clipped to exactly your map extent. When it is off, all coverage features within the map limits will be drawn. As a general rule, use CLIPMAPEXTENT ON for graphics files and CLIPMAPEXTENT OFF for graphics window displays.

Getting Information About Page and Map Settings

That's it for map-to-page transformation. It's not such a difficult subject after all. It does require that you think a little bit before plunging into ARCPLOT commands. A couple of quick calculations such as we did above will reward you with easy graphics files the way you want them the first time.

At any time when you are setting up page and map units, you can get a summary of the unit status with the MAPINFO command. Just type `MAPINFO` at the ARCPLOT command. Here's a sample response:

```
Arcplot: MAPINFO
-- PAGEUNITS  in inches
-- MAPUNITS  in inches
-- PAGESIZE   width   11.000   height   8.500 in inches
-- device size   width   12.968   height   9.018 in inches
-- MAPSCALE     1:       16445.143 AUTOMATIC
-- MAPANGLE    0.00 degrees
-- CLIPMAPEXTENT ON

-- MAPEXTENT       xmin         ymin           xmax           ymax
in MAPUNITS 1712867.000 1684424.000 1741646.000 1713203.000
in PAGEUNITS     8.250       6.250        10.000          8.000

-- MAPLIMITS    xmin          ymin           xmax          ymax
in MAPUNITS 1712867.000 1684424.000 1741646.000 1713203.000
in PAGEUNITS    8.250         6.250        10.000        8.000

-- MAPPOSITION                  X                Y
in MAPUNITS              1712867.0000   1684424.0000   (LL)
in PAGEUNITS                8.2500            6.2500   (LL)
```

Setting Symbol Properties

In this section, we'll discuss how to draw the primary features in a coverage. Because of space limitations, we'll skip how to draw route systems and regions, but the ARCPLOT commands for these extended features are analogous to the ones we discuss below for primary coverage features like arcs and label points.

Setting Symbol Sets

First, we should specify our symbol sets. A symbol set is a special ARC/INFO file that contains graphics for drawing label points, arcs, polygons, and annotation. We'll learn more about symbol sets in the next chapter.

If you want to use multiple symbol sets, you can repeatedly change symbol tables during an ARCPLOT session. These are the four ARCPLOT commands for setting symbol sets:

```
Arcplot: MARKERSET <markerset_file>
```

sets the current markerset file for drawing label points and nodes.

```
Arcplot: LINESET <lineset_file>
```

sets the current lineset for drawing arcs, routes, and sections.

```
Arcplot: SHADESET <shadeset_file>
```

sets the current shadeset for drawing polygons and regions.

```
Arcplot: TEXTSET <textset_file>
```

sets the current textset for drawing coverage annotations as well as text from attributes or map elements.

Specifying Symbols

In ARCPLOT, you can draw features either by a fixed symbol, or by referencing a symbol item. These commands set a fixed symbol for drawing coverage features.

```
Arcplot: MARKERSYMBOL <symbol_number>
Arcplot: LINESYMBOL <symbol_number>
Arcplot: SHADESYMBOL <symbol_number>
Arcplot: TEXTSYMBOL <symbol_number>
```

When you set a symbol number with the LINESYMBOL, MARKER-SYMBOL, SHADESYMBOL, or TEXTSYMBOL command, other symbol properties are also set, such as color and size. For example, LINESET CARTO.LIN and LINESYMBOL 110 set the current line symbol to be a black dashed line with medium thickness. Reference ArcDoc for a depiction of all symbols within symbol sets.

Specifying a New Color

Once you've set a symbol number, you can override the color properties for the current symbol with these commands:

```
Arcplot: MARKERCOLOR <color>
Arcplot: LINECOLOR <color>
Arcplot: SHADECOLOR <color>
Arcplot: TEXTCOLOR <color>
```

There are several ways to specify colors, and we'll cover the two simplest. First, you can use a number from 0 to 7 to define these colors:

0 - background color, as defined by CANVASCOLOR setting
1 - foreground color, the opposite of the CANVASCOLOR setting
2 - red
3 - green
4 - blue
5 - cyan
6 - magenta
7 - yellow

Or, you can type in the name of a color. Here is a partial list of the color names that ARCPLOT accepts. Reference the color illustrations in the ARC/INFO documentation of symbology for the appearance of these colors. The shadeset, COLORNAMES.SHD has many of these colors.

```
snow          GhostWhite      WhiteSmoke  gainsboro
FloralWhite   OldLace         linen       AntiqueWhite
PapayaWhip    BlanchedAlmond  bisque      PeachPuff
NavajoWhite   moccasin        cornsilk    ivory
LemonChiffon  seashell        honeydew    MintCream
azure         AliceBlue       lavender    LavenderBlush
MistyRose     white           black       DimGray
SlateGray     LightSlateGray  gray        LightGray
MidnightBlue  navy            NavyBlue    CornflowerBlue
SlateBlue     LightSlateBlue  RoyalBlue   blue
```

```
DodgerBlue       DeepSkyBlue      SkyBlue          LightSkyBlue
SteelBlue        LightSteelBlue   LightBlue        PowderBlue
turquoise        cyan             LightCyan        CadetBlue
aquamarine       SeaGreen         LightSeaGreen    SpringGreen
LawnGreen        green            chartreuse       GreenYellow
LimeGreen        YellowGreen      ForestGreen      OliveDrab
khaki            LightYellow      yellow           gold
LightGoldenrod   goldenrod        RosyBrown        IndianRed
SaddleBrown      sienna           peru             burlywood
beige            wheat            SandyBrown       tan
chocolate        firebrick        brown            salmon
LightSalmon      orange           coral            LightCoral
tomato           OrangeRed        red              HotPink
DeepPink         pink             LightPink        maroon
magenta          violet           plum             orchid
BlueViolet       purple           thistle
```

You can more precisely define colors with five different color models. You can use these other arguments for any of the color commands in ARCPLOT:

- ❑ HLS <hue> <lightness> <saturation>
- ❑ RGB <red> <green> <blue>
- ❑ CMY <cyan> <magenta> <yellow>
- ❑ CMYK <cyan> <magenta> <yellow> <black>
- ❑ HSV <hue> <saturation> <value>

Unless you have special color needs, you'll probably stick with the first two, simpler methods for specifying color.

Overriding Symbol Scales

When you pick a symbol, by default it will appear on the screen or graphics page with the same size as in the symbol set. This size is fixed in page units. Often, you would like symbols to scale up or down, as you change map scales.

Each of these commands has two usages. When you use them with just a <scale> argument, they define a relative scale factor. For example, a value of 2.0 would cause symbols to draw at twice their normal size. With the MAPSCALE <scale> argument, these commands calculate the factor between your present map scale and the specified scale, and adjust the symbol sizes accordingly.

```
Arcplot: LINESCALE <scale>
Arcplot: LINESCALE MAPSCALE <scale>
Arcplot: MARKERSCALE <scale>
Arcplot: MARKERSCALE MAPSCALE <scale>
Arcplot: SHADESCALE <scale>
Arcplot: SHADESCALE MAPSCALE <scale>
Arcplot: TEXTSCALE <scale>
Arcplot: TEXTSCALE MAPSCALE <scale>
```

NOTE: *While most symbol property commands affect only features drawn with a fixed symbol number, the scaling commands are an exception; they affect ARCPLOT draw commands using symbol items as well as fixed symbols.*

Other Symbol Properties

There are many other ARCPLOT commands that affect the properties of features drawn with fixed symbols. Following are some of the more useful and commonly used commands.

```
Arcplot: LINESIZE <width | *>
```

sets the width of lines with a width in page units, or by interactively picking two points to set a width.

```
Arcplot: MARKERANGLE <angle | *>
```

sets a rotation angle to be applied to markers drawn. You can type in an angle in degrees, or pick two points to specify the angle. A positive angle is counterclockwise.

```
Arcplot: TEXTANGLE <angle | *>
```

sets a rotation angle to be applied to text. You can type in an angle in degrees, or pick two points to specify the angle. A positive angle is counterclockwise.

```
Arcplot: TEXTJUSTIFICATION <LL | LC | LR | CL | CC | CR | UL
         | UC | UR>
```

sets the justification position for text.

These are a few of the ARCPLOT commands that affect the current symbol. In the next chapter, we'll cover how to make many of these changes permanent in a symbol table.

NOTE: *When you change symbol properties in an ARCPLOT session, the new properties remain until changed. Typically, you are frequently setting new symbol properties in an ARCPLOT session.*

Drawing Coverage Features

Following are the basic commands to draw coverage features. These commands draw coverage features either with the current symbol or by the item you specify. There are up to four current symbols; line, marker, shade, and text. If you have a selected set defined for a coverage, all of these commands will draw only the selected features.

In the plotting exercise at the end of the chapter, we will use most of these commands. Most of the coverages supplied with GIScity have feature attribute tables with a symbol item defined. In the exercise, you'll see features drawn with both fixed symbols and symbol items.

Drawing Polygons

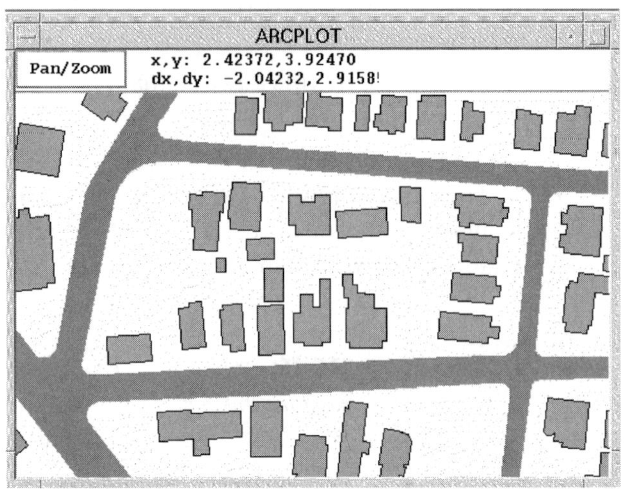

Road and building polygons from the GIScity coverages.

```
Arcplot: POLYGONS <cover>
```

This command draws the outlines of polygons using the current line symbol.

```
Arcplot: POLYGONLINES <cover> {item | symbol}
         {lookup_table} {NOISLANDS}
```

Like POLYGONS, this command draws the outlines of polygons, but it gives you the flexibility to specify a symbol item, an alternate symbol number, or whether to use a *lookup table* (a relate table which has a relate item with the PAT and a numeric item called SYMBOL). The NOISLANDS argument specifies that island polygons will not be drawn.

```
Arcplot: POLYGONSHADES <cover> <item | symbol>
         {lookup_table} {angle_item | angle}
```

This command allows you to shade the interiors of polygons. You can specify a symbol item for the shade, or type in a shade number. As with POLYGONLINES, you can specify a lookup table. Also, you can specify an item to set an angle to be applied to the shade pattern, or type in an angle value.

Drawing Arcs and Nodes

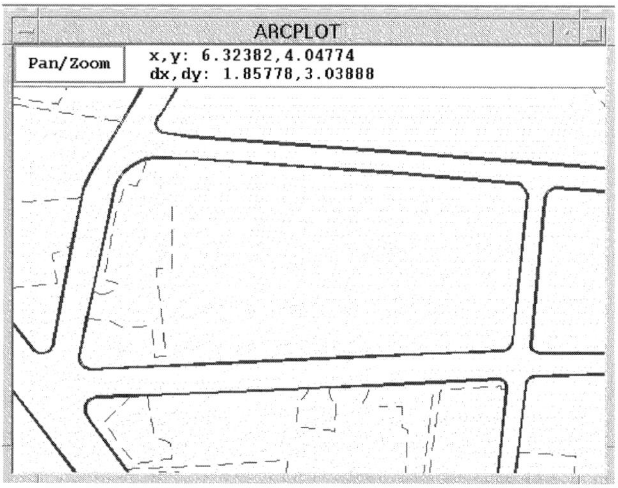

Road arcs from the GIScity coverages.

```
Arcplot: ARCS <cover> {NOIDS | IDS | IDSONLY}
```

This command draws coverage arcs with the current line symbol. You can also draw the arc User IDs if you specify IDS or IDSONLY, but you will rarely, if ever, use this argument.

```
Arcplot: ARCLINES <cover> {item | symbol} {lookup_table}
```

This command draws coverage arcs using a symbol item or specified line symbol number.

```
Arcplot: NODES <cover> {NOIDS | IDS | IDSONLY}
         {angle_item | angle}
```

This command draws nodes with the current marker symbol. You can also specify whether to draw the User IDs for the nodes, as well as rotate nodes.

```
Arcplot: NODEMARKERS <cover> <item | symbol>
         {lookup_table} {angle_item | angle}
```

This command draws nodes using a symbol item you define or an alternate marker symbol number. You can also specify a lookup table, as well as an item with angle values or a fixed angle.

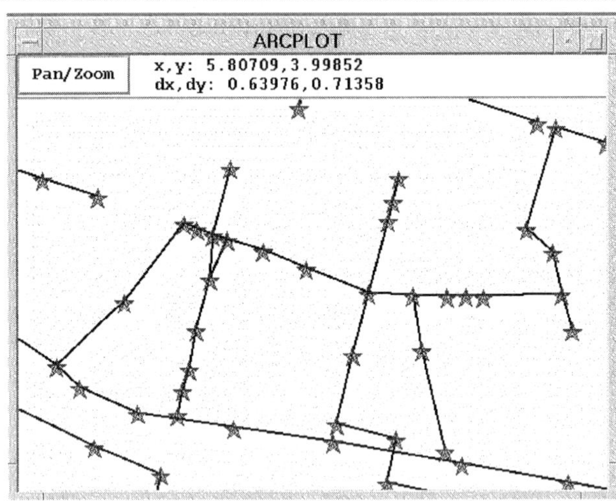

Sewer lines and nodes from the GIScity coverages.

Drawing Label Points

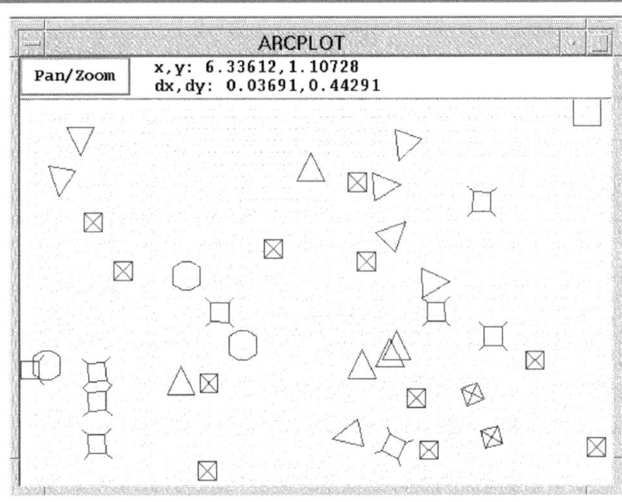

Label points with symbol items at varying angles.

`Arcplot: POINTS <cover> {NOIDS | IDS | IDSONLY}`

This command draws points using the current marker symbol. This is for coverages with points, but no polygons.

```
Arcplot: POINTMARKERS <cover> <item | symbol>
         {lookup_table}
```

This command draws points using a symbol item for markers or alternate symbol. You can also specify a lookup table.

```
Arcplot: LABELS <cover> {NOIDS | IDS | IDSONLY}
```

This command draws label points using the current marker symbol. This is for drawing label points in coverages with polygon topology.

```
Arcplot: LABELMARKERS <cover> <item | symbol>
         {lookup_table}
```

This command draws label points using a symbol item for markers or alternate symbol. You can also specify a lookup table.

Drawing Annotation

Annotation from a coverage.

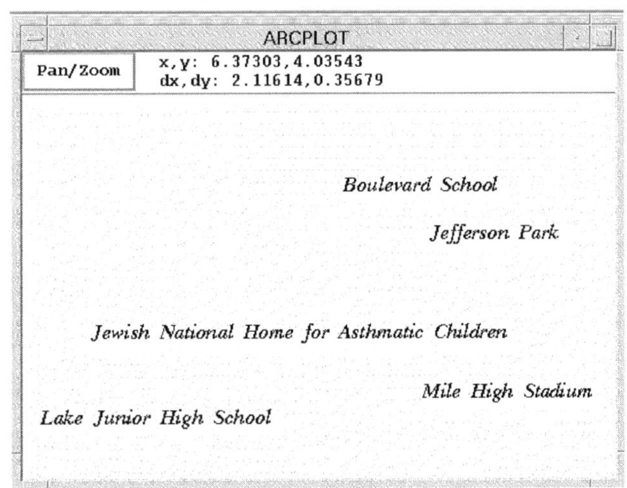


```
Arcplot: ANNOTEXT <cover> {subclass {text_item}
         | ALL} {FLIP} {level...level}
```

This command draws coverage annotations. Annotation properties are defined as pseudo items, which are set with ARCEDIT commands such as TEXTSIZE or TEXTANGLE.

Drawing annotation is different from drawing other coverage features. You can't modify the current text symbol and apply those changes to the drawing of annotation unless you modify the textset while you are in ARCPLOT.

ANNOTEXT has a couple of interesting options. You can specify a subclass and an item from a TAT (annotation attribute table) to replace the text string in the coverage. This dynamically draws coverage annotations based on an item value, instead of the fixed text string.

The FLIP argument can be used with the FLIPANGLE command to ensure that annotation always appears right side up. The ALL argument will draw all annotation subclasses for the coverage.

In the next chapter, we'll cover some more text commands that draw text using item values from other feature attribute tables.

Adding Map Elements

In this section, we'll discuss how to add map elements to your graphics file or screen display. Map elements are separate from coverage features. They are the parts of your map that outline, illustrate, and provide a reference to coverage features. Examples of map elements are neat lines, title blocks, north arrows, title text, and scale bars.

You can review these commands in the Map Elements menu in Display GIScity, which is covered in a later chapter. If you are curious, you can turn Echo On in Display GIScity and run the commands, similar to those we document below. You can then see exactly the steps you might execute in a sample ARCPLOT session.

Adding Lines and Shapes

These commands place lines, circles, points, and shapes by either typing in coordinates in page units or map units, or by selecting points with the screen cursor. All of these map elements are ephemeral: they are not stored in a coverage or anywhere else.

If you want to type coordinates for these commands, check your units. You can tell whether you are using map or page units by typing:

```
Arcplot: SHOW UNITS
```

Also, check the current marker symbol, line symbol, or shade symbol. Here are a few commands that you can use to check out your current symbol status:

```
Arcplot: SHOW MARKERSYMBOL
Arcplot: SHOW MARKERANGLE
Arcplot: SHOW MARKERCOLOR
Arcplot: SHOW MARKERSCALE
Arcplot: SHOW LINESYMBOL
Arcplot: SHOW LINESIZE
Arcplot: SHOW LINECOLOR
Arcplot: SHOW SHADESYMBOL
Arcplot: SHOW SHADECOLOR
```

As you can see, the SHOW command can tell you virtually everything about your current settings in ARCPLOT. Reference ArcDoc, or better yet, just try the SHOW command for any setting you are interested in. Chances are the command will work; there are many, many valid arguments for SHOW.

```
Arcplot: LINE {SPLINE} <* | xy...xy>
```

The LINE command can place a simple line with two endpoints or a compound line with many points. Also, if you use the SPLINE argument, you will get a smoothed line. If you use the cursor for inputting lines, use the 9 input key to return to command level.

```
Arcplot: BOX <* | xmin ymin xmax ymax>
```

This command is a short-cut to placing four lines which make up a rectangle. Specify two opposite corners of a rectangle.

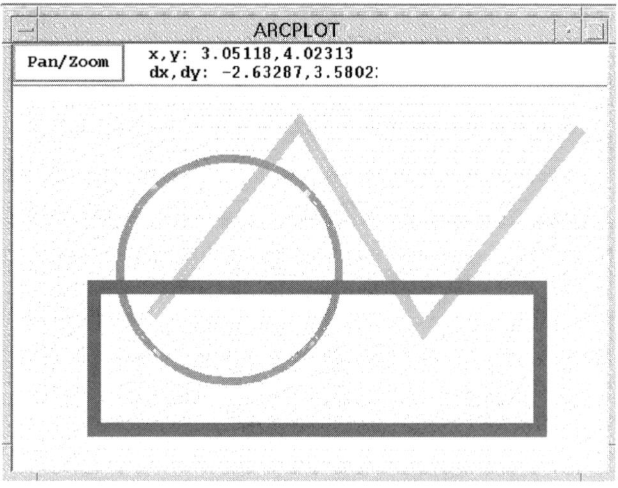

Map elements drawn with LINE< BOX, and CIRCLE.

`Arcplot: CIRCLE <xy radius | * {radius}>`

This command places circles. If you use cursor input, first pick the center point of the circle, and then any point on the circle. Or, you can type in the center point and radius length. Or, you can pick the center point with the cursor and type in a radius length. As an example of the last method, you can interactively place a circle with exactly 10 map units radius by typing `CIRCLE * 10`.

`Arcplot: PATCH <* | xmin ymin xmax ymax> {OUTLINE}`

PATCH places a shaded rectangle using the current shade symbol. This area will obscure whatever is beneath it. If you use the optional OUTLINE argument, the outline of the rectangle will also be drawn with the current line symbol.

`Arcplot: SHADE <* | xy...xy> {OUTLINE}`

SHADE is like PATCH, but instead of a rectangle, SHADE draws a polygon you define by typing coordinates or cursor input. Use input key 9 to close the polygon. As with PATCH, anything beneath the polygon is obscured by the shade symbol.

```
Arcplot: MARKER <* | xy>
```

The MARKER command places the current marker symbol. In the following demonstration plot, we use this command to place north arrows from a markerset containing north arrows.

Adding Text

Before adding text, set your text symbol. There are far more properties for text symbols than any other type of symbol. To see how many commands affect text symbols, type:

```
Arcplot: COMMANDS TEXT
```

You'll see about thirty ARCPLOT commands with names like TEXTALIGNMENT, TEXTSLANT, and TEXTSTYLE.

To check any of these properties, use the SHOW command followed by the name of the text property command.

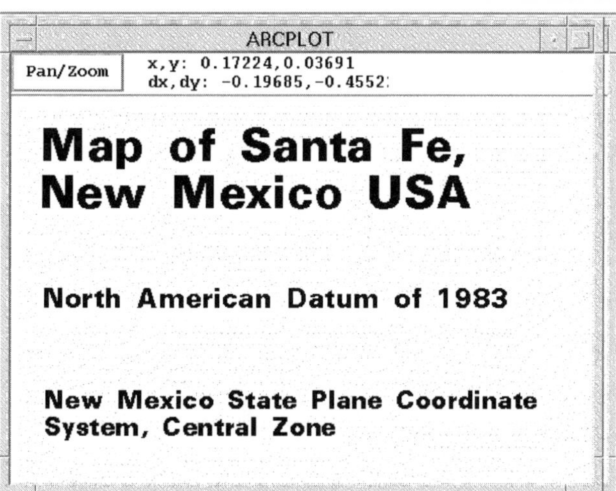

Three text strings drawn with MOVE, TEXTSIZE, and TEXT.

Placing text in ARCPLOT is different from placing other map elements. Most commands to add text require that you first move to the location at which you wish to place text.

```
Arcplot: MOVE <* | xy>
```

The MOVE command specifies the location where the text you are about to place will end up. You can either use cursor input or type in a coordinate in page units or map units.

```
Arcplot: TEXT <text_string> {LL | LC | LR | CL |
         CR | UL | UC | UR}
```

After you have used the MOVE command, you can use TEXT to add a title text. You can optionally set a text justification, which will override the justification you may have set with TEXTJUSTIFICATION.

When you use this command, use single quotes around the text string. This way, ARCPLOT can tell the optional justification argument apart from the text string. For example, if you want to place the text string 'Demonstration Plot' with a justification of lower right, you should enter `TEXT 'Demonstration Plot' LR`.

```
Arcplot: TEXTFIT <text_string> {* | xy1 xy2}
```

Use TEXTFIT instead of TEXT if you want to fit text exactly between two points. You can use cursor input or type in the two points that the text should be fitted to.

```
Arcplot: TEXTFILE <text_file>
```

This command is a powerful way to add a lot of text to your graphics display or graphics file. Any text file on your system can be added. Just prepare a text file with several paragraphs, and then use TEXTFILE to easily add those paragraphs. TEXTFILE is used to add extended descriptions on maps.

Adding North Arrows

There are several ways to add north arrows to your maps. The easiest way is to add a marker from a markerset provided with ARC/INFO called NORTH.MRK. Here are the ARCPLOT commands to do this:

```
Arcplot: MARKERSET NORTH.MRK
Arcplot: MARKERSYMBOL 12
Arcplot: MARKERSIZE 1.5
Arcplot: MARKER *
```

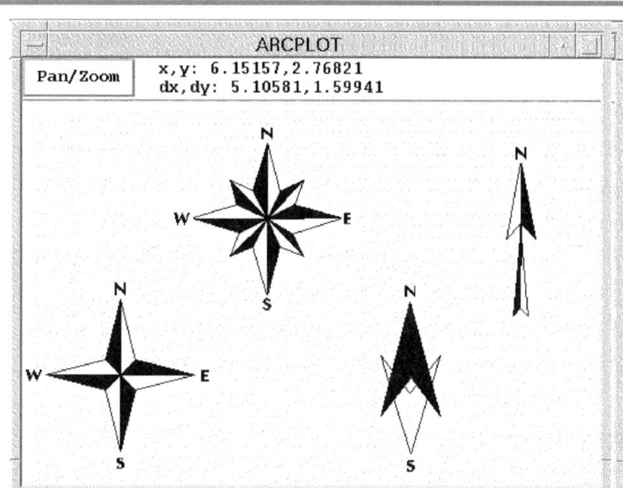

North arrow marker symbols from NORTH.MRK.

These commands place the marker shown in the preceding illustration with a size of 1.5 inches.

While using markers is the easiest way to add a north arrow, you may prefer a north arrow of your own design. You can make a separate coverage just for a north arrow, and then use other ARCPLOT commands, such as MAPLIMITS and ARCS, to add an elaborate north arrow of your design. Frequently, north arrows incorporate agency logos, and this method works well for these north arrows.

Adding Scale Bars

You can add scale bars with the same two techniques as north arrows, but you must add a scale bar with exactly the correct dimensions to provide an accurate scale reference.

The Scale Bar menu in the Map Elements menu in Display GIScity takes the approach of building scale bars in an AML. This AML offers you several choices for number of intervals, interval distances, text symbols, and shade colors. If you are interested, study this AML provided with Display GIScity:

```
$GCHOME/display/disp-element_scalebar.aml (UNIX)
$GCHOME:[DISPLAY]DISP-ELEMENT_SCALEBAR.AML (VMS)
```

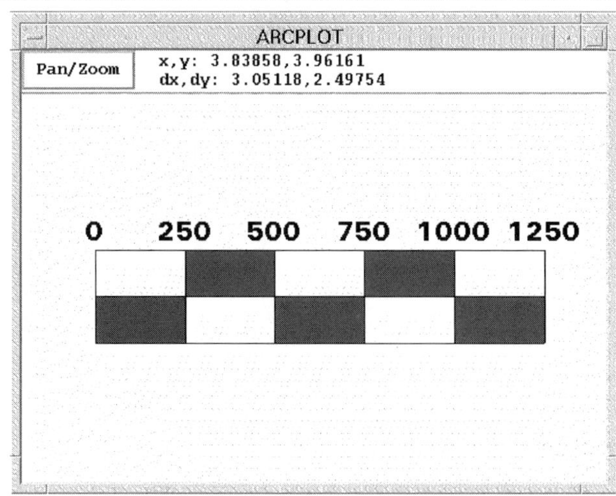

A scale bar generated by an AML supplied with GIScity.

If you are producing a series of maps, you can write a script AML that uses UNITS MAP, PATCH, BOX, and TEXT to create a scale bar. This script AML can be used for any maps at the same scale.

Adding Legends

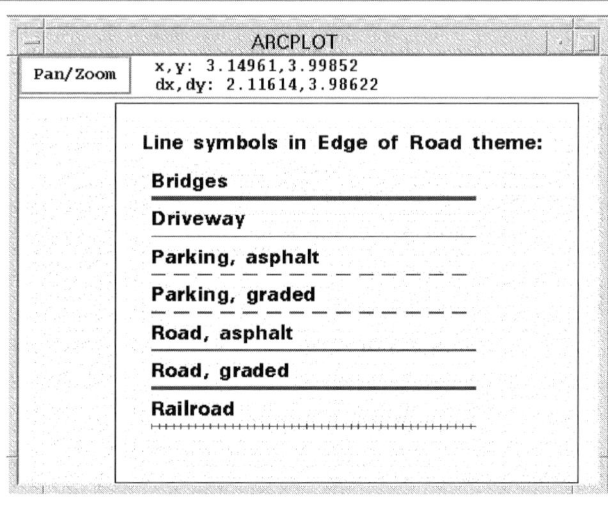

A map legend generated by an AML in GIScity.

ARCPLOT contains a suite of commands designed to streamline the creation of legends. These commands include KEYANGLE, KEYAREA, KEYBOX, KEYLINE, KEYMARKER, KEYPOSITION, KEYSEPARATION, and KEYSHADE. These commands require that you prepare a text file called a *key file*. The key file has a specific format for entering a symbol number and description for selected symbols.

If you'd like to try these commands, reference ArcDoc on these commands and see if the legends these commands create work for you.

Keep in mind that the relative ease of the key legend commands in ARCPLOT leads to limitations in the legends you can design.

Experienced AML users sometimes choose to forgo the key legend commands for the greater flexibility of implementing map legends through an AML. The Map Legend command in Display GIScity takes this approach.

Adding Neat Lines

Neat lines are coordinate grid lines and corners that give you a reference for the coordinate system. For an example, examine the sides of any national survey map, such as the United States Geological Survey quadrangle maps.

Before you use the neat line commands, you must provide a coordinate reference to ARCPLOT. This reference is made through the MAPPROJECTION command. Also, you should have map projection values specified in projection files. Reference the PROJECT command in the ARC program for more information on projection files.

A significant enhancement in ARC/INFO release 7.0 is a set of commands to automatically generate neat lines in ARCPLOT. Reference ArcDoc for these neat line commands: NEATLINE, NEATLINEGRID, NEATLINETICS, NEATLINEHATCH, and NEATLINELABELS.

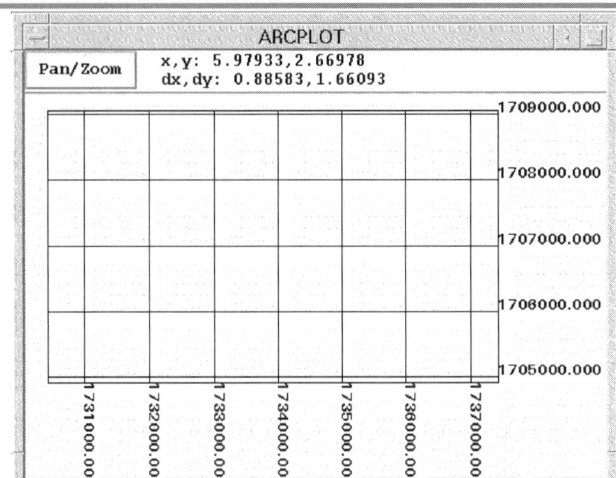

The types of neat lines for maps.

Walking Through a Sample ARCPLOT Session

Now we'll try out many of the basic ARCPLOT commands we've learned in this chapter by stepping our way through the production of the graphics file in the following illustration.

These commands are also contained within an AML supplied with GIScity. ARCPLOT can produce graphics to either a graphics window or a graphics file. To send graphics to a graphics window, type:

```
Arcplot: &run $GCHOME/data/plot_giscity.aml 9999 (UNIX)
Arcplot: &run $GCHOME:[DATA]PLOT_GISCITY.AML 9999 (VMS)
```

To send graphics to a graphics file, type:

```
Arcplot: &run $GCHOME/data/plot_giscity.aml 1040 (UNIX)
Arcplot: &run $GCHOME:[DATA]PLOT_GISCITY.AML 1040 (VMS)
```

A graphics file called PLOTDEMO.GRA will be created in the directory you are in. To see the graphics file, type:

```
Arc: DRAW PLOTDEMO.GRA 9999 3 (UNIX and VMS)
```

Now, let's start the exercise.

Producing a Graphics File

1. To start ARCPLOT and set the graphics environment, type the following commands:

```
Arc: arcplot
Arcplot: display 9999 3
Arcplot: &workspace $GCHOME/data (UNIX or Windows NT) or
Arcplot: &workspace $GCHOME: [DATA]
```

   You'll see a large, blank graphics window.

2. To draw a box outline for the graphics page, type:

```
Arcplot: units page
Arcplot: linesymbol 1
Arcplot: pagesize 11.0 8.5
Arcplot: maplimits 0.75 1.0 7.75 8.0
Arcplot: box 0.5 0.25 10.5 8.25
Arcplot: box 0.0 0.0 11.0 8.5
```

   You will now see two rectangles drawn on the graphics window.

These boxes are the result of commands for Step 2.

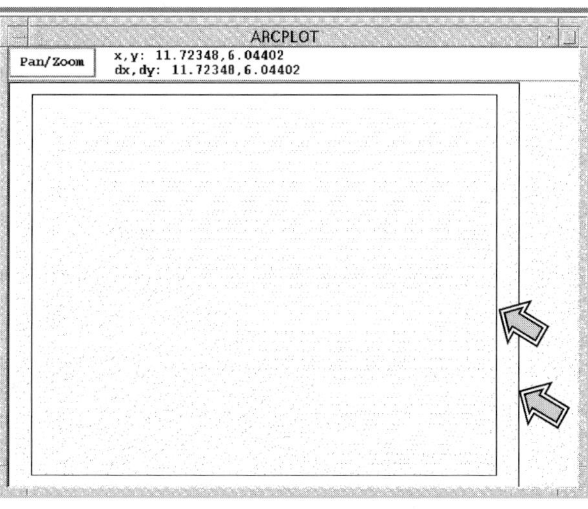

3. To draw downtown Santa Fe, follow the steps listed below.

These commands will draw the main part of the map shown in the preceding illustration. Note that most of the symbol sets used are provided with ARC/INFO, so no directory path is required. However, there is a lineset supplied with GIScity, ELEVRAMP. Because it is not a standard ARC/INFO lineset, you are required to type a path. Both the UNIX and VMS commands are supplied below.

Also, we set a map scale denominator of 7200. Since the MAPSCALE command uses unit-less scales, this is equivalent to a scale of 1":600′.

```
Arcplot: mape 1731200,1704000,1735400,1708200
Arcplot: mapunits feet
Arcplot: mapscale 7200
Arcplot: shadeset colornames.shd
Arcplot: textset font.txt
Arcplot: markerset plotter.mrk
Arcplot: lineset $GCHOME/data/elevramp.lin
         (UNIX or Windows NT) or
Arcplot: lineset $GCHOME:[DATA]ELEVRAMP.LIN (VMS)
Arcplot: arclines contours symbol
Arcplot: polygonshades roadpoly symbol
Arcplot: lineset carto.lin
Arcplot: linesymbol 201
Arcplot: arcs edgeroad
Arcplot: markerset plotter.mrk
Arcplot: markersymbol 68
Arcplot: points manhole
Arcplot: linesymbol 1
Arcplot: linecolor black
Arcplot: linesize 0.04
Arcplot: box 0.75 1.0 7.75 8.0
Arcplot: units map
Arcplot: box 1732603,1705366,1733167,1705930
Arcplot: show mapscale
```

Note how we switched between UNITS PAGE and UNITS MAP for drawing the two boxes above. This technique is used to draw the reference box for an inset map we'll draw in the next step.

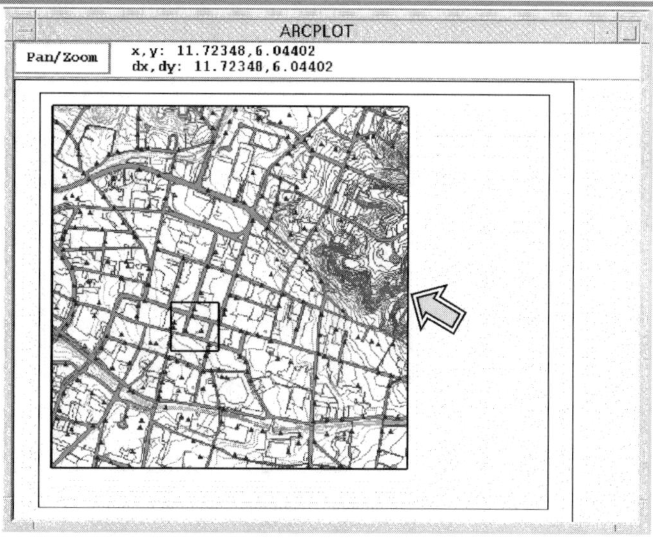

Downtown Santa Fe, as drawn by the Step 3 commands.

4. To draw the plaza inset map, take these steps:

```
Arcplot: mapscale automatic
Arcplot: units page
Arcplot: maplimits 8.25 4.0 10 5.75
Arcplot: mapextent 1732603 , 1705366 , 1733167 , 1705930
Arcplot: lineset $GCHOME/data/elevramp.lin
         (UNIX or Windows NT) or
Arcplot: lineset $GCHOME:[DATA]ELEVRAMP.LIN (VMS)
Arcplot: arclines contours symbol
Arcplot: polygonshades roadpoly symbol
Arcplot: shadeset colornames.shd
Arcplot: polygonshades rooflines 15
Arcplot: lineset plotter.lin
Arcplot: linesymbol 1
Arcplot: arcs rooflines
Arcplot: linesymbol 2
Arcplot: arcs edgeroad
Arcplot: markerset plotter.mrk
Arcplot: markersymbol 68
Arcplot: points manhole
Arcplot: annotext manhole
Arcplot: units page
Arcplot: linesymbol 1
```

```
Arcplot: linecolor black
Arcplot: linesize 0.04
Arcplot: box 8.25 4.0 10 5.75
```

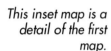

This inset map is a detail of the first map.

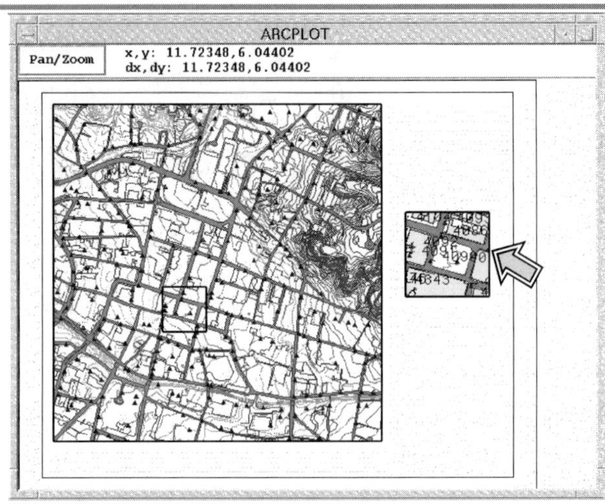

5. This step draws the locator map in the upper right corner of the display. The data used is from the TIGER database provided by the U.S. Bureau of the Census.

```
Arcplot: units page
Arcplot: maplimits 8.25 6.25 10 8.0
Arcplot: units map
Arcplot: mapextent 1712867, 1684424, 1741646, 1713203
Arcplot: lineset plotter.lin
Arcplot: linesymbol 2
Arcplot: linesize 0.04
Arcplot: arcs tgr_road
Arcplot: linesymbol 1
Arcplot: linecolor black
Arcplot: units map
Arcplot: box 1730596,1703978,1735338,1708720
Arcplot: units page
Arcplot: box 8.25 6.25 10 8.0
```

This locator map gives us a reference to a larger area than downtown Santa Fe.

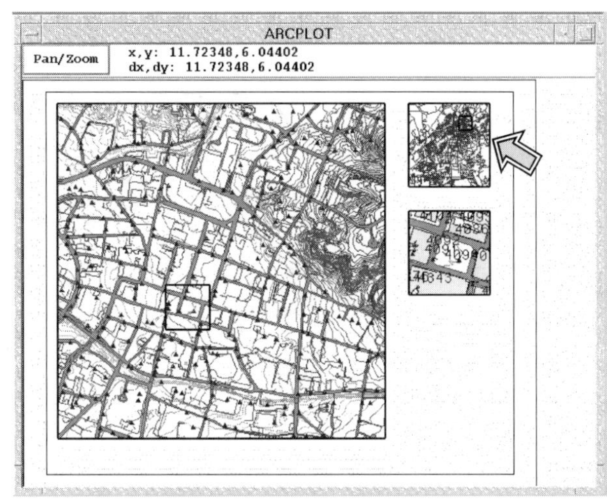

6. This steps uses several of the text commands to place three titles on the graphics display.

```
Arcplot: textset font.txt
Arcplot: units page
Arcplot: textsymbol 3
Arcplot: textsize .35
Arcplot: textjustification cc
Arcplot: move 5.5 0.625
Arcplot: text 'Demonstration Map of Downtown Santa Fe'
Arcplot: textsymbol 1
Arcplot: textsize .20
Arcplot: textjustification cc
Arcplot: move 9.125 3.75
Arcplot: text 'Manholes in\The Plaza'
Arcplot: move 9.125 6
Arcplot: text 'City Locator Map'
```

Now we've added titles to our map.

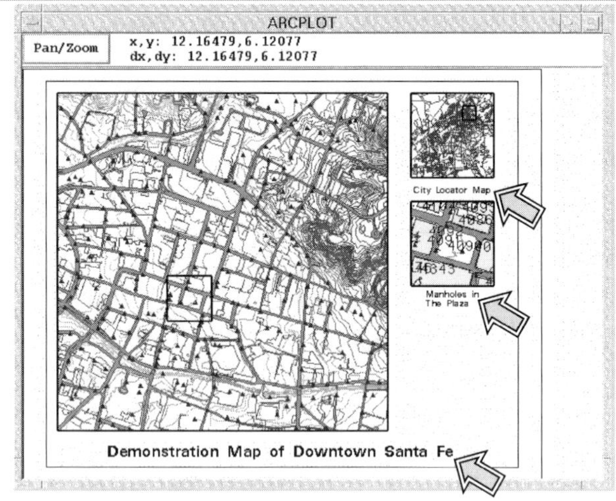

7. We'll place a north arrow using the simplest of the methods discussed in this chapter.

```
Arcplot: markerset north.mrk
Arcplot: markersymbol 12
Arcplot: units page
Arcplot: markersize 1.25
Arcplot: marker 9.125 2.625
```

The north arrow gives us a positional reference.

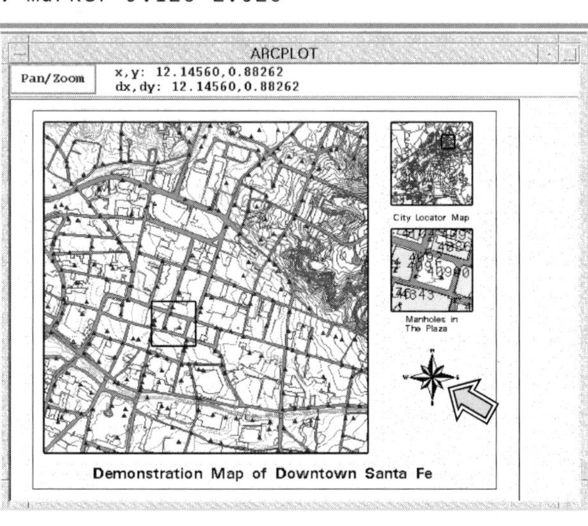

8. In this step, we'll use some of the basic map elements commands to make up a scale bar. Note that the mathematics can be tedious. Display GIScity and other AML applications have commands that can generate a scale bar for you automatically.

```
Arcplot: units page
Arcplot: shadesymbol 1
Arcplot: shadecolor blue
Arcplot: patch 8.375 1.15 8.875 1.30
Arcplot: patch 8.875 1.00 9.375 1.15
Arcplot: patch 9.375 1.15 9.875 1.30
Arcplot: linesymbol 1
Arcplot: linesize 0.02
Arcplot: box 8.375 1.0 9.875 1.3
Arcplot: line 8.375 1.0 8.375 1.4
Arcplot: line 8.875 1.0 8.875 1.4
Arcplot: line 9.375 1.0 9.375 1.4
Arcplot: line 9.875 1.0 9.875 1.4
Arcplot: textsymbol 3
Arcplot: textjustification lc
Arcplot: textsize 12 pt
Arcplot: move 8.375 1.45
Arcplot: text 0'
Arcplot: move 8.875 1.45
Arcplot: text 300'
Arcplot: move 9.375 1.45
Arcplot: text 600'
Arcplot: move 9.875 1.45
Arcplot: text 900'
```

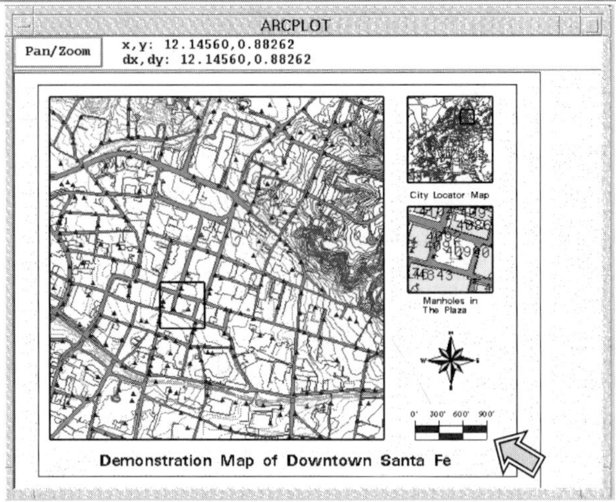

A scale bar, built with ARCPLOT's map element commands.

9. Now we are finished with our ARCPLOT session. After admiring our creation, type:

```
Arcplot: quit
```

10. Now, if you have the stamina, you can type in the following and go back to step 2 to create a graphics file. Aside from the DISPLAY command, all other ARCPLOT commands are identical.

```
Arcplot: display 1040
Enter output filename: giscity.gra
```

After you reach step 9 again, to witness your creation, type:

```
Arc: DRAW GISCITY.GRA 9999 3
```

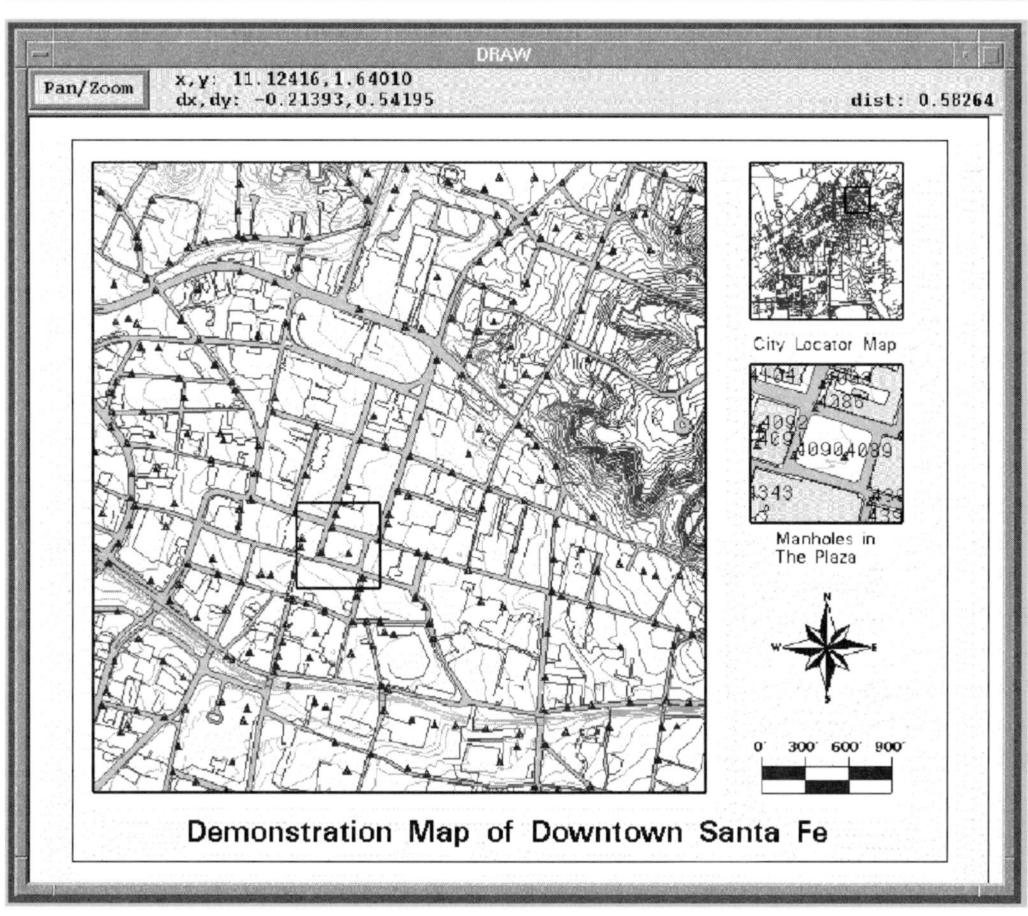

Graphics file called PLOTDEMO.GRA, produced by the exercise commands.

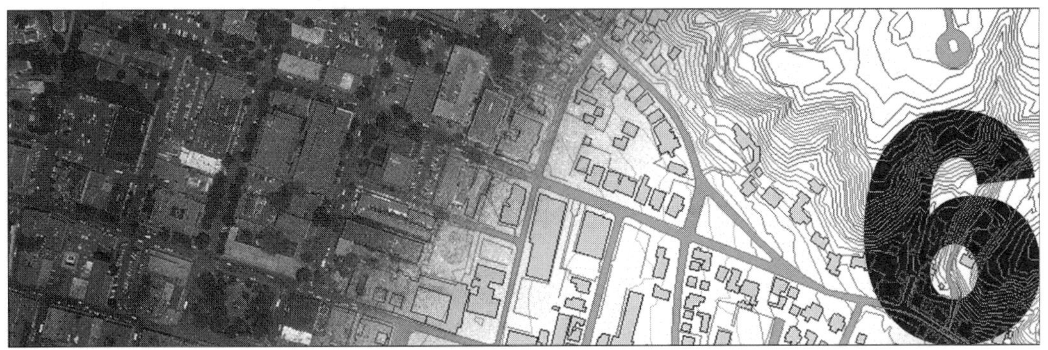

Query, Display, and Symbols in ARCPLOT

Symbolizing Attributes and Editing Symbol Sets

ARC/INFO's unique attribute is its richness of symbolic expression. Every attribute that can be associated with a geographic feature, whether in a feature attribute table, relate table, or external databases, can be portrayed through several visualization techniques: pie symbols, graduated symbols, color ramps, discrete symbols, text on features, and graphs. In this chapter, we will learn these ways to communicate visually in ARCPLOT.

Another important topic for this chapter is how to create and modify the four basic symbol sets in ARC/INFO: markersets, linesets, shadesets, and textsets. With the graphics enhancements delivered with releases 6 and 7 of ARC/INFO, you now have a large palette of controls on symbology. We will go step by step through creating symbols.

Keep in mind that there are many, many commands in ARCPLOT. The commands explored in this chapter are a sampling of the most commonly used commands. You will probably never use a majority of the available commands. But once you understand these several dozen commands, then you can understand and apply other more specialized commands in ARCPLOT.

Advanced Feature Selection in ARCPLOT

We've covered many of ARCPLOT's commands to select features in Chapter 4, Managing Attribute Data. We won't cover the basics for feature selection in ARCPLOT again, but here are two advanced selection commands you might find useful.

The RESELECT OVERLAP command can be used to select wells with 2,000 feet of a selected stream arc.

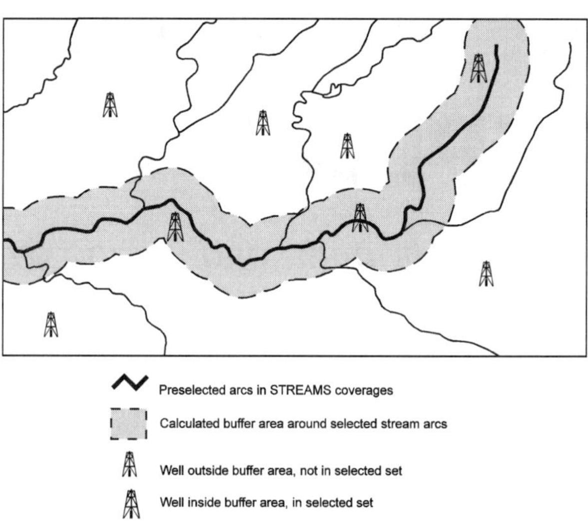
Overlap selection example:
Two coverages, STREAMS and WELLS,
Arcplot: RESELECT WELLS POINTS OVERLAP STREAMS ARCS 2000.0
Preselected arcs in STREAMS coverages
Calculated buffer area around selected stream arcs
Well outside buffer area, not in selected set
Well inside buffer area, in selected set

```
Arcplot: <ASELECT | RESELECT | UNSELECT> <cover>
         <feature_class> OVERLAP <overlap_cover>
         <overlap_feature>  {selection_distance}
         {PASSTHRU | WITHIN}
```

The usage of the ASELECT, RESELECT, and UNSELECT commands lets you select points, lines, or polygons in the selection coverage, which are contained within, overlapping, or within a selection distance of points, lines, or polygons in the overlap coverage.

There are two coverages at work in this command usage. We'll call them the *selection coverage* and *overlap coverage*. In this command, we specify the coverage names and feature classes with the <cover>, <feature_class>, <overlap_cover>, and <overlap_feature> arguments.

If you specify a {selection_distance} in map units, then a temporary buffer is made for you, and all features that overlap with the buffer are selected. {PASSTHRU | WITHIN} applies to line and polygon features in the overlap coverage. Just as in other selection commands, this argument states whether a feature must be completely within a selection area, or if only a part of it need be inside to be selected.

```
Arcplot: <ASELECT | RESELECT | UNSELECT> <cover>
         <feature_class> RANDOM <number> {PERCENT} {seed}
```

This other usage of the ASELECT, RESELECT, and UNSELECT commands lets you modify your selected set by random selection. You can add to, remove, and subset the current selected set.

This command is not often used but is important if you are interested in statistical sampling or simulations on your geographic data sets. You can perform Monte Carlo simulations on a part of a large data set and model environmental scenarios.

Attribute Query and Update

We'll review some ARCPLOT commands that query and update attributes.

Querying Attributes

```
Arcplot: ITEMS <cover> <feature_class>
```

Lists the items defined for the specified coverage and feature class. The coverage and feature class uniquely identify a feature attribute table.

```
Arcplot: LIST <cover> <feature_class> {range} {item...item}
```

Lists the values for the specified coverage, feature class, an optional range of record numbers, and an optional list of items. If you do not specify these last two arguments, then all records in the selected set for that coverage are listed, together with all item values.

```
Arcplot: LISTSELECT {selection_file}
```

Lists all selected sets and their associated coverages and tables.

```
Arcplot: IDENTIFY <cover> <feature_class> <* | xy>
         {item...item}
```

Lists the attributes of one selected coverage feature. Usually, you will use the * argument for interactive feature selection. You can optionally specify a range of items to list.

When you issue this command with the * argument, then the cursor is activated to select a point. If you successfully select a feature in the coverage and feature class, then you will see the item values for that feature appear in your dialog window.

Updating Attributes

In ARCPLOT, there are three primary commands to update feature attributes: CALCULATE, MOVEITEM, and FORMS. These commands are more fully documented in Chapter 4, Managing Attribute Data, so we will just cover them here briefly.

```
Arcplot: CALCULATE <cover> <feature_class>
         <target_item> = <arithmetic_expression>
```

Assigns a numeric value to the target item in the specified coverage and feature class. This operation is done on all features in the selected set.

```
Arcplot: MOVEITEM <cover> <feature_class>
         <'character_string' | source_item> {TO}
         <target_item>
```

Assigns a character string to the target item for the specified coverage and feature class. This operation is done on all features in the selected set.

```
Usage : FORMS <cover> <class> {item...item} {RELATE <relate>}

Usage : FORMS <cover> <class> FORM <form>
```

Invokes a menu to let you walk through a selected set and update attributes.

FORMS is a powerful command that uses the selected set and opens a database cursor to navigate through the features in the selected set.

Measurements

There is one command in ARCPLOT that you can use to make direct measurements of positions, lengths, and areas in your GIS graphics window, MEASURE. This command has three uses we'll cover here: AREA, LENGTH, and WHERE.

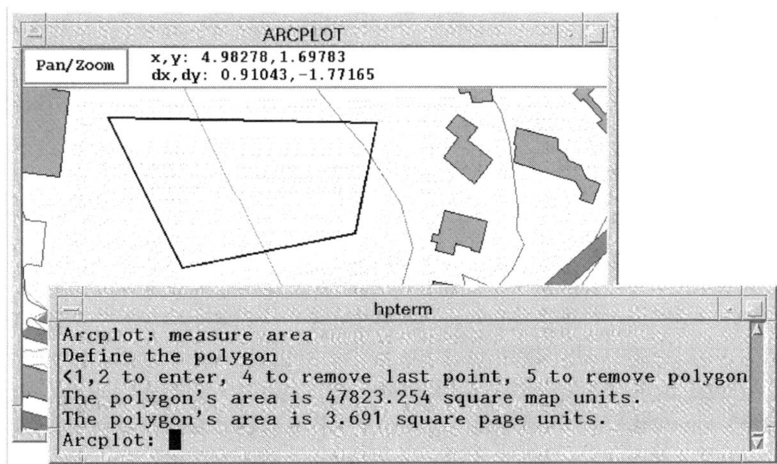

Measuring areas.

```
Arcplot: MEASURE AREA
         Define the polygon
         <1,2 to enter, 4 to remove last point, 5 to
         remove polygon, 9 to end>
```

When you start the MEASURE AREA command, it will engage you in a dialog and activate the cursor. Use the buttons as described in the dialog captured from an ARCPLOT session:

After you close the area with button 9, then the area will be reported to you in the dialog window.

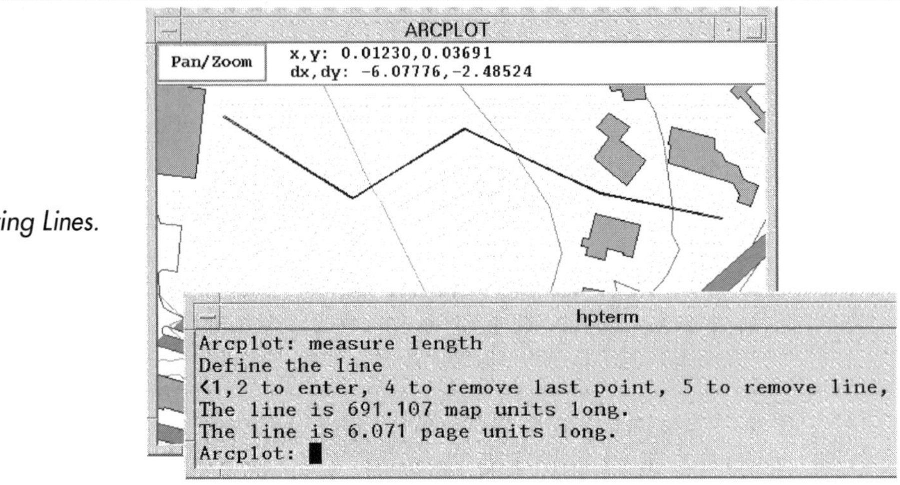

Measuring Lines.

```
Arcplot: MEASURE LENGTH
         Define the line
         <1,2 to enter, 4 to remove last point, 5 to
         remove line, 9 to end>
```

As with MEASURE AREA, MEASURE LENGTH engages you in a dialog to select points. You can use buttons 1 or 2 to mark vertices and nodes, and button 9 to return. The total length appears next in your dialog window.

```
Arcplot: MEASURE WHERE
         Enter point
```

This command reports to you the coordinate position of any point you pick with the cursor. After you select a point with the cursor, you'll see the coordinates in map or page units. The MEASURE WHERE command displays the positions of points you select in the graphics window. Unless you have exceptional aim, it will not be exactly the same coordinate as a point feature you might want to locate, such as a survey marker. In that event, you can do this in ARCPLOT: First, select that point feature and then use the SHOW function with the SELECT argument, as shown below. Insert any coverage name that contains label points.

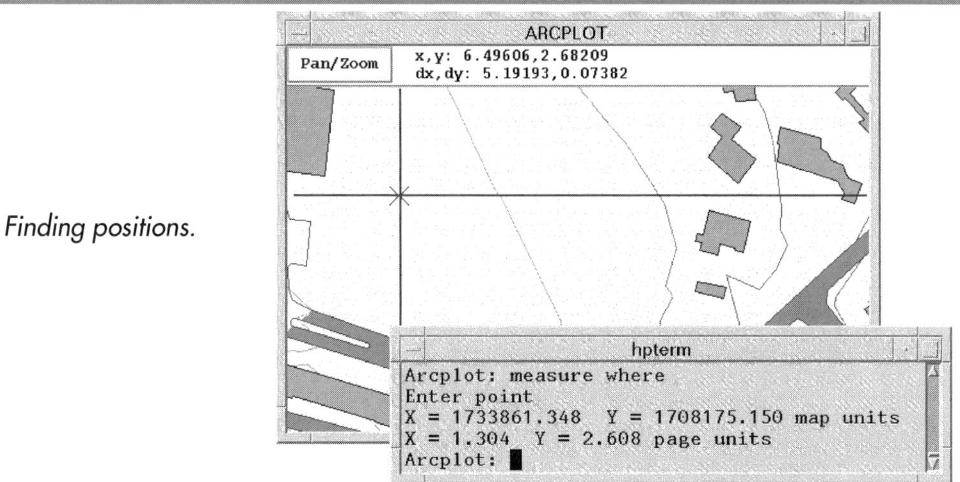

Finding positions.

```
Arcplot: reselect <cover> label one *
Arcplot: &type [show select <cover> point 1 xy]
```

This command displays two real valued numbers. These numbers are the exact coordinates of the selected point feature.

Advanced Display Commands

Now, we'll explore some more of the fundamental commands in ARCPLOT. All of these commands have the same effect whether you are using graphics windows or directing output to a graphics file.

Displaying Attributes as Text

One technique often used in ARCPLOT is to display item values as text next to the associated geographic features. We will cover the ARCTEXT, LABELTEXT, POINTTEXT, NODETEXT, and POLYGONTEXT commands. These commands give you great variety in placing text from attributes on arcs, nodes, points, and polygons.

These item values come from a feature attribute table or a relate table. The command arguments for all these commands are as follows:

```
<cover> <item> {lookup_table} {LL | LC | LR | CL | CC
| CR | UL | UC | UR}
```

<cover> is of course the coverage. <item> is the name of the item to annotate. {lookup_table} is an optional argument to use a simple relate to another table. {LL | LC | LR | CL | CC | CR | UL | UC | UR} is an optional argument to position the text relative to the features. These are the same justifications we have already applied in annotation properties.

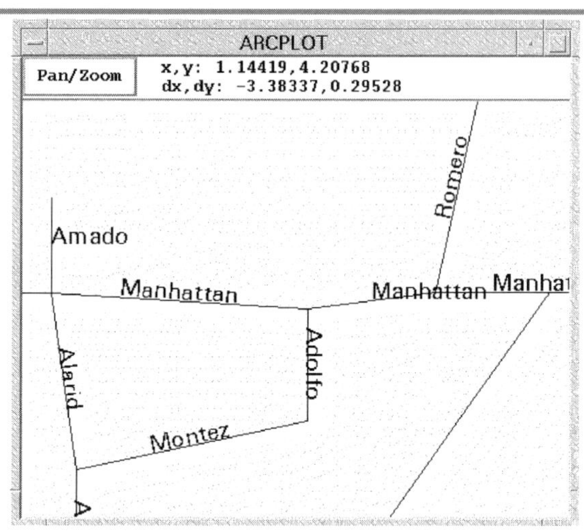

Displaying arc attributes as text.

```
Arcplot: ARCTEXT <cover> <item> {lookup_table}
         {POINT1 | POINT2 | LINE} {offset} {LL |
         LC | LR | CL | CC | CR | UL | UC | UR |
         BLANK} {NOFLIP}
```

This is a general-purpose command to display item values as text from the Arc Attribute Table. The {POINT1 | POINT2 | LINE} choice argument lets you specify whether the text is one point, two points (at an angle), or many points, often used along arcs and routes.

{offset} lets you place LINE text at the specified offset distance. {NOFLIP} ensures that the annotation is displayed right-side up.

Displaying label point attributes and node attributes as text.

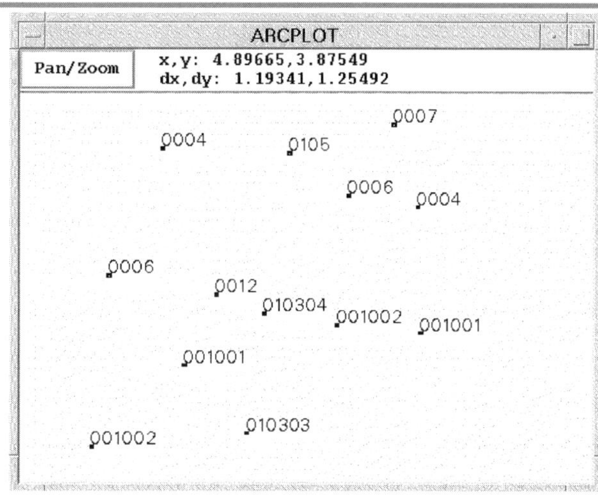


```
Arcplot: LABELTEXT <cover> <item> {lookup_table}
         {LL | LC | LR | CL | CC | CR | UL | UC | UR}
         {NOROTATION}
Arcplot: POINTTEXT <cover> <item> {lookup_table}
         {LL | LC | LR | CL | CC | CR | UL | UC | UR}
         {NOROTATION}
Arcplot: NODETEXT <cover> <item> {lookup_table}
         {LL | LC | LR | CL | CC | CR | UL | UC | UR}
```

These commands display item values as text from a Polygon Attribute Table (LABELTEXT), a Point Attribute Table (POINTTEXT), or a node attribute table (NODETEXT).

```
Arcplot: POLYGONTEXT <cover> <item> {lookup_table}
```

This command displays item values from a Polygon Attribute Table. (The difference between LABELTEXT and POLYGONTEXT is that LABELTEXT places text around the position of the label points, whereas POLYGONTEXT places text to best fit within a polygon.)

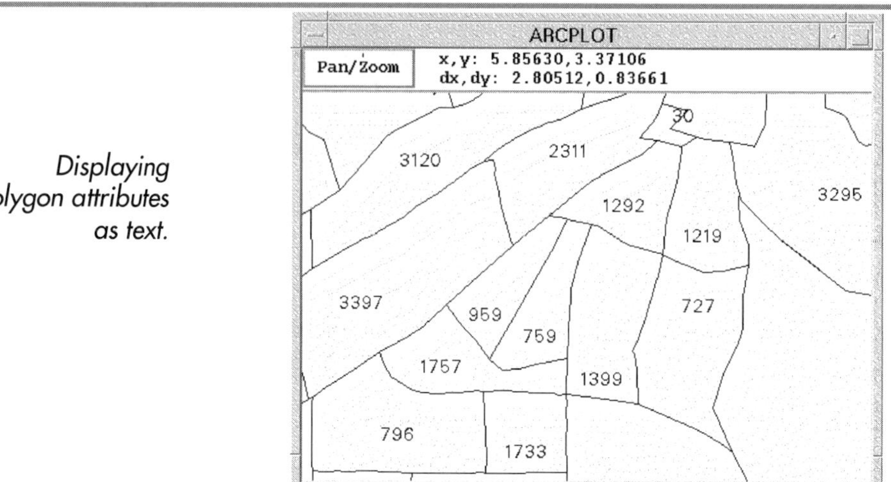

Displaying polygon attributes as text.

Resolving Overlapping Text

When you use the text commands we've just used, ARCTEXT, LABELTEXT, POLYGONTEXT, NODETEXT, and POINTTEXT, the text as drawn on your display frequently overlaps. The OVERPOST is an important new ARCPLOT in release 7 that addresses this problem very nicely.

The OVERPOST command does not directly draw text, but creates an environment that detects overlapping text. Here's how it works: When you use the text commands, each text is drawn one by one. The area subtended by each text is calculated and marked on a bitmap. Any successive text that overlaps with an area already marked on the bitmap is moved to a nearby position that does not overlap with previously drawn text. This is a brilliantly simple solution to a problem that has long plagued GISs and is an impressive enhancement in ARCPLOT.

In practice, you would precede the ARCPLOT text commands with the OVERPOST command. You can enable or disable the overposting environment, as well as set several parameters that affect tolerances. You can use OVERPOST for label markers as well as text, but you will more frequently use it for text.

```
Arcplot: OVERPOST {resolution} {radius} {increment}
```

{resolution} specifies the smallest cell in the bitmap that marks where text is. A smaller value is better, but too small a value increases the time to calculate the overposting environment. The default value is 0.05 inches. {radius} specifies the maximum distance from a feature that a text string can be moved. The default value is 1.0 inches. {increment} sets an interval for the radius by which text is placed. Again, a smaller value is better, but too small a value has no advantage and increases the time for calculation. The default value is 0.05 inches.

```
Arcplot: OVERPOST  | OFF | CLEAR
```

| OFF | CLEAR specifies whether the overposting environment is on or off. CLEAR can be used to keep overposting active, but clears the bitmap. The default value is OFF.

A related command to OVERPOST is LEADERS.

```
Arcplot: LEADERS |ON
```

If you turn leaders on with an active overposting environment, any text strings that are placed away from their associated features can be automatically drawn with a leader line. This too is a nice cartographic enhancement for crowded areas.

Drawing Classifications

Two related sets of commands that let you symbolize by defined classifications are CLASS and a set of color ramp commands: MARKERCOLORRAMP, LINECOLORRAMP, SHADECOLORRAMP, and TEXTCOLORRAMP.

```
Arcplot: CLASS <INTERVAL | QUANTILE> <cover>
         <feature_class> <item> {number_of_classes}

Arcplot: CLASS MANUAL <number_of_classes> <break...break>

Arcplot: CLASS NONE
```

This creates a classification, which is used to assign symbols to features. Once a classification is established in ARCPLOT, it is kept until changed, dropped, or the ARCPLOT session ends. Other ARCPLOT commands, such as ARCLINES and POINTMARKERS, use the classification for drawing graduated symbols.

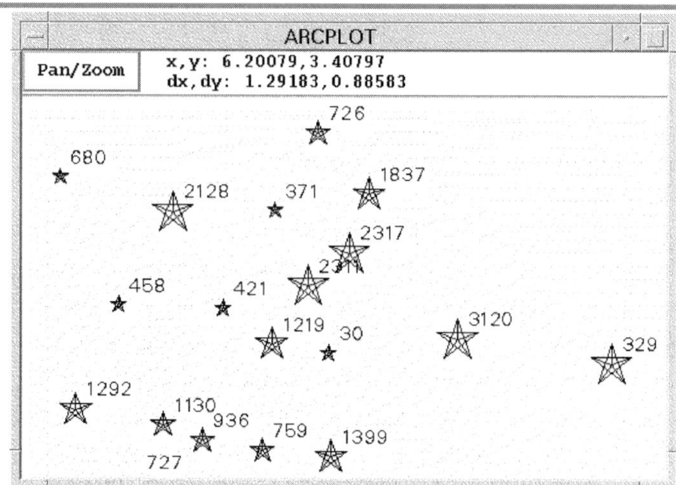

Classifications and color ramps.

You can apply classifications to predefined, discrete symbols, or you can use the following commands to tailor a symbol set for use with the CLASS command:

```
Arcplot: MARKERCOLORRAMP <start_symbol>
         <number_of_symbols> <start_color_spec>,
         <end_color_spec> {linear | nonlinear}
Arcplot: LINECOLORRAMP <start_symbol>
         <number_of_symbols> <start_color_spec>
         <end_color_spec> {linear | nonlinear}
Arcplot: SHADECOLORRAMP <start_symbol>
         <number_of_symbols> <start_color_spec>
         <end_color_spec> {linear | nonlinear}
Arcplot: TEXTCOLORRAMP <start_symbol>
         <number_of_symbols> <start_color_spec>
         <end_color_spec> {linear | nonlinear}
```

These four commands modify a range of symbols in the current symbol set so that colors are graduated between the start and end colors. Here are some examples of how these commands can be applied:

- ❑ You can use marker color ramps to depict accumulation and discharge rates for wells.
- ❑ Line color ramps can color-code electrical attributes, such as green for nominal voltage and red for undervoltage.

- ❑ Toxicity of hazardous waste sites can be mapped with shade color ramps.

Displaying Pie Symbols

In the Display GIScity application, we will explore some commands to depict U.S. Census data as pie symbols. Some of the attributes depicted were level of education, cost of housing, and sex and age distribution. These are the ARCPLOT commands at work behind the Display GIScity pie chart commands:

```
Arcplot: SPOTSIZE <data_value> <spot_size>

Arcplot: SPOTSIZE <data_min> <size_min>
         <data_max> <size_max> {exponent} {CLASS}
```

These commands scale the spot symbols to a defined quantity. For example, you can specify that a label point with attributes summarizing a group of 1000 people be depicted as a half-inch spot. If another label point depicted 4000 people, it would be 1 inch in size. (Spot sizes increase linearly by area, not diameter.) With the exponent argument, you can influence whether features grow linearly or exponentially with increasing item values.

These five spot commands are listed with their most frequently used command usage.

```
Arcplot: ARCSPOT <CIRCLE | BOX | SEMICIRCLERIGHT
         | SEMICIRCLELEFT> <cover> <item | size>
         <item,symbol...item,symbol>
         {halo_distance} {OUTLINE} {INCREASING}

Arcplot: LABELSPOT <CIRCLE | BOX | SEMICIRCLERIGHT |
         SEMICIRCLELEFT> <cover> <item | size>
         <item,symbol...item,symbol> {halo_distance}
         {OUTLINE} {INCREASING}

Arcplot: NODESPOT <CIRCLE | BOX | SEMICIRCLERIGHT |
         SEMICIRCLELEFT> <cover> <item | size>
         <item,symbol...item,symbol> {halo_distance}
         {OUTLINE} {INCREASING}

Arcplot: POINTSPOT <CIRCLE | BOX | SEMICIRCLERIGHT |
         SEMICIRCLELEFT> <cover> <item | size>
         <item,symbol...item,symbol> {halo_distance}
         {OUTLINE} {INCREASING}
```

```
Arcplot: POLYGONSPOT <CIRCLE | BOX | SEMICIRCLERIGHT |
         SEMICIRCLELEFT> <cover> <item | size>
         <item,symbol...item,symbol> {halo_distance}
         {OUTLINE} {INCREASING}
```

So, first you set the scaling for pie chart symbols with the SPOTSIZE commands, then you use the above five commands to display pie symbols. In your $GCHOME/data directory (or $GCHOME:[DATA] in VMS), you will see several AMLs that all begin with the prefix "piechart_." Go into Display GIScity and run the pie chart command with echo turned on. As you watch the dialog window, look for the SPOTSIZE and POLYGONSPOT commands. You'll see how Display GIScity draws these symbols.

Updating Symbol Sets

So far, we've seen how to change default symbols and apply symbology. Now, we will learn how to modify and create new symbols.

ARCPLOT comes with four commands that let you edit symbol tables: MARKEREDIT, LINEEDIT, SHADEEDIT, and TEXTEDIT. Each of these commands invokes a menu that lets you modify the properties of the current symbol. To start any of these symbol editors, just type the command name at the `Arcplot:` prompt. No arguments are necessary.

These four symbol edit commands are really AML macros and menus that are supplied with ARC/INFO to simplify symbol editing. These menus use ARCPLOT's several dozen symbol property commands, of which MARKERSYMBOL, MARKERCOLOR, LINESIZE, SHADESYMBOL, and TEXTJUSTIFICATION are examples.

Whether you use one of the four symbol edit commands or the native symbol property commands in ARCPLOT is a matter of preference. If you have substantial modifications to make on a symbol table, you will find that the symbol edit menus are the quickest way to update symbol tables. For setting just a few symbol properties, using ARCPLOT's symbol property commands can be quick too.

Symbols and Symbol Sets

Before we get started with the MARKEREDIT, LINEEDIT, SHADEEDIT, and TEXTEDIT commands, let's review some basic facts about symbols and symbol sets.

There are four types of symbols and symbol sets in ARC/INFO. Nodes and points are symbolized with marker symbols. Annotation and text are symbolized with text symbols. Arcs, routes, and sections are symbolized with line symbols. Polygons and regions are symbolized with shade symbols.

There is an active marker symbol, line symbol, shade symbol, and text symbol at all times within an ARCPLOT session. You can query the properties of the current symbol with the MARKERINFO, LINEINFO, SHADEINFO, and TEXTINFO commands.

You can change the properties of the current symbol frequently during an ARCPLOT session. One way is to use MARKERSYMBOL, LINESYMBOL, SHADESYMBOL, or TEXTSYMBOL to select a symbol already defined in a symbol set. The other way is to use the symbol property commands.

Each type of symbol (marker, line, shade, and text) spans a set of properties. These properties vary, but some common properties are color, size, and symbol number.

A symbol set is a computer file that stores the properties for that type of symbol. There are four types of symbol sets: markersets, linesets, shadesets, and textsets. They have computer file extensions of .MRK, .LIN, .SHD, and .TXT, respectively.

Each symbol set stores the properties for up to 999 symbols. Symbols within a symbol set are numbered between 1 and 999.

There is always a current markerset, lineset, shadeset, and textset. You can find out what these are with the SHOW MARKERSET, SHOW LINESET, SHOW SHADESET, and SHOW TEXTSET commands in ARCPLOT. These can be changed with the MARKERSET, LINESET, SHADESET, TEXTSET, and SYMBOLSET commands.

To graphically see the symbols in a symbol set, use the SYMBOLDUMP command. This command displays the selected symbol type and symbol range you specify in your graphics window.

NOTE: *When you begin ARCPLOT, the default symbol set is PLOTTER. This means that PLOTTER.MRK, PLOTTER.LIN, PLOTTER.SHD, and PLOTTER.TXT are the default markerset, lineset, shadeset, and textset. ARCEDIT has different default symbol sets: COLOR.MRK, COLOR.LIN, COLOR.SHD, and PLOTTER.TXT.*

More on Symbol Sets

ARC/INFO supplies several symbol sets for your convenience. They reside in the $ARCHOME/symbols directory (or the $ARCHOME:[SYMBOLS] directory in VMS). The predefined symbol sets include CALCOMP, CARTO, COLOR, PLOTTER, VERSATEC, and others. As you might guess from the names, some symbol sets are designed to optimize output from certain types of plotters.

These symbol sets can be separately set with the MARKERSET, LINESET, SHADESET, and TEXTSET commands or can be globally changed with the SYMBOLSET command. If the symbol set you want to use is either in the ARC/INFO directory, $ARCHOME/symbols, or in your current workspace directory, you can just type in the symbol set name after the MARKERSET, LINESET, SHADESET, or TEXTSET commands. Otherwise, you must type in the full path specification followed by the symbol set name.

Applying Symbology in ARCPLOT

All four symbol edit commands have two purposes:

- To change the properties of the current marker symbol, line symbol, shade symbol, and text symbol, and
- To modify symbols and permanently store the changes to a symbol set for later use.

The current symbol is applied when you display coverage features with the ARCPLOT commands POINTS, LABELS, ARCS, and POLYGONS or when you add map elements with commands such as MARKER, LINE, BOX, TEXT, or PATCH.

Symbols in a symbol set can be accessed as a symbol item with the ARCLINES, POINTLABELS, and POLYGONSHADES commands in ARCPLOT.

Editing Marker Symbols

```
Arcplot: MARKEREDIT
```

This command invokes a menu that lets you change the current marker symbol and make permanent changes to markersets.

NOTE: *If you see the error message, "Graphic device not specified," after typing* `MARKEREDIT`*, you have not opened a graphics window. Just type* `DISPLAY 9999 1`*, which will open a small graphics window, then type* `MARKEREDIT` *again.*

To start MARKEREDIT, type the following sequence:

```
Arc: ap
Arcplot: display 9999 1
Arcplot: markeredit
```

This is what you will get:

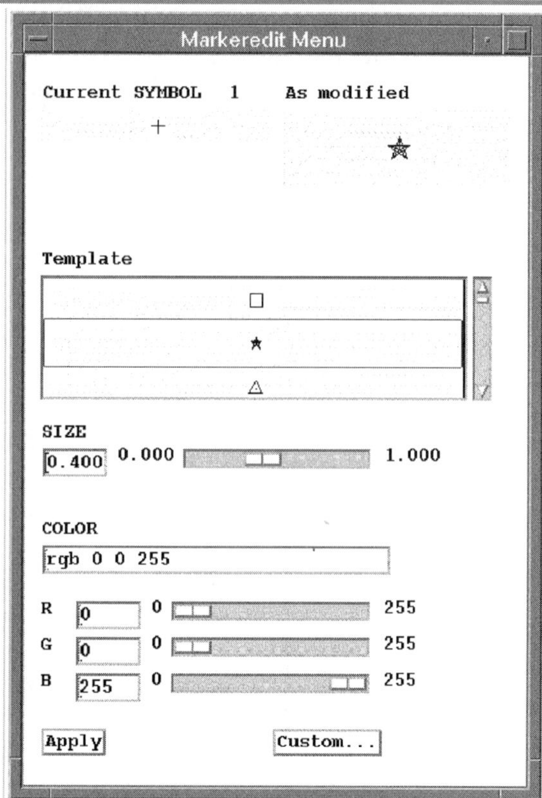

The main MARKEREDIT menu.

This is the main menu for MARKEREDIT. Its purpose is to let you quickly modify the properties of the current marker symbol.

On the top of the menu, there are two symbols depicted, the Current and As Modified marker symbol. You can make changes to the current marker symbol, see the changes, and then commit those changes to the current marker symbol with the Apply button.

In the main MARKEREDIT menu, you can set these marker properties: pick a new template symbol from the scrolling list, specify a new size with a slider or by typing, and change the color by typing in a color name or specifying the RGB (red, green, blue) components of a color. Typing in a color name is the easiest way; see the standard ARC/INFO colors listed in the previous chapter.

NOTE: *The menus MARKEREDIT, LINEEDIT, SHADEEDIT, and TEXTEDIT were not programmed with a Exit or Dismiss button, like GIScity and most GIS applications. Instead, they were designed to be closed with the Motif or OpenLook exit control. In Motif, click on the widget in the upper left corner and select Close from the menu. In OpenLook, click on the pushpin widget.*

If you want to modify other marker symbol properties, or if you want to apply the changes permanently in a markerset, click the Custom... button. This is what you will see:

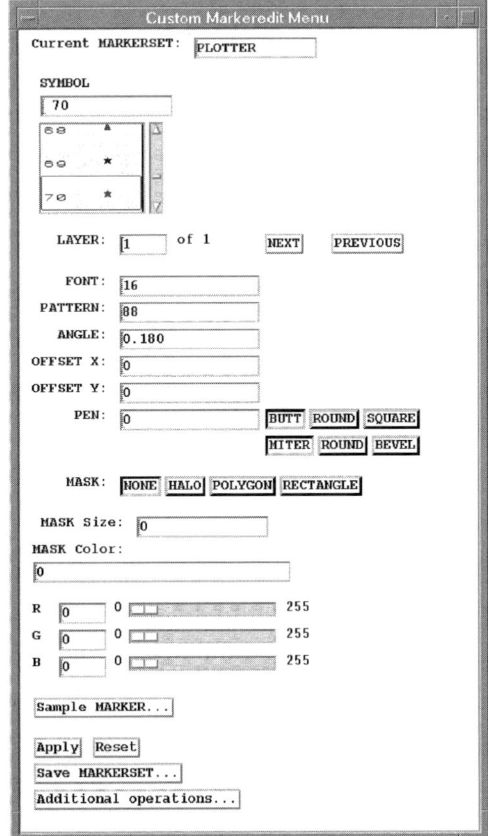

The Custom MARKEREDIT menu.

The Custom MARKEREDIT menu lets you assign these additional marker properties:

- The markerset is the name of current markerset. You can type in a new markerset. The symbol number is the number of the current marker in a markerset. You can look through the scrolling list for a new marker to apply, or you can type in a symbol number.
- Each symbol can be defined to contain several layers. The reason for using layers is to be able to assign multiple patterns, colors, and other properties to a symbol. Use the NEXT and PREVIOUS buttons to navigate through the layers of a symbol.
- Markers are defined from fonts and patterns. A font is a file that stores patterns, which are the lines and areas that make up a symbol. The default font for marker symbols is font 25. User-defined patterns can be placed in new font files, numbered from 26 to 40.
- Markers can be rotated at an angle, which you can set by typing in a geometric angle in degrees. You can also place markers at an offset X and offset Y distance from a point.
- There are several pen properties for marker symbols. First, you can specify a pen size in inches, centimeters, or whichever page units you've set. For thick lines, you can use the BUTT, ROUND, or SQUARE choice to affect the rendering of ends of lines. The MITER, ROUND, and BEVEL choice affects how two lines meet at an angle.
- A mask lets you define a symbol that will opaque any graphics underneath it: NONE, HALO, POLYGON, and RECTANGLE. You can also set a mask size and mask color.

When you set any of these marker properties in this menu, it is equivalent to using these marker symbol property commands:

```
Arcplot: MARKERANGLE <angle | *>

Arcplot: MARKERCOLOR <color_spec>

Arcplot: MARKERFONT <font>

Arcplot: MARKERMASK <HALO | POLYGON | RECTANGLE>
         <size> {color_specification}

Arcplot: MARKERMASK COLOR <color_specification>
```

```
Arcplot: MARKERMASK NONE
Arcplot: MARKEROFFSET <x_offset y_offset | *>
Arcplot: MARKEROFFSET MARKERLAYER <x_offset y_offset | *>
Arcplot: MARKERPATTERN <pattern>
Arcplot: MARKERPEN <size> {BUTT | ROUND | SQUARE}
         {MITER | ROUND | BEVEL}
Arcplot: MARKERSIZE <height | *> {width | *} {PT}
```

On the bottom of the Custom MARKEREDIT menu, there are several actions you can take:

- ❑ The Sample MARKER... button activates the cursor in your graphics window. Pick a point, and a marker with the current symbol properties will be displayed.
- ❑ The Apply button saves any properties you have set to the current marker symbol.
- ❑ The Reset button restores the previous properties in the MARKEREDIT menus to the beginning of the session or last apply.
- ❑ The Save MARKERSET... button lets you save the modified symbol properties to the current markerset file, which is stored on your computer's disk.
- ❑ The Additional operations... button will invoke another menu with actions common to MARKEREDIT, LINEEDIT, SHADEEDIT, and TEXTEDIT. We'll document this menu later.

Patterns and Fonts

So far, we've seen many possible properties that you can set for markers. The most important property of a marker symbol is its pattern. This defines the appearance of the symbol, whether it is round, square, triangular, or another shape.

MARKEREDIT lets you set the pattern number, but MARKEREDIT cannot edit patterns. Patterns begin as arcs and polygons in a special coverage. You would edit that special coverage in ARCEDIT. Next, you can use a couple of ARC commands that let you convert that special coverage into a pattern, which is contained within a font.

Each font has a resolution. All the positions of lines that define a pattern snap to a grid of the specified resolution. The default grid size is (0,0) to (400,400), for a resolution of 400.

You can specify higher resolutions, up to about 4000 by 4000, but a fine resolution will slow down your display time when drawing marker symbols in ARCPLOT or ARCEDIT. You can select a lower resolution, such as (0,0) to (50,50), but because all the points in the pattern will snap to this 50 by 50 grid, you may not get all the detail you want in the finished marker symbols. Symbols that will be displayed at small marker sizes can have lower resolutions, but large marker symbols should have higher resolutions.

Each pattern in a font has a number. You can use any number between 32 and 126 for a pattern number.

Here are some facts about fonts:

- ❑ Font files live in the ARC/INFO directory $ARCHOME/igl63exe (UNIX) or $ARCHOME:[IGL63EXE] (VMS).
- ❑ Typical names for font files are fnt021 and fnt025 in UNIX or FNT021.DAT or FNT025.DAT in VMS. These refer to font files 21 and 25.
- ❑ Fonts 0 through 19 are reserved by ARC/INFO for text fonts. Fonts 20 through 25 are predefined and store all the patterns you see in the markersets supplied with ARC/INFO. You can update fonts 20 through 25, but it is recommended that you create new font files to place your custom patterns.
- ❑ User-defined fonts can be numbered from 26 to 40. Patterns within fonts can be numbered between 32 and 126.
- ❑ Each font file has an implicit resolution. You specify the resolution when you create the font.

Pattern Coverages

Patterns are first defined in a special coverage. After the coverage is ready, you will use the ARCFONT command to convert this coverage into a pattern in a font. Even though this chapter covers advanced commands in ARCPLOT, it is necessary for us to digress to ARC and ARCEDIT for a bit.

Here are the requirements for that special coverage:

Sample coverage for defining a pattern.

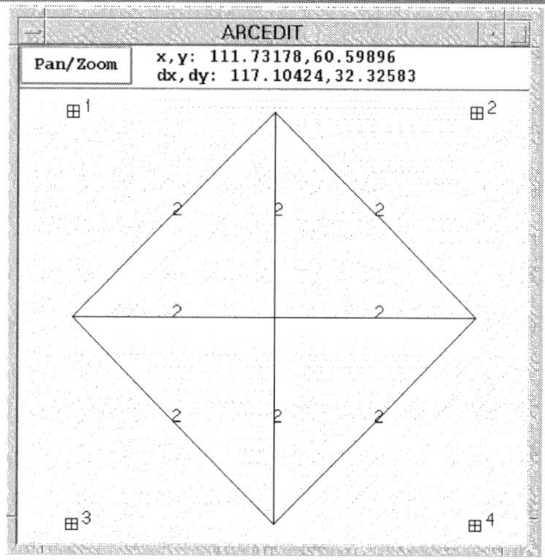

The pattern coverage should only have tics and lines.

- There should be four tics, numbered exactly as shown in the illustration. The tics should bound all the arcs that you want to define your pattern. The coordinates of the tics are not important; you can use arbitrary coordinates like (100,0), (100,100), (0,0), and (0,100), but what is important is that the tics form a perfect square.
- The arcs that define the pattern must have only one of three User ID values: 2, 6, or 7. Most often, you will use a User ID of 2, which specifies a single line for the pattern. A User ID of 6 for an arc specifies that a thick line will be created in the pattern. A User ID of 7 specifies a panel fill. An arc with a User ID of 7 must close on itself clockwise to form a closed area.
- If you need to create several pattern coverages, first create one with just the tics. Make a copy of that coverage and use it as a template for all your pattern coverages.

TIP: *Usually you will edit simple arcs to define patterns in ARCEDIT. Most patterns are fairly simple, but they can be quite detailed. If you want to make a marker symbol for a company logo that is in the AutoCAD DXF format (or other format), you can use the ARCDXF or similar command to import the graphics, then use the GET command in ARCEDIT to import those arcs into your pattern coverage. Set your tics to bound these arcs in a square and to calculate the User ID of the new arcs to 2, 6, or 7.*

Creating Fonts

For your custom patterns, it is a good idea to create a new font, with a number of 26 or greater. It is best to leave alone fonts 20 through 25, ARC/INFO's fonts with pattern cells for the predefined marker symbols.

To create a font, first set your workspace to $ARCHOME/igl63exe (UNIX) or $ARCHOME:[IGL63EXE] (VMS). Use the &workspace directive. This ensures that your new font will be in the same directory as the other fonts supplied with ARC/INFO. From the ARC program, run the FONTCREATE command. Here is how it works:

```
Arc: FONTCREATE <font> {xmin} {ymin} {xmax} {ymax}
```

Create a new font with the specified font number and coordinate range. The font number should be 26 through 40. The resolution can be from (-256, -256) to (3849, 3849). The default resolution is (0,0) to (800,800).

Instead of creating a brand new font, you can make a copy of an existing font and then modify it.

```
Arcplot: FONTCOPY <from_font> <to_font> {from_pattern}
         {to_pattern}
```

This copies an existing font to a new font, with the option to copy a subset of patterns.

Adding Patterns to Fonts

Once you have your pattern coverages and a new font, you can start inserting patterns in the font. To use these commands, you should be in the ARC program and should also be in the same workspace as your font file, which is probably $ARCHOME/igl63exe (UNIX) or $ARCHOME:[IGL63EXE] (VMS).

```
Arc: ARCFONT <in_cover> <out_font> <out_pattern>
```

This updates a pattern in a font file with the arcs and panel fills from the input coverage.

The input coverage should follow the properties outlined previously for pattern coverages. If you get unexpected results, check the tics and User IDs of arcs. For the output font, enter the number of the new or existing font file you wish to add a pattern to. This will usually be between 26 and 40. For the output pattern, enter a number between 32 and 126, which will be the pattern number.

Here's a related command to ARCFONT:

```
Arc: FONTARC <in_font> <in_pattern> <out_cover>
```

This outputs the pattern from the input font file to the output coverage. You won't use this command often, but it can be very useful. FONTARC is the reverse of ARCFONT and is handy if you want to modify a pattern but lost the coverage from which it was defined.

Exercise: Constructing a Marker Symbol

Here is a complete sample session in ARC, ARCEDIT, and ARCPLOT to make a coverage called ARROW_RIGHT with a coordinate range from (0,0) to (100,100). This pattern will appear as an arrow pointing to the right.

NOTE: *This exercise introduces us to ARCEDIT, the ARC/INFO program to edit coverages. We will document ARCEDIT in later chapters. You may choose to either return to this exercise later, or proceed and type in the commands verbatim.*

We will use keyboard input for the coordinates so that we can enter the precise corners of this coordinate range. In this session, we will use these command abbreviations: AE for ARCEDIT, COO for COORDINATE, DE for DRAWENVIRONMENT, CR for CREATE, and EF for EDITFEATURE,

1. Let's set our workspace to a GIScity directory:

```
Arc: &workspace $GCHOME/data (UNIX or Windows NT) or
Arc: &workspace $GCHOME:[DATA] (VMS)
```

2. Next, start ARCEDIT:

```
Arc: ae
```

3. Set coordinate input to keyboard so that we can type in exact coordinates.

```
Arcedit: coo keyb
```

4. Set the draw environment to arcs and tics to draw the User IDs.

```
Arcedit: de tics ids
Arcedit: de arcs ids
```

5. Use the CREATE command to make a new coverage. It will engage you in a dialog to add tics and an initial boundary. Use the keyboard to type in the coordinates shown:

```
Arcedit: create arrow_right
Creating ARROW_RIGHT
Digitize a minimum of 4 tics.
Signal end of tic input with Tic-ID = 0
Tic-ID: 1
Enter X, Y: 100, 0
Tic-ID: 2
Enter X, Y: 100, 100
Tic-ID: 3
Enter X, Y: 0, 0
Tic-ID: 4
Enter X, Y: 0, 100
Tic-ID: 0
```

The initial boundary is the map extent the coverage begins with.

```
Enter initial boundary
Define the box
Enter X, Y: 0,0
Enter X, Y: 100,100
```

6. Now, add arcs as shown. Again, we will type in the coordinates with the keyboard.

```
Arcedit: ef arc
Arcedit: add
```

7. First, we want to use the digitizing options to set a User ID of 2:

```
Enter Key, X, Y: 8
Enter Key, X, Y: 1
New User ID: 2
```

8. Now we digitize the arc:

```
Enter Key, X, Y: 2, 50, 100
Enter Key, X, Y: 1, 100, 50
Enter Key, X, Y: 1, 50, 0
Enter Key, X, Y: 1, 50, 25
Enter Key, X, Y: 1, 0, 25
Enter Key, X, Y: 1, 0, 75
Enter Key, X, Y: 1, 50, 75
Enter Key, X, Y: 2, 50, 100
Enter Key, X, Y: 9
```

9. Let's see our creation:

```
Arcedit: draw
```

Your graphics window should look like the one shown in the following illustration. You have just created a pattern coverage with an arrow pointing to the right.

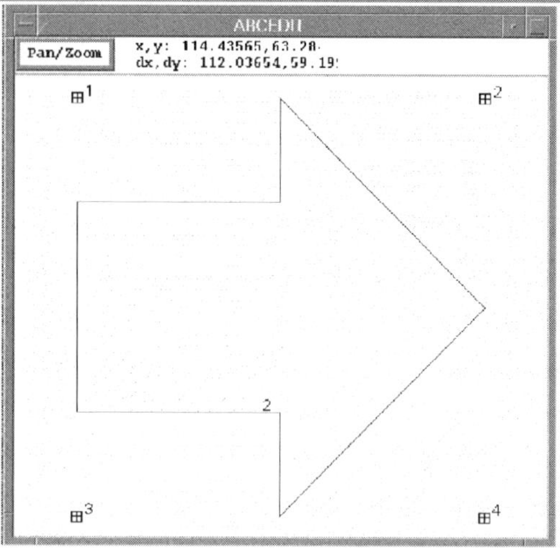

Graphics window after successful execution of user exercise.

10. Let's exit ARCEDIT, saving all our edits, and move on.

```
Arcedit: quit yes
```

11. Now, let's create a new font. We'll make font number 38 and use the default resolution of (0,0) to (800,800).

```
Arc: fontcreate 38
```

If you looked at the files in the directory, you would see a new file called fnt038 in UNIX or FNT038.DAT in VMS.

12. Now we will place the pattern from the ARROW_RIGHT coverage into font 38 as a new pattern, numbered 50.

```
Arc: arcfont arrow_right 38 50
```

That's it! We've created a new font with a new pattern.

13. Now, let's go to ARCPLOT, open a small graphics window, and start MARKEREDIT.

```
Arc: ap
Arcplot: display 9999 1
Arcplot: markeredit
```

14. When the MARKEREDIT menu appears, press the Custom... button. In the Custom MARKEREDIT Menu, type in `38` for the font and `50` for pattern. Look at your main MARKEREDIT Menu. The "As modified" symbol should now appear as the arrow pointing right that you created.

We've now walked through all the steps to create a brand new marker symbol with a pattern you define. Although we had to visit ARC, ARCPLOT, and ARCEDIT to finish all the steps, you can see it's not that hard to create a new pattern from scratch.

TIP: *If you want to use this new font, it is a good idea to copy it to the $ARCHOME/igl63exe directory (UNIX) or $ARCHOME:[IGL63EXE] directory (VMS). This will ensure that the font file will always be recognized, no matter which workspace directory you are in.*

Editing Line Symbols

Now, we will turn to editing line symbols. Type this command to start the menu to edit line symbols:

```
Arcplot: LINEEDIT
```

This command lets you change the properties of the current line symbol and make permanent changes to linesets.

On the top of the LINEEDIT menu, you see two lines: the current line symbol and line symbol as modified. When you start LINEEDIT, they will probably look the same until you start changing line properties.

Just as with MARKEREDIT, the main LINEEDIT menu lets you change the three main properties of line symbols. You can select a new line template, size, and color to apply to your current line symbol.

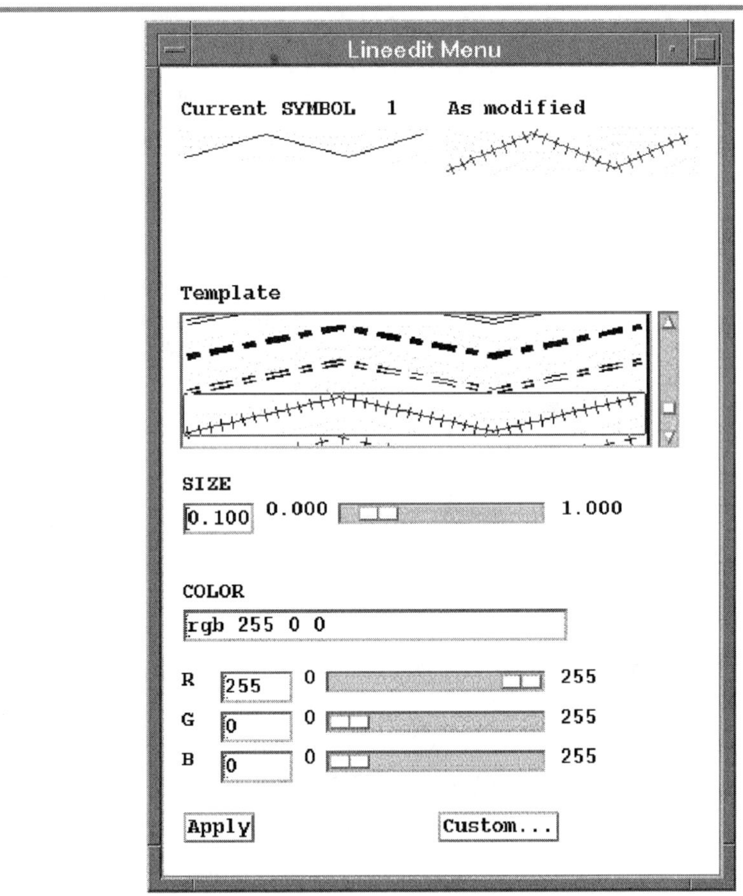

Main LINEEDIT menu.

If you want to modify additional line symbol properties and make changes to a lineset, press Custom.... You'll invoke the two following new menus:

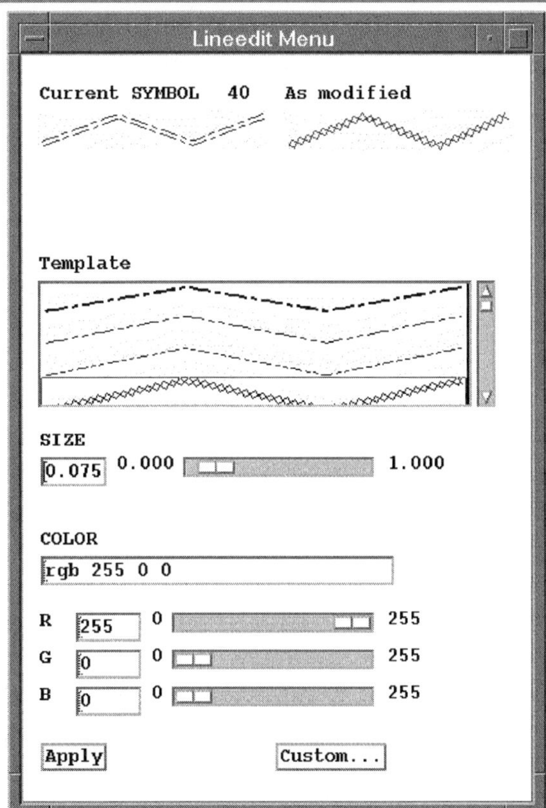

Custom LINEEDIT and Linetype menus.

These are some of the properties you can change in this menu:

- ❑ You can change the current lineset by typing in the name of a new lineset. To change the current symbol, either type in a new symbol number or select a symbol from the scrolling list. Like markers, line symbols can be defined with layers, so that you can overlay different colors and patterns to make a complex line symbol.
- ❑ By typing in a value in page units, you can specify an offset for the line symbol. This means that the line will actually be drawn at a parallel offset according to this value. A positive value defines an offset to the right, relative to the arc, route, or event direction. A negative value defines an offset to the left.

- If your line symbol has embedded markers, you can optionally define a Marker X and Marker Y offset.
- If your line symbol has a repeating pattern, you can set the interval distance for gaps and lines. Gaps and lines are defined with a template, which is a string of integers. For example, a template value of 321 together with an interval distance of .1 inches means that the resulting line will have a line segment of .3 inches, followed by a gap of .2 inches, then a line segment of .1 inches. This repeats for the length of the line.
- You can set a percent for line adjustment, which is used for improving the appearance of line symbols at nodes. You can use the hollow, pen, closure angle, and miter angle properties to control the appearance of lines and how they join at vertices.
- The LINETYPE... button invokes the Linetype menu if you have dismissed it. The Sample LINE... button activates the screen cursor for you to preview the current line symbol by digitizing a sample line. Use buttons 1 or 2 to mark vertices and button 9 to quit this command.
- The Apply button saves the properties you've set in LINEEDIT to the current line symbol. The Reset button restores the line properties to their initial state or the state as of the last Apply command.
- The Save lineset... button stores changes to the current lineset. The Additional properties button invokes another menu with more commands to modify line symbols and layers.

These are some of the line property commands that are run in LINEEDIT when you modify line properties:

```
Arcplot: LINEADJUSTMENT <percentage>
Arcplot: LINECLOSUREANGLE <angle>
Arcplot: LINECOLOR <color_spec>
Arcplot: LINEHOLLOW <width | *>
Arcplot: LINEINTERVAL <length | *>
Arcplot: LINEMITERANGLE <angle>
Arcplot: LINEOFFSET <offset | *>
```

```
Arcplot: LINEOFFSET MARKERLAYER <x_offset y_offset | *>
Arcplot: LINEPATTERN <pattern>
Arcplot: LINEPEN <size> {BUTT | ROUND | SQUARE}
         {MITER | ROUND | BEVEL}
Arcplot: LINESIZE <width | *>
Arcplot: LINETEMPLATE <template>
Arcplot: LINETYPE <HARDWARE | DIAMOND | DOTS | HASH | NONE>
Arcplot: LINETYPE <SCALLOP | SCRUB | SLANT | ZIGZAG> {FLIP}
Arcplot: LINETYPE WIDE {BOXED}
Arcplot: LINETYPE MARKER <font> <pattern>
Arcplot: LINETYPE VERTEXMARKER <font> <pattern>
```

Exercise: Creating a New Line Symbol

You will create a new line symbol that incorporates the marker symbol you made in the previous exercise.

1. If you don't already have a graphics window, open a small window like this:

   ```
   Arcplot: display 9999 1
   ```

2. Invoke LINEEDIT:

   ```
   Arcplot: lineedit
   ```

3. Set the size in the LINEEDIT menu to 0.2 inches. Next, press the Custom... button to invoke the Custom LINEEDIT menu.
4. In the Linetype menu, press the icon on the lower left side of the menu. Then type `38` for the font and `50` for the pattern.
5. Go to the Custom LINEEDIT menu and set these properties: Type `0.15` for INTERVAL and `11` for TEMPLATE. Then press the Sample LINE... button. Digitize a line, and you should see something like the following figure:

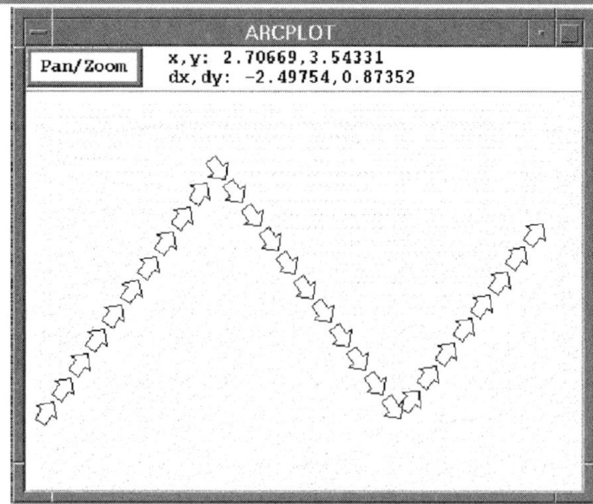

Sample line symbol as modified in the user exercise.

You've now set new properties for the current line symbol. Setting all of the possible properties for line symbols is quite extensive; experiment with LINEEDIT or consult with ArcDoc for more information about line symbols.

Editing Shade Symbols

Shade symbols are used to fill polygons and defined areas.

```
Arcplot: SHADEEDIT
```

The SHADEEDIT command invokes a menu that lets you change the current shade symbol and make permanent changes to shadesets.

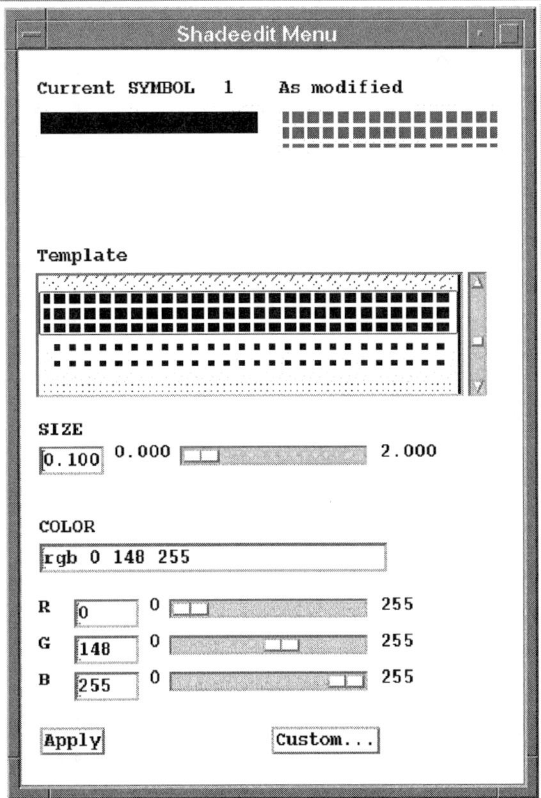

Main SHADEEDIT menu.

The Current shade symbol and the As Modified shade symbol are displayed on the top of the SHADEEDIT menu. You can modify three properties in this menu: template, size, and color. To access additional properties for shade symbols, press the Custom... button. As with LINEEDIT, two menus are invoked:

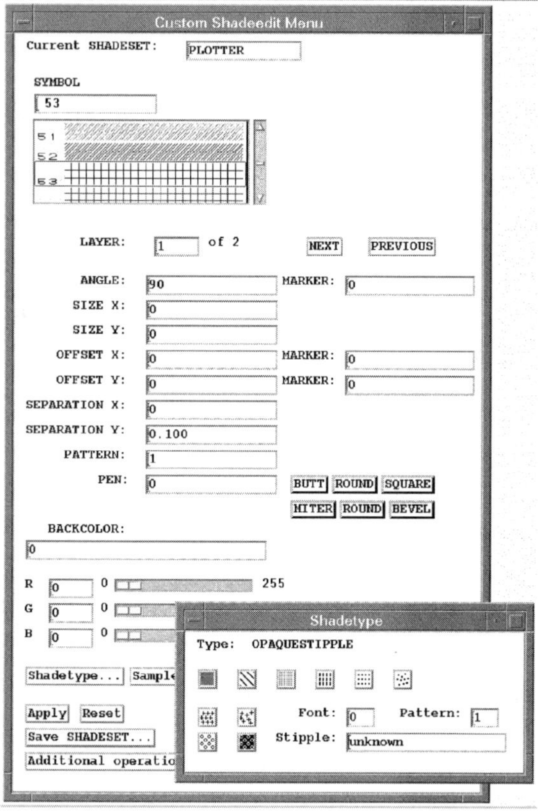

Custom SHADEEDIT and Shadetype menus.

You can set a new shadeset to select or set shade symbol properties. Also, you can type in or select a new shade symbol.

Like other symbols, shades can comprise several layers. You can also specify an angle to be applied to the shade symbol and a marker angle for markers within the shade symbol.

Size X and Size Y set the shade layer size. You can apply Offset X and Offset Y values for shade layer offsets. Also, you can specify offsets for markers that compose the shade.

Separation X and Separation Y specify the separation between elements in a shade layer. The Pattern applies one of the hardware patterns to define the shade symbol.

You can modify several pen properties that affect the stroking of shade symbols: Butt, Round, Square and Miter, Round, Bevel. The Backcolor sets a background color if the SHADETYPE is set to OPAQUESTIPPLE. SHADETYPE COLOR is best for color raster plotters.

You can invoke the Shadetype menu by clicking the Shadetype... button. Also, you can preview the current shade symbol by pressing the Sample shade... button.

The Apply button applies properties you've set to the current shade symbol. The Reset button restores shade properties to their prior status.

Save SHADESET... stores all modified shade symbols to the current shadeset. The Additional operations button will invoke another menu, giving you more options for modifying shade symbols and layers.

If you prefer not to use SHADEEDIT, these are the ARCPLOT commands that set the same shade properties directly on the current symbol:

```
Arcplot: SHADEANGLE <angle | *> {markerangle | *}
Arcplot: SHADEBACKCOLOR <color_spec>
Arcplot: SHADECOLOR <color_spec>
Arcplot: SHADEOFFSET <x_offset y_offset | *>
Arcplot: SHADEOFFSET MARKERLAYER <x_offset y_offset | *>
Arcplot: SHADEPATTERN <pattern>
Arcplot: SHADEPEN <size> {BUTT | ROUND | SQUARE}
         {MITER | ROUND | BEVEL}
Arcplot: SHADESEPARATION <separation1 | *> {separation2 | *}
Arcplot: SHADESIZE <size1 | *> {size2 | *}
Arcplot: SHADETYPE <COLOR | HARDWARE | HATCH |
         RECTANGLE | DOTS | RANDOMDOTS | NONE>
Arcplot: SHADETYPE <MARKER | RANDOMMARKER> <font> <pattern>
Arcplot: SHADETYPE <STIPPLE | OPAQUESTIPPLE> <stipple_name>
```

Editing Text Symbols

Like other features in ARC/INFO, text symbol properties are applied to annotation and text. The properties of the current symbol are used for text in ARCPLOT.

However, unlike other types of symbols, you can't define a symbol item for annotation to control the text symbol of an annotation. Rather, these properties are directly stored with annotation when you create and edit them.

```
Arcplot: TEXTEDIT
```

The TEXTEDIT command invokes a menu that lets you change the current text symbol and make permanent changes to textsets.

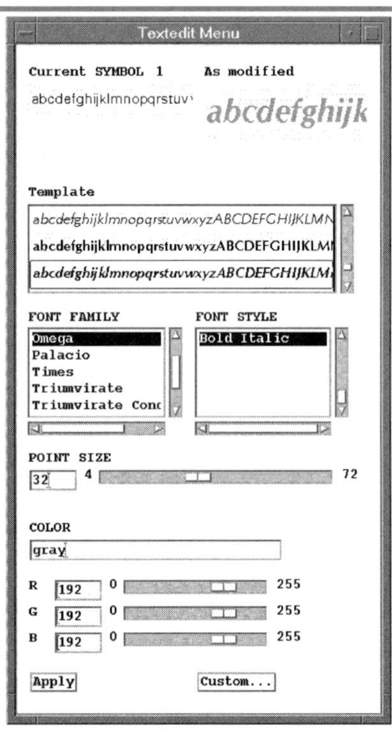

The main TEXTEDIT menu.

The main TEXTEDIT menu has a few more properties than the other symbol edit menus:

The Current symbol and the As Modified symbol are shown on top of the menu. You can select one of the text symbols from the template scrolling list, or you can define a text symbol from the font family and font style. Two more properties are font size and color.

When you click the Custom...button, this is what you'll see:

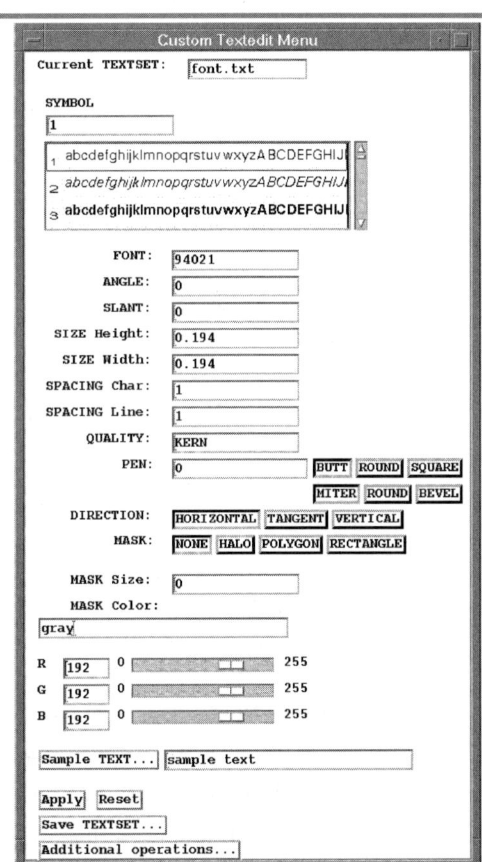

The Custom TEXTEDIT menu.

These are some of the properties you can change in the TEXTEDIT menu:

- ❑ You can change the current textset by typing in a new textset. To change the current symbol, either type in a new number or select a text symbol from the scrolling list.
- ❑ You can specify the font either in the main TEXTEDIT menu or by typing in a font number. Reference ArcDoc for a listing of all fonts and their numbers.
- ❑ Annotation can be drawn at an angle. Type in an angle in degrees. Also, you can specify a slant angle. Some other properties you can set are size height and size width in page units. These apply to individual letters, not the entire text string.
- ❑ Spacing Char sets the spacing between characters, and Spacing linesets the spacing between lines of text. Quality sets one of several spacing modes.
- ❑ As with the other symbol edit menus, you can set these pen properties: Size sets the thickness of the pen strokes; Butt, Round, and Square affect the ends of each line; Miter, Round, and Bevel affect how lines join at an angle.
- ❑ There are three values you can set for direction: horizontal, tangent, and vertical. These affect the individual characters in a text string, so that you can have a text string that may be read vertically or at an angle.
- ❑ An important enhancement for text in ARC/INFO 7.0 is the ability to mask text symbols. This means that you can opaque an area around an annotation. The purpose is to reduce visual clutter on a map. You can define a mask size, which is like a buffer distance around annotation, and a mask color.
- ❑ The Sample text... button activates the cursor so that you can add a sample text on the graphics window.
- ❑ The Apply button applies any text properties you set to the current text symbol. The Reset button restores text properties to the current text symbol to what they were before.
- ❑ The Save textset... button stores all modifications to the symbol set.

If you prefer not to use TEXTEDIT, these are the ARCPLOT commands that set the same properties for the current text symbol:

```
Arcplot: TEXTALIGNMENT <LEFT | CENTER | RIGHT | AUTOMATIC>
Arcplot: TEXTANGLE <angle | *>
Arcplot: TEXTCOLOR <color_spec>
Arcplot: TEXTDIRECTION <HORIZONTAL | TANGENT | VERTICAL>
Arcplot: TEXTFILE <text_file>
Arcplot: TEXTFIT <text_string> {* | xy1 xy2}
Arcplot: TEXTFONT <font>
Arcplot: TEXTJUSTIFICATION <LL | LC | LR | CL | CC |
         CR | UL | UC | UR>
Arcplot: TEXTMASK <HALO | POLYGON | RECTANGLE> <size>
         {color_specification}
Arcplot: TEXTMASK COLOR <color_specification>
Arcplot: TEXTMASK NONE
Arcplot: TEXTOFFSET <x_offset y_offset | *>
Arcplot: TEXTPEN <size> {BUTT | ROUND | SQUARE}
         {MITER | ROUND | BEVEL}
Arcplot: TEXTPRECISION <AUTOMATIC | number_of_decimal_places>
Arcplot: TEXTQUALITY <CONSTANT | PROPORTIONAL |
         KERN | TIGHTKERN>
Arcplot: TEXTSIZE <height | *> {width | *} {PT}
Arcplot: TEXTSLANT <angle | *>
Arcplot: TEXTSPACING <character> {line}
Arcplot: TEXTSPLINE <text_string> <* | xy...xy>
Arcplot: TEXTSTYLE <SIMPLE | TYPESET>
```

Additional Operations for the Symbol Edit Menus

The MARKEREDIT, LINEEDIT, SHADEEDIT, and TEXTEDIT menus all have a button on the bottom that says Additional Operations. They all invoke the same menu, which gives you a suite of commands that will help you manage symbols and layers in a symbol set.

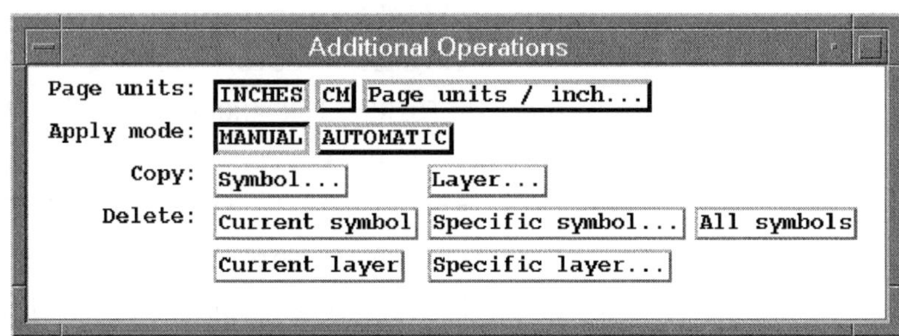

The Additional Operations menu.

These are the properties you can set in the Additional Operations menu:

- ❑ You can set page units to inches, centimeters, or other units.
- ❑ Apply Mode lets you set manual or automatic application of edits. This lets you make changes immediately, without having to use the Apply button in the other MARKEREDIT menus.
- ❑ You can copy symbols and layers with the Copy Symbol... and Copy Layer... commands. Likewise, you can use Delete Current Symbol, Delete Specific Symbol, Delete All Symbols, Delete Current Layer, and Delete Specific Layer to run those operations.

The equivalent ARCPLOT commands to manage symbols in symbol sets and their layers are MARKERCOPY, MARKERCOPYLAYER, MARKERDELETE, MARKERDELETELAYER, LINECOPY, LINECOPYLAYER, LINEDELETE, LINEDELETELAYER, SHADECOPY, SHADECOPYLAYER, SHADEDELETE, SHADEDELETELAYER, TEXTCOPY, and TEXTDELETE.

If you choose to modify symbol tables through ARCPLOT commands rather than the symbol editing menus like MARKEREDIT, you need to be aware of these additional ARCPLOT commands:

```
MARKERPUT <symbol>

LINEPUT <symbol>

SHADEPUT <symbol>

TEXTPUT <symbol>
```

saves the current symbol properties into a symbol number slot.

All of these commands place the current symbol definition for marker, line, shade, and text into a symbol number location from 1 to 999. These remain current for the remainder of your session and are useful for ARCPLOT commands like ARCLINES.

```
MARKERSAVE <markerset_file> {TEMPLATESONLY}
LINESAVE <lineset_file> {TEMPLATESONLY}
SHADESAVE <shadeset_file> {TEMPLATESONLY}
TEXTSAVE <textset_file> {TEMPLATESONLY}
```

saves all the symbol number locations into a symbol set file.

When you make modifications on symbols in an ARCPLOT session, these save commands can permanently store any changes you make to the symbol set file on your computer's disk. Use these commands with care, because they will overwrite the previous symbol set files.

Gathering Features in Display GIScity

Selection, Themes, Views, and Relates in ARCPLOT

A GIS is an instrument for the placement, retrieval, and delivery of information through geography. In other words, a GIS is a geographic database manager.

GIS is a powerful enabling technology because the majority of information we use relates to geographic locations. Businesses are vitally interested in the demographic and geographic distribution of their customers, as well as in the proximity of competitors. Utilities must manage a large infrastructure and, by necessity, they rely upon geography for inventory, maintenance, and emergency response. State environmental agencies use geography extensively because environmental quality directly relates to the proximity of pollution sources.

The starting point of geographic analysis is retrieving the data that you are interested in. This chapter explores the following general methods for selecting geographic features:

- **Spatial selection.** You will learn to use the mouse cursor to pick one or many geographic features. You will also collect features

within boxes, circles, and polygons that you can define interactively.

- ❑ **Logical selection.** You can build simple or compound logical queries to select geographic features. An example would be "Find all the buildings with an area greater than 3,000 square feet and that are within the Guadalupe Historical District."
- ❑ **Overlap selection.** An important advantage of GIS is that you can select geographic features that are adjacent to, or within a buffer distance of, other geographic features. For example, you may try to locate wells within a buffer distance of a stream, buildings within a census tract, or businesses near selected streets.

In this and the two succeeding chapters, we revisit the Display GIScity application which is supplied on the companion CD-ROM. Display GIScity is written in AML and operates within ARCPLOT.

In the first chapter, we learned how to start Display GIScity and how to zoom into the graphics display. In the first part of this chapter, we will learn how to highlight selected features. In the second part of this chapter, we'll see how GIScity further organizes geographic data into themes and views. We'll also learn more about how ARC/INFO manages geographic information through relates.

These are some ways that GIScity organizes coverages and other information:

- ❑ **Themes.** Geographic features are grouped into themes, such as roads, building, and contours. Each theme contains a *feature attribute table*, which is a database in which information about the features is stored. In Display GIScity, there is always a theme that is active for interaction. A very important property of a theme is its *class*. Examples of classes are points, lines, and polygons.
- ❑ **Views.** In a graphics window, you need the ability to easily specify which themes are to be drawn, and the order of their appearance. Views are lists, or groups, of themes with additional information about how the themes are set for display within your graphics window.
- ❑ **Relates.** While each theme contains its own feature attribute table, in a GIS you will want to explore data relations between different

themes with other tables. The definition of logical connections between tables is called a *relate*, and the use of relates is central to maximizing the potential of a GIS.

This chapter covers the six Select commands, which are outlined.

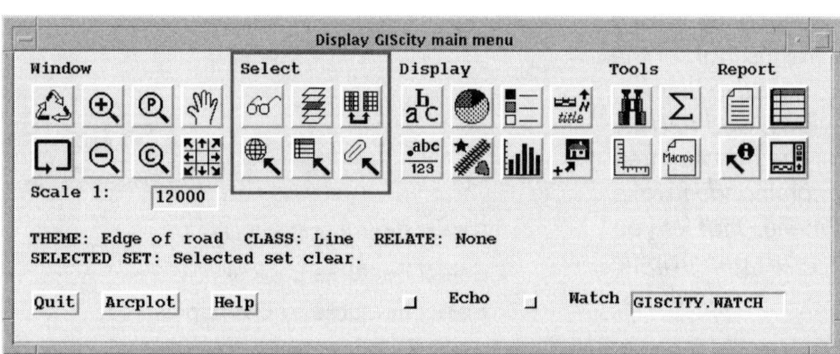

The Select Commands

The functions of selecting geographic features and managing themes, views, and relates are all served by the Select commands in the main menu of Display GIScity. Each of these six commands invokes another menu, which will be displayed below.

These icons are designed to help you intuitively choose the Select command you want to use. The eyeglass icon signifies Views. The icon of overlaying sheets is a common image used to denote Themes within a GIS. The next icon shows an arrow between two tables, which is an accurate image of a Relate.

The six Select commands within the Display GIScity main menu. The top three commands invoke menus that set the active themes, views, and relates. The bottom three commands invoke menus that let you select geographic features.

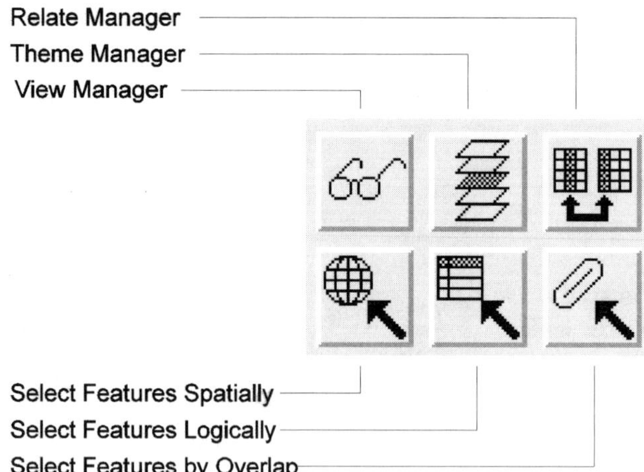

The bottom three icons all have a diagonal arrow, which denotes selection. The globe depicts Spatial Selection. The table depicts Logical Selection. The line with a surrounding polygon, or buffer, depicts Overlap Selection.

NOTE: *The author invented the majority of the icons used in the Display GIScity applications. The remainder are taken from ArcTools or based upon icons from ArcView. The following criteria were used in the creation or selection of icons through GIScity: (1) To visually communicate the function as clearly as possible. (2) To emulate icons with similar functions in common Windows applications such as Microsoft Word, CorelDraw, and other popular programs. (3) To emulate icons with like functions in ArcView or ArcTools.*

Selected Sets in ARCPLOT

In the first chapter, you performed the Window commands to zoom in, out, and around Santa Fe. Certain features turned off and on when you changed scales, such as the street names that appeared above a specific scale threshold. But the content of the GIS display in the graphics window

was essentially fixed. We'll now look at some ways to create changing displays.

Definition and Uses of Selected Sets

A central concept within ARC/INFO is the *selected set*. Many commands in both ARCPLOT and ARCEDIT interact through selected sets. Some of the purposes of selected sets in ARCPLOT are

- ❑ To display geographic features selectively in the graphics window.
- ❑ To perform visualization commands, such displaying text from attributes.
- ❑ To choose features to include in tabular reports.
- ❑ To control the content of graphics files output to a plotter or printer.

In this section, we'll use the three Select Features menus—Spatially, Logically, and Overlap—within Display GIScity to learn the primary techniques for selecting features in ARCPLOT.

Selected Set Basics

Before we get to the specific commands in the Select Features menus, let's review some general characteristics of selected sets.

- ❑ A *selected set* is a virtual list of geographic features within a theme.
- ❑ When you begin a session, the selected set for a theme is *clear*. A clear selected set is equivalent to having all features in a theme selected. This might seem a bit counterintuitive, but ARCPLOT works this way for your convenience. Otherwise, when you issue draw commands at the beginning of a session, no features would display until you established a selected set. Frequently, you are working with the entire theme, and having clear selected sets at the beginning will save you the step of specifying a selected set.

- Many commands in ARCPLOT and ARCEDIT operate directly on selected sets. These commands, especially in ARCEDIT, will not execute until a selected set is defined. They will give you a warning message if a selected set is not defined. Other commands are less fussy; they operate on both clear and selected sets. The commands that draw features in your graphics window are a good example.
- In ARCPLOT, you can define multiple selected sets, one for each theme. When you use the Select Features commands, you are working on the current active theme. You can see the active theme and the feature class in the Display GIScity main menu.
- You can also store and retrieve selected sets for use at a later time. Be aware, though, that if you have edited themes in Edit GIScity or ARCEDIT, the stored selected sets are no longer valid because they no longer match what was in the theme when they were saved.

NOTE: *When we cover Edit GIScity and ARCEDIT later in the book, you will see that most of the techniques we are now using to select features in ARCPLOT are the same, but there are some important differences. These differences will be outlined and explained later.*

The Three Select Features Menus

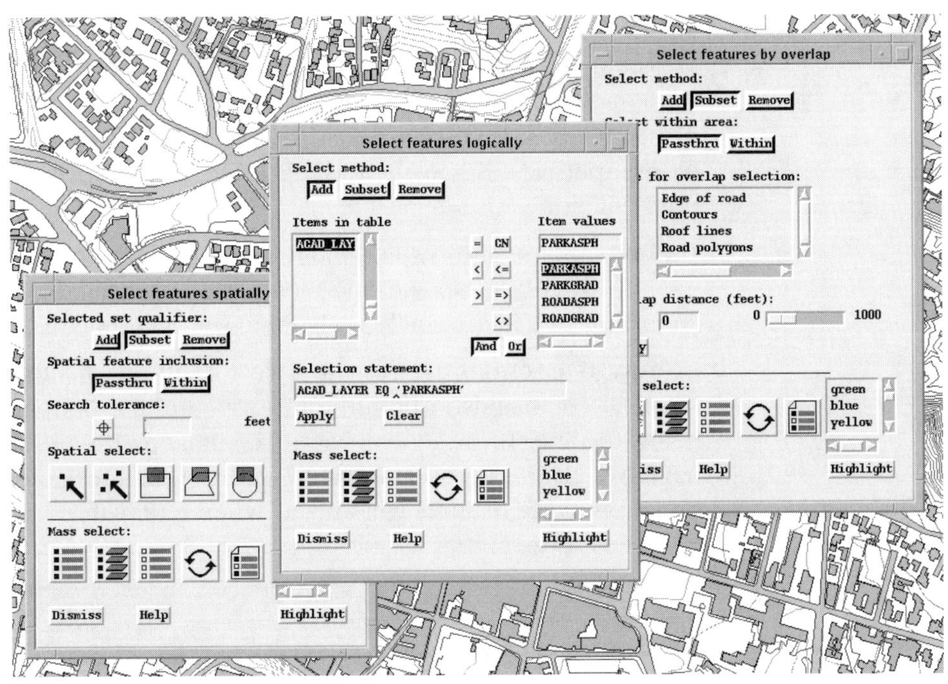

The three menus for selecting geographic features: Select Features Spatially, Select Features Logically, and Select Features by Overlap.

The three feature selection menus can all be used simultaneously. You can also use them in coordination with one another to pose compound queries. For example, you may want to search for all features matching a criterion within a certain area. You would use the Select Features Logically menu to define the criterion, and the Select Features Spatially menu to define the area.

Common Characteristics of the Select Features Menus

When you use the three Select Features menus, you will notice several common characteristics:

- ❑ **Message area.** At the bottom of each menu, you will frequently see a message indicating how many features are currently selected. At other times, if you have just issued a command that requires more input, a message such as "Pick two points for selection in box" will appear.
- ❑ **Updates.** Some choices and values are shared among the three menus. When you update one menu, the corresponding choices and values are immediately updated in the other menus.
- ❑ **Common commands.** All the Select Features menus contain identical commands in both the top row and the bottom section below the line. These general control commands, discussed below, have two main purposes: mass selection of features, and control over how these features will appear when redrawn.

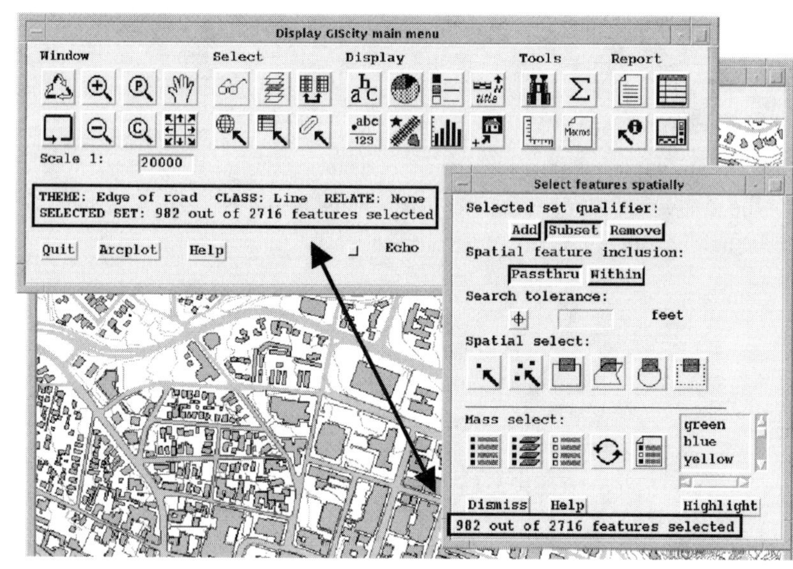

The Display GIScity main menu always displays the current theme, class, relate, and the present status of your selected set—in this example, "982 out of 2716 features selected." Each of the Select Features menus also displays the present status of your selected set.

NOTE: *All three Select Features menus could have been consolidated into a single menu, but this would have resulted in a large and unwieldy menu. Three menus were created for Display GIScity because this is a logical subdivision of the main feature selection functions in ARCPLOT.*

The Selected Set Qualifier

These two choices in the Select Features menus specify two options for selecting features. The Selected Set Qualifier is present in all three menus. The Spatial Feature Inclusion is present in two of the three menus.

Selected set qualifier:

Add Subset Remove

Spatial feature inclusion:

Passthru Within

The Selected Set Qualifier choice specifies how the selected set is to be modified. You can add to the selected set, remove from the selected set, or select a subset of the selected set. This choice does not perform any immediate action; it sets the selection qualifier that the other selection commands will use.

In the following illustration, 6 of the 17 squares were previously selected (left, black squares). The large box defines the area for the next selection operation. If the Add Selection Set Qualifier is used (top right), all squares within the selection box will be added to the set. If Subset is used as a qualifier (middle right), the previously selected squares in the selection box are retained, and the squares outside the box are not selected. If Remove is used as a qualifier (bottom right), all the squares in the selection box are removed from the set.

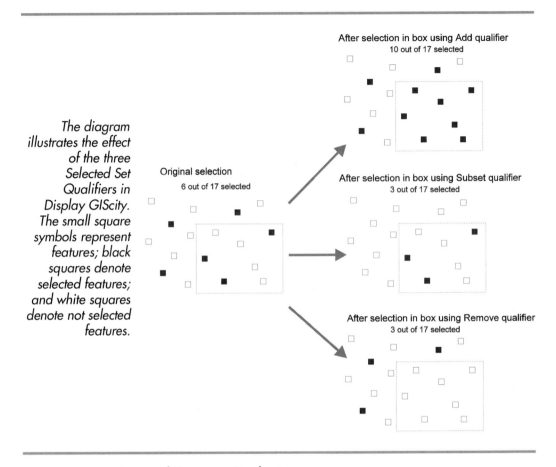

The diagram illustrates the effect of the three Selected Set Qualifiers in Display GIScity. The small square symbols represent features; black squares denote selected features; and white squares denote not selected features.

Original selection
6 out of 17 selected

After selection in box using Add qualifier
10 out of 17 selected

After selection in box using Subset qualifier
3 out of 17 selected

After selection in box using Remove qualifier
3 out of 17 selected

Spatial Feature Inclusion

Spatial Feature Inclusion is the second item on the Select Features Spatially and Select Features by Overlap menu. This choice lets you determine whether features to be selected must be completely, or only partially, within an area such as a box, circle, or polygon. This Passthru/Within choice is active when you select line or area features from themes.

If your current themes contain lines and areas, Spatial Feature Inclusion becomes active when you use the Select by Box, Select by Circle, Select by Polygon or Select Screen commands in the Select

Features Spatially menu, or any type of selection in the Select Features by Overlap menu.

The Passthru/Within inclusion choice is ignored if your active theme contains a feature class of points. Although a point appears to have a discrete size on the screen, in the GIS it is considered to have just a single X,Y coordination position.

This diagram shows how the Within and Passthru inclusion choices work. The inclusion choice applies to selecting lines or areas; it has no effect on selecting points. The Within choice selects all features completely within the selection shape. The Passthru choice selects all features partly within the selection shape.

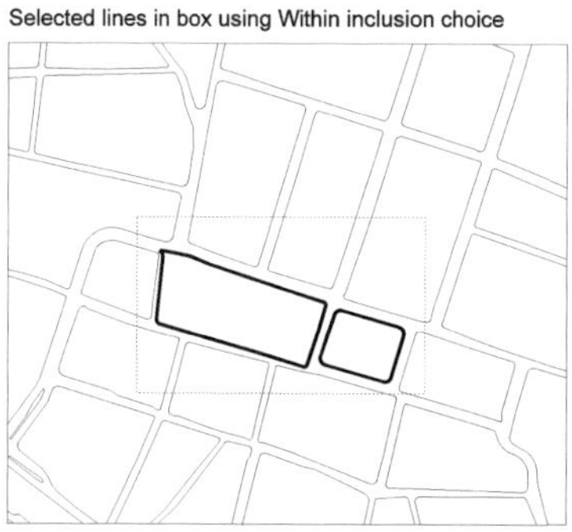

Selected lines in box using Passthru inclusion choice

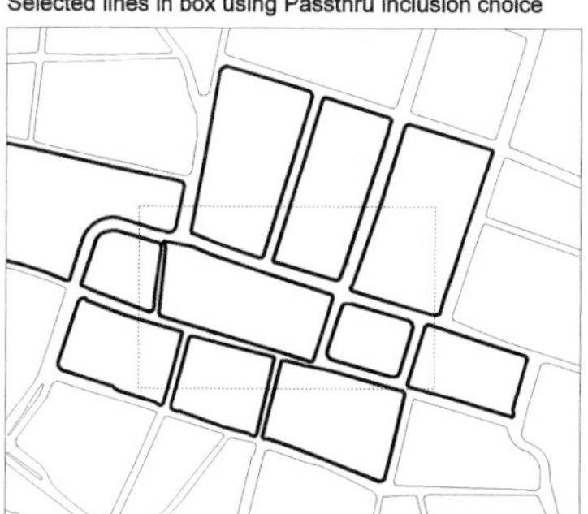

The Mass Select Commands

These five commands, present in all Select Features menus, perform immediate actions on your selected set. Selected Set Qualifier and Spatial Feature Inclusion have no effect on these commands.

Mass select:

Clear Selection
Clear Selection, All Themes
Select None
Reverse Selection
Read/write Selected Sets

Clear Selection will select all features in the current theme. Display GIScity does this by clearing the selected set in ARCPLOT. Remember, a clear selected set in ARCPLOT has all features selected.

Clear Selection, All Themes will clear all selected sets in all themes. This command is handy when you have selected sets in more than two themes and want to quickly clear all of them. You could also change active themes and clear them one by one, but that would be time consuming.

Select None will select none of the features in the current theme. This command is usually followed by other selections that add to the selected set.

Reverse Selection will make all selected features not-selected and all not-selected features selected. If you have 900 out of 1000 features selected, when you execute this command, you will have the other 100 of the thousand features selected. Although it may seem an odd command, it is quite useful.

Read/Write Selected Sets will let you save the current selected sets to a system file. This is useful if you are performing complex analyses, or want to save selections for creating map plots later. This command will invoke another menu which lets you save or retrieve selected sets.

You can save or retrieve selected sets to system files with this menu. In the data list, you'll find several prepared selected sets.

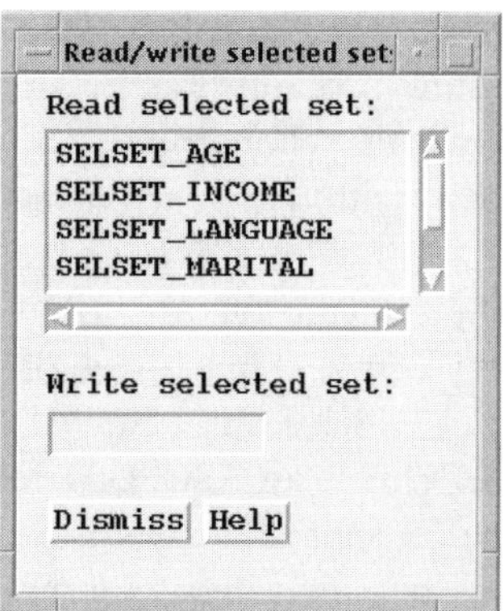

NOTE: *Saved selected sets are valid only as long as you don't edit their theme. Although you cannot edit themes in Display GIScity, in Edit GIScity, you have the full ability to edit every aspect of a theme. Once you do so, you will find your saved selected sets are no longer valid. Consequently, saved selected sets are useful primarily for immediate processing, or on relatively static themes.*

Display Control Commands

You can choose a color to highlight selected features at any time. This data list and button are present in each Select Features menu.

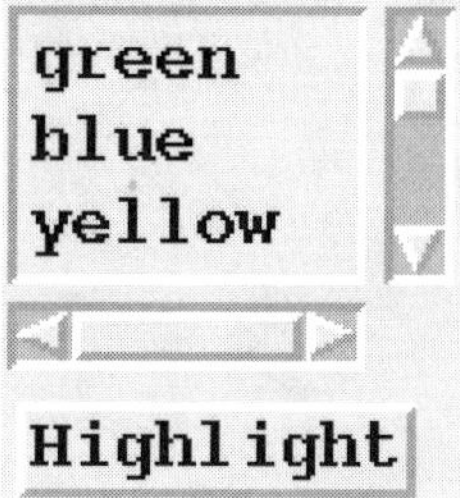

The display control commands have two parts. First, select a highlight color from the data list. Then, when you select the Highlight command, the selected features are drawn with the highlight color.

This Highlight command gives you quick feedback verifying which features you have just selected. The command will act instantly and will

not redraw the graphics window. The selected features will be drawn with simple point, line, or area symbols.

However, when you use a Window command to redraw the graphics window, within the current theme you will see only the selected features drawn. In fact, if you have selected sets for other themes, you will only see the selected features drawn for those themes.

You will see that most of the Window and Display commands in Display GIScity draw only the currently selected features within each theme. This is an ARCPLOT feature designed to control the appearance of final output and to assist analysis.

The mass selection command Clear Selection will restore the display of all features within a single theme. The Clear Selection, All Themes command will restore the display of all features within every theme.

At times you will want to apply more sophisticated symbols to show information about selected features. You will learn some versatile commands to symbolize selections in the next chapter.

The Select Features Spatially Menu: Getting Features from the Graphics Window

Now that we have covered the preliminaries of general selection, let's start with some selection commands. On the Display GIScity main menu, press the globe and arrow icon for Select Features Spatially.

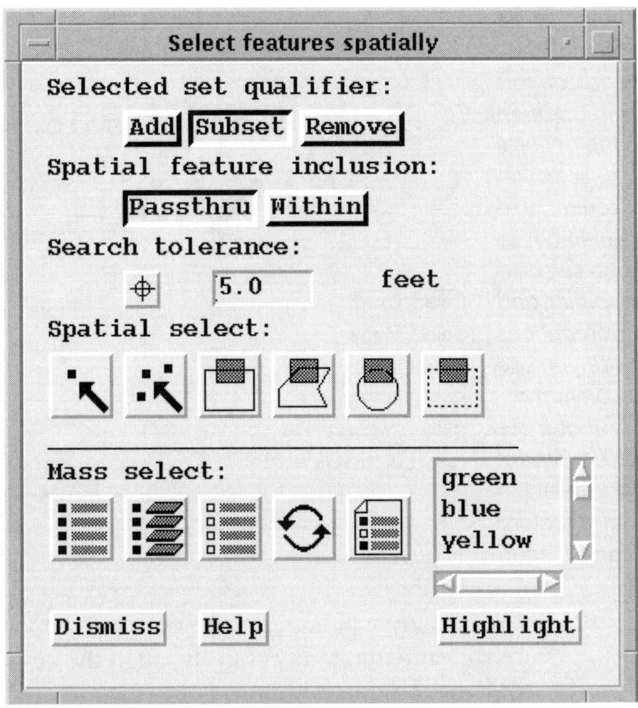

The Select Features Spatially menu provides you with quick graphical selection of map features.

The Spatial Select Commands

Select One will activate the cursor for selection. Move the cursor inside your graphics window, position it above a feature that is contained in the active theme, and press mouse button 1. (If you are picking among closely spaced features, you might select more than one feature because of the search tolerance, discussed below. If you are having this difficulty, try making your search tolerance smaller, or use another selection method, such as Select in Box.)

There are six Spatial Select commands on this menu, in addition to the general selection commands. Remember to check the selected set qualifier and inclusion options before you use these commands. All of these commands, except for Select in Screen, require cursor input.

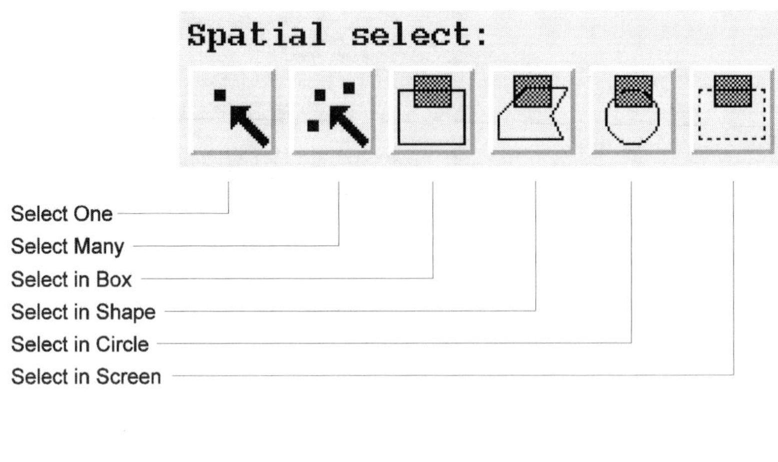

Select Many will activate the cursor to repeatedly select many features. To select features successively, align the cursor over each desired feature and press mouse button 1, in turn. To terminate the selection, press mouse button 9. (See the input keys on mouse button diagram in chapter 1).

Select in Box will select all the features within a box that you specify. Move the cursor inside the graphics windows and press mouse button 1 once to locate one corner, then again to mark the opposite corner of a box. The order or direction of the corners is not important; what is important is that the corners be diagonally opposite corners which include the features you want. A box will appear in your graphics window. It will disappear when you refresh or redraw your graphics window.

Select in Shape will select all the features within an area that you interactively define. Move the cursor inside the graphics windows and successively press mouse button 1 to locate the vertices of the containing shape. When you are finished, press mouse button 9 while still inside the graphics window. The area outline will appear in your graphics window. The area will disappear when you refresh or redraw the graphics window.

Select in Circle will select all the features within a circle that you interactively define. With your cursor, press mouse button 1 to locate the center point of a circle. Press mouse button 1 again to locate a point anywhere on the circumference of the circle. A circle will be drawn and every feature in the circle will be selected. The circle will disappear when you refresh or redraw the graphics window.
Select in Screen will select all the features within the current screen. This command will execute immediately, without cursor input. The Passthru/Within inclusion option applies to this command.

Troubleshooting

Try using each of these commands with different settings for the selected set qualifiers of Add, Subset, and Remove, as well as the Passthru/Within inclusion option. Here are some things to watch out for when using these commands in combination:

- ❑ Nothing seems to happen when you set the selected set qualifiers or inclusion options. The choices will not take effect until you execute another selection command.
- ❑ If your selected set is clear, the Add qualifier will not change your selected set. This is because a clear selected set is like a full selected set, and you cannot add to something that is already full. (Yes, this is confusing!) You have to specify the Subset or Remove qualifiers to affect clear selected sets.
- ❑ A clear selected set is distinct from an empty selected set. You can end up with an empty selected set by trying to select features and missing. When this happens, the theme will seem to disappear when you redraw the screen. This is a good time to either execute the Clear Selection command or choose features with the Add qualifier.
- ❑ If your active theme class is *image*, then none of the selected set commands will have an effect. There are no discrete geographic features that you can select from an image in ARCPLOT. (An optional ARC/INFO module, called GRID, lets you interact with an imagelike data set called a grid. GRID is not covered in this book.)

- ❑ If your active theme class is *point*, the Passthru/Within inclusion option will be ignored.

Other Tips

- ❑ Keep an eye on the message areas of both the Select Features menu and the Display GIScity main menu. You'll get current information on your active theme and selected set.
- ❑ Try the highlight commands. They will take effect immediately. They are a handy way to see if the selection you intended was successful.
- ❑ Even if a theme is not visible, you can still select features from it. If you are zooming to scales where certain theme features no longer display, you can still select the features. The highlight commands will show them for you.
- ❑ Likewise, you can select features that are not presently visible because they are not in the selected set and you have redrawn the screen. Again, the highlight command will reveal them to you, and another redraw will add them to your display.

Later in this chapter, we'll cover how to change your active theme. When you have done so, you may want to revisit some of these selection commands and try them with different active themes.

A view of buildings during selection. The selected set is clear, the Selected Set Qualifier is Subset, and Spatial Feature Inclusion is set to Passthru. You have just used the Select in box command...

And this is the view of the same building theme after the graphics window is redrawn with the Redraw window command. Clear selection will restore all buildings in the graphics windows after a redraw.

Search Tolerance

When you select features with the Select One or Select Many commands, a search tolerance is applied around the point you select. Any features within that tolerance are selected.

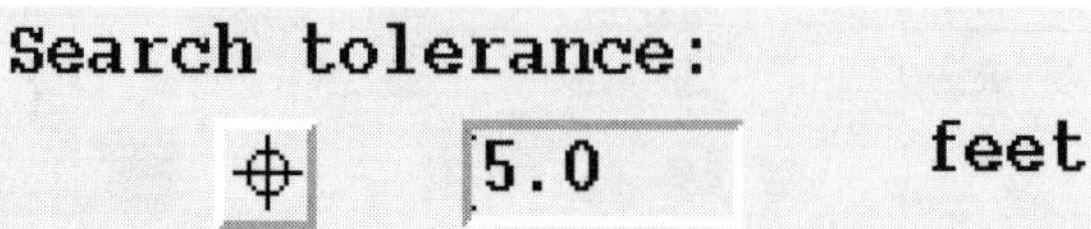

Select One and Select Many in the Select Features Spatially menu use a search tolerance. When you pick a point on the graphics window, ARCPLOT uses the search tolerance to find all features within a circle that has a radius distance equal to the search tolerance.

When you begin ARCPLOT, a default search tolerance is set. With the Search Tolerance command, you can type a new tolerance in the text input field or pick the cross-hair icon, which will activate the screen cursor for you to pick two points. The distance between the two points is used to update the search tolerance, and you will see the new updated value.

Try setting the search tolerance to a large value, and then try the Select One command in a dense part of your graphics window. You might collect a whole slew of features with the Select One command! After you've convinced yourself that search tolerances that are too large are usually as undesirable as tolerances that are too small, reset a reasonable value with the screen cursor.

NOTE: *The cross-hair icon you see in this menu is used in many Display GIScity and Edit GIScity menus. It will usually be coupled with a text input field or slider. All the cross-hair icons have a consistent meaning: You can either set a value, or place something with the screen cursor. Whenever you press the cross-hair icon, the screen cursor is activated. For setting values, you will usually need to select two points to set a distance.*

TIP: *You won't need to change the search tolerance often. If you are selecting among closely spaced features, you can zoom in to get a better separation. Or, if you have trouble picking one from several adjacent features, you'll find that using Select Box with either the Passthru or Within options is an easy way to pick the feature you want.*

The Select Features Logically Menu: Building Logical Expressions

Within ARCPLOT, you can select geographic features through *logical queries.* These are sometimes called Boolean expressions.

Each theme in Display GIScity has a *feature attribute table* associated with it. Each feature attribute table has a set of *database fields,* which are called *items* when you refer to feature attribute tables. (Another industry term for database fields or items is *columns.*)

This database feature allows you to pose queries such as:

- In the road centerline theme, "Find all line features with the street name of Paseo de Peralta."
- In the contour theme, "Find all contours above 7100 feet."
- In the building theme, "Find all buildings with greater than 4000 square foot area."

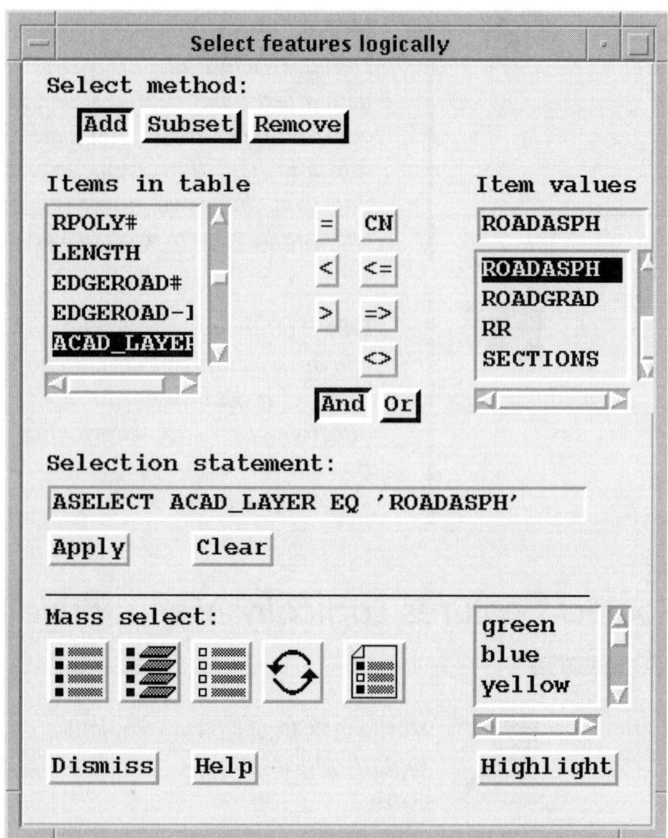

The Select Features Logically menu lets you select map features by their attributes.

ARC/INFO has a standard syntax for posing queries. Display GIScity makes it easy for you to build queries by using the Select Features Logically menu.

As you use the menu, you will see the query being built for you in the Selection Statement text input field. Once you've learned ARC/INFO's syntax for queries, you can move your cursor directly into the text input field and type in your own queries.

The syntax and some of the terminology in ARC/INFO, such as items, are borrowed from the INFO database management system.

NOTE: *A common misconception in the GIS community is that ARC/INFO relies on the INFO database management system. It does not. This confusion is understandable because ARC/INFO preserves some terminology and query syntax from INFO. In fact, the name ARC/INFO came from the combination of "ARC," a geographic term which refers to the line features that are the basis of GIS data structures*

ARC/INFO uses a table format that is compatible with the INFO table format, but it does not use INFO internally for table management. The INFO database management is supplied with ARC/INFO licenses for the convenience of users who choose to use it. You can elect to not install INFO on your workstation and still enjoy full functionality in ARC/INFO.

Increasingly, ARC/INFO users are using the Database Integrator (DBI) functions within ARC/INFO. DBI allows you to connect feature attribute tables to commercial databases such as Oracle, Ingres, Sybase, Informix, and others. DBI is particularly attractive if you are using one of these commercial databases for other nongeographic applications, because you can then enjoy powerful and seamless data relationships between all your corporate databases.

The following steps outline the procedure for constructing a logical query in the Select Features Logically menu. Read through them and then try the exercise that follows.

Step 1. Begin by determining whether you want to add, remove, or select a subset of the existing selected set. Make the appropriate choice from the Select Method widget. If your selected set is clear, you probably will want to specify the Subset choice.

As you begin to build a query with the Select Features Logically menu, you will see a query verb precede the statement. This is ARC/INFO syntax for query commands that correspond to your Select Method choice. When you specify Add, you will see the selection command ASELECT. When you specify Subset, the selection command is RESELECT. When you specify Remove, the selection command is UNSELECT. Watching the query as it is being built will give you a head start on the ARC/INFO query syntax.

The main area of the Select Features Logically menu. Certain portions are outlined and connected by arrows. You can read this illustration as a flow chart for using this menu.

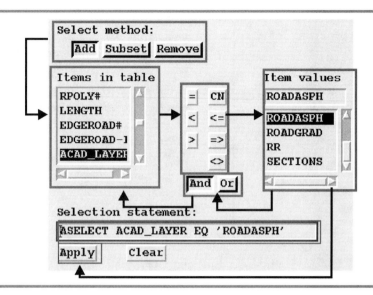

Select method:
Add Subset Remove
Items in table
RPOLY#
LENGTH
EDGEROAD#
EDGEROAD-I
ACAD_LAYER
= CN
< <=
> =>
<>
And Or
Item values
ROADASPH
ROADASPH
ROADGRAD
RR
SECTIONS
Selection statement:
ASELECT ACAD_LAYER EQ 'ROADASPH'
Apply Clear

Step 2. Next, select from the Items in Table data list. Once you have selected an item, you will see a list of possible values appear in the Item Values data list. These are all the possible items, but don't choose one yet!

Step 3. After you have specified an item from Items in Table, choose an operator such as =, <, or >. These symbols are consistent with what you learned in high school algebra:

- ❑ = means "equals"
- ❑ < means "less than"
- ❑ > means "greater than"
- ❑ <= means "less than or equal to"
- ❑ => means "greater than or equal to"
- ❑ <> means "not equal to"

The other operator, CN, means "contains." This operator is used for alphanumeric character items, such as street names. With this operator, you don't have to type in a complete character value, only enough to specify what you want.

NOTE: *Certain combinations of items, operators, and values don't make sense. Selecting an integer value by "CN" is an example. If the Select Features Logically menu detects an inconsistency at any time, it will give you a message in the message area at the bottom of the menu. Don't worry, nothing adverse will happen if your query is invalid. You can always begin again by pressing the Clear button.*

Step 4. Either select from the Item Values data list, or type a value in the text input field under the Item Values data list.

Step 5. You now have a simple query. You should see a selection command typed for you in the text input field labeled Selection Statement. Now, you can select all the feature specified by this query by pressing the Apply button at the bottom of the menu. Or, you can append the query by selecting from the And/Or choice. If you select And or Or, then you must start another selection clause by repeating steps 2 through 4.

At any time, if the selection statement is incorrect, you can get a fresh start in this menu by pressing the Clear button. If you have already applied a selection and the result is not what you desired, then you can use the Clear Select button in the Mass Select command group at the bottom of the menu and start over.

Once you become proficient in the ARC/INFO query syntax, you can bypass all of the above steps and type the query directly in the Selection Statement text input field. You can also partially build an expression with the menu, and then type modifications within the Selection Statement text input field.

As with the other Select Features menus, the Mass Select commands and Highlight commands are available in the Select Features Logically menu for your convenience. You will find yourself using these commands to clear or modify your selected set quickly and simply.

You will also encounter circumstances where you need to use two (or rarely, three) Select Features menus simultaneously. This happens when you need to specify a combined spatial and logical query to select the features you are interested in. These menus are designed to support simultaneous updating of selected sets, and they all interact with the identical selected sets in your ARCPLOT session.

Exercise 1: Selecting Asphalt Roads

1. First, you'll set an active theme. On the main menu, pick the Theme Manager icon—the stack of papers in the Select group. At the top of the Theme Manager menu is the Select New Active Theme data list. From it, pick the Edge of Road item. Click the Dismiss button.
2. On the main menu, pick the Select Features Logically icon—the chart with an arrow pointing to it.
3. On the Select Features Logically menu, first clear your selection set by picking the first icon (Clear Selection) in the Mass Select group.
4. At the top of the menu, pick the Subset button in the Select Method group.
5. Now you're ready to build your query. From the Items in Table data list, choose the ACAD_LAYER item.
6. The Operator buttons are located to the right of the Items in Table data list. Pick the **=** ("equals") operator.
7. From the Item Values data list, choose the item ROADASPH.
8. You have just selected all the asphalt roads in the Edge of Road theme. The Selection Statement text input field should read
   `RESELECT EDGEROAD LINE ACAD_LAYER = 'ROADASPH'`
9. Press the Apply button to activate the selection on your display. You can press the Clear Selection button in the Mass Select group to get rid of the selection set.

NOTE: *Some of you are users of commercial databases that support the Standard Query Language (SQL). You may notice some general similarities to SQL in the ARC/INFO query syntax, but ARC/INFO more closely resembles algebraic notation. The use of ARC/INFO with the Database Integrator supports SQL queries, and can be used as an alternative to this form of query, but that is beyond the scope of this book.*

The Select Features by Overlap Menu: Locating Features by Areas and Buffers

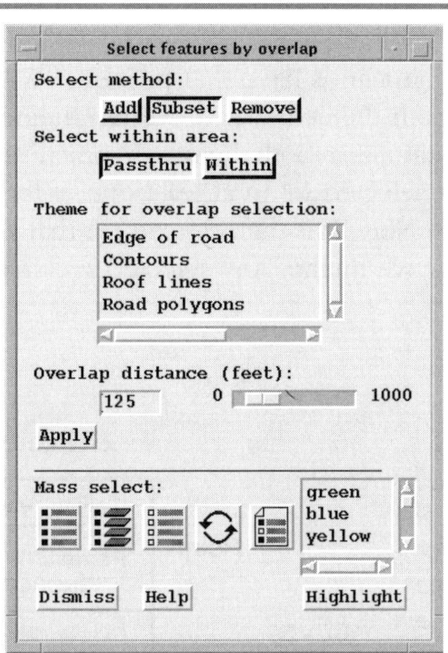

The Select Features by Overlap menu lets you select map features contained within, or adjacent to, other map features.

The Select Features by Overlap menu begins to demonstrate some of the power of a GIS. Almost every CAD or computer graphics program supports selecting features with a screen cursor. Many graphics programs even support selecting features through a logical query. A GIS differentiates itself by its powerful techniques for selecting features that are inside other features, intersect other features, or lie within a buffer distance of other features.

Until now, we have been selecting features from a single active theme. (The active theme is always displayed in the lower part of the Display GIScity main menu.) With this command, we'll introduce overlap themes.

It is very important now to pay attention to theme classes. Remember, each theme has a class. The classes of themes in GIScity are Point, Line, Polygon, Annotation, Node, Image, and Route. Other classes are possible in ARC/INFO, but these seven well represent the classes needed, at least until you start using the optional modules such as TIN and GRID.

The power of this menu is the ability it gives you to choose features in one theme that have a geographic relationship to features in another theme.

The Select Features by Overlap command is unusual in that it can simultaneously operate on two distinct selected sets. Few ARCPLOT commands have that property. When selected sets are not specified in both the active and overlap themes, this command will operate on all features in both themes. In practical use, you will find yourself specifying a selected set in at least one of the themes, and often in both.

Note that the selected set that is modified is always the one for the active theme. Any selected sets in the overlap theme are unchanged.

Not all combinations of theme classes for the active theme and overlap theme will use the values you set for the Passthru/ Within choice, or Overlap Distance.

Active theme class	Overlap theme class: Point	Overlap theme class: Line	Overlap theme class: Polygon
Point	Overlap distance APPLIED Passthru/Within IGNORED	Overlap distance APPLIED Passthru/Within IGNORED	Overlap distance IGNORED Passthru/Within APPLIED
Line	Overlap distance APPLIED Passthru/Within IGNORED	Overlap distance APPLIED Passthru/Within IGNORED	Overlap distance IGNORED Passthru/Within APPLIED
Polygon	Overlap distance IGNORED Passthru/Within APPLIED	Overlap distance IGNORED Passthru/Within IGNORED	Overlap distance IGNORED Passthru/Within APPLIED

Since there are three basic classes for both overlap and active themes, nine combinations are possible. Here's a list of the combinations, with some notes on each.

1. **Active theme class: Point; Overlap theme class: Point.** An example is selecting all wells near point pollution sources. To be useful, this option requires you to specify an overlap distance. An interesting use would be to find all liquor stores near churches or schools.

2. **Active theme class: Point; Overlap theme class: Line.** An example is selecting all businesses (which are represented by points in GIScity) within an overlap distance of specified streets. Again, an overlap distance should be specified.
3. **Active theme class: Point; Overlap theme class: Polygon.** This combination could be used to select all businesses within tax districts. This time, an overlap distance would be ignored.
4. **Active theme class: Line; Overlap theme class: Point.** We can select all streams within an overlap distance of specified wells. This combination requires an overlap distance to give you a reasonable answer.
5. **Active theme class: Line; Overlap theme class: Line.** This is an interesting combination of overlap and active theme classes. If you set the Passthru/Within inclusion option to Passthru, you will select every feature that intersects features within another coverage. For example, you can find every road that crosses a set of streams. Or, you can use this option with an overlap distance. Then you could select every stream within a certain distance of selected roads. Both options are active.
6. **Active theme class: Line; Overlap theme class: Polygon.** This can be used for finding all sewer lines within city boundaries or service districts, or electrical lines contained within tax districts. Utilities can utilize this combination for their tax assessment requirements.
7. **Active theme class: Polygon; Overlap theme class: Point.** You can find all city blocks that contain a well, or all voting districts that have a school within them.
8. **Active theme class: Polygon; Overlap theme class: Line.** With this combination, you can locate all voting districts along a given street. Or, you could find all lakes that feed, or are fed by, a selected stream system.
9. **Active theme class: Polygon; Overlap theme class: Polygon.** You can find polygons that overlap other polygons, or polygons that wholly contain other polygons. An example is finding all islands within selected lakes. The Passthru/Within inclusion option is useful

with this combination. The overlap distance is ignored.

You will probably use this command most frequently with an overlap theme class of polygon because the command is so handy for finding features within polygons. Select Features by Overlap begins to distinguish GIScity from CAD programs, and offers you some simple but powerful ways of selecting geographic information.

Exercise 2: Using Select Features by Overlap Menu

In this exercise, you'll select manholes that are within 100 feet of the street named Guadalupe.

1. First, use the Zoom commands to display the area of downtown Santa Fe.
2. Pick the Theme Manager icon in the main menu. In the Theme Manager menu, set your active theme to "TIGER Roads" by picking that name from the Select New Active Theme data list. Pick the Dismiss button.
3. Now you'll select the specific roads you want from the active theme. On the main menu, click on the Select Features Logically icon (the arrow pointing to a table.) When the menu appears, click on the Clear button to make sure that you have no prior selected set. Then click on the Subset button in the Select Method group.
4. From the Items in Table data list, select the item FNAME, and then click on the = ("equals") operator button. In the Item Values area of the menu, either click on the text input field and type `Guadalupe`, or scroll through the data list and pick each item named Guadalupe. Click on the Apply button. Your Selection Statement should read `RESELECT TGR_ROAD LINE FNAME EQ 'GUADALUPE'`.
5. Next, return to the Theme Manager menu (by way of the main menu) and use the Select New Active Theme scrolling list to change your active theme to Manhole. (The Manhole theme has the class of points.)

6. From the main menu, choose the Select Features by Overlap icon (the arrow pointing to an oval with a line in it.) Your selected set (of roads named Guadalupe) in TIGER Roads is still in effect.
7. In the Overlap menu, choose TIGER Roads from the Theme for Overlap Selection data list, and set the Overlap Distance to 100. (You don't need to worry about the Passthru/Within setting because the Manhole theme has the class of points.)
8. Pick the Apply button. After thinking a bit, Display GIScity will give you the answer in the selected set.

In the preceding exercise, although Manholes is the active theme and TIGER Roads is the overlap theme, it was first necessary to select TIGER Roads as the active theme in order to specify your selection set. The same general procedure is used in the following exercise.

Exercise 3: Using Contours as the Overlap Theme

In this exercise, you'll select all the buildings in Santa Fe at an elevation greater than 7,100 feet.

1. Choose the Theme Manager menu and select "Contours" from the Select New Active Theme data list. (Remember, Contours will be your overlap theme, but first you must make it the active theme in order to specify your selection set.)
2. Choose the Select Features Logically menu. Click on the Clear button to get rid of any previous selection set, and choose the Subset button in the Select Method group.
3. From the Items in Table data list, choose ELEV. Now pick the > ("greater than") operator button. Type `7100` in the Item Values text input field. Pick the Apply button.
4. Use the Theme Manager to change your active theme to Buildings, which has a class of polygons.
5. Choose the Select Features by Overlap icon (the arrow pointing to an oval). Click on the Passthru option in the Select Within Area group, and choose "Contours" from the Theme for Overlap Selec-

tion data list. Click on the Apply button, and in a while, you will have selected many buildings above 7,100 feet. Caution: This is not a foolproof way to select buildings above a certain height, but works pretty well here because the contour interval in the Santa Fe data is dense and nearly every building is intersected by a contour line.

NOTE: *This is the first command in GIScity that may take more than a few seconds to run. It may even take several minutes if your themes or selected sets are substantial. When you initiate this command, you will see a pop-up window notifying you that this command is in progress. When the command is complete, the pop-up window will automatically disappear, prompting you to resume your work.*

Managing Features Through Themes, Views, and Relates

The Theme Manager Menu

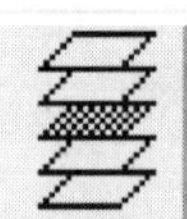

In a GIS, we would like to logically group similar features together. These groupings are called `themes`. Here are some of the reasons to establish themes in a GIS:

- ❑ Themes are easily turned on and off in a display, so that we can see only the geographic information that we desire.
- ❑ We can set *scale thresholds* for themes, so that as we zoom in the graphics window, features can turn on and off automatically. For example, when we see the map index, only major features such as streets and rivers appear. As we zoom in, other features such as buildings and utility lines begin to appear.
- ❑ Themes often connect directly to other databases. By making features in a theme directly correspond to rows in a database table, we simplify the process of collecting and managing that data.

The theme manager lets you set the active theme, scale thresholds, and symbology for each theme.

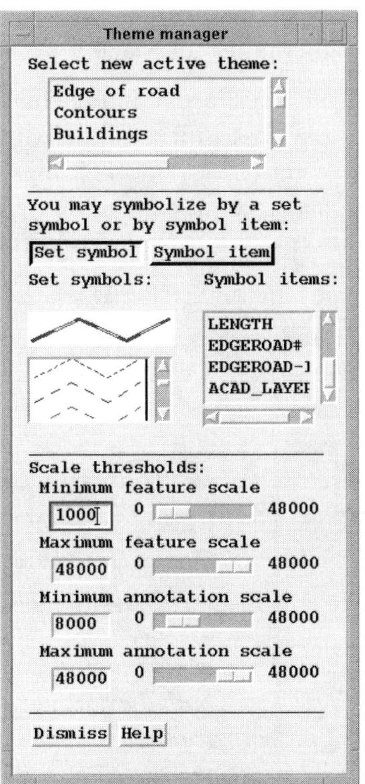

In a typical GIS application, the number of themes defined can range from a dozen to several dozen.

The following are some characteristics of themes:

- Theme definitions may change from one application to another. For instance, a general purpose GIS application may contain all electrical features within a single theme, but an electric utility will want to separate that single electrical theme into a dozen or so more finely subdivided themes.
- Themes are (or appear to the user as) a continuous map. In a well-designed GIS application, you will not see map borders as you move from one area to another.
- One theme can be a subset of another theme. This is a very powerful idea. For instance, from a single geographic dataset for

roads, we can define one theme for primary roads and another theme for all roads.

- ❑ Themes are very useful for symbolizing the same geographic features differently in various situations. For example, you may want to display buildings as point symbols when your display is at a low magnification and to display buildings by their roof lines when your display is zoomed to a high magnification.

Themes are similar to *layers* in CAD systems, but themes in ARC/INFO are much more powerful because they are actual databases. Themes, as well as layers, can be turned on or off. Themes can be used to explore data relationships with each other, and with external databases.

NOTE: *Themes are not an intrinsic type of data organization within ARC/INFO. The basic unit of data storage in ARC/INFO is called a coverage. Themes, however, are an intrinsic type of data organization within many GIS applications, including an ESRI GIS software product, ArcView. The GIScity application, through data definitions and AML modules, has built the data structure of themes to generally emulate themes within ArcView.*

Setting the Active Theme

The top part of the Theme Manager lets you set a new active theme by selecting a choice from a data list.

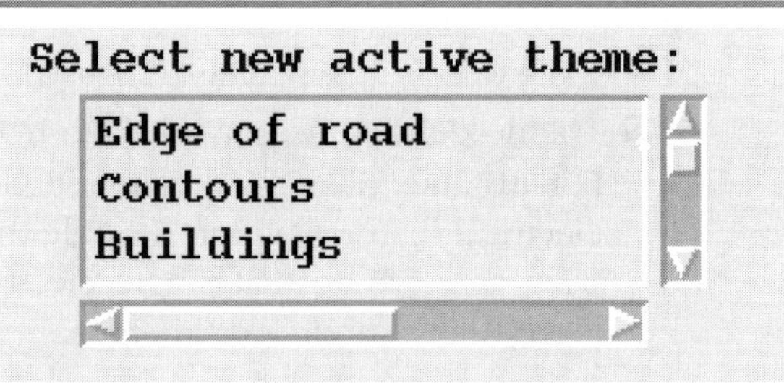

You've already used this data list in the preceding exercises, and you'll continue to use it frequently to quickly change the theme you're working with to another theme. You will also use this command extensively as a prelude to using the Display commands.

Whenever you use the Select New Active Theme command, watch the informational lines in the Display GIScity main menu. They will tell you the new active theme you have just selected and its theme class. The selected set should be clear unless you previously selected features within this theme.

Using Symbols

You can display a theme's features using either a Set Symbol, which is the same for every feature in the theme, or by differentiating between subclasses of features with Symbol Items.

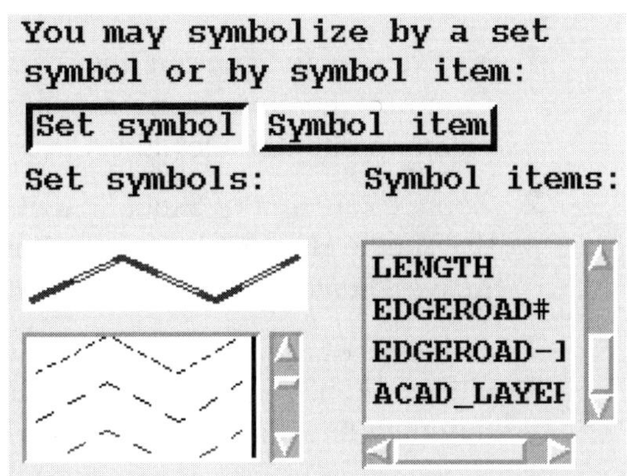

ARC/INFO uses two types of commands for drawing features: Set Symbols and Symbol Items. For instance, if you're drawing roads, you can use Set Symbols to draw every feature in the road theme with the same graphic symbol. If you choose the Symbol Item option, various subclasses of features within the theme will each be drawn using a different symbol. In Display GIScity, the Symbol Items subgroups have been assigned predefined graphic symbols, which all have numeric item names.

Most themes are displayed with Set Symbols. That is because Display GIScity can redraw them that way more quickly, and your graphics window is not visually cluttered. However, themes such as TIGER Roads have a line Symbol Item defined for it, called SYMBOL. This line symbol uses a thick line style for highways and major roads, and other line styles for secondary roads. These symbols generally match the United States Geological Survey (USGS) quadrangle map symbols.

To use this part of the menu, first decide whether you want a set symbol or a symbol item. Make your choice from the Set Symbol/Symbol

Item buttons. Next, select one of the elements in the respective data list below. You can change back and forth from Set Symbol to Symbol Item, or change the data list value at any time.

Exercise 4: Changing Symbol Types

Here are a couple of ways to manipulate the symbols for a theme.

1. You've seen the contour lines draw with a graduation of colors; the Contours theme is using a Symbol Item, which assigns different colors to different elevations.
2. Zoom to an area of Santa Fe that shows lots of contour lines. Pick the Theme Manager icon, and choose Contours from the Select New Active Theme data list.
3. Pick the Set Symbol button, to disable the current Symbol Item choice. From the Set Symbols data list, choose a symbol, such as a thin gray line, and Redraw your screen.
4. You will see the contours lose their color graduation and be replaced with the new line symbol you chose.
5. Next, change your active theme to Buildings. This theme uses a Set Symbol to shade the buildings, and the Set Symbols data list now displays shade colors instead of lines. (You may need to zoom in further to display the buildings as polygons.)
6. Click on the Symbol Item button to disable the set symbols. From the Symbol Items data list, choose the item called BLDG-TYPE. Redraw the graphics window. You will see commercial and residential building types clearly distinguished by color.
7. Now, from the Symbol Items data list, select BLDG-HIST and Redraw the graphics window. The display will change to show buildings inside and outside of historical districts in downtown Santa Fe with different colors. The effect is striking.

A key advantage of a GIS is complete flexibility of symbolization. We have just used stored symbol values. In the next chapter, we will see how to change symbology with dynamically calculated attributes.

NOTE: *The Symbol Items data list shows every available item for the active theme. Most of the listed items, if selected, would give you a nonsensical result when you redraw your graphics window. Pick only items with the name SYMBOL, or with "SYM" as part of the item name. The data list within Display GIScity cannot distinguish between these types of items, and it relies on you to make a reasonable choice.*

Setting Scale Thresholds

Scale thresholds control which features are displayed at any given scale.

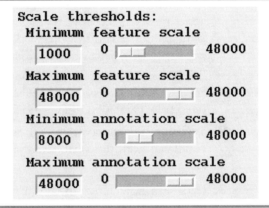

```
Scale thresholds:
 Minimum feature scale
  1000    0  48000
 Maximum feature scale
  48000   0  48000
 Minimum annotation scale
  8000    0  48000
 Maximum annotation scale
  48000   0  48000
```

The bottom part of the Theme Manager menu contains four sliders with interesting values. These are the scale thresholds within Display GIScity. This is how they work: whenever you issue a Window command that either zooms in or out in your graphics window, Display GIScity checks the scale thresholds to determine whether or not to draw the theme's various features or annotations.

You've already seen the effect of these scale thresholds when you've seen road names appear as you zoom in. The edge-of-road theme has annotation that is triggered to display once you've zoomed into a scale of 1:8000 or closer. With these sliders, you can modify the scale thresholds which have been set for you by default.

You'll also notice two general sets of themes:

- ❑ Themes for the central part of Santa Fe, mapped at a high level of detail meant for publishing maps at a scale of about 1:2400.

- ❑ Themes from the Census TIGER digital maps, which are created from maps at a scale of 1:100 000.

You will find some overlap between the scale thresholds for these two sets of themes, but they will usually appear independently of each other.

Exercise 5: Understanding Scale Thresholds

1. In the Scale box on the main menu, type in a scale of `200 000` and press enter. You will zoom way out of downtown Santa Fe and your view will extend into the surrounding countryside. You now see a new collection of themes. These are the small-scale themes principally derived from the U.S. Census TIGER data. They didn't appear before because their scale thresholds did not take effect at the scales you've been using. Also, the familiar large-scale themes, such as Edge of Road and Buildings, are suppressed, because we have now exceeded their scale threshold.
2. Type a scale between 36 000 and 48 000 in the Scale box. Now you'll see both the small-scale and large-scale themes at the same time.
3. Choose a new active theme in the Theme Manager menu. Check the theme's current Scale Thresholds, and try zooming in and out.
4. Now, use the Minimum and Maximum Feature Scale text input fields (or sliders) to change the scale thresholds. Repeat your zooming experiments to verify the new scale thresholds. Try modifying the scale thresholds for various themes, and note the different effects when using themes with classes of point, line, and polygon.

ArcView and many GIS applications use scale thresholds also. You can now see that this is a technique to control screen clutter and prevent slow display of geographic features. Published maps contain only information that is appropriate to that scale. Quality GIS applications also employ scale thresholds because doing so applies the same cartographic principles mapmakers have used since the earliest maps.

NOTE: *Scale thresholds are not an intrinsic part of ARC/INFO, but Display GIScity has implemented them through the ARC Macro Language. Look for scale thresholds in the AML applications you build or purchase.*

NOTE: *Many themes have no annotation. If this is the case, setting the scale thresholds for annotation will have no effect.*

Using the View Manager to Control Display

A view is a list of the themes that are active for display. A view was automatically set for you when you began GIScity, and you can dynamically modify views.

Views are defined in GIS applications so that the applications will be easier to use for different types of users. For example, a city engineer, a city planner, and a visitor would all like to see distinctly different views. These different users can easily specify their preferred views and see the geographic data appropriate for their particular interest.

Two characteristics of a view are:

- Views define which themes are currently displayed.
- Views determine the order in which themes are drawn.

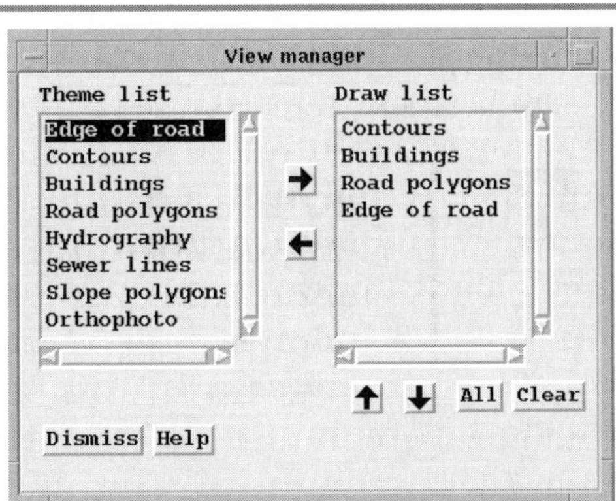

The View Manager menu gives you control over which themes are displayed, and in which order.

There are two data lists in this menu: Theme and Draw. The Theme List does not change; it is the list of all themes supplied with GIScity. The Draw List is the list of all themes that are currently specified for display in the graphics window. The themes are listed in the order they will appear when you redraw the graphics window.

To add themes to your draw list, select a theme in the Theme List, and press the arrow icon pointing right. To remove themes from your Draw List, select a theme in the Draw List, and press the arrow icon pointing left.

To move a theme up in the order of drawing, first select the theme in the Draw List, and then press the arrow icon pointing up. To move a theme down, select a theme in the Draw List and press the arrow icon pointing down.

The All button will move every theme from the Theme List into the Draw List, maintaining the same order as in the Theme List. The Clear button will remove every theme from the Draw List.

Normally, you should first draw themes with a class of polygons, followed by themes with lines, and lastly themes with points. The reason is simple: polygons tend to cover up line and point features. However, sometimes polygons that obscure line features are desired. For example, in Display GIScity, the Contour theme comes first, because you often want to suppress contour lines so that they don't obscure man-made features.

Try selecting the Contour theme in the Draw List. Press the down arrow icon. The Contour theme will move down a position. Next, redraw your graphics window. You will now see contours overlaid on the theme they were just swapped with.

NOTE: *As with themes, views are not directly defined as part of the ARC/INFO system, but many GIS applications, including GIScity and ArcView define views to better organize GIS data. GIScity has modeled views and themes through the Arc Macro Language.*

Accessing External Data with the Relate Manager Menu

The Relate Manager lets you establish logical connections between a theme's attribute table and other tables.

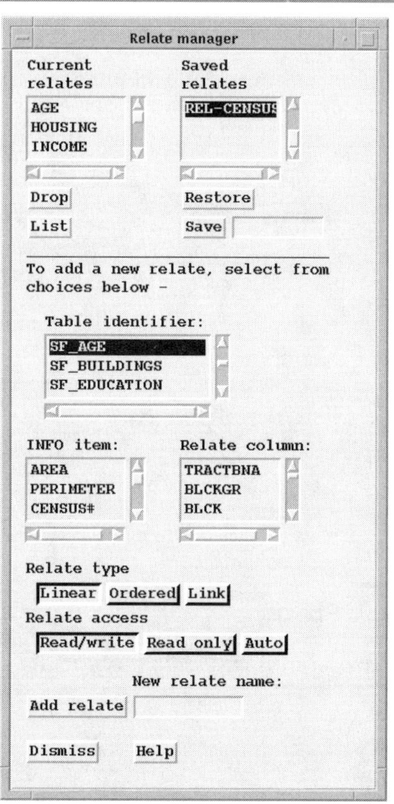

A powerful feature of ARC/INFO is the ability to relate geographic features to other tables or databases. A relate is a logical connection between the attribute table for a theme and any other table, provided there is a common link between the two.

Relates will be discussed in greater detail later, but we will use this menu to quickly cover relates already established for you in Display GIScity.

This diagram illustrates the georelational model that is key to ARC/INFO. Each theme has a feature attribute table (usually point, line or polygon) associated with it. The attributes table contains database fields called items. Each attribute table can optionally have an item which is a link to another table. The other table can be another feature attribute table, an INFO table, or a table from another database management system.

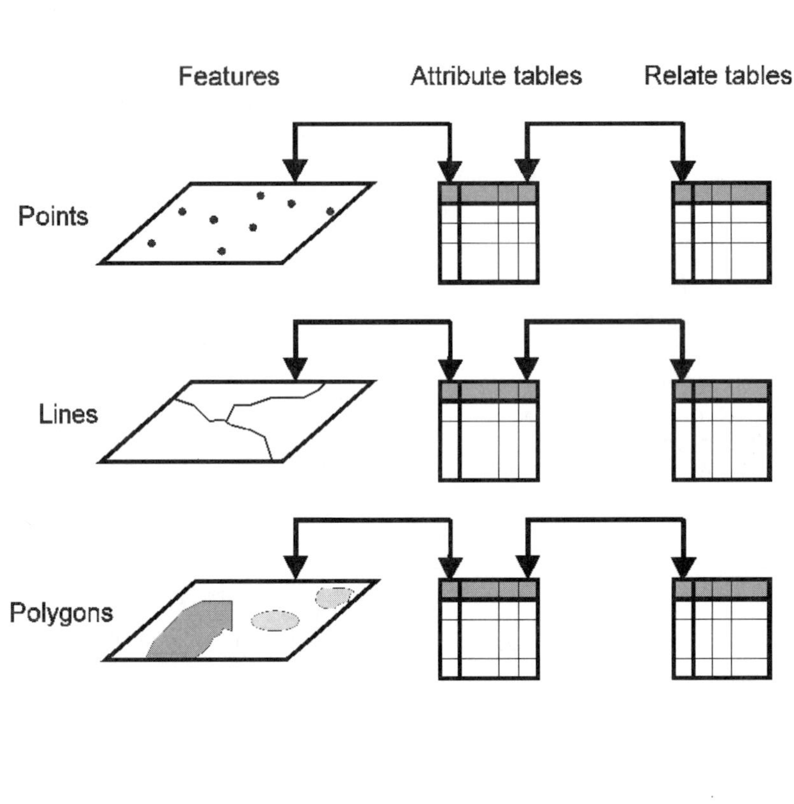

Saving Relates

The Current Relates data list contains the relates for the active theme that you have previously defined and saved, and new relates that you define for that theme. You can remove a relate from the Current Relates data list by selecting it with the mouse and then pressing the Drop button.

The definition of a relate (or several relates) can be stored within an INFO table. These are called *Saved Relates*. You can make a selection from the Saved Relates data list and press the Restore button. You should see one or several relates being added to the list of Current Relates.

The top part of the Relate Manager manages your list of current relates and lets you save and retrieve relates.

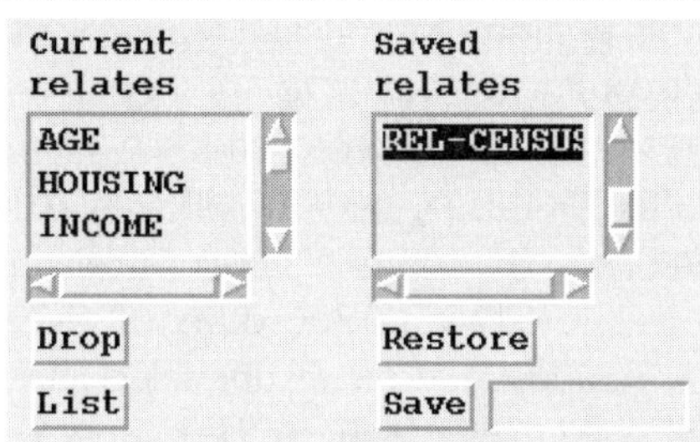

After you have defined some relates, you can save them for another day by first typing a name for your new relate in the Save text input field and then pressing the Save button. The new name will appear in the Saved Relates data list.

At any time, you can press the List button to get a detailed listing of all the information about a relate: what type it is, what items connect the two sides, and what the connected tables are.

Establishing New Relates

The lower part of the Relate Manager lets you establish new relates.

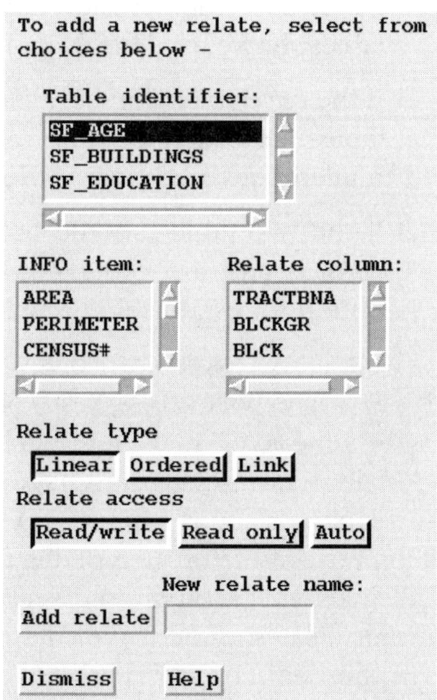

This portion of the Relate Manager menu connects the feature attribute table in your active theme to another table. Before establishing a new relate, look at your main menu to check what your feature attribute table presently is. Often, at this point, you will want to use the Theme Manager menu to reset the active theme. We'll cover relates in more detail in later chapters, but for now the following procedure will give you an overview of creating new relates.

To add a new relate, first select a table in the Table Identifier data list. These are the tables which can be connected to your current feature attribute table.

Next, select from the INFO Item data list. This selection should be an item which has a value also present in the table selected from the Table Identifier data list.

Then, pick from the Relate Column data list. (Frequently the names in the INFO Item list and in the Relate Column list are identical, but this is not required.)

You have two choices for structuring the relates: Relate Type and Relate Access. The Relate Type choice gives you the option of Linear, Ordered, or Link relates.

- ❑ If in doubt, pick Linear. The choice of Linear makes no assumptions about the order of the entries in the two tables; with very large tables, however, Linear choices take longer to process.
- ❑ If the two tables don't exactly match, but item values in the related tables are ordered in the same way, such as alphabetically or in numeric order, choose Ordered.
- ❑ If the two tables should have exactly the same number of table records and are ordered identically, choose Link—the quickest option for establishing a relate.

The Relate Access choice allows you to specify the type of access to the relate table: Read and Write access, Read Only access, or Auto access. The Auto choice sets the same access to the relate table as to the feature attribute table.

Finally, a new relate must have a name. Pick a short, descriptive name (no blanks or funny characters) and type it in the New Relate Name dialog box. After you've typed in the name, press the Add Relate button. If all goes well, you should see the new relate appear in your list of Current Relates at the top of the menu.

In the next chapter, we'll cover several commands that exploit the relates we can establish with this menu.

Visualizing Information in Display GIScity

Text, Symbology, and Graphics in ARCPLOT

GIS software is endowed with a rich set of commands to depict information graphically. Its unique character is the ability to overlay information upon and through maps. ARCPLOT spans the majority of visualization commands within ARC/INFO programs. Some of the capabilities we will explore in this chapter have counterparts within ARCEDIT, but ARCPLOT is designed to deliver geographic information to you graphically through almost any means you can imagine. Visualization is where ARCPLOT shines.

In this chapter, we will apply visualization techniques to the interactive graphics window on your workstation. All of the commands we will explore can also be applied to graphics files, which can be sent to your plotter or printer. In the next chapter, we'll learn how to create graphics files and incorporate the effects of most of these commands within plotted or printed maps within the Display GIScity application.

Introducing the Display Commands

A GIS can deliver information to you through a variety of means:

- **Annotation.** You can select any attribute table item or relate item and annotate its values next to geographic features.
- **Pie chart symbols.** Statistical information within attribute tables or relate tables can be displayed as pie charts. A number of macros have been predefined for you to easily portray some interesting demographic and housing statistics from the 1990 United States Census enumeration.
- **Color ramps and graduated symbols.** You can select any numeric item from an attribute table or relate table, specify a *quantile* (which is a subdivision of the range of values for an item), choose a start color and finish color and/or a start size and finish size, and redisplay the features for that item graduated by color and size.
- **Map legends.** For any theme, you can display a map legend of the feature classifications and their symbols. Map legends are also available for pie chart symbols, color ramps, and graduated symbols.
- **Graphs.** You can select any two items and prepare a graph of their relation to each other. Four types of graphs are supported: point, bar, line, and shade.
- **Images referenced to features.** A couple of buildings within the downtown area of Santa Fe have scanned images referenced to them. You will be able to select one of those buildings and invoke a display of that image.

In the previous chapter, we covered selected sets, themes, and relates. You may have wondered why we spent quite some time with those concepts. The reason, quite simply, is that most of the interesting commands in ARCPLOT work directly upon the selected sets within your active theme. Before using many of the Display commands, you must first specify a new theme and modify the selected set.

Also, much of the interesting information in the GIScity data set is only accessible through relates. Again, a relate is a logical connection between a theme attribute table and another database table. In the GIScity application, all relates are set to INFO tables. Relates take some getting used to, but are well worth the effort because much of the data you want to access will not be within the ARC/INFO attribute tables, but in relate tables.

In your real working environment, you'll probably establish relates to both INFO tables and a database management system (DBMS) that your company or agency uses for other applications. Examples of commercial DBMSs are Oracle, Ingres, Sybase, and Informix.

Wherever you see a data list in the Display command menus, you may see relate items within that list. You will see the relate items as well as the items for the theme attribute table. This will happen only when you have an active relate set. Look at the message area of the Display GIScity main menu to see if you have an active relate. In the exercises in this chapter, we'll walk through several examples that use relate items, and then the utility and power of relates will become clear to you.

Most of the Display commands we are about to cover use the selected sets that you learned about in previous chapters. As with the Window commands, all geographic features will be visible with a clear selected set; otherwise only selected features will be symbolized.

Unlike the symbol settings that you can control with the Theme Manager menu, the Display commands operate immediately. Their visual effect disappears when you perform one of the Window commands such as Redraw or Zoom In. (In the next chapter, we will learn how you can preserve the effect of some of these commands in a map plot.)

Display GIScity makes many ARCPLOT visualization commands very easy for you to use. In this chapter, you will learn some principles of visualization and the capabilities of ARCPLOT that may not be obvious at the command level.

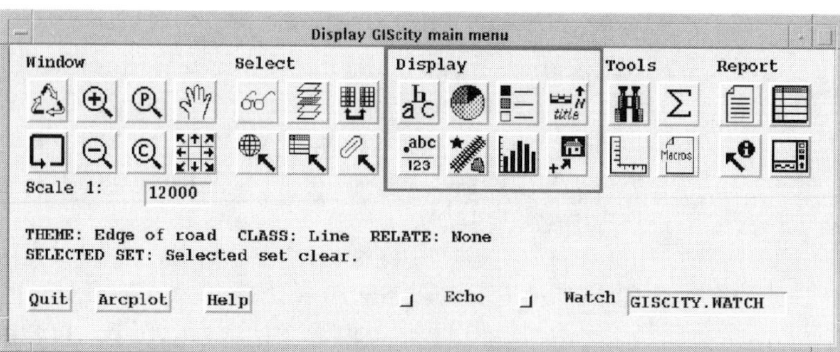

This chapter covers the outlined Display commands within the Display GIScity control panel.

While you use the Display commands, be aware of the current theme, selected set, and active relate, all displayed on the main menu. Every Display command operates on one or more of the contexts that you have specified through the Select commands on the control panel. None of the GIScity Display commands will modify any of those selection contexts for you.

You will discover that underneath these Display commands in Display GIScity are some substantial macros. For evidence of the underlying complexity necessary for these simple commands, toggle on the Echo check box on the main menu and watch the dialog window as you try these commands. Some of the display commands perform several dozen ARCPLOT commands to achieve their result. A few commands may execute a couple of hundred ARCPLOT commands.

NOTE: *Turning on the Echo check box will give you insight into ARCPLOT commands, but will also slow your workstation's performance somewhat. Normally the Echo check box should be off unless you want to study the underlying ARCPLOT commands while you use Display GIScity.*

The eight Display commands within Display GIScity.

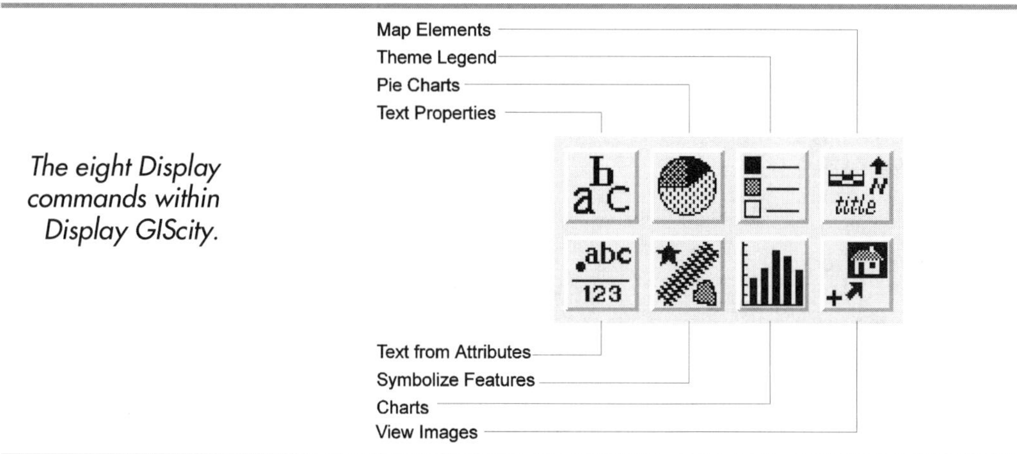

While these eight commands by no means exhaust the possibilities for displaying information graphically, they offer an instructive and attractive subset of visualization techniques within ARCPLOT.

Setting Text Properties

The Text Properties menu is different from the other Display commands because it does not immediately change the display in your graphics window. Rather, it sets the default text characteristics that most of the other Display commands use. Because ARCPLOT offers you many options to control the appearance of text, setting text characteristics warrants a menu of its own. This menu controls most of the following text characteristics possible in ARCPLOT:

- Font
- Color
- Text spacing
- Text size in points
- Position of text relative to point features
- Position of text relative to line features

The Text Properties menu, invoked by pressing the Text Properties icon button. Whenever you modify a text property, you will see the sample text at the bottom of the menu updated with the new property.

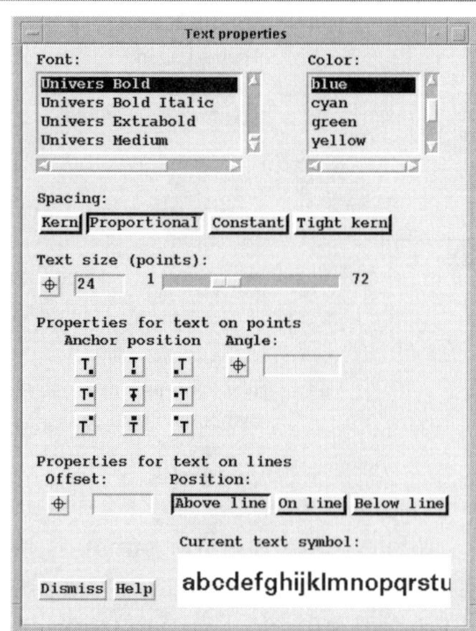

Fonts

These text properties apply to text in most situations.

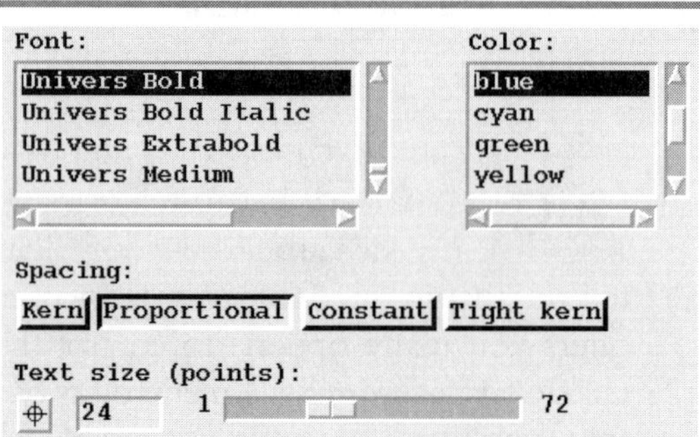

These text properties in the upper part of the menu are your most frequently used text characteristics. First, you can choose from a variety of fonts. Use the data list to select from the several dozen that are available.

The main typefaces delivered with ARC/INFO.

Univers
Triumvirate
Times
Omega
Palacio

Within the data list, you will notice variants of these typefaces with characteristics of bold, italic, or other text properties. By choosing a font from the data list, you are choosing several text properties (typeface, bold, italic) at once.

As you try out these fonts, you will see that some fonts have a prefix of IGL. These are basic fonts provided with ARC/INFO. Other fonts have names such as Univers, Triumvirate, Times, Omega, and Palacio. These are typefaces provided with ARC/INFO from the Agfa Compugraphics font library and are called *Intellifont typefaces.*

NOTE: *ARC/INFO comes to you with a suite of fonts. If you prefer to use a font not available in this data list, you can purchase additional fonts from Agfa Compugraphics and add them to the font library, where they will be accessible from within ARC/INFO. Reference the ARC/INFO system dependencies documentation for information on acquiring additional fonts.*

The IGL fonts were the original fonts delivered in early versions of ARC/INFO. While they are not as attractive as the Intellifont typefaces (added in release 6.0 of ARC/INFO), the IGL fonts can be useful for pen plotters, because text using IGL fonts can be plotted much more quickly.

For raster output devices (electrostatic plotters, ink jet plotters, and laser printers) there is no performance penalty in selecting the Intellifont typefaces. This is one reason why raster devices are becoming more popular and steadily replacing pen plotters.

While redrawing your graphics window, you will seldom notice a difference in speed between drawing text with the IGL and the Intellifont fonts, unless you have many text items to draw.

Locating Text on Lines

These text properties apply to text placed along line features.

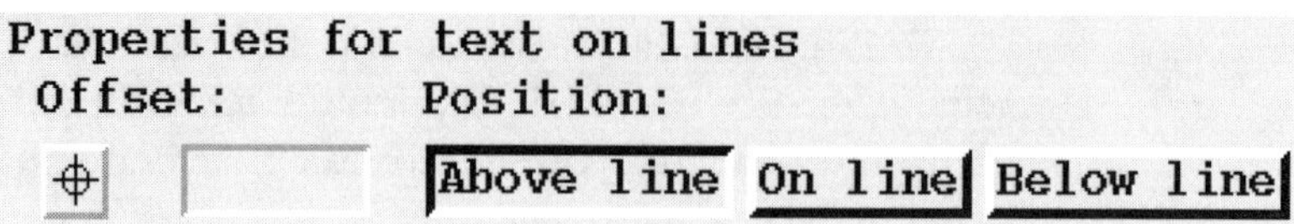

Offset and position are two text properties in this menu that control text placement along lines. The offset can be specified in two ways. If you pick the cross-hair icon, the screen cursor will be activated for you to select two points on the graphics window. ARC/INFO will calculate the distance between the two points and set the offset for you.

Or, you can type in a value in map units for an offset position. In GIScity, the map units are in feet. If you are zoomed to a scale of 1:24 000, typing in an offset of 100 feet would result in an offset distance on your workstation screen of 1/20 of an inch. (This result is calculated by dividing 100 feet by 24,000, and multiplying by 12 inches to a foot to derive a screen offset of 1/20 of an inch.)

The second property for text on lines is the position choice, Above Line, On Line, or Below Line. When we cover the Text from Attributes menu in the next section, you will see examples of Above Line and Below Line positions.

Locating Text on Points

These text properties specify options for text placed on point features.

Properties for text on points
Anchor position Angle:

There are nine possible Anchor Positions for text on point features. Each of the nine icons illustrates one of these options with a letter T to symbolize text and a small square to signify the relative placement of the point feature. The top left Anchor Position icon sets a text justification of lower left for text, which is the default setting if you set no other.

Normally, text on point features will be placed horizontally. Occasionally, you might choose an angle for text. This is sometimes done to resolve closely spaced text along a horizontal line. The cross-hair icon and text input field for text angles can be used to set a new default angle for text on point features. If you type in a value of 45, all future text on points will be placed at that angle.

Most angles you can type into ARC/INFO are measured from the East azimuth. Called geometric angles, these are the same angles that you probably learned in school.

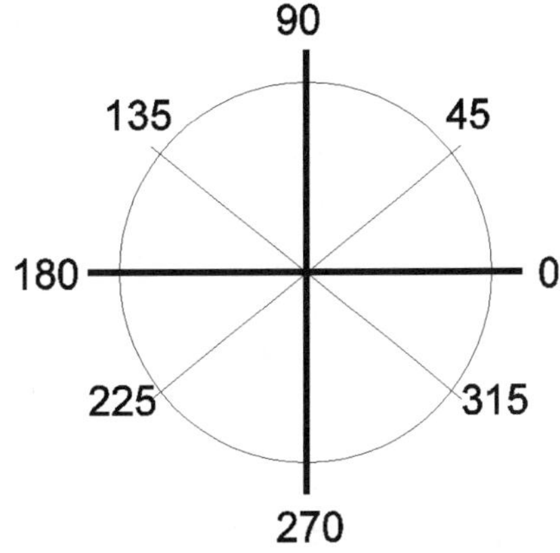

NOTE: *The Angle command will have a limited effect for text on lines. Most text on lines will be placed along line segments, either above, on, or below. However, text that is larger than the line segment is placed anchored at the midpoint of the line and at whatever text angle you have set.*

Drawing Text from Attributes

On the graphics window, you can draw any information included in a theme attribute table or relate table using the Text from Attributes menu.

First, select the Text from Attributes icon from the Display GIScity main menu. On the menu that appears, scroll through the data list, and left-click the mouse button on the item you want to draw as text on the graphics window.

Immediately, you will see text appear on your graphics window. If your active theme has a class of Line, most of the text will be oriented above, on, or below its associated line feature. If your active theme has a class of Point, the text will be oriented according to the anchor position specified in the Text Properties menu. And if your active theme has a class of Polygon, the text will be centered within the polygons.

The text will be drawn with the applicable text characteristics set in the Text Properties menu. If your screen is cluttered with many text items overwriting each other, you can decrease the annotation size.

If the Text from Attributes item list does not have the information you wish, you can switch active themes using the Theme Manager, or establish an active relate with the Relate Manager.

The Refresh Window and Text Properties buttons at the bottom of the menu perform the same actions as the icon commands with the same name in the Display GIScity main menu. They are placed here for your convenience because they are frequently needed when drawing text from attributes.

This graphic illustrates the effect of clicking on the ACAD_LAYER item from the Edge of Road theme. The Above Line position choice was set in the Text Properties menu.

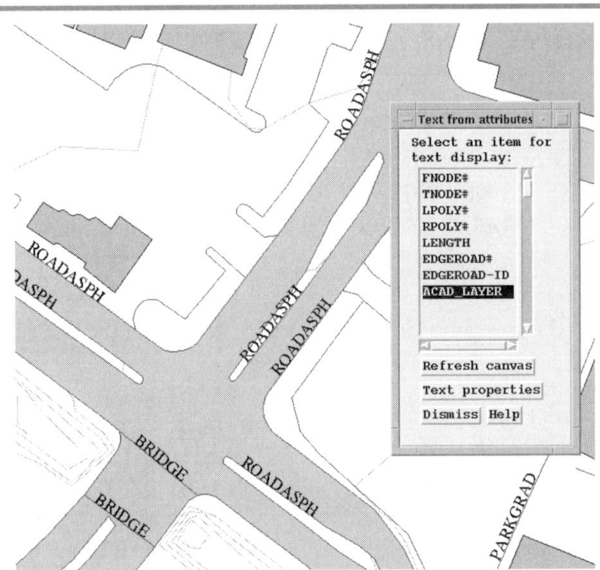

This graphic illustrates the same selection, but in this instance, the Text Properties menu was used to set the Below Line position choice.

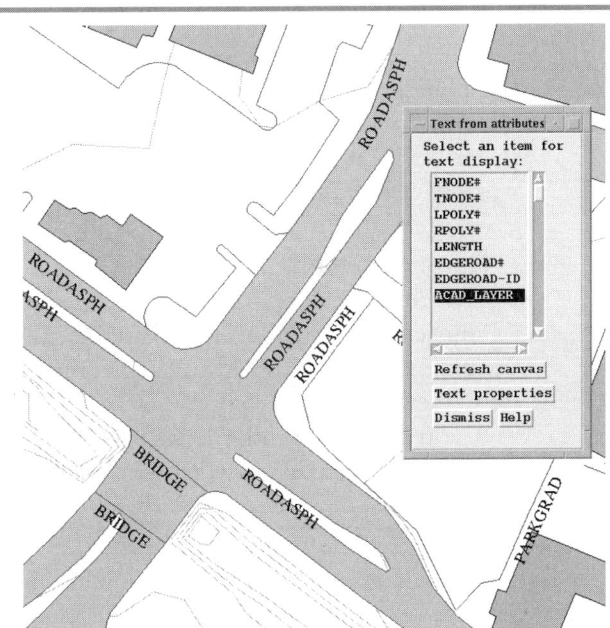

Resolving Overlapping Text

An important enhancement in release 7 of ARC/INFO is the new capability to resolve overlapping text. This feature is important, because displaying text from attribute values is a common and useful task in GIS, but displayed text in close proximity typically overlaps itself.

Commands that affect the resolution of overlapping text.

```
Overpost status:
 [On] [Off] [Clear]

 Resolution (inches)
0.050  0.010  [slider]  0.200
 Radius (inches)
1.150  0.100  [slider]  2.000
 Increment (inches)
.050   0.010  [slider]  0.200

[ ] Leader lines
```

On the lower part of the Text from Attributes menu is a group of commands that enable and control this feature, which is called overposting.

The On, Off, and Clear choice turns this feature off and on. Once enabled, overposting text can successfully resolve text from different coverages. Clear keeps overposting active, but lets you clear the overposting status.

The three sliders for Resolution, Radius, and Increment control tolerances for how far text can be moved. Reference the ArcDoc documentation for the ARCPLOT command OVERPOST for details on these factors.

If you click the Leader Lines check box on, then leader lines can be automatically drawn between features and their text.

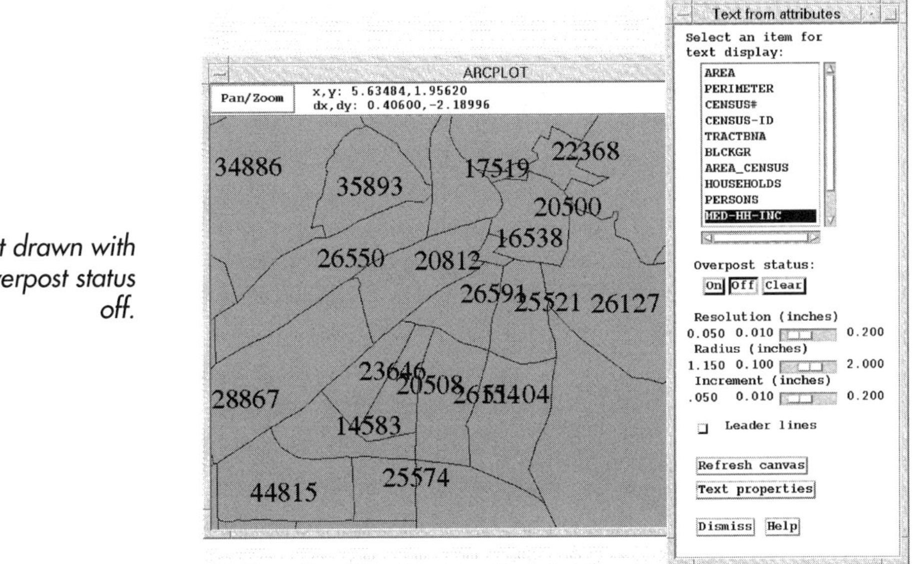

Text drawn with the overpost status off.

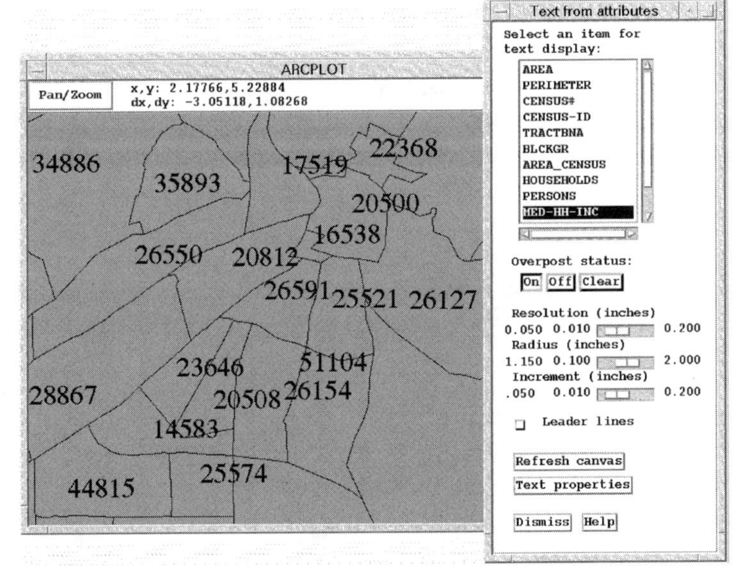

Same text drawn with the overpost status on. Note how the same text in the previous picture no longer overlap.

Exercise 1: Duplicate the Text in the Illustrations

In this exercise you will....

1. Click the Theme Manager icon from the main menu, and select the Edge of Road theme from the active theme data list. Click the Dismiss button.
2. Zoom in to any area within downtown Santa Fe, to a scale of about 1:1200.
3. Click the Text Properties icon. Select the text font of your choice from the Font data list, and select the On Line position choice. Click the Dismiss button.
4. Click the Text from Attributes icon. From the menu that appears, click the item named ACAD_LAYER. You should see a display similar to the ones in the illustrations.
5. Click the Refresh Window button to clear the text from your graphics window. Experiment by setting different text properties in the Text Properties menu, and then click the ACAD_LAYER item again on the Text from Attributes menu.
6. For another interesting example, use the Theme Manager to set Contour as the active theme. Choose ELEV from the Text from Attributes data list, to draw elevation values on the contour lines. You may have to try some different display scales to get annotated contours such as you might see on topographic maps.

Exercise 2: Draw Text on Polygon Features and by Relate Items

In this exercise, you'll use the Text from Attributes menu to automatically place text in polygons, and to add further data to your display with the relate function.

1. Use the Theme Manager to set your active theme to Census. Using the Scale Box on the main menu, zoom to a scale of 300 000. You will see an area that is a good part of Santa Fe county, and you will

also see some new themes from the U.S. Census Bureau TIGER digital maps. The detailed themes you saw in downtown Santa Fe will no longer be visible, due to scale thresholds for those themes.

2. On the Text from Attributes menu, click the item PERSONS. For each of the polygon areas in your graphics window, you will see a value appear, which is how many persons reside within that area.
3. Next, click the Relate Manager icon. Click the relate AGE from the Current Relates data list to become your active relate. Verify that this relate is active by looking at the Display GIScity main menu. Click the Dismiss button.
4. Go to the Text from Attributes menu. Select the relate item AGE//AGE_15-19. You will now see the count of all 15- to 19-year olds drawn within the same polygons.

NOTE: *Whenever you see a double slash in an item list, it means that is a relate item. The name to the right of the double slash is the name of the relate, and the name to the left is the relate item.*

Displaying Statistics with Pie Charts

If you have statistical data in your database tables, and if you can relate that data to a GIS theme, you have the potential for powerfully displaying that data in ARCPLOT with pie chart symbols. Pie charts are useful for organizing this statistical data when you have a range of database columns (or items in ARC/INFO).

A good example of this type of data is census demographics. Each item is a tabulation for a class of census attributes, such as age, race, income, and other characteristics. U.S. Census data from 1990 for all of Santa Fe County, New Mexico, is included with the data on the companion CD-ROM. Display GIScity also comes with some predefined macros that make it very easy for you to visualize this data with pie charts.

The Pie Chart menu offers you several prepared macros, size controls for the pie charts, and an optional legend for displaying class ranges and colors.

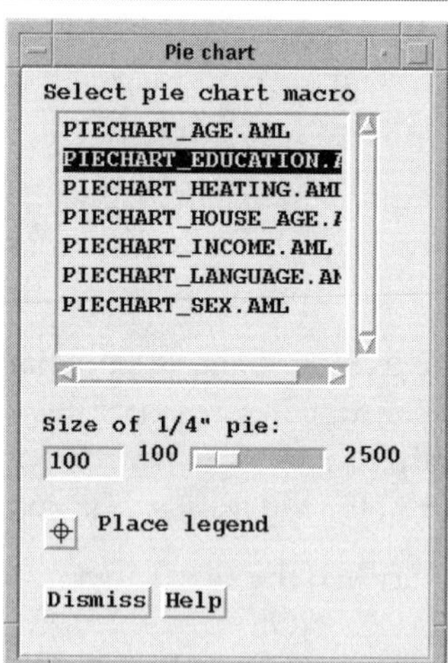

The Display GIScity macros provide you with pie charts to quickly and easily symbolize the following information available from the U.S. Census STF-3 data. All of this data is tabulated by block groups. A block group comprises several hundred people. You will see block group boundaries in Display GIScity only when you zoom to a scale of 1:48 000 or more. (You won't see block group boundaries if you can see the detail themes for downtown Santa Fe.)

PIE_CHART_AGE.AML. This pie chart macro summarizes the age distribution of the population. The classifications are ages 0-4, 5-9, 10-14, 15-19, 20-24, 25-29, 30-34, 35-39, 40-44, 45-49, 50-54, 55-59, 60-64, 65-69, 70-74, 75-79, 80-84, 85 and over.

PIE_CHART_EDUCATION.AML. Educational levels within Santa Fe county are depicted with this macro. The following graphic shows the effect of this macro. The classifications are: less than 9th grade education; from 9th to 12th grade, no diploma; high school diploma; some college education, no degree; associate (2 year) degree; bachelors degree; and graduate degree.

PIE_CHART_HEATING.AML. Primary household heating sources are displayed when you pick this macro. The classifications are: utility gas, LP gas, electricity, fuel oil, coal or coke, wood, solar energy, other energy source, and no fuel used.

PIE_CHART_HOUSE_AGE.AML. This macro shows the distribution of housing by its age. Classifications are: since 1989, 1985 to 1988, 1980 to 1984, 1970 to 1979, 1960 to 1969, 1950 to 1959, 1940 to 1949, and before 1939.

PIE_CHART_INCOME.AML. This pie chart macro tabulates families whose income falls within specified ranges. The classifications are: $0 to $20,000, $20 to $40,000, $40 to $60,000, $60 to $75,000, $75 to $100,000, and more than $100,000. Income ranges are shown in the second graphic illustration below.

PIE_CHART_LANGUAGE.AML. This shows the primary language spoken. Compared with most areas of the United States, a significant portion of the population in New Mexico speaks primary languages other than English. The classifications are: English, Spanish, Native American languages, and Others.

PIE_CHART_SEX.AML. Sex and marital status are shown with this macro. Classifications are: Male, never married; Male, married; Male, separated; Male, spouse absent; Male, widowed; Male, divorced; Female, never married; Female, married; Female, separated; Female, spouse absent; Female, widowed; and Female, divorced.

These prepared macros provide you with selected access to several dozen database fields contained within the U.S. Census STF-3 data. Over 2,000 database fields exist within this data!

To create pie chart symbols, first click the Pie Chart icon from the Display GIScity main menu. When the Pie Chart menu appears, click one of the macros described above. You will see pie charts appearing throughout your graphics window.

If the pie charts do not appear, use the Scale Box in the main menu to zoom to a scale between 1:50 000 to 1:800 000.

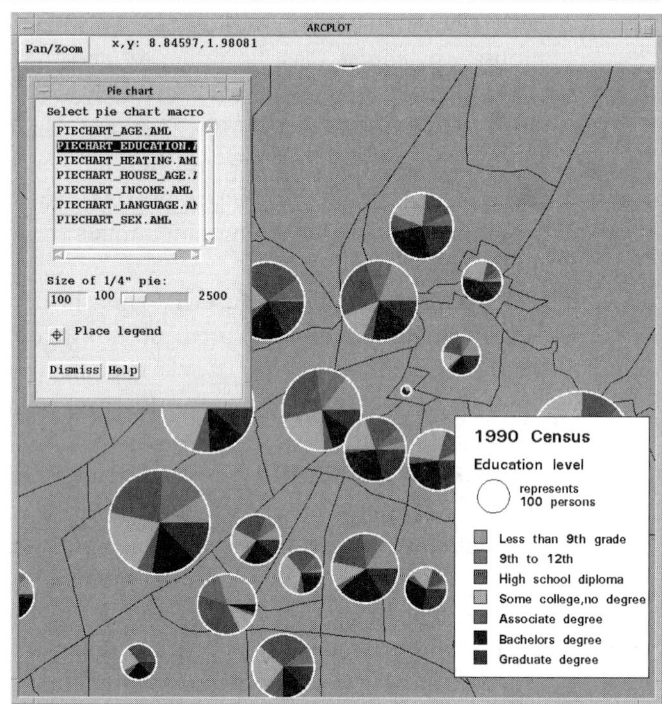

This graphic depicts the range of education levels within Santa Fe. The size of the pie chart signifies the size of the tabulation.

If the pie chart symbols are too big or too small, you can change the size of the pie by increasing or decreasing the value in the Size of 1/4" Pie text input field or slider. Alternately, you can increase or decrease your scale.

Once you have created pie chart symbols that look right, click the Place Legend command. Move your cursor within the graphics window and select a lower left corner for a map legend. (This legend will use the text properties you have set currently.)

This graphic shows the range of household incomes in Santa Fe. Can you find a correlation between household income and education level?

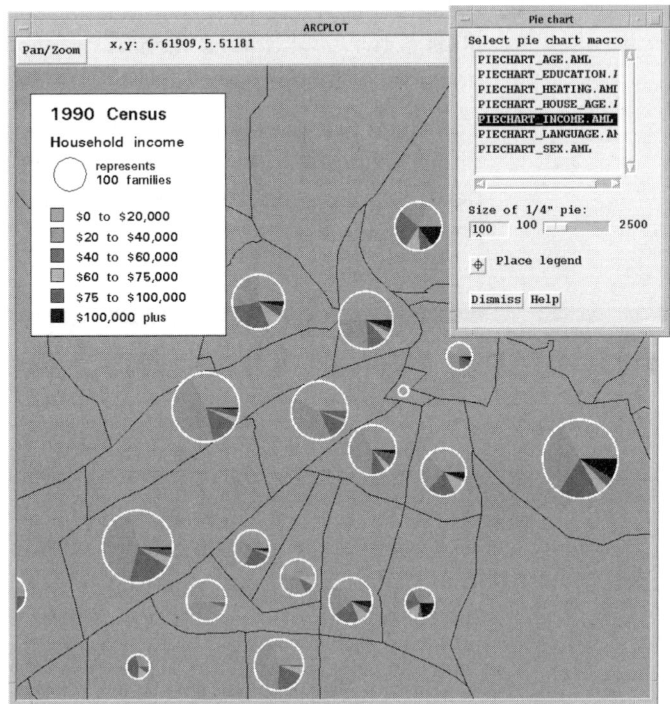

Exercise 3: Explore All the Census Pie Chart Macros

1. Go to the Theme Manager and set the active theme to Census.
2. Go to the Scale Box in the main menu and type in a scale between 50 000 and 800 000.
3. Click the Pie Chart icon in the main menu.
4. Click any of the pie chart macros in the data list.
5. Change the size of the reference 1/4" pie and click the pie chart macros again.
6. Click the Place Legend command, and then pick a part of your graphics window that is relatively empty.
7. Experiment with different settings in the Text Properties menu, different scales, and different pie chart macros.

8. Using the different pie chart macros, see if you can detect correlations between attributes such as primary house heating and age of housing, or income level and educational level.
9. You'll see that pie chart symbols can communicate spatial patterns of distribution very well. Map presentations like these can provide persuasive evidence for patterns and trends in populations, housing, and businesses.

Symbolizing Point, Line, and Area Features

Often, you need to portray information graphically by color or size. This type of map is called a *thematic* or *chloropleth* map. Newspapers and magazines commonly use this type of map to portray national patterns relating to social issues, such as cost of housing, money spent on education, or election results.

The Symbolize Attributes menu provides you with two useful techniques to make thematic maps: *Color ramps*, which use a gradation of colors to symbolize increasing values, and *graduated symbols*, or symbols which increase in size to denote increasing values. These two display techniques can be applied to points, lines, and polygons. Color ramp thematic maps are usually most striking when used for symbolizing polygons.

NOTE: *If you've carefully looked at the contour lines in Display GIScity, you will notice that a color ramp is already applied to contours. This has been done permanently through assigned line symbol numbers corresponding to darkening colors at higher elevations. In the lower elevations, contours are light green, and range to dark green in the foothills within the downtown area. Later, we'll cover making this kind of permanent color ramp assignment.*

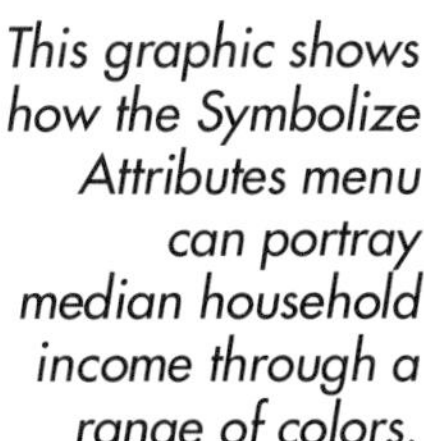

This graphic shows how the Symbolize Attributes menu can portray median household income through a range of colors.

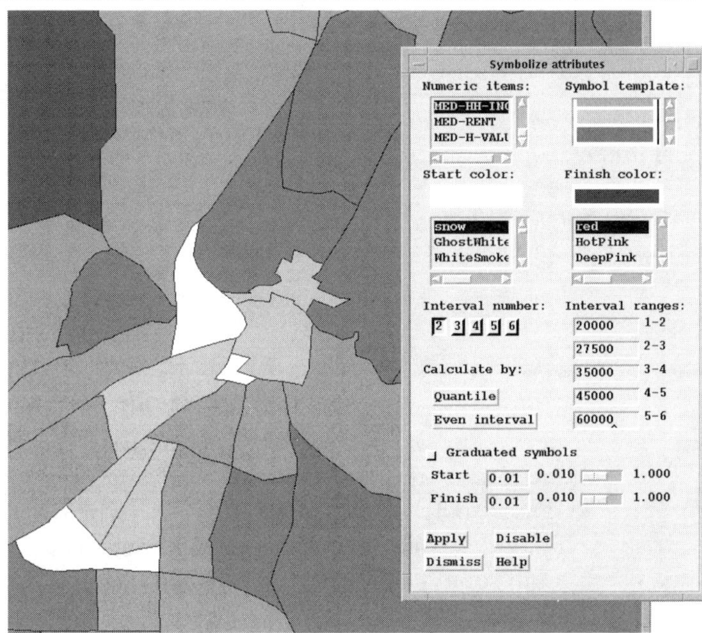

In the preceding example, the item MED-HH-INC (median household income) was selected. The item to be symbolized must be numeric. A color ramp from snow white to red is set, and 6 intervals are chosen, with interval ranges from 0 to 20 000, 20000 to 27 500, 27 500 to 35 000, 35 000 to 45 000, from 45 000 to 60 000, and from 60 000 on up. The colors for the intermediate intervals are automatically blended from white to red.

To produce a thematic map, first click the Symbolize Attributes icon from the Display GIScity main menu. The Symbolize Attributes menu will appear. You must then select a numeric item to be symbolized. The contents of the Numeric Items list are determined by your active theme and relate.

The Symbol Template data list specifies the type of symbol you wish to depict. What's interesting about this menu is that, depending on the theme class you've specified, point symbols, line symbols, or shade symbols will appear in the data list.

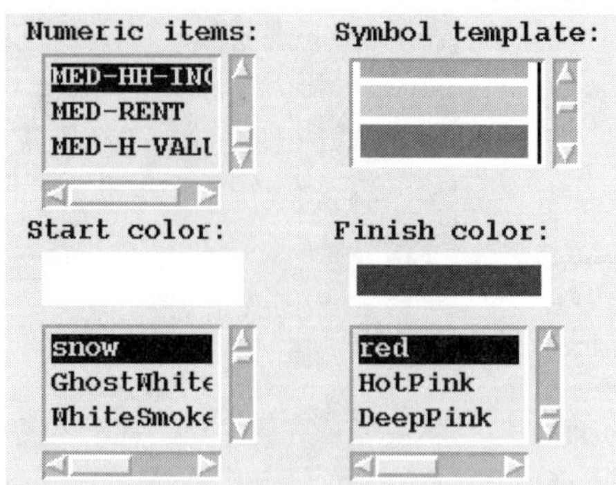

A detail from the upper part of the Symbolize Attributes menu.

For shades, this is not a particularly interesting choice because you'd nearly always like a solid shade symbol. But this scrolling list provides interesting choices for points and lines. For points, you can choose a star, triangle, square, or circle for the symbols. For lines, you can choose among several solid lines of different thicknesses, or you can choose from some lines that have symbols applied.

Defining Symbol Colors

The Start Color data list specifies the display color of the lowest value class of the color ramp. In the case of the preceding illustration, we selected a color of snow (white) for the class of $0 to $20,000. Likewise, the Finish Color data list specifies the display color of the highest value class interval. In our illustration, we selected red for the highest class, which is $60,000 and above.

Between these two colors, Display GIScity automatically calculates a color ramp for you. Using the color definitions and the number of intervals (discussed below), a gradual palette of colors is provided that transitions from the start color to the finish color.

NOTE: *If you are an ArcView user, you will recognize most of the capabilities in this menu. This menu purposefully emulates ArcView within ARC/INFO to both illustrate some ARCPLOT capabilities and to show you how some ArcView functions can be provided in ARC/INFO through application macros.*

Defining Intervals

Interval number: Interval ranges:
2 3 4 5 6
20000 1-2
27500 2-3
Calculate by: 35000 3-4
Quantile 45000 4-5
Even interval 60000 5-6

A detail of the middle section of the Symbolize Attributes menu.

Once you have selected an item, you can choose an interval number. The interval number defines how many discrete ranges the total range of values will be subdivided into. This menu supports between 2 and 6 intervals.

The menu offers two methods for structuring the contents of intervals, Quantile and Even Intervals. The Quantile button will scan the item values from your selected features and calculate interval ranges so that the same number of values falls within each interval. While roughly the same number of values will be present in each interval, you will see that the spread of values in some interval ranges is larger than in others.

The Even Interval button will scan the item values for a minimum and maximum value, and based on the interval number, will calculate the interval ranges to be equally sized. Though the spread of values in each interval will be similar, the number of values in each interval will likely be uneven, and perhaps significantly uneven.

Here's an analogy to help you understand the difference between the Quantile and Even Interval calculations: In news reports about housing costs, you will often read about the median and the average price of a house. The median price is usually lower than the average price. If you selected only two intervals, the Quantile calculation would effectively give you the median price and the Even Interval calculation would give you a number closer to the average price.

The Interval Ranges text input fields have two functions: When you choose the Quantile or Even Interval calculations, they will update and show you the calculated interval ranges. Or, you can type interval ranges

of your choosing directly in the Interval Ranges text input fields. For example, you can perform the Even Interval calculation, and then modify the Interval Ranges to become rounded numbers.

NOTE: *The Interval Ranges text input fields show you only the intermediate interval values. The first interval always contains all features with a value less than the first interval range break, and the last interval always contains all features greater than the last interval range break.*

NOTE: *If you choose an interval number less than six, the lower interval range breaks will not be applicable and will become blanked out on your menu.*

Defining Graduated Symbols

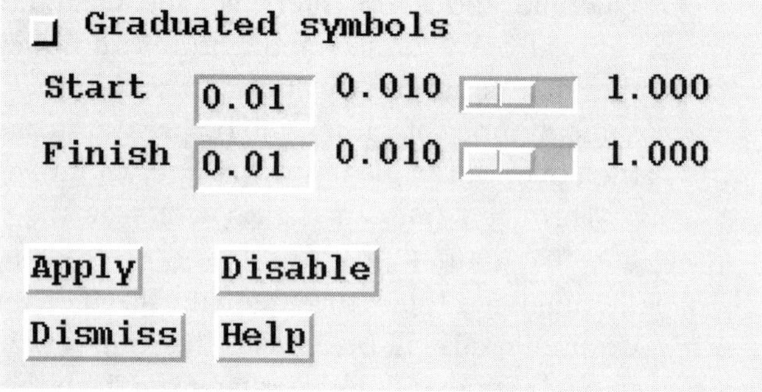

A detail from the lower section of the Symbolize Attributes menu.

You can optionally define symbols that are graduated by size. These graduated symbols do not necessarily replace the color ramps associated with point, line, or polygon features, but add another dimension to how you can visualize information. You can use graduated symbols for added emphasis.

Alternately, you can symbolize by graduated symbols only, using a fixed color. To do this, select identical Start and Finish colors, but choose different start and finish size symbols for point and line classes.

Exercise 4: Symbolizing Polygon Features

1. Select the Census theme with the Theme Manager. Go to a scale of about 1:300 000.
2. Choose the Symbolize Attributes icon. From the Numeric Items data list, select MED-HH-INC (median household income).
3. Try selecting different start and finish colors. Also, try the Quantile and Even Interval range calculations. You will see the Interval Ranges values update. Watch how the ranges vary with each type of calculation. Click the Apply button to see the result on your graphics window. Use the Refresh Window command to clear the experiment from your screen.
4. Click the MED-RENT (median rent) and MED-H_VALUE (median house value) items in succession, and repeat your experiments.
5. Try this: Use the Select Features Spatially menu to select a group of polygons in the center of your screen. Click the Apply button again and note that only the selected polygons are symbolized.

Exercise 5: Symbolizing Line Features

In this exercise, you'll notice the difference in symbolizing polygon and line features.

1. Select the Contour theme with the Theme Manager. Zoom to a scale of about 1:10 000, making sure to include an area of the map that contains contour lines.
2. In the Symbolize Attributes menu, select the ELEV (elevation) item. Try setting different interval numbers and ranges.
3. You can replace the permanent color ramp within the Contour theme (light green to dark green for higher elevations) with other colors. Choose a new start and end color. (Don't worry; you won't destroy the permanent color assignment.)

4. Next, turn on the Graduated Symbols check box. Try a Start size of 0.01 and a Finish size of 0.20. When you click the Apply button, not only will contours change color as they increase in elevation, they will also thicken.

Drawing Map Legends

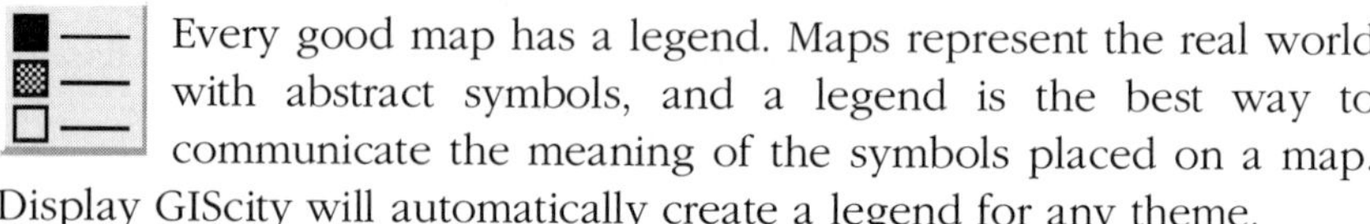

Every good map has a legend. Maps represent the real world with abstract symbols, and a legend is the best way to communicate the meaning of the symbols placed on a map. Display GIScity will automatically create a legend for any theme.

Choose the active theme you want, and click the Theme Legends icon. From the Theme Legends menu, choose a background color, click the Place Legend button, and select an area on the screen for the legend's position.

The illustration shows a legend for the Edge of Road theme.

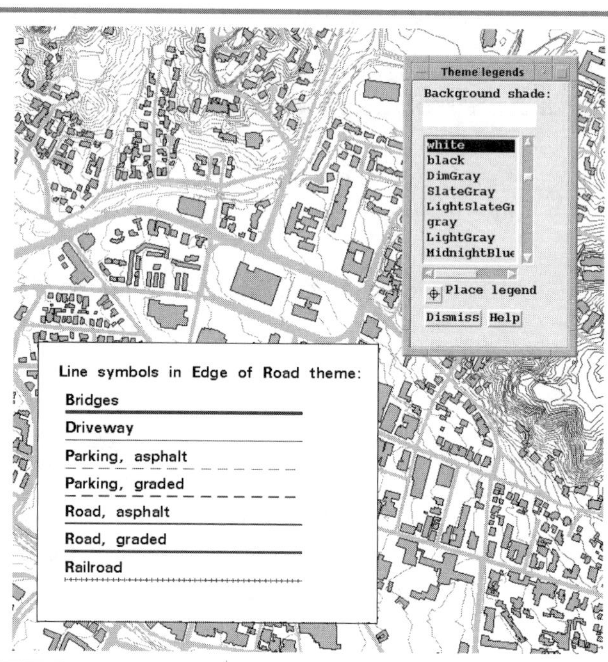

Graphing Statistics with Charts

ARCPLOT comes with a suite of commands specifically designed for creating charts. Like most of the other Display commands, the underlying commands are quite numerous, but Display GIScity contains macros that greatly simplify the generation of charts.

Creating a chart with Display GIScity is quite easy. By clicking the Chart icon in the main menu, you will invoke the Graph Attributes menu. All you have to do to make a chart is to pick an item for the X-axis of the chart, and another item for the Y-axis. Next, pick the Point, Bar, Line, or Shade option. The appropriate chart will appear for whichever of the four types you selected.

What's actually more difficult is finding two different items that will produce an interesting and meaningful graph when combined. In the following illustrations, the active theme is Buildings, and the two items picked are Polygon areas (square feet) and Perimeters (lineal feet). In the user exercises, you'll be guided to some other items that will make interesting graphs.

NOTE: *When the Graph Attributes command is used, graphs are sent to a brand new graphics window, which has the name GRAPHWINDOW in the title bar. This command shows how certain ARCPLOT commands can be directed to graphics windows other than your main graphics window with its standard map view.*

Point charts (also called *scatter point charts*) are used for numerous data points that fall within a discernible pattern, but with loose correlations. With a point chart, you can nicely see the concentration of data points, in this case near the graph origin.

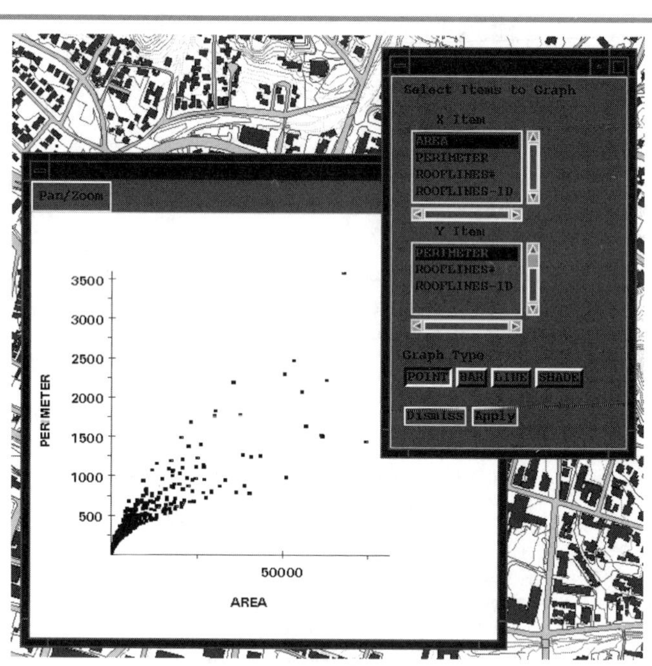

A point chart of the Area and Perimeter items within the Buildings theme.

Bar charts are useful for smaller data sets. They are most effective when the data has an equal interval along the X-axis. Like point charts, bar charts can also indicate the areas where the data is densest, but the example we're using is not well represented by a bar chart.

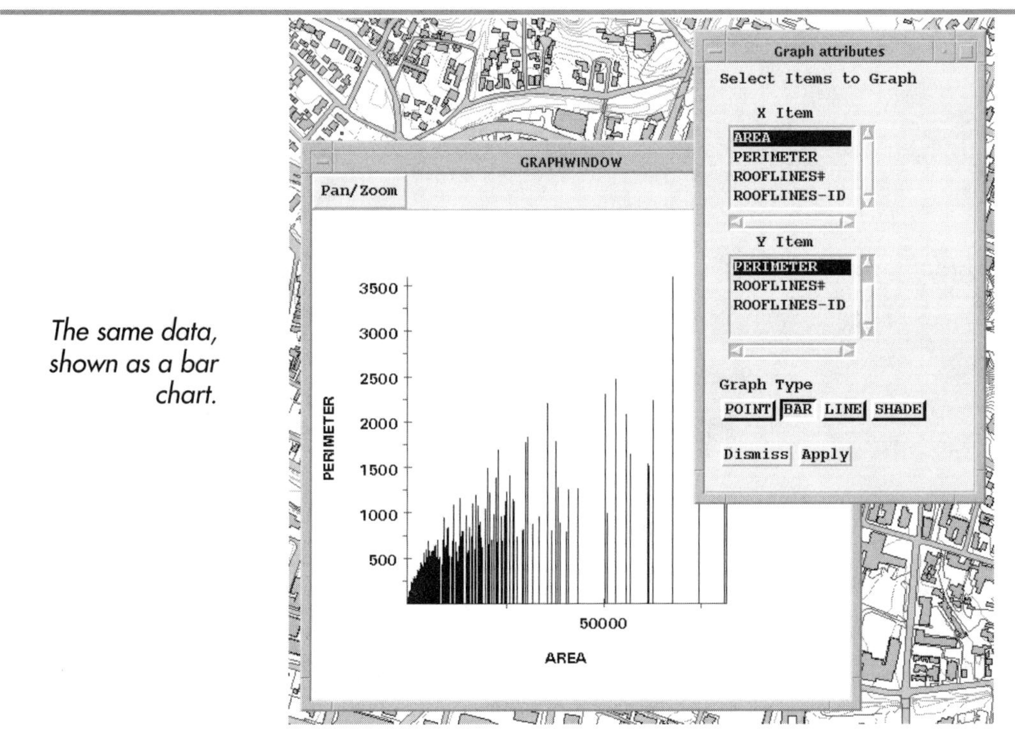

The same data, shown as a bar chart.

A *line chart* has the advantage of showing linear trends and reducing clutter in a chart, but it does not clearly indicate where data is densest. Also, as demonstrated in this illustration, line charts can contain spikes from certain data points.

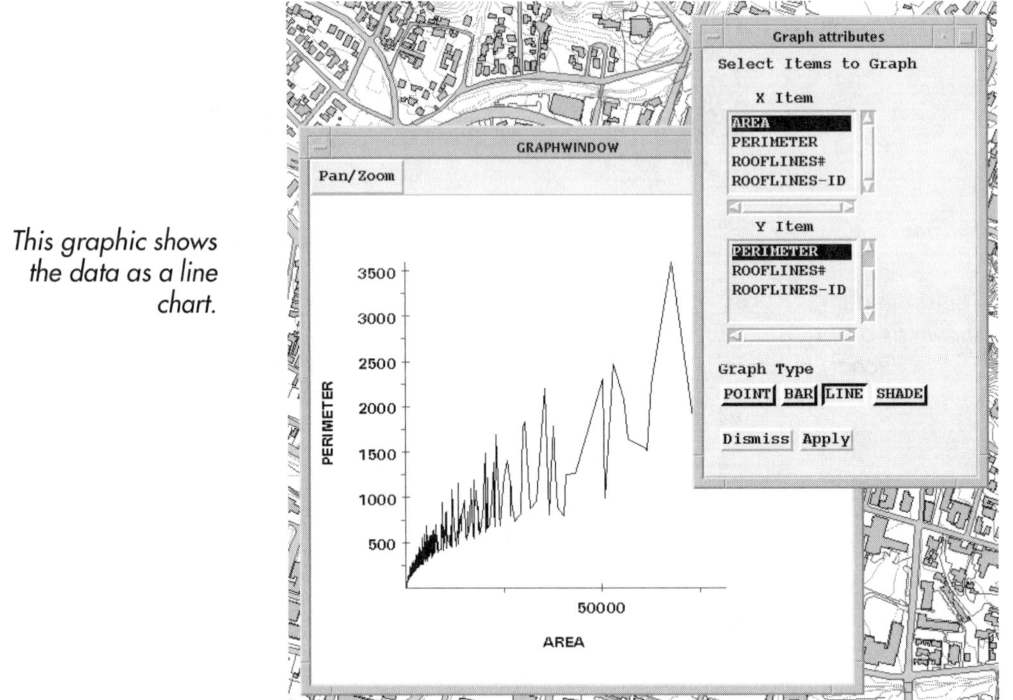

This graphic shows the data as a line chart.

The *area chart* is similar to the line chart, but the distribution pattern stands out more clearly. For this particular example of the AREA item versus the PERIMETER item, clearly the point chart works best. As you use charts, you will learn to select chart types based on the character of the items, their distribution and density, and how strongly the data is correlated.

Lastly, the data shown as a shade (or area) chart.

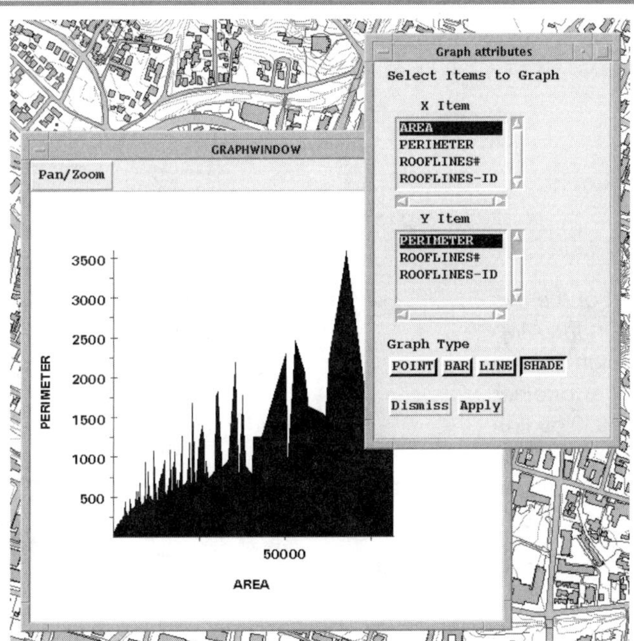

All charts use the characteristics you set in the Text Properties menu, except for size. The Graph Attributes command will calculate the text size appropriate for the size and scale of each chart. This is a great convenience for you, because it is difficult to predict what size text would be right for charts that can vary greatly.

Placing Map Elements

No map is complete without some finishing cartographic elements, such as neat lines, north arrows, title blocks, and title text. The Map Elements menu lets you add many of these map elements to produce a finished map.

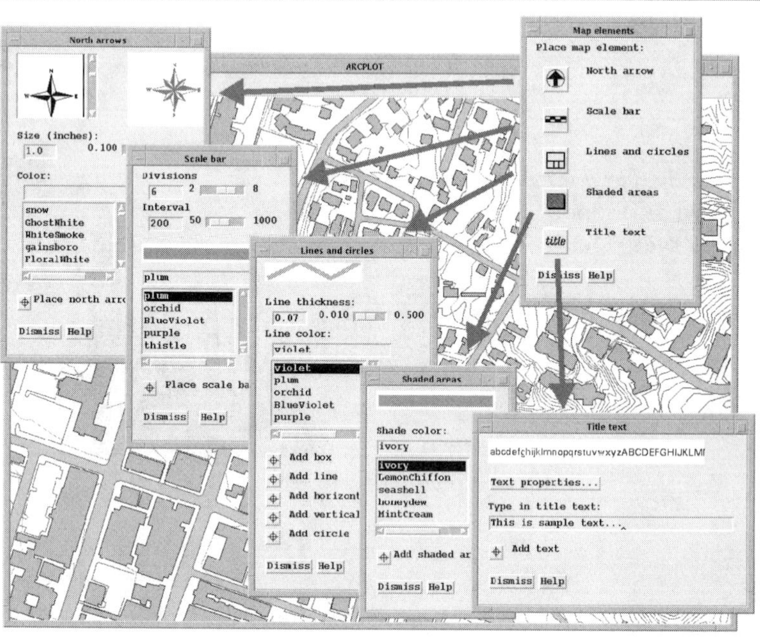

Each of the five icons in the Map Elements menu invokes another menu. You can bring up these menus as you need them.

The Map Elements menu contains five icons. When you press any of these icons, an additional menu will appear, with the same title that appears next to the icon. The five menus are North Arrows, Scale Bar, Lines and Circles, Shaded Areas, and Title Text.

As you use these menus to place map elements in your graphics window, the elements will obscure whatever is beneath them. All map elements in Display GIScity are ephemeral; they vanish whenever you redraw or refresh your graphics window. At any time, you can immedi-

ately restore the normal appearance of the graphics window with the Refresh Window command.

Adding map elements does not change any features or attributes within the GIS databases. This is by design. In a GIS database, you should never add map elements to your GIS database as geographic features. Map elements would disrupt the relations between geographic features. Network traces, or finding features by overlap, would yield erroneous results. At times you need to incorporate some nongeographic features within a map, such as leader lines connecting annotation to a feature. When this is necessary, it is usually best to place them in a separate theme.

In the next chapter, when we create graphics files for output to plotters and printers, we'll see how to record and capture these map elements in hard copy maps.

NOTE: *The map elements in Display GIScity are kept simple because this book is an introduction to ARC/INFO. Look at some quality, commercial GIS applications which support these further enhancements for map elements: snapping to a user-defined grid, saving map elements in a map layout which can be repeatedly used in map production, index maps, which show the entire territory (city, county, utility, or other) with the present map sheet highlighted, automatic neat line and trim line generation, multi-level divisions in the scale bar, and automatic insertion of text from an attribute table within a title block. ArcTools contains some, but not all, of these enhancements.*

We'll now step through each of these menus and create some simple map elements.

Defining Shaded Areas

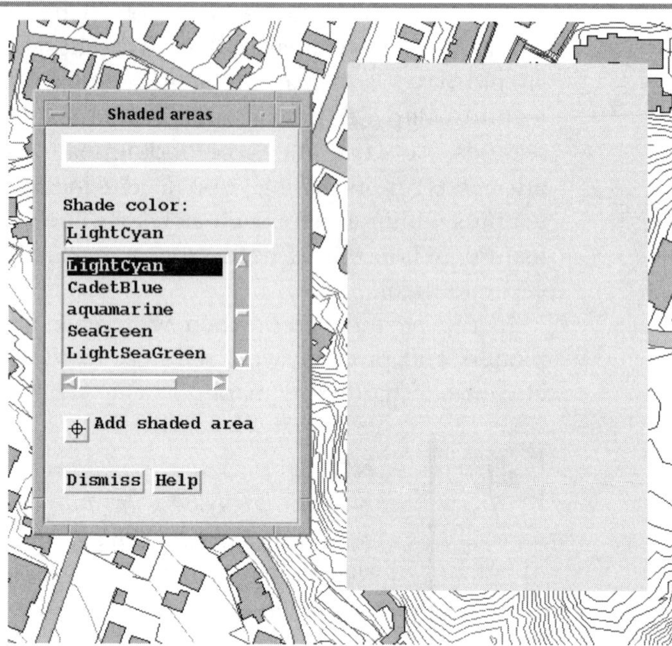

The Shaded Areas menu, invoked from the Map Elements menu.

The logical place to begin is shaded Areas, because other map elements are built upon them. With this menu, you can place a shaded area anywhere within your graphics window.

First, select a color from the data list, or, optionally, type in a color name. When you choose a color, the actual color will appear on the top part of this menu. This gives you a chance to preview the selected color.

Next, click the Add Shaded Area cross-hair icon; your screen cursor is activated to receive two points within your graphics window. Click mouse button 1 twice to mark two opposite corners. Immediately after you click the second point, the marked area will be shaded. If the result is not what you wish, you can restore the screen with the Refresh Window command.

Adding Lines and Circles

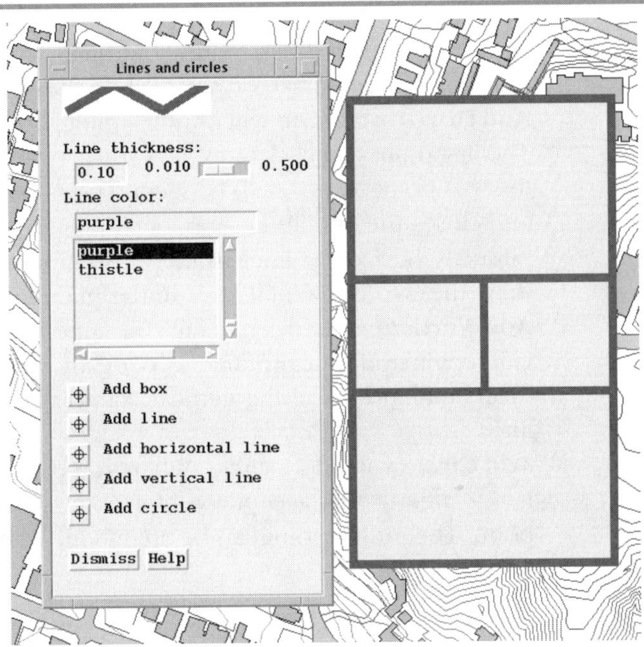

The Lines and Circles menu, invoked from the Map Elements menu.

You can outline and subdivide your shaded area with the commands in this menu.

First, select a line thickness. You can either type in a value directly, or move the slider to the thickness you desire. As you change the thickness, you will see the display of the sample line change.

Next, select a line color. You can type in the color or pick it from the data list. Again, when you pick a color, the display of the sample line at the top of the menu changes. After you have picked these two line characteristics, you are ready to add lines.

There are five commands for adding lines and circles. When you pick the cross-hair icon next to each of these commands, your screen cursor will be activated for you to add that element on the graphics window. **Add Box** will activate the cursor for you to click two opposite corners of a box. In the illustration, this command was used to draw the outline of the shaded area.

Add Line enables you to draw simple or compound lines. When you click this command, all the points you pick with the cursor will be connected by line segments. When you are finished entering lines, press button 9. (Either press mouse button 3 while simultaneously pressing [Ctrl] on the keyboard, or press [9] on the keyboard.)

Add Horizontal Line will draw a simple, horizontal line with two points. The first point you click fixes the vertical position of your line, and defines the start point of the line. The second point uses the same vertical position and closes the line at the horizontal position you choose. The result is a perfectly horizontal line. In the illustration, this command was used to draw the two horizontal lines in the middle of the shaded area.

Add Vertical Line operates in the same fashion as the Add Horizontal Line command, but produces a perfectly vertical line. In the illustration, this command made the vertical line joining the two middle horizontal lines.

Add Circle will add a circle. When you click this command, your cursor is activated to first accept a center point of the circle, and then a radius point. The radius point can be anywhere along the circumference of the circle you want to create.

Adding North Arrows

ARC/INFO comes with some predefined north arrow point symbols. With this menu, you can choose from a variety of north arrows and place them wherever you want. The arrow point marked "N" always points directly up because the GIScity base map is oriented with north at the top.

There are three properties for you to choose. First, you can go through the data list of north arrow symbols and pick a style. As soon as you pick a north arrow symbol, it will appear in a display field to the right of the data list.

Next, you can modify the size of the north arrow. You can type in a new size or use the slider to change the size. As soon as you change the size, the symbol in the display will change.

And you can specify a color for the north arrow. As in many other menus in Display GIScity, you can either type in or select a color from the data list.

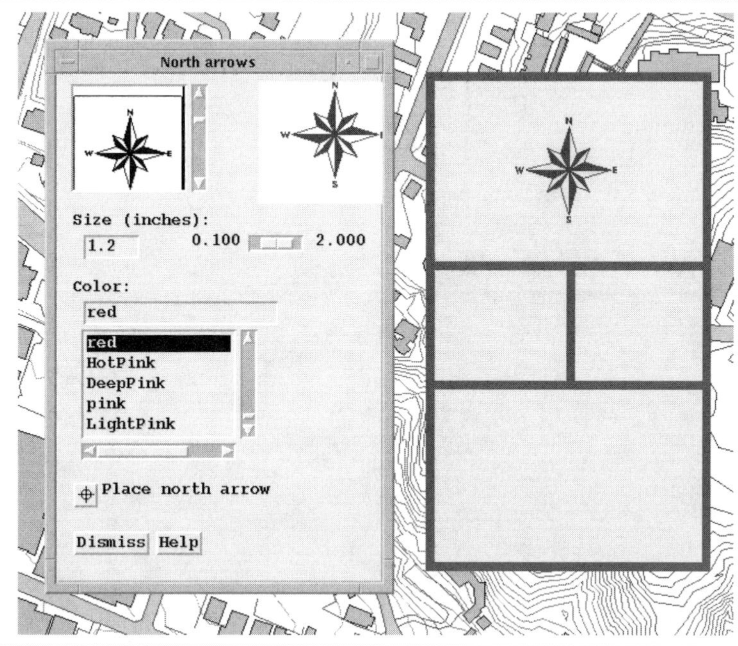

The North Arrows menu, invoked from the Map Elements menu.

Once you have set these three properties for a north arrow, you can pick the Place North Arrow command and pick a point anywhere on your graphics window. The north arrow will be centered at that point.

TIP: *North Arrows can be formed several ways: as a symbol; through a set of map elements; or even through a separate GIS dataset. In Display GIScity, it is quickest to use the symbols ARC/INFO provides. However, in production environments, GIS applications usually have a standard map layout that includes a north arrow which is permanently edited as a set of lines, together with neat lines, trimlines, and title block layout. If you prefer a north arrow other than the ones provided as point symbols in ARC/INFO, it is best to incorporate the north arrow this way in a map layout. By doing so, you will have greater flexibility in creating and refining the appearance of your north arrow.*

NOTE: *Later, we will discuss point symbols in greater detail. Point symbols are cataloged in files called* marker sets*. For your reference, the north arrow symbols you see here are in the marker set called NORTH.MRK.*

Creating Scale Bars

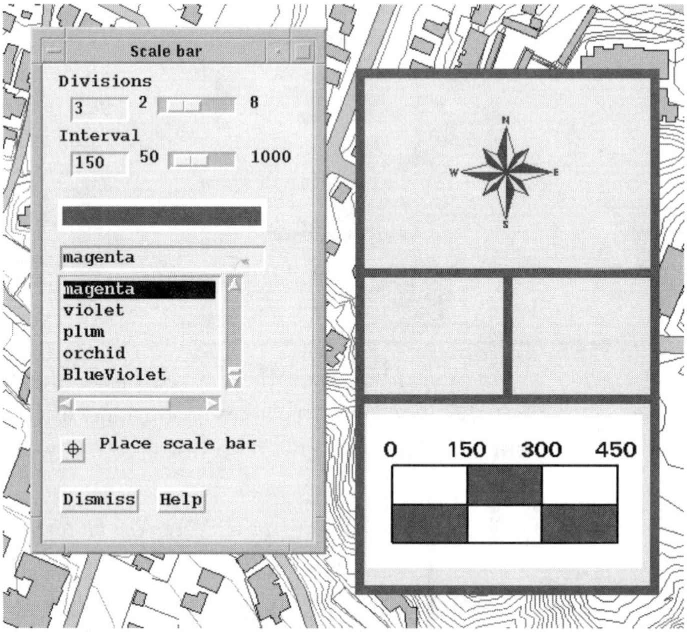

The Scale Bar menu, invoked from the Map Elements menu.

Scale bars are common on maps and are made up of shaded areas, lines, and annotation. Display GIScity makes creating scale bars very easy for you.

First, select the number of Divisions that you want to use in the scale bar. In the illustration, the scale bar has three divisions.

Next, choose an Interval value. This is a distance of feet, in map units. It is the horizontal distance each division of the scale bar will span.

NOTE: *Sorry, Non-Americans. In GIScity, the GIS database has internal units of feet, and so distance units throughout GIScity are in feet. Slowly, the United States government and American industry are adopting the metric system, but the English system will predominate for some time.*

Finally, choose a color. This color will be applied as the shade color in the alternating bars within the scale bar. (In Display GIScity, the line color in scale bars is always black. Black is a good outline color and the menu can be kept simpler that way.)

Once you have set these three properties for scale bars, you can click the Place Scale Bar command. With your cursor, click a lower left corner position for the scale bar. All of the map elements will be drawn automatically for you, and your scale bar will be correct to the scale you are presently using. The annotations above the scale bar will be the correct real-world distances in reference to your graphics window.

This is a visually striking command. If you have the Echo check box turned on in your main menu, you can watch the dialog window to see all of the ARCPLOT commands triggered by this simple Display GIScity command.

NOTE: *The text that appears in your scale bar will inherit all the characteristics you have set in your Text Properties menu, except one: size. The Scale Bar command automatically calculates an appropriate text size for your scale bar, but does not change the size set in the Text Properties menu.*

Adding Title Text

Maps need titles. The Title Text menu lets you place any titles you wish on the graphics window.

At the top of the menu is a display of sample text with the current properties. The text you see here appears the same as it does in the Text Properties menu. If you want to modify these properties, you have a

shortcut in the Title Text menu: simply press the Text Properties... button to invoke that menu.

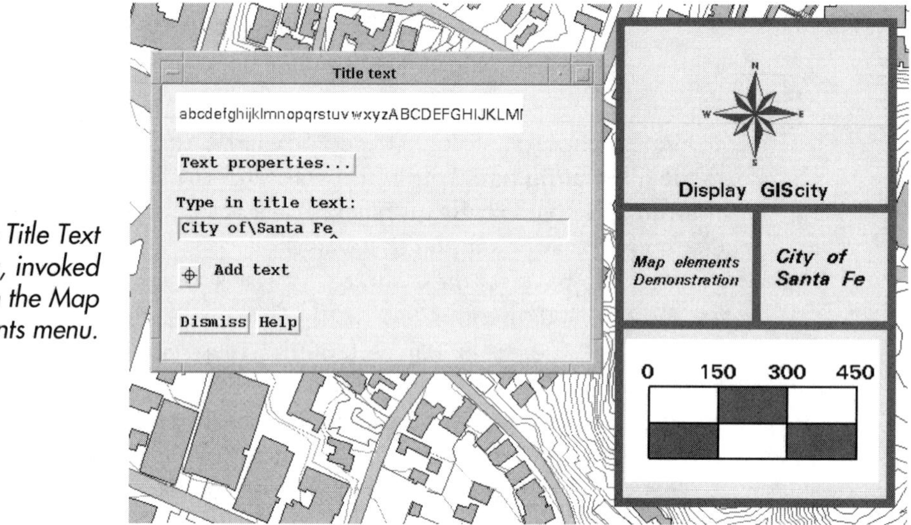

The Title Text menu, invoked from the Map Elements menu.

Next, use the Type in Title Text input field to key in the text you want. When you have finished typing a line, pick the Add Text command, and select any point on the graphics window. The text you typed will appear on the screen with a lower left anchor position.

In between placing separate text items, you can modify text characteristics. In the illustration, the "Display GIScity" text uses the Univers Bold font set. The "City of Santa Fe" text is in Univers Bold Italic. And the text that says "Map elements Demonstration" uses a smaller text size.

If you look carefully at the illustration, you'll notice something curious: the text appears in the text input field as "City of\Santa Fe" and it appears on the graphics window on two separate lines. This is a standard within ARC/INFO. If you're inputting text, wherever you insert a backward slash ARC/INFO will separate the line at that point into two (or more) lines. You can place complete paragraphs on the graphics window in this way.

Exercise 6: Add a Legend and Map Elements to the Graphics Window

1. Add several map legends to your graphics window. Use the Theme Manager to select an active theme. Then click the Theme Legends icon and use that menu to select a background shade and position for each legend.
2. Look around your workplace for a sample map that has map elements you would like to duplicate.
3. Click the Map Elements icon, and from it, choose the Shaded Areas menu to add to your graphics window a shaded area similar to that on your sample map. White is a good color for a shade color. Click the Dismiss button to get back to the Map Elements menu.
4. Next, use the Lines and Circles menu to add a border around the shaded area. Set the Line Thickness slider to a thickness of 0.05 inches and pick a color for the lines. Use the Add Box cross-hair to outline the shaded area. Think a moment about what elements you want to add to the shaded area, and then use the Add Horizontal Line and Add Vertical Line cross-hairs to divide the shaded area into smaller areas.
5. Add a scale bar. Try different values for intervals and divisions. Also, try placing scale bars in a different area of your graphics windows, away from your shaded area. Scale bars take some experimenting to find the right settings.
6. Place a north arrow. Scroll through the list and you'll find several attractive north arrows. Click one and try setting its size to 1.5 inches and its color to darkGray.
7. Finally, add some title text. Go back to the main menu and press the Text Properties icon. Select a different font and click the Dismiss button. Choose the Map Elements icon and then the Title Text icon. Type in the text and choose a position for it with Add Text.
8. Now, try changing properties between different text items, but this time, use the Text Properties... button on the Title Text menu. Try embedding backslashes in your text to make multiple-lined text.

Reporting GIS Information in Display GIScity

Tools and Report Commands in Display ARCPLOT

In addition to its visualization capabilities, a GIS has the ability to serve information through methods common to other database management systems. Examples include query, form update, and reports. It has these capabilities because a GIS is a true database management system, albeit one tailored to managing information through geography.

Reporting information from database management systems comprises two steps: *select* and *report*. First, you select (and optionally sort) the data records of interest, and then you choose tabular formats and headings for a printed report.

ARC/INFO produces standard tabular reports in this way but also transcends other database management systems by offering a fully graphic method for reporting information through geography. In this chapter, we will learn how to prepare a *graphics file* in Display GIScity.

A graphics file is a geographic report from ARC/INFO. It can be saved on disk and output to a plotter or printer at any time. Like a tabular report, a graphics file is not a database, but a snapshot in time of selected GIS information. A graphics file, when printed, is a map.

NOTE: *CAD systems also produce maps of high cartographic quality. If your primary need is for maps that have a relatively constant appearance, a quality CAD system is probably the right tool for you. For example, publishers of national road atlases use CAD with great success.*

However, if you perform analysis with maps and would like to produce a variety of maps from one map database, you cannot escape selecting a GIS for your mapping requirements. In a database management system, you can produce many types of reports from one database. Likewise, in a GIS, you can produce many types of maps from one GIS database.

This chapter will present a suite of general purpose tools for ARCPLOT through several commands for reporting information in Display GIScity.

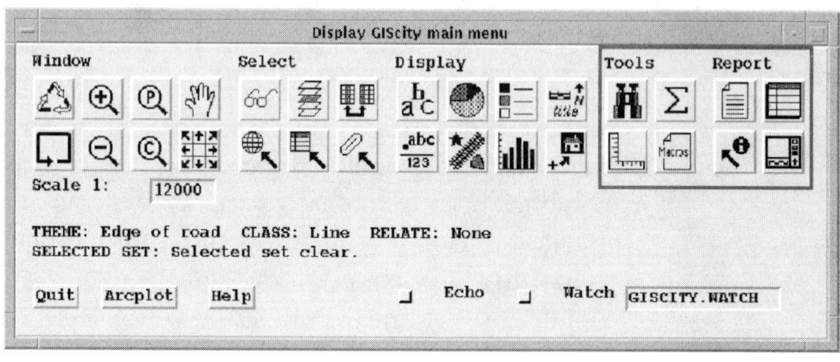

This chapter explores more techniques for presenting information in ARCPLOT through the Tools and Report commands in the Display GIScity main menu.

The Display GIScity Tools Commands

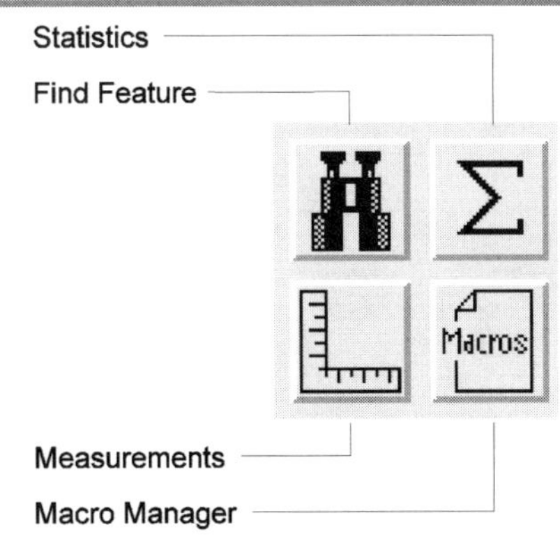

The Tools commands in the Display GIScity main menu.

The Tools commands in Display GIScity provide you with these useful utilities:

- **Find Feature.** This command lets you quickly find any feature within any theme by typing in any characteristic that uniquely identifies that feature.
- **Statistics.** You can easily perform several statistical calculations upon the selected features of your active theme.
- **Measurements.** You can measure coordinate positions, lengths, and areas within your graphics window.
- **User Macros.** You can augment Display GIScity with macros that you make yourself. Chapter 18 provides you with the basic techniques for creating simple macros. This menu is an entry point for you to access them.

The Find Feature Menu

The first Tools command in Display GIScity, Find Features is a quick command for locating any feature in any theme by an item value. As you'll see, it works best with features for which

an attribute value is unique, such as an occupant name or phone number.

Actually, everything this command can do can also be done with other commands in Display GIScity, but not as quickly and easily. Find Feature works like an interesting subset of other commands we've already tried: Select Features Logically and Theme Manager. It has much in common with Identify Feature, which we'll cover later in this chapter.

The Find Feature menu is streamlined to find features, anywhere, by attribute.

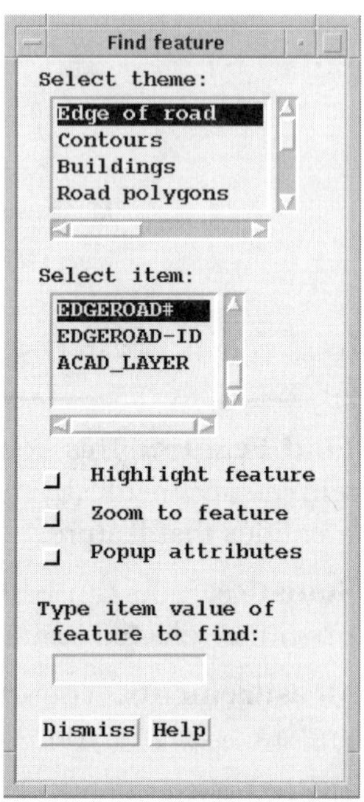

To find a feature, first select a theme and item. When you select a new theme, you'll see the item list change. When you select an item, the scrolling list of item values will be updated with the possible item values for that theme. After you've specified theme item, and item value, you can then select any of three options:

- The **Highlight Feature** check box specifies that the selected feature(s) will be highlighted. It uses the same highlight color you've set in any of the three Select Features menus.

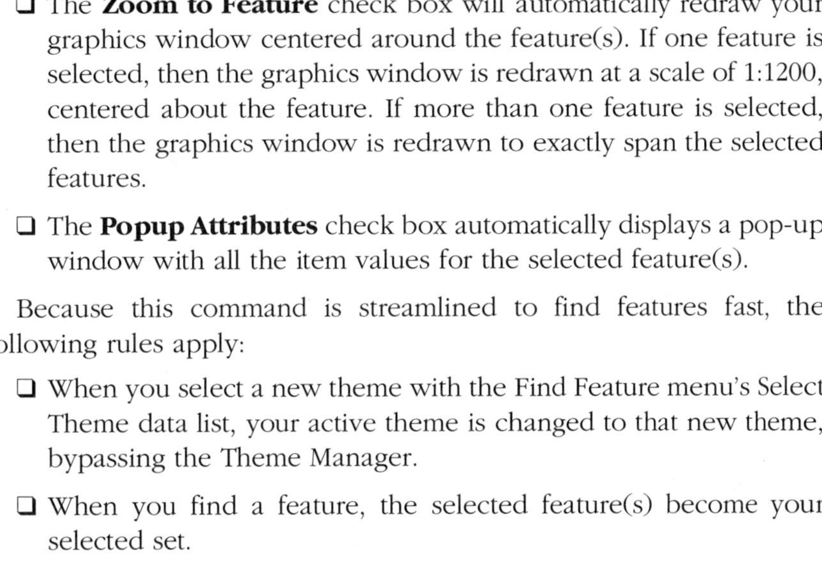

- The **Zoom to Feature** check box will automatically redraw your graphics window centered around the feature(s). If one feature is selected, then the graphics window is redrawn at a scale of 1:1200, centered about the feature. If more than one feature is selected, then the graphics window is redrawn to exactly span the selected features.
- The **Popup Attributes** check box automatically displays a pop-up window with all the item values for the selected feature(s).

Because this command is streamlined to find features fast, the following rules apply:

- When you select a new theme with the Find Feature menu's Select Theme data list, your active theme is changed to that new theme, bypassing the Theme Manager.
- When you find a feature, the selected feature(s) become your selected set.
- If your selection fails, you will see an error message and your selected set will be clear.

Exercise 1: Finding Elevations

1. Click the binoculars icon to display the Find Feature menu. Choose Contours from the Select Theme data list, and choose ELEV from the Select Item data list.
2. Click the Highlight and Popup Attributes check boxes.
3. Now select 7100 in the item value data list and click Apply. Repeat with 7110 and 7120. You can see the direction in which the elevation is rising. Click the Dismiss button.

The Measurements Menu

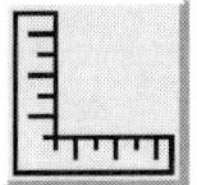

In a GIS, map features are stored in a true map projection coordinate system. As you zoom in and around the GIS display, it is easy to lose a sense of real-world distances. The Measure-

ments commands help orient you to the dimension and coordinates of points within your graphics window.

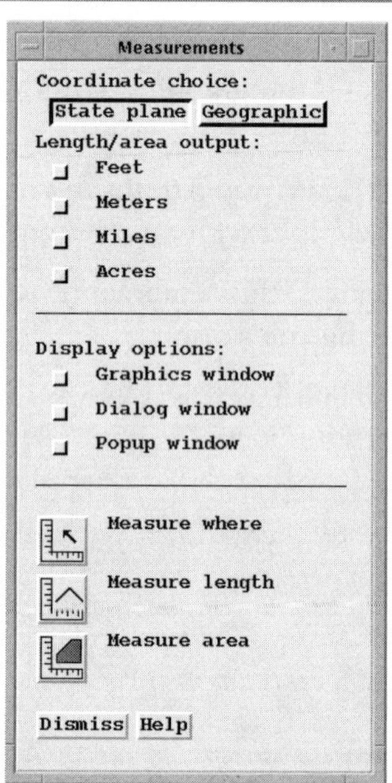

With the Measurements menu, you can perform and present a variety of measurements on your graphics window.

To perform a measurement with this menu, first decide what type of measurement you need and then follow these general steps:

Step 1. Use the Length/Area Output check boxes and the Coordinate Choice buttons to select the measurement units you want Display GIScity to calculate and display:

- For coordinate measurements, use only the State Plane/Geographic choices.
- For length measurements, use the Feet, Meters, and Miles check boxes.
- For area measurements, use the Feet, Meters, Miles, and Acres check boxes. (The area results are in square feet, square meters, square miles, and acres.)

Step 2. Select any (or all) of the Display Options:

You can send measurements to the dialog window. (Make sure that the Echo check box in the main menu is turned off; otherwise the measurements will be lost in the dozens of lines the Display GIScity will execute.)

- ❑ You can send measurements to the graphics window, where they will be displayed as text. The characteristics you set in the Text Properties menu will be used to draw the measurement text in the graphics window.
- ❑ You can display the measurements on a pop-up window. (You will have to Quit the pop-up window before you can return to other menus and commands in Display GIScity.)

Step 3. Choose one of the three measurement commands, Measure Where, Measure Length, or Measure Area.

Step 4. When you choose one of the measurement commands, your screen cursor becomes active. Select as many points in your graphics window as you want. When you are finished, click button 9 (click mouse button 3 while holding down the [Ctrl] key, or press [9] on the keyboard.)

TIP: *The Measurements commands provide you with the positions, lengths, and areas between the arbitrary points you select on the graphics window.*

To measure the exact length and area of existing features as stored in the database, use the Identify Feature command, discussed later in this chapter. For line features, look at the LENGTH item for the exact length of that feature in map units (feet in the GIScity database). For polygon features, look at the PERIMETER and AREA items for those respective measurements (in GIScity, linear feet and square feet).

Using the Measure Area Command

The Measure Area command gives you line segment, perimeter, and area measurements.

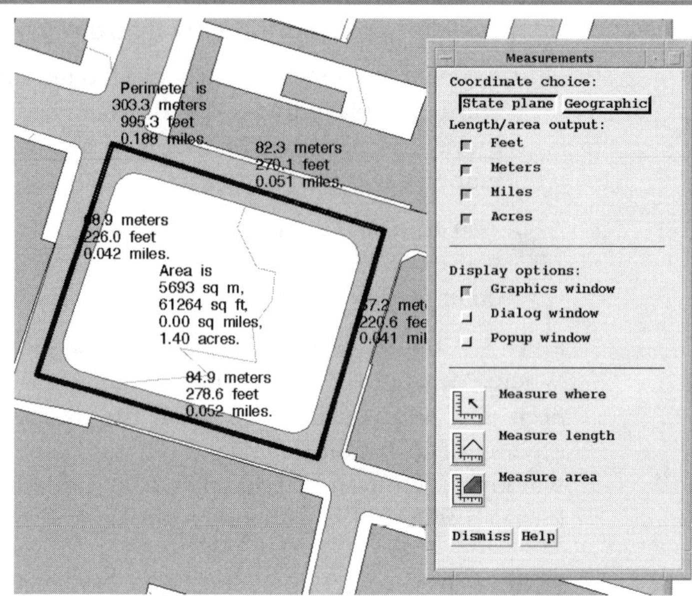

When you use Measure Area, three types of measurements are presented: lengths of each line segment, length of total perimeter, and total area.

Lengths for each line segment and the perimeter are returned to you in feet, meters, and/or miles, as you've specified in the Length/Area Output check boxes. The total area is returned in square feet, square meters, square miles, and/or acres, again as you've specified in the check boxes.

When you use this command, the line segments are automatically drawn for you as you locate points. The line segment lengths are annotated at the midpoint of each segment.

When you want to close the area, do not move the cursor to the beginning point of your area. Instead click mouse button 9 (or key [9])to automatically close the area. When you close the area, the perimeter length is displayed at the starting point of the area. The area measurement is displayed at the center of the area.

TIP: *Some of the text items may overwrite one another when you're sending measurements to the graphics window. You cannot completely eliminate this clutter, but you can mitigate its effect by trying the following: zoom to a higher scale; decrease your text height in the Text Properties menu; limit the Output Measurement units you choose; or choose the Dialog Window or Popup Window Display Options instead.*

Using the Measure Length Command

The Measure Length command gives you line segment and total length measurements.

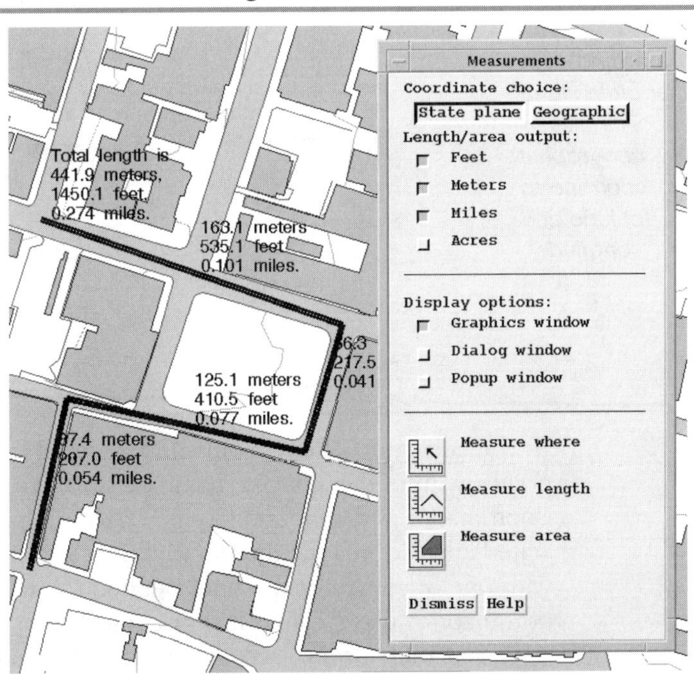

You can measure the total distance between successive points in your graphics window with the Measure Length command.

After you click on this command, your screen cursor is active to accept points. The first point marks the starting point of the total measurement. Beginning with the second point, each line segment will be drawn for you, and the length for that segment is annotated at the midpoint of each segment in the length measurement units you've set. When you

click button 9, the total length measurements are annotated at the endpoint.

Using the Measure Where Command

The Measure Where command gives you coordinates in either State Plane coordinates or geographic coordinates (latitude and longitude).

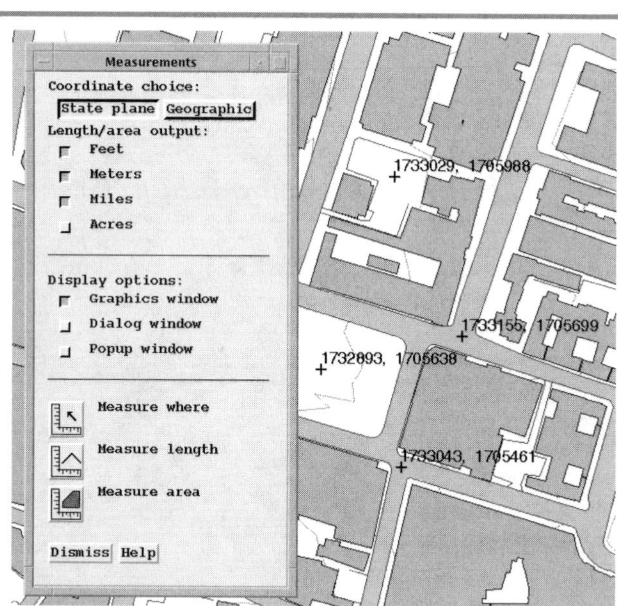

The Measure Where command gives the geographic locations of points within the GIScity database in two formats: the State Plane Coordinate System or Geographic Coordinates (latitude/longitude). The State Plane Coordinate System (SPCS) is an American coordinate system with one or several map zones defined for each of the fifty states. Most of the zones use a cylindrical or conical map projection.

Casual map users don't need to understand the details of geodesy (the science of accurately measuring the Earth) and surveying technology to use the State Plane Coordinate System. Map users in the U.S. gain at least a familiarity with SPCS because most local maps in the United States use that system.

NOTE: *The internal coordinate system in the GIScity database is the New Mexico Central state plane zone, North American Datum of 1983 (NAD83), with units of feet. While this is a typical coordinate definition for American GISs, there is a growing trend toward adoption of meters as the unit of measurement in the United States.*

Before you use this command, choose either the State Plane or Geographic options. When you pick this command, your screen cursor is active. Press button 1 successively in the graphics window. For each point, a crosshair appears and the coordinate is annotated in the choice of coordinates.

Exercise 2: Measuring Santa Fe

1. Use the Zoom In command to zoom to the center part of downtown Santa Fe.
2. Choose the Text Properties menu. Use the Text Size input field or slider to choose a font size between 8 and 16 points.
3. Choose the Measurements menu. Click on the Feet check box in the Length/Area group, and click on the Graphics Window check box in the Display Options group.
4. Click on the Measure Area icon and use the cursor to select the corners of a city block. (Close the last segment of the perimeter by pressing the [9] key.) If you have text items overwriting each other on the graphics window, try the tips mentioned previously for resolving text.
5. Now, zoom out to a scale of 1:800 000. Leave the Length/Area Output setting at feet and pick the Measure Length icon. Pick points on your screen to establish a series of segments. (Press [9] to end the selection mode.) Try measuring single and multiple lengths.
6. Zoom back in and measure some more lengths. You should see quite a difference in the length values.

The Statistics Menu

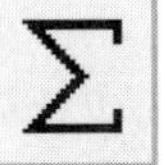

In ARCPLOT, you can perform some basic statistical analyses on your selected set. You can perform means, averages, minimum and maximum values, and standard deviations.

The Statistics command uses two basic modes. You can apply statistics directly to a *summary item* on the selected features in your active theme. Or you can break down statistical summaries into subdivisions by a *case item*.

Suppose you want to run statistics on the lengths of road segments. If you use the summary item mode, the statistics reported will include all road segments in your selected set, regardless of the type of road. If you use the case item mode, your report will be broken down into subdivisions corresponding to whatever division applies to the chosen item. In the following illustration, the case item designation has separated the statistical analysis into sections corresponding to asphalt parking, graded parking, asphalt road, and so on, and reported the statistics separately for each group.

Here's how to get statistics with summaries by case on your selected set:

Step 1. Select an item from the Summary Item list. Because you can perform statistics only on numeric values, this data list will show you only those items that are numeric (integer or real value).

Step 2. Select an item from the Case Item list. Because this data list presents you with every available item, including numeric items, be careful with your selection. You want to chose an item with a only a few values, such as a classification.

Step 3. Check the Perform Analysis On boxes for the statistical results you want to see. The check boxes control which values applicable to the selected set the command calculates and presents the sum of all the selected items, the mean value, minimum value, maximum value, and/or the standard deviation.

Step 4. Once you have specified the statistics you'd like reported, click the Apply button. After some calculations which usually take a few seconds, you will get a pop-up window with the statistics you requested. The Reset button will clear both the case and summary items.

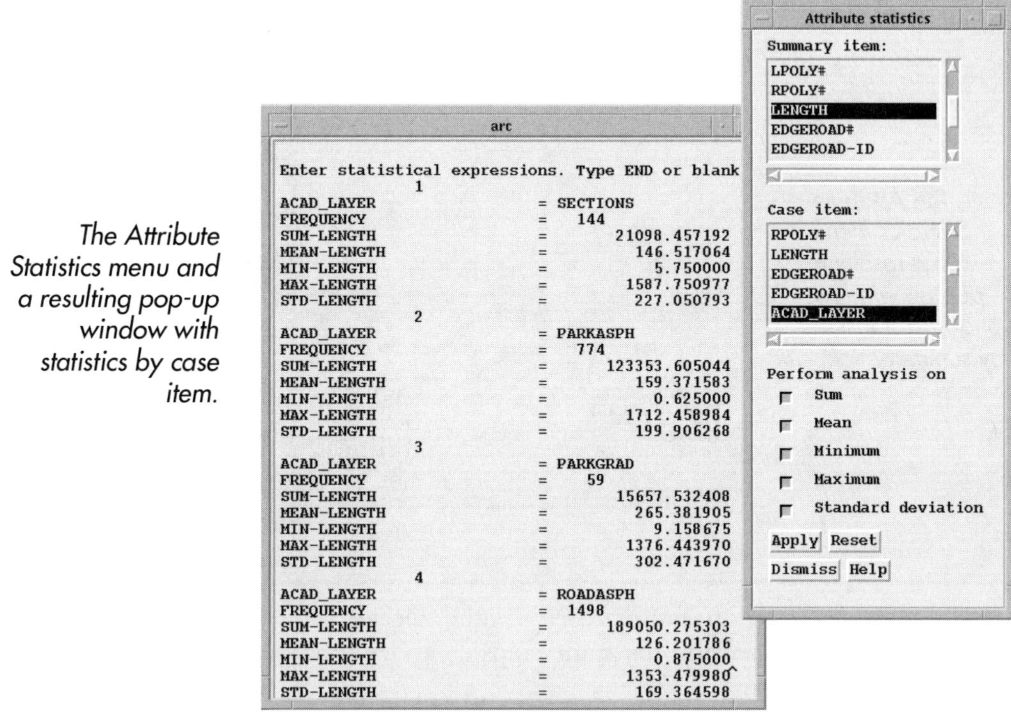

The Attribute Statistics menu and a resulting pop-up window with statistics by case item.

NOTE: *Some of the statistical summaries with case subdivisions that you might require can take quite a bit of time. If the case item has many values, you will end up with that many subdivisions. In general, select only case items that have limited and discrete sets of values. A good example is an item such as ACAD_LAYER in the Edge of Road theme. Performing statistics on items that have many values not only takes much more time, it yields the least interesting statistics.*

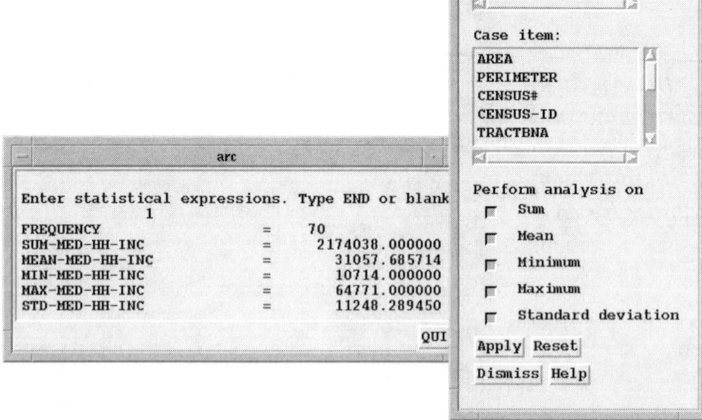

The Attribute Statistics menu with a resulting pop-up window showing statistics by summary item.

To report statistics on the entire selected set, without subdivision into cases, perform the same steps as above, but skip the selection of the case item.

NOTE: *This menu has a Reset button and most other menus do not. In other menus, you can undo many actions or preferences with the buttons or check boxes in that menu. But in the Attribute Statistics menu, once you've selected a case item, you can't deselect it without using the Reset button, or dismissing the menu.*

Exercise 3: Running Statistical Analyses

The Edge of Road and Census themes have the most interesting information for statistical analysis.

1. Make the Edge of Road theme your active theme.

2. Click the Attribute Statistics icon from the Display GIScity main menu and pick LENGTH for the Summary Item. Pick several or all of the check boxes for analysis. Then click the Apply button. You'll see the statistical summaries on the lengths of all the road features.
3. Keeping LENGTH as the Summary Item, click on ACAD_LAYER in the Case Item data list. Click on the Apply button again. You'll now see statistical summaries broken down by type of road.
4. Pick the Select Features Spatially menu. Use the first Mass Select icon to clear all the selected sets, then click the Select in Box icon and select just a few road segments.
5. In the Attribute Statistics menu, click the Reset button, select LENGTH from the Summary Item data list, and click Apply. You'll now see the statistical summaries only on the selected road features.
6. Invoke the Theme Manager and change the active theme to Census. Zoom to a scale of 1:250 000. From the Summary Item data list, pick any of the values for household income, housing value, or rent. Click on the Apply button to perform summary statistics.
7. Now use the Select Features Spatially menu to select only the polygons in the downtown Santa Fe area (where the polygons are smaller and more tightly packed). Pick the Apply button again to perform the same summary statistics calculations, only this time on the limited area. What can you infer from the statistics on the distribution of wealth between the urban and rural areas of Santa Fe county?

The Macro Manager Menu

The introduction to this book states that you can best use ARC/INFO through macros written in the ARC Macro Language (AML). This is because ARC/INFO itself is designed as a geographic toolkit, instead of an off-the-shelf application. While a full discussion of AML is beyond the scope of this book, it's not hard to make simple and useful AML modules.

This menu gives you access within GIScity to macros that you can create yourself.

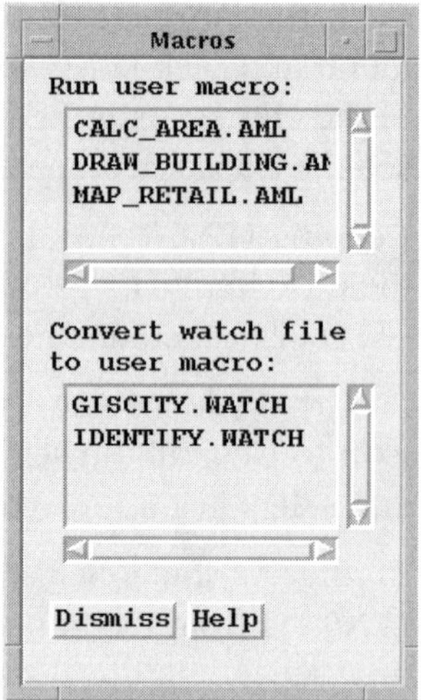

In Chapter 18 of this book, we will learn how to create simple macros in ARC/INFO. If you try making some simple macros for ARCPLOT later in this book, you can quickly execute them with this menu.

For beginning AML programmers, there is a shortcut to writing macros: you can convert *watch files* to macro files. While there are some limitations to look for when using this shortcut, it can be both a useful and instructive technique. We'll reserve a discussion on making simple AML modules for later in the book. You can return to this menu later to try AMLs you've made.

The Display GIScity Report Commands

In the previous chapter, we learned a variety of means for getting ARCPLOT to deliver information to you through visualization. ARCPLOT also supports the delivery of information through these techniques:

- **Direct query.** You can pick and identify any geographic feature in Display GIScity. With this command, a pop-up menu with all attributes for the chosen feature will appear.

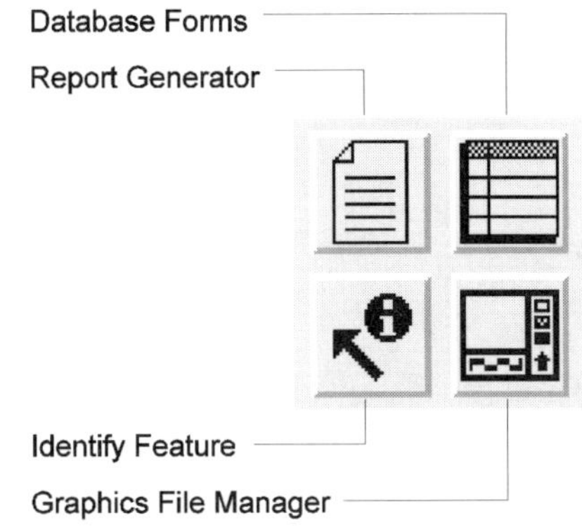

The Report commands in the Display GIScity main menu.

- ❑ **Database forms.** While you cannot edit geographic features in ARCPLOT, you can edit their attributes (item values). With this command, you can easily invoke a database form menu and walk through your selected set, updating attributes.
- ❑ **Reports.** You can create tabular reports from your active theme with this command. While not intended to compete with commercial report generators, this menu is a quick and handy way to make simple reports.
- ❑ **Graphics files.** You can direct any of the map displays you've composed on your graphics window to a graphics file. This file is stored and can be sent to your plotter and printer anytime. This chapter presents an extended discussion of the ways to control the extent, size, scale, and format of graphics files.

The Report Generator Menu

While you purchased ARC/INFO specifically for its rich geographic analysis and visualization tools, you can also use ARCPLOT to produce tabular reports. The Report Generator will make a simple report on the selected features in your active theme. With this menu, you can choose the items you want in your report, and their order of appearance. After you have made the report, you can preview it in a pop-up window, and then send it to your printer.

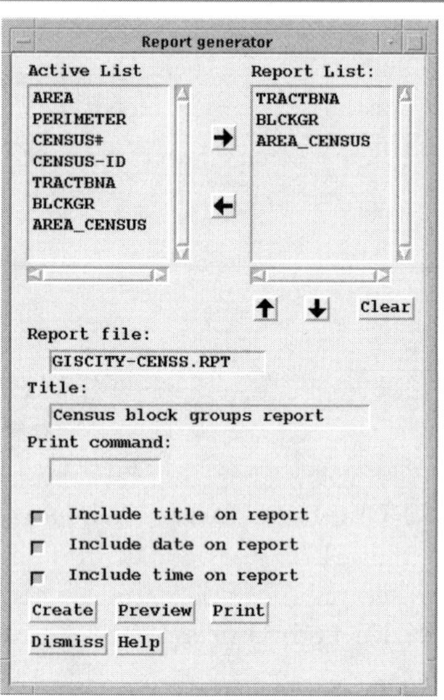

The Report Generator menu gives you a simple and flexible tool to make tabular reports.

NOTE: *Commercial database management systems offer sophisticated techniques to produce tabular reports. These enhancements include page breaks, level breaks, summations, output masks, aliases, calculated columns, and other techniques. An important design philosophy of many GIS applications, including GIScity, is not to attempt to duplicate the functionality of other software tools, but to provide clean links to transfer information to other tools (such as graphics programs, spreadsheets, word processors, and relational database managers).*

If you need to produce elaborate tabular reports, ARC/INFO's Database Integrator technology can produce an extract of selected tables. Another commercial database management system can then take those extracted tables or files and make a professional report. Adhering to that principle, GIScity provides only rudimentary report generation.

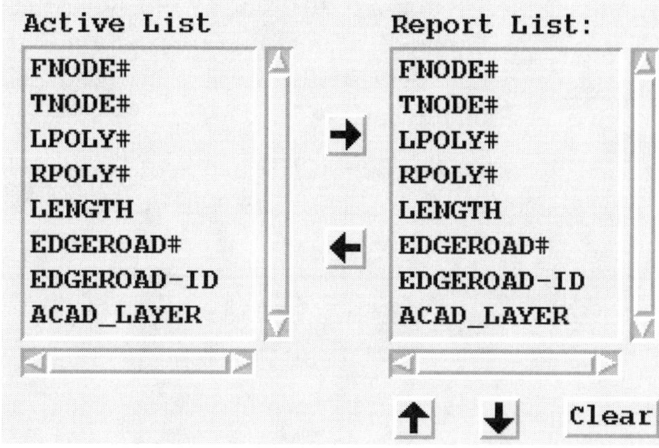

The top portion of the Report Generator menu.

The starting point in making a report is specifying the items that you want to see.

On the left side of the menu is a list of all the items in the attribute table for the active theme class. This list is called the Active List in your menu. If you have an active relate, those relate items will also appear in this list. The Active List cannot be changed; it is just used for picking items to move to the Report List.

On the right side of the menu is the Report List, a list of all items that will be sent to the report, in the order in which they will appear. There are four ways you can change this list: adding an item, removing an item, promoting an item in the list, and demoting an item in the list. Here's how you perform these four actions:

- To add an item to the Report List, select an item in the Active List and then click the arrow icon that points to the right. (If that item is already in the Active List, there will be no change.)
- To remove an item from the Report List, select that item from the Report List and click the arrow icon that points to the left. You will see that item removed from the Report List.
- To promote an item in the Report List (which has the effect of moving that column one position to the left in the report), select that item in the Report List and click the arrow icon pointing upward.

- ❑ To demote an item in the Report List (which has the effect of moving that column one position to the right in the report), select that item in the Report List and click the arrow icon pointing downward.

To clear the Report List and start over, click the Clear button. After this command, the Report List will be empty and ready for you to add items from the Active List.

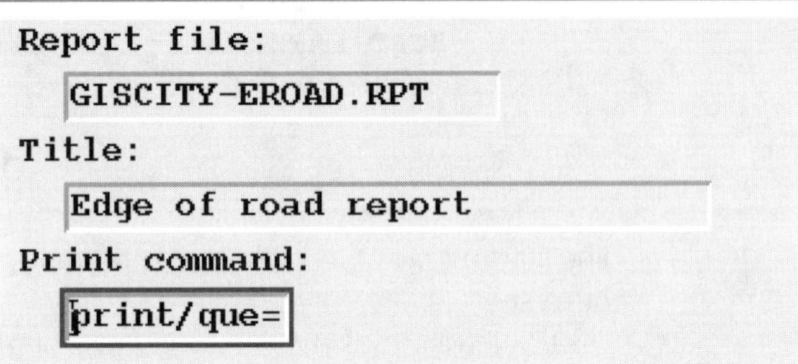

The middle portion of the Report Generator menu.

You have three characteristics to set for your report: its file name, a report title, and instructions for sending that report to whichever printer you have on your computer system.

The Report Generator menu creates a default file name for reports that you produce. That name is "GISCITY-" followed by a theme abbreviation in GIScity and a file extension of .RPT. You can accept this default name, or you can move your cursor to the Report File text input field and type in a new name.

This menu also prepares a default title for you: the descriptive name of your active theme with the word "report." Again, you can accept the default title or type another in the Title text input field.

Lastly, this menu lets you specify the system level command to send this report to your printer. Because you can run GIScity on a variety of workstation makes and operating systems, this menu does not try to predict your system command for printing text files.

NOTE: *If you are not certain of the correct print command for your system, consult with your system administrator. On UNIX systems, that command may start with "lp". On VMS systems, that command may start with "PRINT/QUEUE=." Typing this print command is only necessary if you want to print it to a line printer or laser printer.*

The lower portion of the Report Generator menu.

Include title on report
Include date on report
Include time on report
Create Preview Print
Dismiss Help

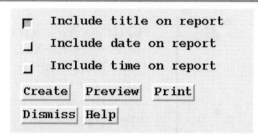

You have three additional options for your report.

- You can add the report title to the top of your report by enabling the Include Title on Report check box. By default, this check box is enabled.
- You can have today's date printed below the title by enabling the Include Date on Report check box. By default, this check box is disabled.
- You can include the current time (with an am or pm suffix) by enabling the Include Time on Report check box. By default, this check box is disabled.

On the bottom of the menu, you have five buttons. The bottom two buttons, Dismiss and Help, work the same as they do in every other menu.

- The Create button will generate the report with all the items in your Report List. This report will become a text file on your computer disk, using the file name you've selected, the report title, and other report options.

- The Preview button will present the new report in a pop-up window so that you can preview it before sending it to a printer. (You need to create a report before you can preview it.)
- The Print button will send the report file to the printer you specified with the Print Command text input field. If you have not entered a correct Print Command for your system, you will see a system error message in your dialog window.

NOTE: *For this Report Generator, it is best to keep the list of items short enough that its total length is 80 characters or less. If the sum of the item lengths (with single space separators) is greater than 80 characters, the underlying ARCPLOT command (the LIST command) will automatically reorganize the report so that only one item value appears on each line of the report. While this will present every item value you asked for, it makes for a very long report listing!*

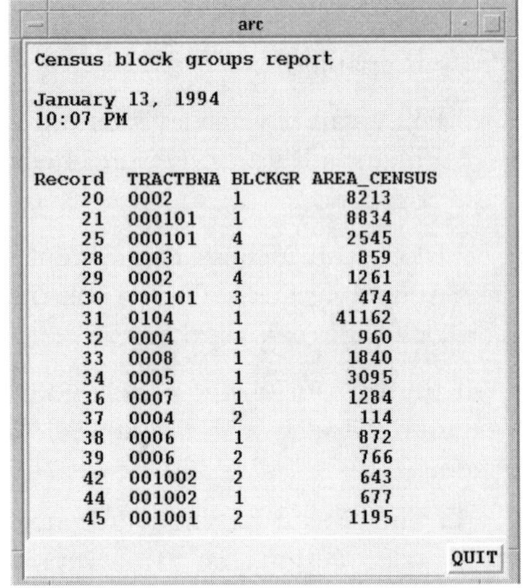

arc

Census block groups report

January 13, 1994
10:07 PM

Record	TRACTBNA	BLCKGR	AREA_CENSUS
20	0002	1	8213
21	000101	1	8834
25	000101	4	2545
28	0003	1	859
29	0002	4	1261
30	000101	3	474
31	0104	1	41162
32	0004	1	960
33	0008	1	1840
34	0005	1	3095
36	0007	1	1284
37	0004	2	114
38	0006	1	872
39	0006	2	766
42	001002	2	643
44	001002	1	677
45	001001	2	1195

QUIT

A sample report from the Census theme, previewed in a pop-up window.

Exercise 4: Generating a Report

We will go through all the steps to create a report.

1. Use the Theme Manager to choose Census as your active theme.
2. Invoke the Report Generator menu. Select the TRACTBNA item from the Active List and click the right arrow icon. Select the AREA_CENSUS item from the Active List and click the right arrow icon. Select the BLCKGR item from the Active List and click the right arrow icon.
3. Now, select the BLCKGR item in the Report List and click the upward arrow icon. See how BLCKGR and AREA_CENSUS are switched in the Report List.
4. Click the Create button, and then click the Preview button. You should see a report very similar to the preceding illustration.

The Identify Feature Menu

This menu lets you quickly select any feature and display all attributes for that feature.

To invoke a listing of all attributes for any feature, first determine the theme a feature is contained within. If necessary, use the Theme Manager to change the active theme.

Next, click the Identify Feature icon. From the menu, just select the cross-hair icon, move your screen cursor within the graphics window, and select a feature by clicking the left mouse button. When you've selected a feature, a pop-up window will appear displaying all the values for the attribute table for that theme.

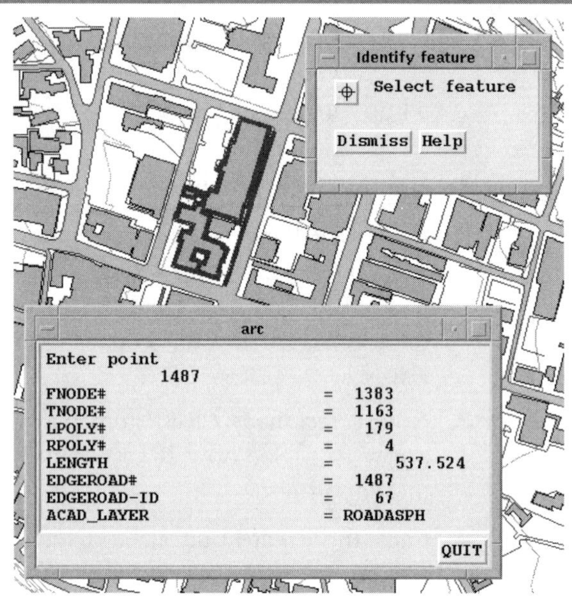

The Identify Feature command highlights the selected feature and reports all attributes in a pop-up window.

NOTE: *If you have trouble selecting a feature, try the following: First, try selection again, with greater care to align the cursor precisely on the feature. Next, make sure the feature is within your active theme. You can verify this by turning on only that theme in the View Manager and redrawing the screen. If the features remains, it is in the current theme. Failing those two steps, you can try increasing the Search Tolerance setting in the Select Features Spatially menu.*

Exercise 5: Identifying Contours

1. Use the main menu Scale box to zoom to a scale of less than 1:48000 within Downtown Santa Fe.
2. Use the Theme Manager to set your active theme to Contours. Click on the Identify Feature icon.

3. Click on the cross-hair icon and then click on a contour line in the graphics window. You'll see the pop-up window appear with attributes for that line. Click on the Quit button.

4. Click on several other contour lines and note how the ELEV and SYMBOL values change in the pop-up windows. Pick the Dismiss button on the Identify Features menu.

Updating Item Values with the Database Forms Menu

The Form menu lets you step through selected features and update their values. This example is for the Census theme in Display GIScity.

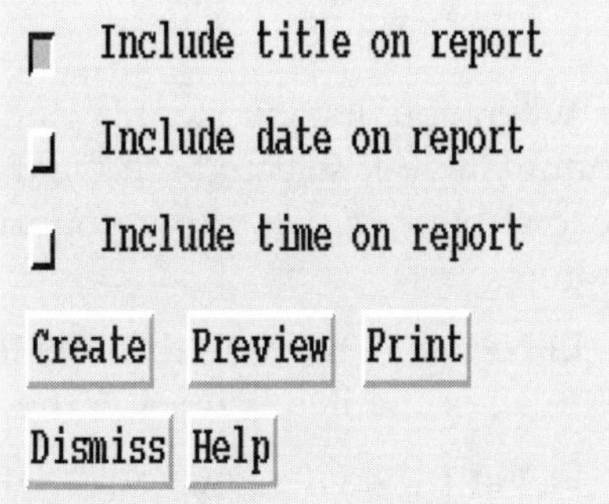

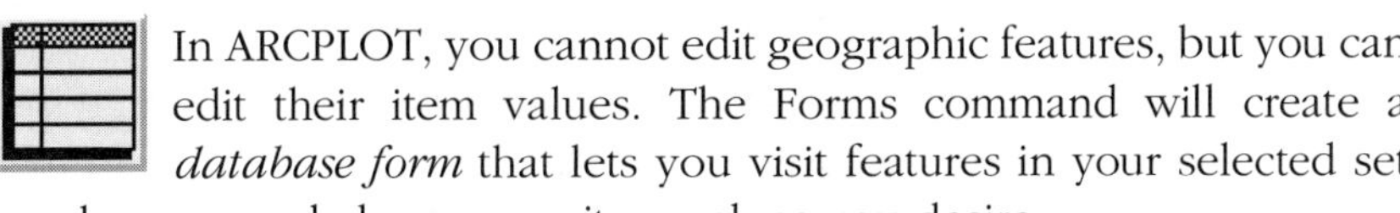

In ARCPLOT, you cannot edit geographic features, but you can edit their item values. The Forms command will create a *database form* that lets you visit features in your selected set one by one, and change any item values you desire.

Before you start this command, take a look at your active theme and selected set.

When you click the Form command, ARCPLOT will automatically construct a custom form menu for you with exactly the items that exist in your active theme. This may take a bit of time; when done, the menu will appear over your graphics window.

Some item values cannot be modified. These are items that ARCPLOT manages internally for you. In the illustration above, these items are AREA, PERIMETER, and CENSUS#.

For polygon themes, ARC/INFO automatically creates, calculates, and manages two items, AREA and PERIMETER. Those items are self-describing and are calculated without your intervention. For line themes,

ARC/INFO also creates and stores a LENGTH item, which is likewise calculated without your intervention.

The next four items have text input fields associated with them. These are items you can modify. In the illustration, these are CENSUS-ID, TRACTBNA, BLCKGR, and AREA_CENSUS. To modify any of these values, just move your screen cursor inside the text input field and type in a new value.

WARNING: *Be very judicious in the use of this menu. When you modify any item values, those changes take place immediately in the theme table. They cannot be undone.*

When you first invoke this menu, you will see item values for the first feature in your selected set. On the bottom of the Form menu, there are several buttons that allow you to navigate among menus for different features.

- ❑ **Next** will move to the next feature in the selected set. You will see all of the item values in the menu change immediately.
- ❑ **First** will take you back to the first feature in the selected set. This is useful when you suspect you've made a mistake and want to revisit the previous selected features.
- ❑ **Who** will flash the current feature for you on the screen.
- ❑ **Reselect Box** will let you narrow your selected set by selecting features in a box. This works just like the Select Box command in the Select Features Spatially menu with the Subset choice.
- ❑ **Cancel** will dismiss this menu.
- ❑ A **Relate** button will appear if you have an active relate. If you click this button, another page of items will appear, with all the relate items available for your inspection and updating.

If your theme table has more than 20 items, you will see a button called Page2. Selecting this button gives you another page of item values to inspect. (None of the themes in GIScity exceeds 20 items.)

NOTE: *Forms navigates through your attribute tables through the use of cursors. This is a built-in capability within ARC/INFO that lets you walk through a selected set. With a cursor, you can start with the first feature in the selected set and then step through successive features until you reach the end of the selected set. For each feature reached through the cursor, you can inspect and modify all item values for that feature. We'll cover cursors in detail later in this book. ARC/INFO cursors are very similar to cursors in commercial database management systems.*

TIP: *For aspiring AML programmers, analyzing how the Database Forms command works is very interesting. Turn the Echo check box on and execute this command. This command comprises a set of macros that construct an AML menu on demand. If you study the watch file, source macro code, and resulting menu, you can learn a lot about cursors, form menus, and updating item values.*

The Graphics File Manager Menu

A *graphics file* is a report from the GIS database. It is a graphical snapshot of any portion of the area spanned by your GIS database. When a graphics file is printed or plotted, it is a map.

An important attribute, and powerful advantage, of a GIS is the versatility and variety of maps that you can produce from one set of GIS themes. As you become proficient with ARC/INFO, you will be struck by your flexibility in creating maps that present any information present in the themes or relate tables.

Maps produced with CAD software are relatively static. While you can turn layers off and on in a CAD map, you don't have the richness of the GIS toolkit to portray information graphically.

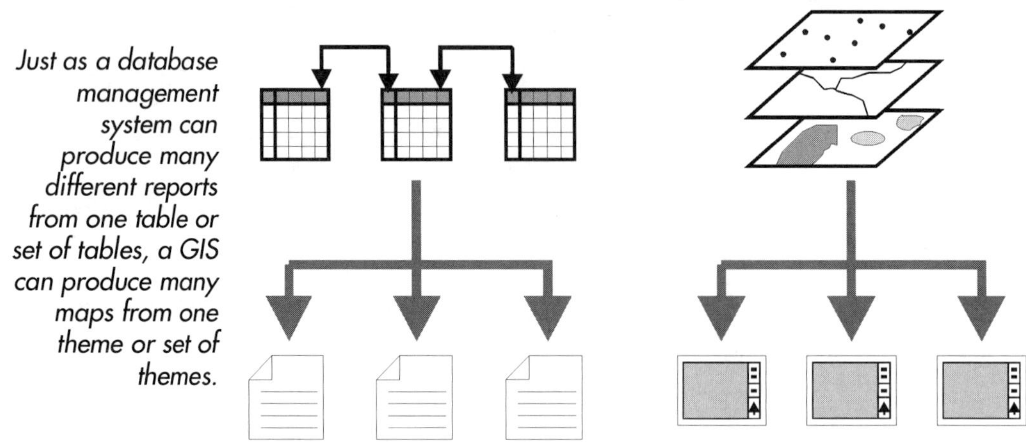

Just as a database management system can produce many different reports from one table or set of tables, a GIS can produce many maps from one theme or set of themes.

Any and all of the information you can visualize in Display GIScity, you can direct to a graphics file. This includes turning themes off and on, adding map elements, symbolizing features by attribute, and drawing text from attributes.

A graphics file can also be sent to every plotter in common usage. Graphics files can be sent to raster output devices (electrostatic and ink jet plotters and laser printers) and vector output devices (pen plotters).

The format of a graphics file is independent of the type of plotter it is sent to, or the type of computer workstation you used to create it. Graphics files can be transferred from one type of computer system to another, and output to a panoply of graphics devices.

Another advantage of graphics files over similar files in HPGL or Calcomp binary formats is that graphics files are, by design, more compact in size. Graphics files can be copied from one make of computer to another without losing any graphical intelligence. Graphics files are platform-independent.

Lastly, you can alternately produce Encapsulated PostScript, Adobe Illustrator, or CGM formatted files instead of the ARC/INFO graphics file format. Many ARC/INFO users output graphics to these formats for final touchup in other popular graphics programs.

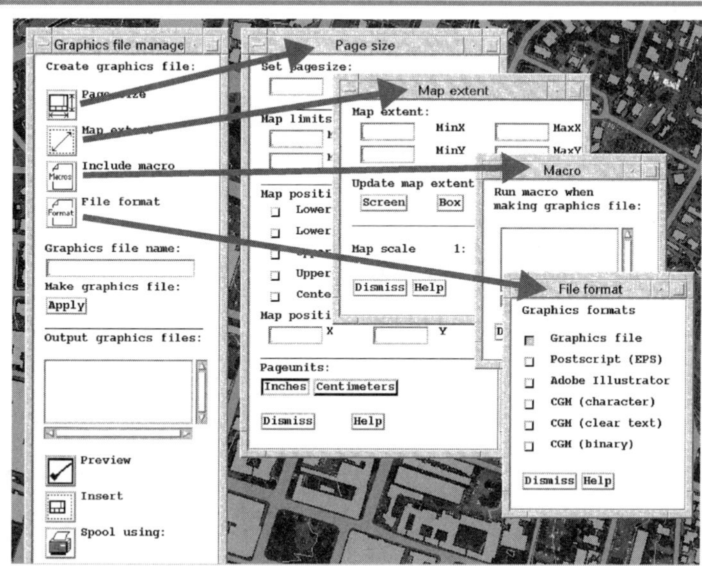

The Graphics File Manager provides you with many ways to specify the definition and output of graphics files. Four additional menus are invoked by the first four icons in this menu.

The **Page Size menu** specifies the physical dimensions of the output medium (paper, mylar, vellum, or other) you will use for your graphics file.

The **Map Extent menu** controls which portion of the GIScity database you will output.

The **Include Macro menu** lets you add an AML module that you create into the generation of a graphics file. This is how you can include the map visualization commands into a graphics file.

The **File Format menu** gives additional choices of formats to output your graphics. For almost any plotter, keep the default graphics file format as the selected format. For popular desktop publishing and graphics programs on IBM-compatible PCs or Macintoshes, select another format that your software can import.

All of these menus have default values. In fact, to produce a graphics file, you don't have to invoke any of them. The defaults set within these menus are as follows:

- The Page Size is set to 24 inches wide by 18 inches long. The map limits (area left blank on the page) is a 2-inch border all around the sheet. The map will be centered on the sheet.
- The default Map Extent is exactly the map extent you are currently zoomed into in Display GIScity. The map scale is adjusted so that

the map extent in the graphics window fills an area of 20 inches by 14 inches.

- By default, no Macros are run when making a graphics file.
- The default File Format produces an ARC/INFO graphics file.

When you want to modify these default values, just invoke the appropriate menu. The two menus you'll use most often are Page Size and Map Extent.

Specifying the Page Size

The Page Size icon in the Graphics File Manager invokes this menu to manage the page size of your graphics file.

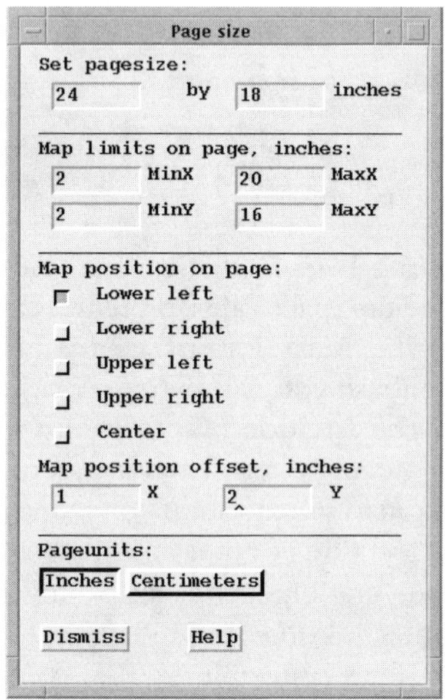

The first step in setting up a graphics file is to specify the actual size of the medium, paper or otherwise, on which you are plotting.

The Set Pagesize text input fields let you set the page size by typing in a width followed by a height.

The Map Limits on Page text input fields specify that portion of the page which will contain graphics. Nearly all maps have a margin of blank space around the edges. The map limit can be used to specify this

standard margin, or to further limit the area of the page available for drawing geographic features.

Once you've typed in a new page size, a default 2-inch margin is calculated within the page size to determine a new map limit. If you set a page size of 36 by 24, your default map limit will be 32 by 20 (2 inches off left and right edges = 36 - 4 = 32; etc.).

There are five choices for Map Position on Page: lower left, lower right, upper left, upper right, and center. With each of these choices, you can optionally enter a Map Position Offset value.

If you enter a map position offset, the map is shifted by a further distance in the X and Y directions. A map position offset is usually entered with the lower left map position.

Finally, you can toggle between inches and centimeters for Pageunits to measure page size. All further calculations will take into account which page size units you've set.

Setting Map Extents

The Map Extent menu is invoked by the second icon in the Graphics File Manager.

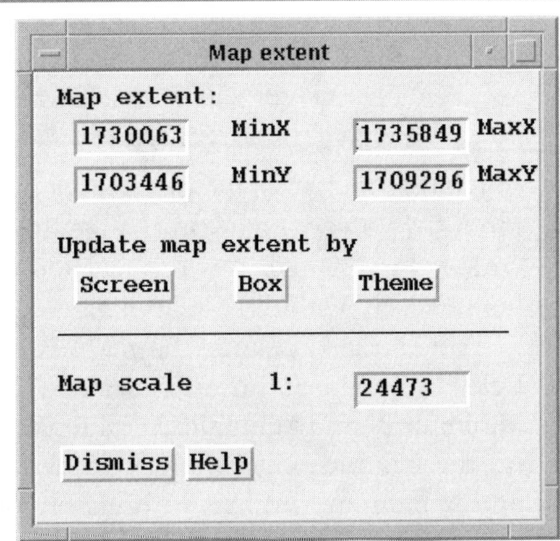

Once you've set a page size and orientation, you need to select which part of the GIS database you want to send to the map. A *map extent* is an imaginary rectangle that defines an area on the map that you want to plot. Every time you zoomed in or out with a Window command in the Display GIScity main menu, you changed the map extent.

Map extents are expressed in the real world coordinate system of the GIS database. In the Display GIScity database, those coordinates are in the New Mexico Central state plane zone.

When you first invoke this menu, you'll see a map extent predefined for you. It actually is the current map extent you've set in Display GIScity. You can accept that map extent and move on, or you can change the map extent in the following ways:

- You can type new values in the four Map Extent text input fields. This method controls exactly the range of the map you want to print, and using overlayed graticules, you can make a very neat map.
- By just clicking the Screen button, you can set the map extent to match exactly the map extent in your graphics window.
- You can click on Box to interactively define a map extent by a box displayed in your graphics window.
- You can click on Theme to specify a map extent that spans the active theme. If your selected set is clear, the extent will include the entire theme. If you have selected features in the active theme, the map extent will be the same as a bounding box defined around the selected features.

In the bottom half of the map extent menu, you will see a Map Scale text input field. Every time you change the map extent, you will see the map scale value change. It is recalculated each time so that the map extent fits precisely within the map limit area you defined.

You can also change the map extent with the Map Scale text input field, by typing in an exact scale. This menu will use the Map Extent Minimum X and Minimum Y values you've already set (which, in effect, set the left and bottom edges of the map) and update the Maximum X and Maximum Y values to precisely fit the Scale value and Map Limit area. For production mapping, this fifth option is the most useful method for setting a map extent because it can produce maps at an exact size and scale.

NOTE: *Most of the time, the relative height and width dimensions of your map extent won't match the relative height and width dimensions of your page size. These relative dimensions are called aspect ratios. For example, the size of your map extent can be 5000 feet by 6000 feet (an aspect ratio of 5:6), and your page size can be 20 inches by 30 inches (an aspect ratio of 2:3). Regardless of which option you choose to set a map extent, the Map Extent menu will automatically adjust your maximum X and Y values so that the two aspect ratios match. The page size and map limits will not change.*

Including Macros

The menu is invoked by clicking "Include Macro" in the Graphics File Manager menu.

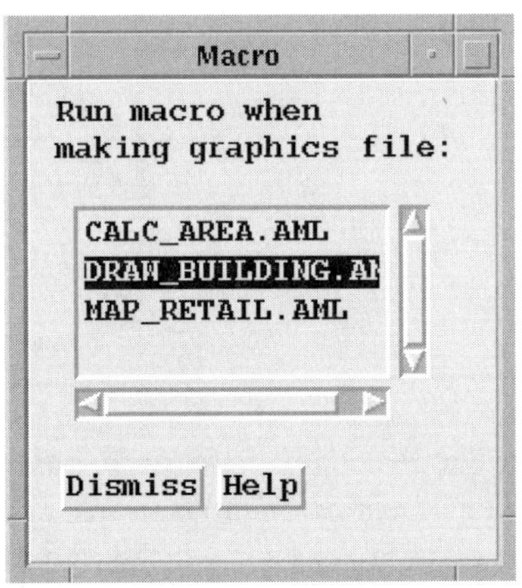

Any AML module you create can be added in the course of building a graphics file.

Remember, most of the Display commands are ephemeral; as soon as you redraw the graphics window, they disappear. Because the graphical elements from the Display commands discussed in the last chapter are not permanently stored in a GIScity theme, we need a way to get them into a graphics file.

The Macro menu provides that way. The graphics are not stored in the macro, but the ARCPLOT commands to draw them are. Again, refer to Chapter 18 for instructions on creating simple macros with the Arc Macro Language.

TIP: *After you've read about making simple macros, here are three rules for successfully creating macros to incorporate into graphics files. (1) Do not invoke any menus while recording a macro. (2) Keep the recording session short and just perform the visualization commands whose effects you want to record. (3) Do not perform any Windows commands.*

Selecting a File Format

The File Format menu presents six choices for output file formats.

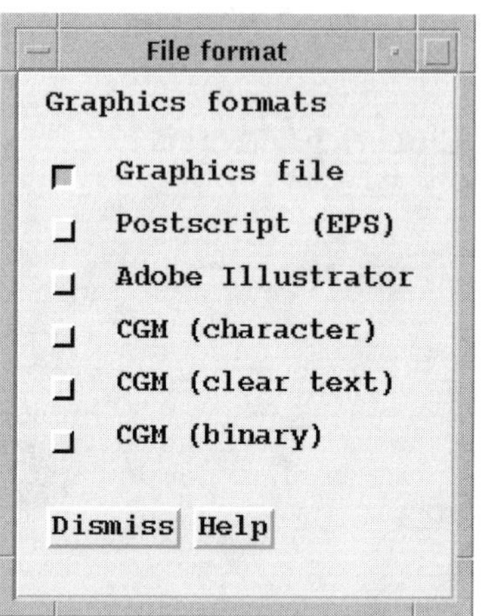

The File Format menu is simple to use, and most often, you'll just accept the Graphics File default, which is the ARC/INFO graphics file format.

Options are provided for using other graphics programs. The Encapsulated PostScript (EPS) format is a very common page format understood by many laser printers. The Adobe Illustrator format is a common standard used by many desktop publishing and graphics programs; it is easier to work with than EPS. The CGM format (constructive geometry metafile) has three different flavors, and is also widely accepted by graphics programs. You should be able to simply copy these alternate graphics files formats from a UNIX or VMS workstation to an IBM-compatible PC or Macintosh.

These six check boxes are mutually exclusive. When you choose one, all the other file format choices are disabled. (These six check boxes are implemented to work like the choice widget.)

Producing a Graphics File

Once you've used the five menus that specify settings for a graphics file, to create the file, just do the following:

Step 1. Type a name in the Graphics File Name text input field. Use only the characters that are valid for file names in your system, such as "A" to "Z", "1" to "9", "-", and "_". Don't use funny characters like "@", "\", or "&". Also, do not type in a file extension. The Graphics File Manager will automatically provide a file extension for you. (The file extension for graphics files is .GRA.)

Step 2. Next, just click the Apply button. You will see your graphics window temporarily disappear. Making a graphics file may take from a few seconds to several minutes. When it is done, you will see the new graphics file become available to you in the Output Graphics Files list.

The lower part of the Graphics File Manager lets you preview and output graphics. The three icons perform commands directly.

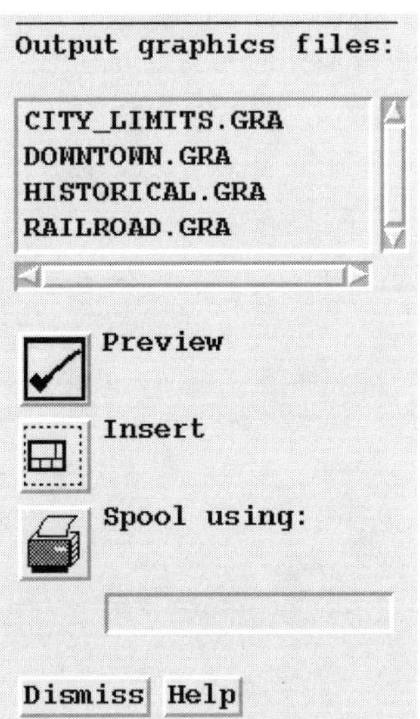

Once you've made some graphics files, you can view them or send them to a plotter or printer with the three output commands in the Graphics File Manager:

The **Preview** button will draw the graphics file you've selected in the Output Graphics File list in your graphics window. When you are ready

to return to your graphics window, use any Window command in the Display GIScity main menu.

The **Insert** button adds the graphics file element to your graphics display by activating your screen cursor to accept two points in the graphics window. Immediately after you click the second point, the graphics file will be drawn to fit between those two points.

Lastly, with the **Spool Using** button and text input field, you can spool your graphics file to a printer or plotter. On your system, you may have a plot queue set up to easily send graphics files to your plotter. Consult with your system administrator on plot queues. If the administrator pleads ignorance, bring along the ARC/INFO system dependencies manual.

You can also send a graphics file directly to your plotter or printer. You can send graphics files to the plotter with ARCPLOT commands like CALCOMP and HPGL. You can type those plotter commands in this text input field, but this method has the disadvantage of tying up your Display GIScity session until the plot is done.

Part III

ARCEDIT

UNIVERSITY OF GLAMORGAN
PRIFYSGOL MORGANNWG
Learning Resources Centre

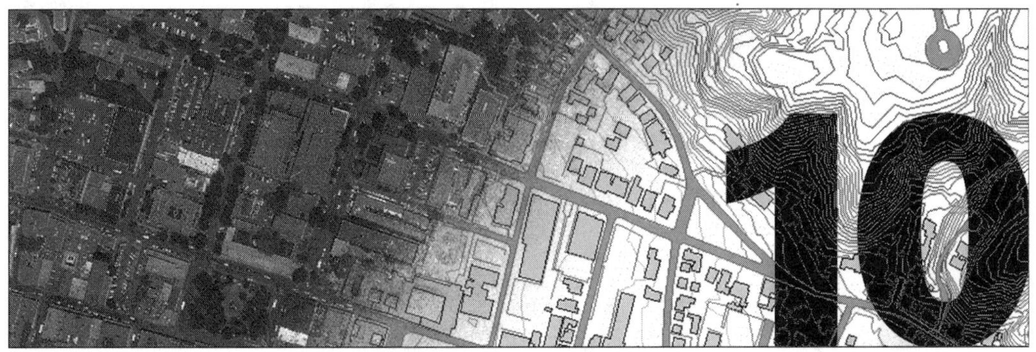

Editing Coverages in ARCEDIT

Steps to Automating Geographic Data Sets

ARCEDIT is the ARC/INFO program for editing coverages. With ARCEDIT, you can add or modify every type of feature within a coverage.

As you start using ARCEDIT at the command level, you'll appreciate the ease of use of ARCEDIT AML applications like Edit GIScity. You should seek out (or write) AML applications for your routine work, but an application designed for ease of use cannot span all the functions of ARCEDIT.

These are two circumstances when experienced users may enter ARCEDIT's command level: making a series of simple edits on a coverage, or jumping to the ARCEDIT command level from within an AML application to execute commands not included in the AML application. Otherwise, the use of an AML application is preferred. Clicking a single command in an AML application may be the equivalent of typing several or many ARCEDIT commands at the command level.

The Six Basic ARCEDIT Steps

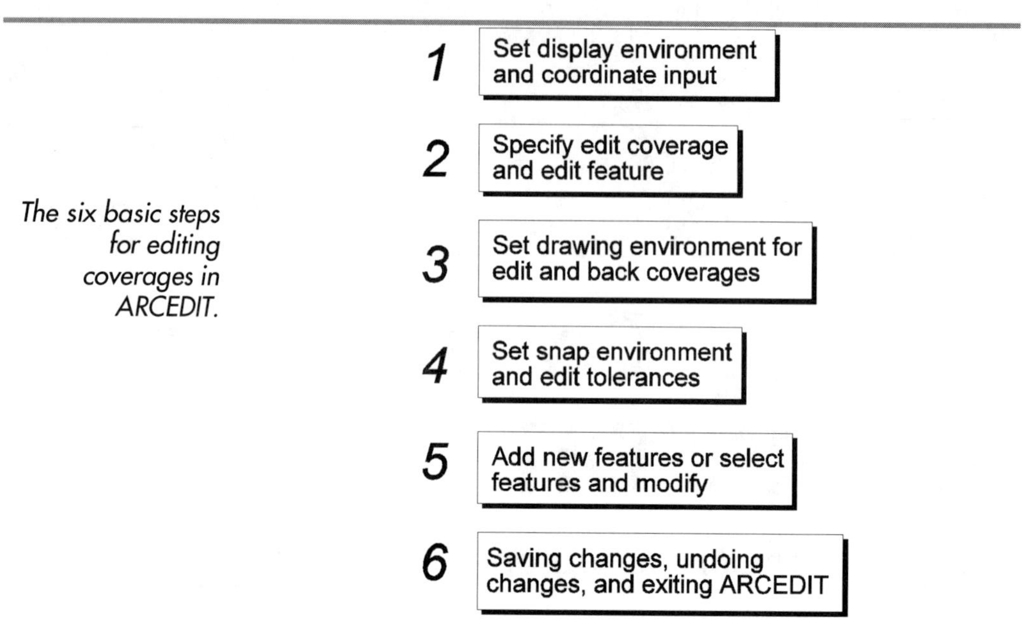

1
Set display environment and coordinate input
2
Specify edit coverage and edit feature
3
Set drawing environment for edit and back coverages
4
Set snap environment and edit tolerances
5
Add new features or select features and modify
6
Saving changes, undoing changes, and exiting ARCEDIT

The six basic steps for editing coverages in ARCEDIT.

When you edit a coverage in ARCEDIT, you interact with several environments: the display environment, the edit coverage and feature, the drawing environment, the snapping environment and edit tolerances, and the selected set. A general characteristic of these environments is that, once set, they remain active until changed. Also, most of these environments have default values which are set at the beginning of an ARCEDIT session or stored with the coverage. This chapter will acquaint you with these environments in ARCEDIT.

NOTE: *You have another alternative for editing coverages: ArcTools, the graphical user interface to ARC/INFO. Much of the design of GIScity follows ArcTools, but GIScity is simpler and ArcTools is more comprehensive. ArcTools, by necessity, is a graphical interface to nearly every possible ARCEDIT command and usage. GIScity streamlines the use of ARC/INFO and teaches you only the most important commands. To begin ArcTools, type ARCTOOLS at the ARC, ARCEDIT, or ARCPLOT command prompt.*

The steps in the preceding illustration outline the basic procedure for editing coverages in ARCEDIT. Though some are optional and the order of executing them can be modified, these steps comprise the main elements of most ARCEDIT sessions.

This chapter covers steps 1 through 4, part of step 5, and step 6. The fifth step, adding and modifying features, is the main content of an ARCEDIT session and is covered in the next chapter.

Before we proceed with the first step, let's review how to enter the ARCEDIT program. Simply type `ARCEDIT` at the `Arc:` command prompt. (A valid command abbreviation for ARCEDIT is AE.)

Step 1: Set the Display Environment and Coordinate Input

To start an edit session, first turn on your digitizer if you're using one, and then start ARCEDIT. After starting ARCEDIT, the usual next step is to specify your workstation's display environment and the coordinate input.

The Display Environment

The DISPLAY command creates the graphics window in which you see your coverage. Although the following usage appears quite substantial, in common practice you usually specify only the first few arguments.

```
Arcedit: DISPLAY <device> {option} {baud} {tty_line}
         SIZE {FRAME | CANVAS} <width> <height>}
         {POSITION <xy>}
         {POSITION <xy | UL | UC | UR | CL | CC | CR | LL | LC | LR>
```

```
{SCREEN | THREAD <thread>}
{xy | UL | UC | UR | CL | CC | CR | LL | LC | LR}
POSITION <ABOVE | BELOW | LEFT | RIGHT | CENTER>
{SCREEN | THREAD <thread>}
```

Reference the type of graphics display with <device>. For this argument, nearly all users will type 9999, which denotes an X-windows device. Virtually all workstations today are X-windows devices. Other numbers are used for alternate graphics devices which were common in the past, such as the Tektronix graphics displays.

The four options for the DISPLAY command.

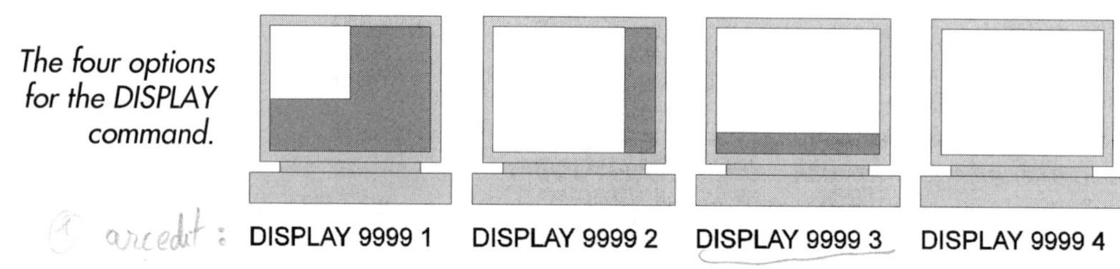

If you type in one of these four usages of the DISPLAY command, you will see a graphics window appear with one of these sizes and positions. The most commonly used option is DISPLAY 9999 3. It creates a graphics window that spans most of the workstation display, yet leaves space for a dialog window at the bottom.

The other arguments, such as SIZE and POSITION, are generally not used at the ARCEDIT command level, because you can avoid a lot of typing by using the windowing system on your workstation (Motif or OpenLook) to resize the graphics window. These arguments are more frequently used in AMLs to precisely set window sizes.

After you have created a window with one of the options, such as DISPLAY 9999 1, move your mouse to one of the edges or corners of your graphics window and drag the margin of the graphics window to a new position. When you let go, the window will be resized. When you resize a graphics window by stretching it, whatever coverage features are displayed will be immediately redrawn.

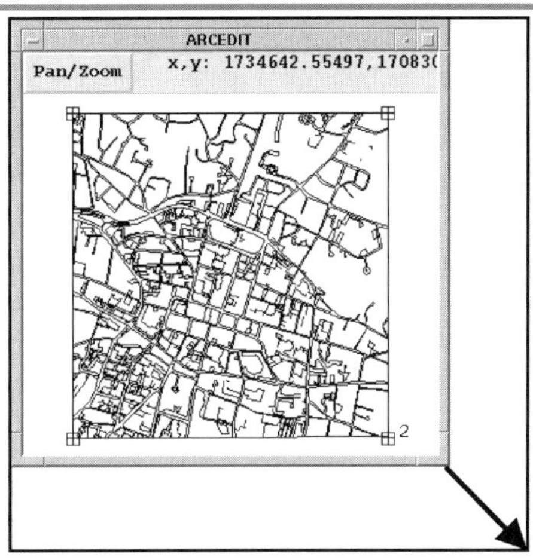

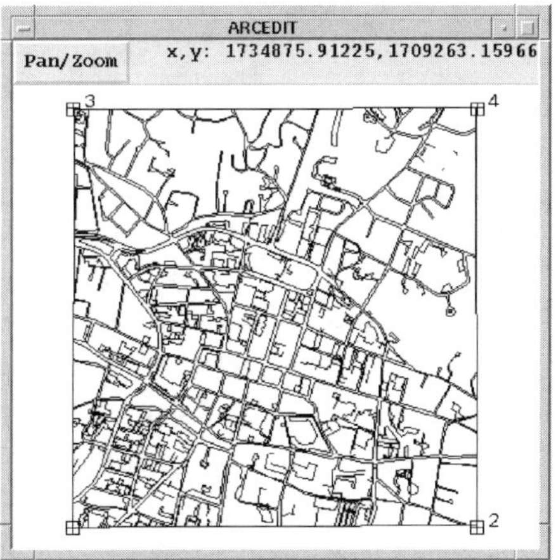

Resizing a graphics window with Motif. The graphics window is redrawn right after you resize the window.

NOTE: *It is not mandatory to create a graphics display in ARCEDIT. If you enter ARCEDIT to perform strictly non-graphic changes, you can interact with ARCEDIT solely through the dialog window.*

The Terminal Setting

If you will be using any AML menus in your ARCEDIT session, you need to issue the &TERMINAL command to specify the type of workstation and the input device for menus. As with the DISPLAY command, the device setting is nearly always 9999, which denotes an X-windows workstation or display.

```
Arcedit: &TERMINAL <device> {&CURSOR | &TABLET |
         &MOUSE | &KEYPAD}
```

The common usage for this command is

```
Arcedit: &TERMINAL 9999 &MOUSE
```

&TERMINAL appears different from other commands we use in ARCEDIT because it is preceded with an ampersand (&). Later, we will learn more about commands beginning with an ampersand, which are called AML directives.

Digitizer Registration

```
Arcedit: DIGITIZER <digitizer_type> {<tty_line>:{option1}:
         {option2}:{option3}} {POINT | STREAM}
```

This command sets up communications between your digitizer and workstation. Because the options and <tty_line> designations vary from one hardware system to another, consult your ARC/INFO System Dependencies Guide for the arguments to specify your operating system and digitizing equipment.

The POINT and STREAM choice lets you specify whether you digitize discrete points by pressing a button on the digitizer's cursor (POINT mode) or continuously digitize points as you trace with the digitizer's cursor (STREAM mode).

The STREAM choice is recommended because it gives you the most flexibility. Also, keep in mind that your current weed tolerance controls how close vertices in new arcs can be.

TIP: *Troubleshooting digitizer connections can be tricky. The ARC command DIGTEST will help you isolate the possible cause of the failure to communicate. A common reason for a communication failure is turning on the digitizer after the ARCEDIT session was begun. Whenever you use a digitizer, always turn it on before you begin ARCEDIT.*

Coordinate Input

Next, you should use the COORDINATE command to specify the type of coordinate input for your ARCEDIT session. During the course of an ARCEDIT session, you can repeatedly change the coordinate input. While you are digitizing from a digitizing table, you may find it helpful to temporarily switch coordinate input to a mouse and then back to the digitizer. By default, the coordinate input is set to the mouse. If you intend to digitize exclusively with the mouse, you won't need to use the COORDINATE command.

```
Arcedit: COORDINATE <CURSOR | TABLET | MOUSE>
```

specifies that a terminal cursor, tablet, or mouse is set for coordinate input. For X-windows devices, all three arguments have the same result—activating your mouse for coordinate input.

```
Arcedit: COORDINATE KEYBOARD {XY | POLAR | RELATIVE}
```

specifies that the keyboard is the active device for input. By default, the XY argument is specified, which means positions are typed in as real-world coordinates. POLAR specifies that a polar angle and distance are to be typed in. RELATIVE specifies that delta X and Y distances are typed in. While digitizing with a keyboard may seem strange, it is a useful option for entering precise survey data or for adding features at a precise offset from another position.

```
Arcedit: COORDINATE DIGITIZER {cover} {AFFINE | PROJECTIVE}
Arcedit: COORDINATE DIGITIZER {NOTRANS | DEFAULT}
```

Sets the coordinate input to a digitizing table. When you use the COORDINATE DIGITIZER command with the {cover} argument to specify a coverage, a new registration is set. The AFFINE and PROJECTIVE choice specifies a mathematic correction applied to errors in registration. The usual choice is the default AFFINE. The PROJECTIVE choice is used for digitizing from aerial photographs of flat terrain.

When you use the COORDINATE DIGITIZER without the {cover} argument, coordinate input is set to the digitizer using the registration previously set. This allows you to switch between mouse input and digitizer input. The rarely used NOTRANS argument accepts straight coordinates from the digitizing table. The DEFAULT argument specifies that the digitizer setting is returned to the most recent registration you've set.

Tic registration with COORDINATE DIGITIZER.

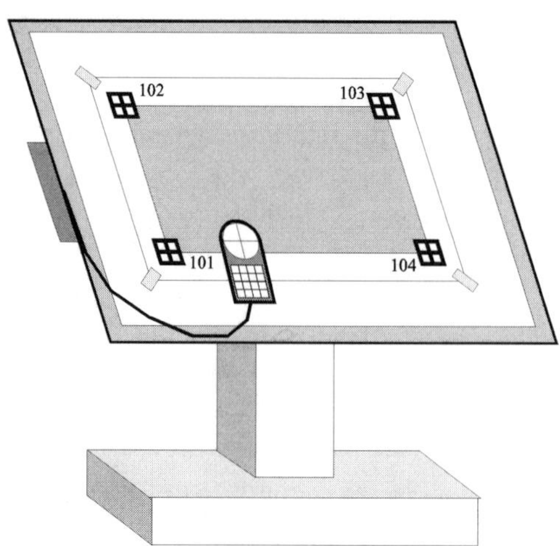

When you use a digitizer, first define a geometric relation between a map manuscript taped flat upon the digitizer surface and the tics in the specified coverage. The coverage used for registration does not need to be your edit coverage.

Following is a dialog from a sample registration. The arguments for the DIGITIZER command depend on the type of digitizer and operating system you are using. (This example uses a typical UNIX port designation.) Also, this dialog sets the registration for four tics in a coverage named EDGEROAD, with Tic ID values of 101, 102, 103, and 104.

```
Arcedit: DIGITIZER 9100 /dev/ttya/ POINT
Arcedit: COORDINATE DIGITIZER EDGEROAD
DIGITIZER TRANSFORMATION
Digitize a minimum of 4 tics.
Signal end of tic input with Tic-ID = 0
Tic-ID:  101*
Tic-ID:  102*
Tic-ID:  103*
```

```
Tic-ID:  104*
Tic-ID:  0*
```

The tic IDs are entered with the keys on the digitizer cursor. They vary from one manufacturer to another, but most have sixteen keys numbered 0 through 9 and A through E. When you see the Tic-ID prompt after starting the COORDINATE DIGITIZER command, these are the steps to follow:

1. Carefully place the cross-hair of the cursor over the point on the map manuscript corresponding to a tic on your coverage. Press cursor key 1.
2. Next, press the keys corresponding to the tic ID. After you've entered all the numeric keys, press the A key (on most digitizers) to indicate that the tic number has been entered.
3. Repeat steps 1 and 2 until at least four tics are entered. When done, press the 0 key followed by the A key to complete the registration process.

You can enter as many tics as you need. An important consideration is entering tics that are widely spaced. The four outermost corners of a map are ideal. Survey points are also good positions for tics.

Inevitably, there will be some error in registration when you register between a map manuscript taped on the digitizer and the tics already in your coverage. After you finish tic registration with the COORDINATE DIGITIZER command, you will see a message informing you of the registration error found.

You will see several values. The important one is the *RMS error* (Root mean square error.) This is a statistical measurement of the average discrepancy between each tic and point you've marked on the digitizing table. Ideally, you should keep this value below 0.004. Sometimes you cannot achieve this value because of distortions or inaccuracies in your map manuscript. If you cannot get an RMS error under 0.004, try registration several times to determine the best value you can achieve. Make sure that the map manuscript is taped down tight.

The RMS error is your measure of the best accuracy you can expect from digitizing. It does not guarantee accuracy—careful digitizing is a necessary follow-up—but good registration is a prerequisite for accurate digitizing.

Step 2: Specify the Edit Coverage and Edit Feature

An essential step in editing coverages is specifying which coverage and which type of feature you are about to edit. These are called your *edit coverage* and *edit feature*. During an ARCEDIT session, you can change both frequently.

Changing Workspaces

First, make sure that you are in a workspace that contains the coverage you wish to edit. Before you begin ARC/INFO, you can use your computer's operating system command to move to that directory with the cd command in UNIX or SET DEFAULT in VMS. Just type:

```
UNIX: cd $GCHOME/data
VMS: SET DEFAULT $GCHOME:[DATA]
```

Moving to a workspace directory through the operating system command or through ARC/INFO's &WORKSPACE command is a matter of preference. If you want to move to that same directory from within the ARCEDIT program, type:

```
Arcedit: &WORKSPACE $GCHOME/data (UNIX)
Arcedit: &WORKSPACE $GCHOME:[DATA] (VMS)
```

Whenever you need to check which workspace you are in, type the following command:

```
Arcedit: SHOW WORKSPACE
```

If you want to see which coverage(s) are in your current workspace, type:

```
Arcedit: DIRECTORY COVER
```

Setting the Edit Coverage

One of the first commands issued in an ARCEDIT session sets the edit coverage and, optionally, the edit feature:

```
Arcedit: EDIT <cover> {feature_class}
```

You can have up to ten edit coverages within an ARCEDIT session. The most recently specified edit coverage is the active one. In the event that you reach the limit of ten edit coverages and want to specify another, you should use the REMOVEEDIT commands. When you remove an edit coverage, if you have not saved the edit changes you will be asked if you want to do so. This is the usage for the REMOVEEDIT command:

```
Arcedit: REMOVEEDIT ALL {YES | NO}
Arcedit: REMOVEEDIT <cover> {YES | NO}
```

The REMOVEEDIT command with the ALL argument will remove all your edit coverages. With the <cover> argument, this command removes selected edit coverages. The YES argument specifies that you do want all edit changes saved. If you don't specify the last argument, you will be asked if you want to save changes.

There are two ways to ask ARCEDIT what the edit coverage(s) are:

```
Arcedit: SHOW EDITCOVERAGE
```

will show you the current edit coverage.

```
Arcedit: STATUS EDIT
```

will show you all the edit coverages and the current edit coverage.

Setting the Edit Feature

During the course of an ARCEDIT session, you will probably change the edit feature a number of times.

```
Arcedit: EDITFEATURE <NONE | ARC | NODE | LABEL | TIC
         | LINK | POLYGON | ANNO.subclass |
         SECTION.subclass | ROUTE.subclass |
         REGION.subclass | GROUP.subclass>
```

The EDITFEATURE command lets you change to any feature type possible in a coverage. You will become very familiar with this command, which can be abbreviated as EF. If you forget to set the edit feature, you will get an error message from many ARCEDIT commands that require an edit feature.

To see what the current edit feature is at any time, type:

```
Arcedit: SHOW EDITFEATURE
```

Step 3: Set the Drawing Environment for Edit and Back Coverages

There are three ways to draw coverages in ARCEDIT. You can draw the current edit coverage, you can specify *back coverages* to draw, or you can execute an AML to draw any coverage through ARCPLOT graphics commands.

ARC/INFO's Graphic Display Environment

Before you enter ARC/INFO, you can specify at your operating system prompt several system settings that affect graphics display in both ARCEDIT and ARCPLOT. There are about a dozen in all. To find out more, refer to the "The ARC/INFO Display Environment" section of ARC/INFO's on-line documentation system, ArcDoc. Following are three useful settings:

```
ARCDYNAMICPAN <OFF | ON>
```

If you set this value to ON, all DRAW commands in ARCEDIT will store into memory features in an area around the current map extent. While this slows down the display of features a bit, it gives you a benefit: when you use the DYNAMICPAN command, features adjacent to the map extent will be drawn very quickly.

```
CANVASCOLOR <BLACK | WHITE>
```

By default, the background for graphics windows in both ARCPLOT and ARCEDIT is black. You can set white as the background color instead.

NOTE: *GIScity applications have a light gray background, which is not one of the options for the CANVASCOLOR setting. Edit GIScity and Display GIScity use a trick—an ARCPLOT command which shades the entire map extent with a light gray color every time. Light gray was chosen because it is pleasing to the eye and complementary to all colors.*

```
CROSSHAIRCOLOR <color_name>
```

This setting controls the color of the screen cursor in ARCPLOT or ARCEDIT.

Specifying these three commands varies between UNIX and VMS. In UNIX, use the system command setenv. In VMS, use the system command DEFINE. Here are a few examples:

```
UNIX: setenv ARCDYNAMICPAN ON
UNIX: setenv CANVASCOLOR BLACK
VMS: DEFINE CANVASCOLOR WHITE
VMS: DEFINE CROSSHAIRCOLOR RED
```

To set ARC/INFO's graphics display environment in Windows NT, go to your desktop, right-mouse click on the My Computer icon, and select Properties. In the tabbed dialog that appears, click on Environment. To set CANVASCOLOR to white in Windows NT, enter CANVASCOLOR as the variable and "white" as the value.

The CANVASCOLOR and CROSSHAIRCOLOR settings apply to both ARCPLOT and ARCEDIT. The ARCDYNAMICPAN setting applies only to ARCEDIT.

The Draw Environment for the Edit Coverage

ARCEDIT contains several commands designed to depict and highlight potential editing errors. The DRAWENVIRONMENT command is a key command designed for interactive use in ARCEDIT. For example, you can highlight arcs which appear to join, but which are actually dangling arcs in close proximity.

```
Arcedit: DRAWENVIRONMENT
         ALL {ON | OFF}
         ARC {ON | OFF | IDS | ARROWS | INTERSECT}
         NODE {ON | OFF | IDS | ERRORS | DANGLE | PSEUDO}
         LABEL {ON | OFF | IDS}
         TIC {ON | OFF | IDS}
         ANNO.subclass {ON | OFF | level...level}
         LINK {ON | OFF}
         POLYGON {ON | OFF | IDS | FILL}
         REGION.subclass {ON | OFF | IDS | FILL}
```

```
SECTION.subclass {ON | OFF | IDS | ARROWS |
   MEASURES |        POSITIONS}
ROUTE.subclass {ON | OFF | IDS | ARROWS |
   ROUTEERRORS | MEASUREERRORS}
GROUP.subclass {ON | OFF}
```

DRAWENVIRONMENT affects the display of only the current edit coverage. Refer to the Set Draw Environment command in Edit GIScity as documented in later chapters. You will find most of the arguments for DRAWENVIRONMENT there.

By default, the drawing environment for all features is turned off. This is one of the first ARCEDIT commands you will use in an editing session. This command can be abbreviated as DE.

To find out what the current draw environment is, type:

```
Arcedit: SHOW DRAWENVIRONMENT
```

Drawing Coverages in ARCEDIT

The graphic display of features in ARCEDIT is primarily designed to give you feedback about potential editing errors. Also, feature drawing in ARCEDIT is dynamic: when you add or delete features, you will see the result immediately in your graphics window. To draw coverages in ARCEDIT, just type:

```
Arcedit: DRAW
```

Whenever you resize the graphics window, the DRAW command is implicitly executed.

If you have set back coverages or specified an AML with the AP command, those will draw as well.

If no features are drawn when you issue the DRAW command, here are the most likely of the possible causes:

- ❑ You have not set the draw environment for edit coverages or back coverages.
- ❑ Your coverage may be empty.
- ❑ Your current map extent may be outside the range of your coverage(s).
- ❑ You current map extent may be set to an empty part of your coverage(s).

Setting the Map Extent

The map extent is the range of X and Y coordinates spanned by your graphics window. The coordinates are the real-world coordinates set for your coverage: Universal Transverse Mercator, State Plane Coordinate System, or other. The units are nearly always meters or feet.

There are some other usages for MAPEXTENT, but these are the most commonly used ones. The star (*) argument will activate the screen cursor to select two points:

```
Arcedit: MAPEXTENT {* | DEFAULT | SELECT}
Arcedit: MAPEXTENT <xmin ymin xmax ymax>
```

Coverages normally have a stored map extent which envelopes all features in that coverage. (If the stored map extent is not current, the Arc: REBOX command can restore it.) The DEFAULT argument sets the map extent to the edit coverage's stored map extent. The SELECT argument will set the map extent to span exactly the selected features in the edit coverage and edit feature. Be aware that if you use this argument choice with only one point feature selected, the map extent will be very, very small.

The <xmin ymin xmax ymax> argument is very useful, but it is normally employed in an AML rather than used at ARCEDIT's command level. For example, many of the zoom commands in Edit GIScity, such as Zoom to Previous and Zoom Out, take the current map extent, calculate a new X and Y coordinate range, and then use this argument to set a precise map extent.

Whenever you issue a MAPEXTENT command, nothing will seem to happen. However, when you redraw your screen, it will be at the new map extent. (The zoom commands in Edit GIScity combine the two basic steps of setting new map extents and drawing the coverages.)

The MAPEXTENT command can be abbreviated as MAPE. Two short ARCEDIT commands you will find yourself using frequently are MAPE* and MAPE DEF.

NOTE: *If you use the EDIT command to set an edit coverage and do not already have a map extent set, ARCEDIT sets the map extent to the edit coverage's stored map extent, as in its bnd file.*

Setting Symbol Tables

If you use symbols to display coverages, as Edit GIScity and Display GIScity do extensively, you can set symbol tables. There are four types of symbol sets:

- **Linesets** symbolize arcs, routes, and sections. The standard file extension is .lin.
- **Markersets** symbolize label points. The standard file extension is .mrk.
- **Shadesets** symbolize polygons and regions. The standard file extension is .shd.
- **Textsets** symbolize annotation and text. The standard file extension is .txt.

When you begin ARCEDIT, several symbol tables are set as the default. They all reside in the directory $ARCHOME/symbols and are plotter.txt, color.lin, color.mrk, and color.shd. Refer to the ARC/INFO documentation for a pictorial summary of the contents of these and other symbol tables supplied with ARC/INFO. In Chapter 19, we will learn how to create and modify symbol tables.

These are four commands to set symbol tables:

```
Arcedit: LINESET <lineset_file>
Arcedit: MARKERSET <markerset_file>
Arcedit: SHADESET <shadeset_file>
Arcedit: TEXTSET <textset_file>
```

These symbol tables can reside in either the $ARCHOME/symbols directory or in any directory path you specify. If you do not supply a directory path (such as $GCDATA/symbols) in the argument to these commands, ARCEDIT will assume they are in either the $ARCHOME/symbols directory or in your current working directory.

Symbolizing the Edit Coverage

When you start ARCEDIT, the features in your edit coverage are drawn without the cartographic symbols you saw throughout the Display GIScity and Edit GIScity applications. Without symbols, label points appear as little crosses, and arcs are drawn as simple lines. You can optionally set point, line, and other symbology to the features in your edit coverage.

```
Arcedit: SYMBOLITEM {cover} <ARC | LABEL | TIC | NODE
         | POLYGON | ANNO.subclass | REGION.subclass |
         SECTION.subclass | ROUTE.subclass> <NONE |
         item | constant>
```

The first argument is optional and lets you specify a coverage. If no coverage is entered, SYMBOLITEM assumes that you are applying the command to the edit coverage.

The second argument specifies the feature class. You can use this command several times to specify symbology for different feature types within the edit coverage.

The third argument is how you specify the symbol. Normally you will enter the name of an item which contains symbol numbers. This item must be an integer item. The NONE option disables the application of symbols to features in your edit coverage.

Arcs drawn in a coverage with and without a symbol item specified.

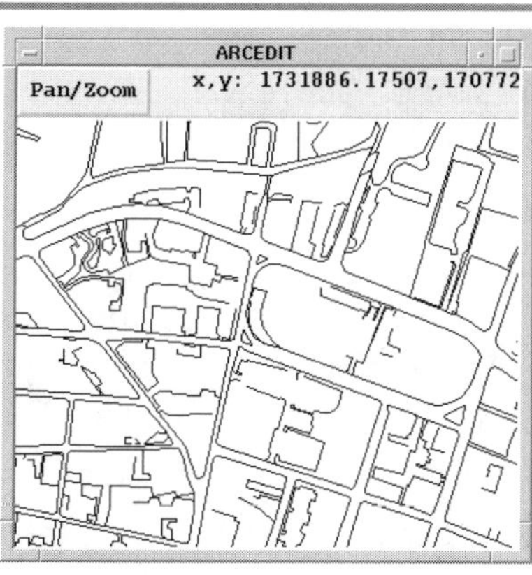

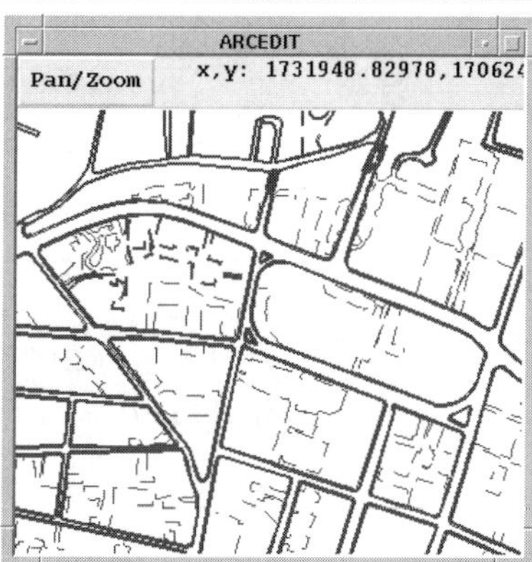

All of the GIScity coverages have a predefined item called SYMBOL—an integer item with an item width of three bytes. It is by no means required that a coverage contain an item reserved for symbol numbers, but it is a good practice for most applications. The name of an item used for symbols can be anything you want. Also, you can represent alternate symbols by multiple items in a feature attribute table or by using a relate item.

Although using a relate item to a lookup table best adheres to a relational database design standard called *normalization*, it has one drawback. Symbolizing features using a relate item will noticeably slow down your display time for large coverages.

Adjusting the Display Scale

Normally, ARCEDIT and ARCPLOT display feature symbols on the graphics window at the same sizes as on a hard-copy map. If a certain symbol is to be 1/20″ on a published map, it will measure exactly that size on the graphics window, no matter what the display scale is.

Although this type of fixed size display for symbols presents a uniform display, often (and particularly when cartographic relations are important), you might prefer symbols that size up and down with the current display scale. Following are a suite of commands that allow you to specify symbol scales:

```
Arcedit: LINESCALE <scale>
Arcedit: MARKERSCALE <scale>
Arcedit: TEXTSCALE <scale>
Arcedit: SYMBOLSCALE <scale>
```

These four commands will accept a scale factor and base the display of features on it. For instance, if you type in `2`, the symbols will appear twice as large after the next time you use the DRAW command.

The SYMBOLSCALE command conveniently sets the line scale, marker scale, and text scales simultaneously.

```
Arcedit: LINESCALE MAPSCALE <scale>
Arcedit: MARKERSCALE MAPSCALE <scale>
Arcedit: TEXTSCALE MAPSCALE <scale>
Arcedit: SYMBOLSCALE MAPSCALE <scale>
```

Using MAPSCALE with the scale commands will accept a unitless map scale. (Remember, a unitless map scale is a scale setting without any units, such as 1:48 000, instead of 1″:2000′.)

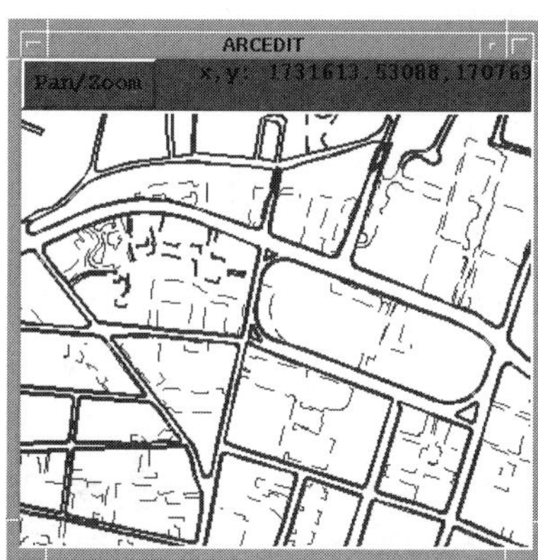

An example from Edit GIScity before and after setting a line scale factor of 2.

Drawing Background Coverages with ARCPLOT

ARCEDIT offers a variety of options and commands to depict coverage features, but it doesn't come close to ARCPLOT in the depth of display and visualization functions. However, here's a neat ARCEDIT command: the AP command can use ARCPLOT display commands within an ARCEDIT session with each draw. To take advantage of this capability, you must have an AML written for ARCPLOT and designed to display coverage features.

```
Arcedit: AP <AML_file | NONE>
```

If you set the AML_file argument choice, AP will use that file to draw the background features with ARCPLOT commands. The NONE argument disables the use of an AML file with ARCPLOT commands.

NOTE: *GIScity contains an AML for use with AP which controls the general display of coverage features in both Edit GIScity and Display GIScity. The intrepid among you may want to take a look at this AML: $GCHOME/tools/tool-draw.aml (UNIX) or $GCHOME:[TOOLS]TOOL-DRAW.AML (VMS).*

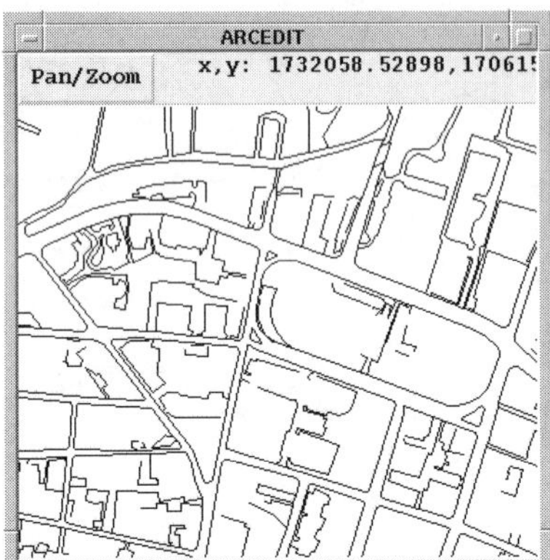

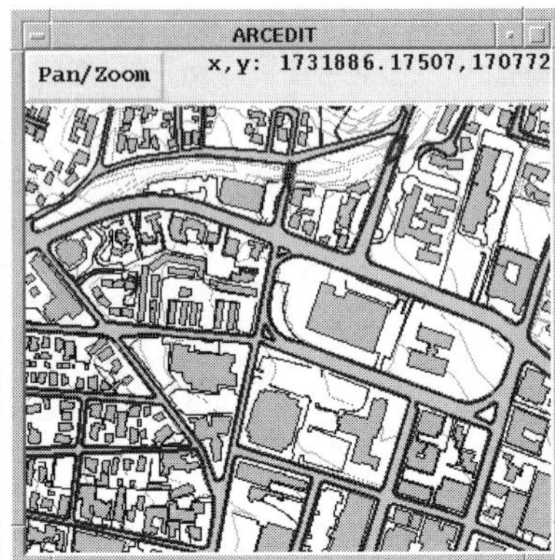

An example from GIScity of coverages drawn without (left) and with (right) the AP command.

NOTE: *Any AML file written to be used with AP should be constructed with care. It can be just a simple script of ARCPLOT commands, but serious applications demand programming logic and validation. For instance, the AML file should have the intelligence to suppress the display of edit coverages and edit features. Otherwise, edit coverages will be drawn twice, and deleted features may reappear after you thought you removed them.*

Setting the Back Environment

With ARCEDIT, you can optionally set backcoverages, which are drawn together with the edit coverage whenever you issue the DRAW command.

Back coverages are for display rather than editing. However, you can change a back coverage into an edit coverage whenever you want.

```
Arcedit: BACKCOVERAGE <cover> {symbol}
```

This command adds a new back coverage. When you issue a draw command, features set with the BACKENVIRONMENT command will draw together with features from the edit coverage.

```
Arcedit: BACKENVIRONMENT {back_cover}
         ALL {ON | OFF}
         ARC {ON | OFF | IDS | ARROWS | INTERSECT}
         NODE {ON | OFF | IDS | ERRORS | DANGLE | PSEUDO}
         LABEL {ON | OFF | IDS}
         TIC {ON | OFF | IDS}
         ANNO.subclass {ON | OFF | level...level}
         LINK {ON | OFF}
         POLYGON {ON | OFF | IDS | FILL}
         REGION.subclass {ON | OFF | IDS | FILL}
         SECTION.subclass {ON | OFF | IDS | ARROWS |
             MEASURES | POSITIONS}
         ROUTE.subclass {ON | OFF | IDS | ARROWS |
             ROUTEERRORS | MEASUREERRORS}
         GROUP.subclass {ON | OFF}
```

Nearly all the settings in this command match the settings in the Set Draw Environment menu in Edit GIScity. The BACKENVIRONMENT command can be abbreviated as BE.

```
Arcedit: REMOVEBACK <cover | ALL>
```

This command removes a back coverage.

```
Arcedit: BACKSYMBOLITEM <backcover> <ARC | LABEL | TIC
         | NODE | POLYGON | ANNO.subclass |
         REGION.subclass | SECTION.subclass |
         ROUTE.subclass | item | constant>
```

This command is just like the SYMBOLITEM command, but for back coverages instead of edit coverages.

To find out what the current back environment is, type:

```
Arcedit: SHOW BACKENVIRONMENT (or SHOW BE)
```

NOTE: *If you use an AML file with the AP command, setting a back environment is really not necessary. The AP technique is more powerful, but requires you have an AML file with ARCPLOT commands prepared. The back environment can be easier to establish if you are using ARCEDIT at the command level.*

Labeling Features with Item Values

One last drawing environment command displays item values as text. The TEXTITEM command specifies a feature type and item. When you redraw the screen, the actual item values for each feature will be drawn. The text that appears when you use this command is temporary; it is not the same as annotation, which you can edit and store in a coverage.

```
Arcedit: TEXTITEM {cover} <ARC | LABEL | TIC | NODE |
         POLYGON | ANNO.subclass | REGION.subclass |
         SECTION.subclass | ROUTE.subclass> <NONE |
         item | 'string'>
```

By default, the TEXTITEM command works on the current edit coverage; you can use the optional {cover} argument to specify another coverage.

The second argument specifies the feature class. For this command to display any text, a feature attribute table must be present for the chosen feature class.

The third argument controls the output of the command. If you type in an item name, item values for that item will appear the next time you issue a DRAW command. To disable the display of item values, use TEXTITEM with the NONE choice.

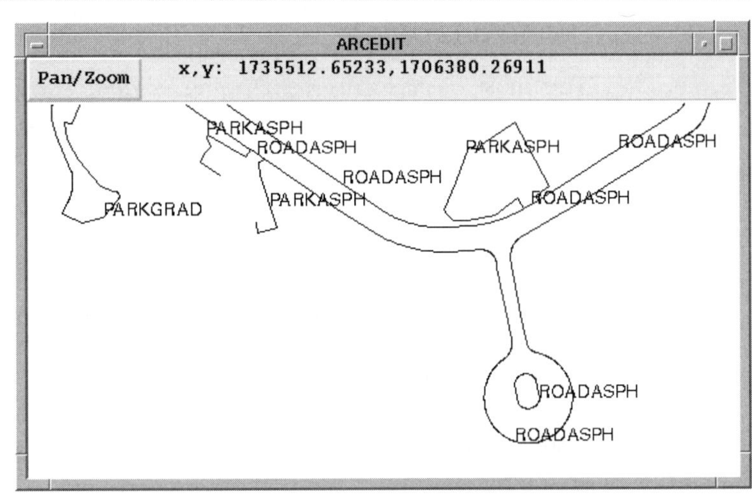

Sample item values displayed with the TEXTITEM command.

Step 4: Set the Snap Environment and Edit Tolerances

The snap environment positions new or modified features to exactly match the positions of other features. The snap environment is important in maintaining topology among features. For instance, the arcs along a polygon must close exactly for the polygon to be defined, and arcs along a network must meet at exact common points to enable network tracing functions. Snapping can also be useful when digitizing coincident features, such as label points depicting values placed at vertices along arcs depicting gas lines.

There are three types of snapping, any of which can be active simultaneously:

- Node snapping
- Arc snapping
- General snapping

To get information about the snap environment at any time, type:

```
Arcedit: STATUS SNAP
```

Node Snapping

The most basic form of snapping is *node snapping*, which is used to ensure that nodes snap together exactly when you digitize one arc near another. As mentioned before, accurate node snapping is critical for building polygons and networks.

```
Arcedit: NODESNAP <FIRST | CLOSEST | OFF> {DEFAULT
         | * | distance}
```

By default, node snapping is enabled when you begin an ARCEDIT session. The default distance is 1/1000 of the height or width of your coverage's map extent, whichever is greater.

The first argument specifies whether the FIRST node found within the distance is snapped to, or the CLOSEST node. OFF disables node snapping.

The second argument sets the distance within which snapping will occur. DEFAULT returns the node snap distance to the default mentioned above. The star choice prompts you to set a snap distance interactively by picking two points. Or you can type in a distance in your coverage units (usually meters or feet.)

NODESNAP will not snap features already digitized; it affects newly added or edited features only.

A new node is snapped to an existing node if it is within the node snapping distance.

Arc Snapping

Arc snapping lets you snap the nodes of new arcs to create intersections with existing arcs. Arc snapping is used to correct *undershoots* and *overshoots.*

```
Arcedit: ARCSNAP ON {* | distance}
Arcedit: ARCSNAP OFF
```

When you specify ARCSNAP ON, you can set the distance with the screen cursor or by typing in a distance in coverage units. The default, ARCSNAP OFF, disables arc snapping. Normally you will use ARCSNAP in conjunction with the INTERSECTARCS command. If you want to perform arc snapping, you probably want the arcs to be intersected as well.

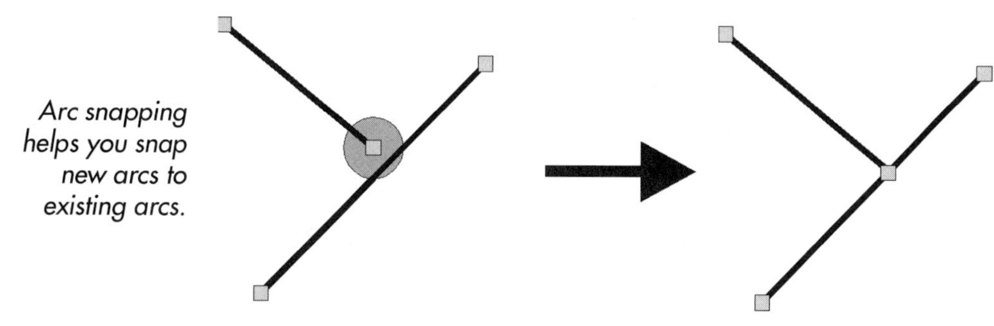

Arc snapping helps you snap new arcs to existing arcs.

General Snapping

This most general type of snapping is not typically used for maintaining topology, but to ensure that features which should align precisely to others are snapped. Here are a couple of examples where general snapping is useful:

- ❑ Survey points can be added as label points. They are also used as tics for registration. You can use general snapping to guarantee that the label points and tics are at exactly the same position.
- ❑ Utility maps contain point features and line features. General snapping can set the addition of point features such as transformers to lie exactly upon an electrical line.

```
Arcedit: SNAPCOVERAGE <cover>
Arcedit: SNAPFEATURES <NONE | ARC | NODE | LABEL |
         TIC | ANNO.subclass> <NONE | ARC | NODE |
         LABEL | TIC | ANNO.subclass>
```

The two preceding commands are required for general snapping. SNAPCOVERAGE specifies the coverage containing the features to snap to. It can be the edit coverage or another coverage. SNAPFEATURES specifies which type of feature in the edit coverage should be snapped to which type of feature in the snap coverage.

```
Arcedit: SNAPPING <OFF | FIRST | CLOSEST> {* | DEFAULT
         | distance}
```

Once you've set the snap coverage and features, SNAPPING activates the general snapping environment. OFF disables general snapping. FIRST and CLOSEST enable general snapping and specify whether the first feature found is snapped to, or the closest one. You have three options to set the distance for general snapping: * activates the screen cursor, DEFAULT sets the distance to 1/1000 of the coverage height or width, or you can type in a distance in coverage units.

Snapping Features Already Present

Normally, features are snapped as they are digitized. When you set the SNAPPING command, all new digitized or modified features can be snapped. If you have features already digitized that weren't snapped, you can use this command:

```
Arcedit: SNAP
```

This command applies general snapping on the selected features. SNAP can be a little tricky if the snap coverage is the same as the edit coverage. SNAP uses the selected set in ARCEDIT, but SNAP won't snap together features that are within the selected set. The trick is to select just the features you want to snap, and not the features you want to snap to.

Edit Tolerances

There are three other settings that affect your editing environment.

```
Arcedit: EDITDISTANCE {* | DEFAULT | distance}
```

The EDITDISTANCE command specifies how close the screen cursor must be to a target feature to successfully select it. You can set this distance either with the screen cursor, as a default value of 1/100 of the coverage height or width, or as a distance you type in coverage units.

```
Arcedit: GRAIN {* | DEFAULT | distance}
```

The grain tolerance sets the closest distance between adjacent arc vertices. This tolerance is used for modifying existing arcs with ARCEDIT commands such as SPLINE and for adding curved arcs with ADD.

```
Arcedit: WEEDTOLERANCE {* | DEFAULT | distance}
```

The weed tolerance also sets the closest distance between adjacent arc vertices. This tolerance is used for new arcs you digitize with the ARCEDIT ADD command and when modifying selected arcs with GENERALIZE.

To find the tolerances you have set, type:

```
Arcedit: STATUS TOLERANCES
```

Step 5: Add New Features or Select and Modify Features

ARCEDIT contains a variety of commands to add and modify features. We will defer a discussion of these commands for the next chapter. In the Edit GIScity application, we explore how to specify and modify the selected set. Refer to the Select Features Spatially and Select Features Logically menus from that chapter; they parallel the actual selection commands in ARCEDIT closely.

A common charateristic of commands that add and modify features is that they modify the *selected set.*

Selection Commands

These are all the selection commands in ARCEDIT.

- ❑ SELECT makes a new selected set.

```
Arcedit: SELECT {ONE | MANY | ALL | OUTLINE | DANGLE |
         CONNECT}

Arcedit: SELECT PATH {from-to_impedance_item}
         {to-from_impedance_item}

Arcedit: SELECT <BOX | POLYGON | CIRCLE | SCREEN>
         {WITHIN | PASSTHRU}
```

```
Arcedit: SELECT {FOR} <logical_expression>
```

- ❑ ASELECT adds to the selected set.

```
Arcedit: ASELECT {ONE | MANY | ALL | OUTLINE |
         DANGLE | CONNECT}

Arcedit: ASELECT PATH {from-to_impedance_item}
         {to-from_impedance_item}

Arcedit: ASELECT <BOX | POLYGON | CIRCLE | SCREEN>
         {WITHIN | PASSTHRU}

Arcedit: ASELECT {FOR} <logical_expression>
```

- ❑ RESELECT selects a subset of the selected set.

```
Arcedit: RESELECT {ONE | MANY | ALL | OUTLINE | DANGLE}

Arcedit: RESELECT PATH {from-to_impedance_item}
         {to-from_impedance_item}

Arcedit: RESELECT <BOX | POLYGON | CIRCLE | SCREEN>
         {WITHIN | PASSTHRU}

Arcedit: RESELECT {FOR} <logical_expression>
```

- ❑ UNSELECT removes from the selected set.

```
Arcedit: UNSELECT {ONE | MANY | ALL | OUTLINE | DANGLE}

Arcedit: UNSELECT PATH {from-to_impedance_item}
         {to-from_impedance_item}

Arcedit: UNSELECT <BOX | POLYGON | CIRCLE | SCREEN>
         {WITHIN | PASSTHRU}

Arcedit: UNSELECT {FOR} <logical_expression>
```

- ❑ NSELECT reverses the selected set.

```
Arcedit: NSELECT
```

Drawing Selected Features

ARCEDIT contains a couple of commands to highlight selected features.

```
Arcedit: SETDRAWSYMBOL <symbol> {color}
```

This command lets you select a symbol number (from the current symbol table set with the MARKERSET, LINESET, SHADESET, or TEXTSET command) and, optionally, a color.

```
Arcedit: DRAWSELECT {ONE | ALL | OFF}
```

This command enables the highlighting of selected features. ONE highlights individual features as they are selected with the MANY option in the selection commands. ALL highlights all selected features. OFF disables highlighting of selected features.

Step 6: Saving Changes, Undoing Changes, and Exiting ARCEDIT

During an ARCEDIT session, you are working on temporary files when you add, modify, or delete features. The actual coverage is not updated until you save the changes. This design allows you to change your mind about whether to keep edit changes. Here's a trio of important session control commands in ARCEDIT:

```
Arcedit: OOPS
```

Many users consider OOPS to be the most useful ARCEDIT command. If you have just made an incorrect edit, you can undo it by typing OOPS. You can also type OOPS repeatedly to step back as many edits as you want, unless you have committed changes with the SAVE command. In that case, you can only OOPS back to the last SAVE.

```
Arcedit: SAVE {new_cover}
Arcedit: SAVE {old_cover} <new_cover>
Arcedit: SAVE {ALL} {YES}
```

Events such as power outages can cause you to lose edits. When you've spent several hours editing a coverage, you'd like to make sure your edits are kept. The SAVE command will write the information in the temporary files to your coverage, making the edits permanent. You cannot OOPS the SAVE command.

Normally, SAVE is used without arguments. The coverage arguments let you save to a new coverage, which is also a way of making a copy of a coverage within ARCEDIT.

```
Arcedit: QUIT {YES | NO}
```

QUIT is the last command in an ARCEDIT session. After you issue this command, you are returned to the ARC program. QUIT with the YES argument saves all edit changes. QUIT with the NO argument exits without saving changes, either from the beginning of the ARCEDIT session or the most recent SAVE.

Getting Information in ARCEDIT

At several points in this book we've used the SHOW and STATUS command to find out about our environment in ARCEDIT.

```
Arcedit: STATUS {CURRENT | EDIT | BACKGROUND | ADD
         | DRAW | WHERE | MISC | ALL | SYMBOLSET |
         SNAP | TRANSFER | LINK | TOLERANCES | VTRACE
         | GRIDEDIT}
```

We've used several of the options for the STATUS command. If you want to see all of the above, use the ALL option.

```
Arcedit: SHOW <arguments>
```

The SHOW command can give you information about almost anything in ARCEDIT. This command has too many options to cover here. Reference ArcDoc for a listing of everything you can find out with SHOW. You'll use this command more and more as you gain experience with ARC/INFO.

It is impossible to memorize every command in ARC/INFO. Here are two more commands to get information about the available commands in ARCEDIT:

Arcedit: COMMANDS{prefix | wildcard}

will give you a listing of all ARCEDIT commands. You might be overwhelmed by the result, so you can narrow your search by using a wildcard, such as the letter A to find all the ARCEDIT commands that begin A.

```
Arcedit: USAGE <command>
```

gives you a listing of all the possible arguments for any ARCEDIT command. All the command descriptions in this chapter were obtained with this command.

Exercise 1: Stepping Through an ARCEDIT Session

We've covered the six main steps for editing coverages, although we've left the actual feature edit commands for the next chapter. In this exercise, you'll type in the commands for a simple ARCEDIT session, using coverages from GIScity.

This ARCEDIT session does not set the snap environment or edit arcs, but does select an edit coverage and features, zooms in, and zooms back to the coverage map extent.

1. Although not all ARCEDIT commands can be abbreviated, the following common abbreviations are used in this exercise.

Command	**Abbreviation**
MAPEXTENT	MAPE
DRAWENVIRONMENT	DE
COORDINATE	COO
EDITFEATURE	EF

2. Type in the following commands and watch the screen to see their result.

```
Arc: ARCEDIT
Arcedit: &term 9999
Arcedit: display 9999 3
Arcedit: coo mouse
Arcedit: &workspace $GCHOME/data
Arcedit: edit EDGEROAD
Arcedit: ef arc
Arcedit: de arc on
Arcedit: draw
Arcedit: mape *
Arcedit: draw
```

```
Arcedit: mape def
Arcedit: draw
```

Exercise 2: Using ARCEDIT Commands

Let's conduct another simple ARCEDIT session, but this time, you'll choose the commands and arguments to type in. Using the commands documented in this chapter, perform the following steps. Try using abbreviations.

1. At your operating system prompt, enter the ARC program.
2. From the ARC program, enter ARCEDIT.
3. Set your workspace to $GCHOME/data (UNIX) or $GCHOME:[DATA] (VMS).
4. Set the terminal setting to an X-window device.
5. Open a graphics window that covers all but the bottom of your workstation's display.
6. Set your edit coverage to the coverage containing buildings called ROOFLINES.
7. Set the edit feature to label points.
8. Zoom in.
9. Draw.
10. Zoom back to the coverage extent.
11. Set the markerset to PLOTTER.MRK.
12. Set the symbol item to SYMBOL.
13. Draw again.
14. Set the draw symbol for selected features to 1 and the color red.
15. Select label points with a box defined with your screen cursor.
16. Highlight the selected features.
17. Zoom to the selected features.

Adding and Modifying Features in ARCEDIT

Editing Arcs, Label Points, Polygons, and Other Features

In the previous chapter, we laid the foundation for editing coverages by discussing six general steps in an ARCEDIT session. Now, we'll explore the specific commands to add, move, copy, delete, and otherwise change features in a coverage.

As we go through each feature type, refer to the Edit GIScity feature editing menus. You'll find that most of the commands covered at the command level in this chapter are present in those menu choices.

Many feature editing commands do not have any arguments. Instead, they work on your selected set. The outcome of the these commands may also modify your selected set.

A special and powerful command in ARCEDIT is the ADD command, which creates new features. ADD has a special submenu for editing each type of feature. You can invoke the choices in the submenu through the input keys on your current coordinate input device, whether mouse, digitizer or keyboard. Because of its utility, we'll pay particular attention to the ADD command.

Editing Feature Attributes

This chapter concentrates on editing the positional information of coverage features. To review, two important ARCEDIT commands are used to update item values. CALCULATE is used for numeric items and MOVEITEM is used for character items. Both commands operate upon the current edit feature and selected set.

Another way to easily edit feature attributes is with the FORMS command in ARCEDIT.

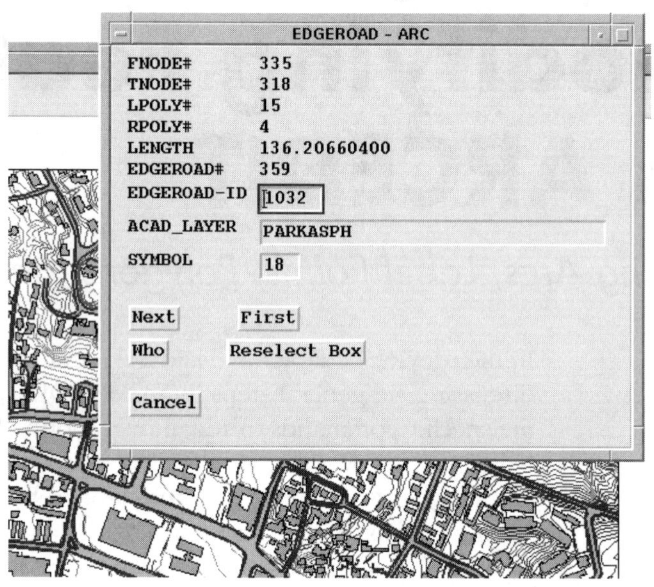

A sample menu invoked by the FORMS command.

When you are editing features with a feature attribute table, you can select a number of features and type in the FORMS command. ARCEDIT will construct a menu for you to edit item values interactively. This command uses database cursors. The buttons on the menu, Next and First, use cursor processing to navigate through the selected set.

```
Arcedit: FORMS {item...item} {RELATE <relate>}
```

The preceding usage creates a new form menu. If you don't use any arguments, FORMS will construct a menu with all the items in your feature attribute table. To select only certain items to be updated, enter a list of item values. You can also specify a relate by name for updating item values in a relate table. The following usage of the FORMS command invokes a pre-existing form menu.

```
Arcedit: FORMS FORM <form_name>
```

An important function for many applications is editing a symbol item. You can do this in a couple of ways: You can update an item, such as a feature classification, and use a lookup table for symbol display, or you can directly update an item in your feature attribute tables for the purpose of symbol display.

NOTE: *Most feature types contain a pseudo-item for symbols, called $SYMBOL. Except for annotation, these pseudo-items affect the symbol assignment only for the duration of the ARCEDIT session. Therefore, you are best advised to add symbology by assigning a feature attribute table item that you define (call it SYMBOL, or whatever you like), and then use the SYMBOLITEM command.*

Moving Features to Other Coverages

Nearly all ARCEDIT commands work only on the current edit coverage and edit feature. The PUT and GET commands are designed to transfer features between coverages:

```
Arcedit: PUT <coverage>
```

The PUT command takes the features in your edit coverage, edit feature, and selected set, and writes them out to the specified coverage. If the coverage specified does not yet exist, this command will create a new coverage. Otherwise, the new features will be appended to the existing features.

WARNING: *You cannot undo the PUT command. Use this command with care.*

```
Arcedit: GET <coverage>
```

The GET command imports all the features from the specified coverage. Any snapping environment or edit tolerances currently set are applied on the imported features.

NOTE: *You will achieve the best results with GET and PUT when the item definitions in the feature attribute table in the sending and receiving coverages are identical. Otherwise, if there is a mismatch in the item definitions, you will lose item values.*

Arcs and Nodes

The arc is the fundamental feature type in ARC/INFO. Accordingly, there are more ARCEDIT commands to edit arcs and their nodes and vertices than for any other feature type. The majority of ARCEDIT commands that modify arcs are described in this section.

Any of these commands which add or modify coordinates of arcs will use the snapping environment you've set. All three snapping methods—node, arc, or general snapping—can affect the placement of nodes, vertices, and arcs.

All but two of the following commands require that you first set the edit feature to arc. Two commands, ALIGN and MOVE, work with the edit feature set to node as well. Nodes are a special feature type because they cannot exist without arcs. Although there are only a couple of commands in ARCEDIT that operate directly on nodes, many commands that edit arcs will also affect the placement or creation of nodes.

Routes and sections are compound features built upon arcs. Editing route systems is an advanced topic. Refer to ArcDoc for information on editing route systems.

The Arc Draw Environment

As you use these commands, remember to set your edit feature to arcs (the abbreviated command is EF ARC) and the draw environment for arcs and nodes. Some useful settings for arcs and nodes are as follows:

- DRAWENVIRONMENT ARC ARROWS lets you see the direction of arcs, which can be important for digitizing networks but not important for polygons.
- DRAWENVIRONMENT ARC INTERSECT will display a square marker with an X inside for arcs that cross without intersections.
- DRAWENVIRONMENT NODE DANGLE displays a square marker for dangling nodes.
- DRAWENVIRONMENT NODE PSEUDO displays a diamond marker for pseudo nodes.
- DRAWENVIRONMENT NODE ERRORS has the same effect as DRAWENVIRONMENT NODE DANGLE and DRAWENVIRONMENT NODE PSEUDO combined.

Remember, DE is a handy abbreviation for DRAWENVIRONMENT.

In general, when you are editing arcs to form radial networks like stream systems or utility lines, you should edit with these goals:

- Arcs should be digitized with a consistent direction, usually downstream.
- Dangling nodes should only occur at the ends of lines, and not at junctions.
- Pseudo nodes are acceptable if an attribute changes at that point.

When you are editing looped networks like streets,

- ❑ The arc direction is not important.
- ❑ Dangling nodes should only occur as endpoints like cul-de-sacs or dead-ends.
- ❑ Pseudo nodes should only be at a point where an attribute, such as a street classification or name, changes.

When you are editing arcs that form polygons, such as parcel lots,

- ❑ Arc direction is not important.
- ❑ There should be no dangling nodes at all.
- ❑ Pseudo nodes should only appear for "island polygons," which are formed by a single arc without any intersections.

It is important to watch for unwanted dangling nodes, because they can cause a gap in a network and prevent a polygon from being correctly closed. Pseudo nodes aren't as critical; unnecessary pseudo nodes don't prevent the topological relations you want to set, but they can clutter up a coverage and affect how line symbols appear on your display or map.

Most, but not all, of the following commands operate upon your selected set.

Arc Intersection

One important setting for arcs is whether or not intersections are created when you add, move, or change coordinates of arcs.

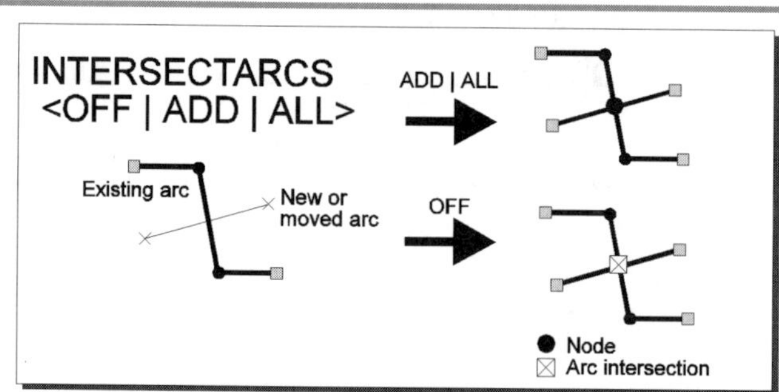

By default, the arc intersection setting is off when you begin an ARCEDIT session. For many types of coverages, especially coverages with polygons, you should use the ADD or ALL setting for INTERSECTARCS. The ADD argument only intersects arcs created with the ADD command. The ALL argument intersects arcs that are modified with any ARCEDIT command that moves nodes or vertices.

When you use the INTERSECTARCS command, no changes will be immediately apparent. It is only after you use the ADD, COPY, MOVE, or other commands that the intersections are calculated and added (or not added).

In the preceding diagram, the arc intersection symbol denotes that two arcs cross without a node at the intersection. This is the symbol you see with DRAWENVIRONMENT ARC INTERSECT.

Arc Editing Commands

The ADD command for arcs has many options and submenus.

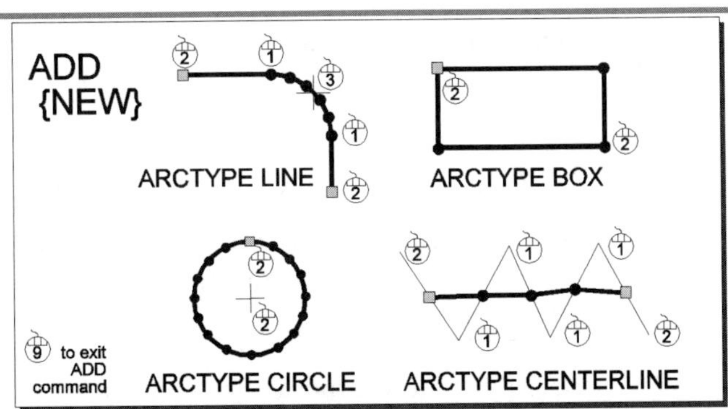

Because you are adding new features, the ADD command disregards the selected set. All arcs you add with the ADD command become the new selected set.

There are four ways to enter arcs: as lines, circles, boxes, and centerlines. By default, arcs as added as lines. If you want to preset the different ways of entering arcs, use the following command before you add arcs.

```
Arcedit: ARCTYPE <LINE | BOX | CIRCLE | CENTERLINE>
```

The command name, ARCTYPE, is a misnomer because there is only one type of arc. There is no difference between an arc added with ARCTYPE BOX or that same arc added with ARCTYPE LINE with exactly the same four corners. The four "arc types" are just short-cuts for editing. Most of the time, you will use the line option for arcs.

When you type the ADD command, you will get one of four submenus, depending on the setting from the ARCTYPE command. These four submenus have options from one to nine. These options use the current coordinate input: mouse, keyboard, or digitizer.

Adding Arcs as Lines

```
Arcedit: ADD
1) Vertex            2) Node                3) Curve
4) Delete vertex     5) Delete arc          6) Spline on/off
7) Square on/off     8) Digitizing options  9) Quit
```

When you add an arc as a freeform line, you will get the submenu above.

Arcs always begin with a node. Most arcs are digitized in this sequence:

1. Use key 2 at the begin point.
2. Then use key 1 successively to digitize vertices.
3. You can optionally add circular segments within an arc by using key 3 to mark a point on the circular segment.
4. Finish the arc with key 2.
5. After you finish one arc, you can either add a new arc with key 2, or return to the ARCEDIT command level by using key 9.

When you use keys 1, 2, and 3, your cursor should be carefully placed over a location on your graphics window or digitizer. Keys 4 through 9 don't require that you place the cursor at any location.

The 4 through 9 keys have special functions you'll find handy:

- ❑ Key 4 removes the most recent vertex you've added.
- ❑ Key 5 removes the most recent arc you've added.
- ❑ Key 6 turns splining off or on. Splining is an automatic smoothing of arcs.
- ❑ Key 7 turns squaring off and on. This is an aid to help you digitize 90 degree corners for features such as building.

- Key 8 invokes a new submenu for digitizing options, documented below.
- Key 9 returns you to the ARCEDIT command level.

NOTE TO WINDOWS NT USERS: *If you are using a two-button mouse, issue this command first:*

```
Arcedit: 2BUTTON
```

This will remap the key input so that you can use a two-button mouse instead of a three-button mouse. See ArcDoc for more details on the 2BUTTON command.

Adding Arcs as Boxes

```
Arcedit: ARCTYPE BOX
Arcedit: ADD
2) Box corner  5) Delete box     8) Digitizing options
9) Quit
```

To add an arc as a box, move the cursor to one corner of a box, click key 2, move the cursor to the opposite corner, click key 2 again. Key 5 deletes the most recent arc added as a box.

Adding Arcs as Circles

```
Arcedit: ARCTYPE CIRCLE
Arcedit: ADD
2) Center/radius point  5) Delete circle   8) Digitizing options
9) Quit
```

To add an arc as a circle, move the cursor first to the center point of the desired circle. Click input key 2. Move the cursor to any point on the circumference of the desired circle. Click 2 again. You'll see a circle drawn, which is actually made up of many short line segments separated by vertices. If you zoom into the circle, you'll see the line segments. Key 5 deletes the most recent arc added as a circle.

Adding Arcs as Centerlines

```
Arcedit: ARCTYPE CENTERLINE
Arcedit: ADD
1) Shape point    2) Start/Stop    5) Delete centerline
8) Digitizing options   9) Quit
```

Adding arcs as centerlines is strikingly different from the other methods, because you are indirectly digitizing the arc. You would use this option if you want to digitize a street centerline from a map with the edges of roads. This method adds nodes and vertices that connect the midpoints of each line segment you enter. To start, use key 2 over a point on the edge of the road. Then use key 1 and zig-zag across the road. To finish, use key 2 to mark the last point on the edge of road. Key 5 deletes the arc most recently added as a centerline.

Digitizing Options

```
1) New User ID            2) New Symbol          3) Autoincrement OFF
4) Autoincrement RESUME   5) Arctype line          6) Arctype box
7) Arctype circle         8) Arctype centerline 9) Quit
```

With each of the four submenus we just discussed, you can invoke the digitizing options submenu with key 8. Keys 1 through 4 are infrequently used. Keys 5 through 8 can be used as an alternate to the ARCTYPE command.

When you digitize arcs, a User ID is automatically assigned for you. For the majority of applications, the actual values of the User ID item is unimportant. You can build a complete and comprehensive GIS model without using User IDs at all. As discussed in previous chapters, if you want to assign identifiers to arc features, creating a special item in your Arc Attribute Table for this purpose is often preferable, because you can more flexibly define this item as character, integer, or other item type.

Key 2 assigns a line symbol for arcs, but this option is rarely used because it does not permanently assign a line symbol value for that arc.

Keys 3 and 4 enable and disable the automatic incrementing of User ID values as you digitize.

The ADD command, by itself, does not assign any user-defined items in your Arc Attribute Table. You must follow the ADD command with the CALCULATE or MOVEITEM commands to assign numeric or character items. Using CALCULATE on an item defined for symbology is the preferred method for assigning symbology. The NEW command also can be used before you use the ADD command to preassign symbol numbers or any item values.

The ALIGN arcs command.

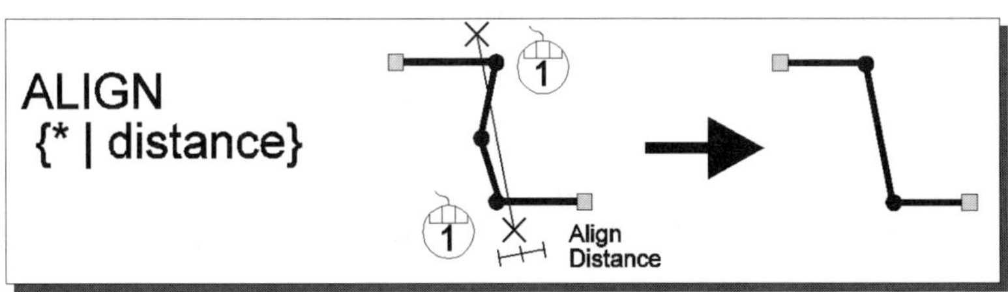

The ALIGN command lets you make corrections to arcs by snapping nodes and vertices to a simple line with two endpoints that you interactively define. The align distance is a tolerance for selecting nodes and vertices to move. If you don't type in a distance, the cursor will be activated for you to specify the align distance between two points.

The ALIGN command does not use your selected set for input; it will work on any arcs within the align distance. It will, however, set any modified arcs to be the new selected set.

When the align distance is specified, the cursor is activated to receive two points. When you initiate this command, you will see a short menu:

```
1 = Enter points        2 = Show line      9 = Quit
```

Use key 1 to pick the two points. After the first point, you can use key 2 to dynamically display the line you are about to set, or you can use key 9 to quit without aligning. If you finish the ALIGN command and the result is not what you wanted, use OOPS to undo this command.

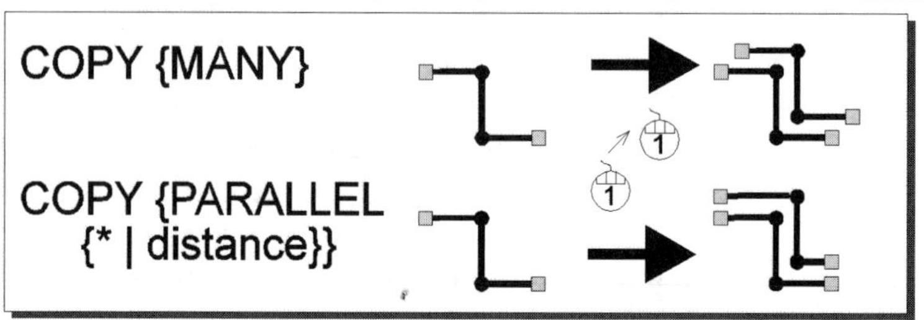

The COPY arcs command.

The COPY command will copy the arcs that are in your selected set. Without the PARALLEL argument, any number of arcs can be selected. With the PARALLEL argument, only one arc can be selected.

If you simply type `COPY`, you will be prompted to pick two points with your cursor. Use any input key except 9 to select these two points to move the selected arcs. Use input key 9 to abort this command.

COPY MANY lets you repeatedly copy the same selected arcs. Use input key 9 to abort.

NOTE: *For the COPY command and several other ARCEDIT commands, this method of interactively picking two points with the screen cursor does not mean that the arcs (or other features) will go to the second point. It means that an X and Y shift is calculated and that shift is used to copy or move the selected features.*

COPY PARALLEL will copy the single arc you have selected to a parallel offset as shown in the illustration. This is very useful for digitizing a feature that parallels another, such as a telephone line that is 50 feet to the right of a street centerline.

With the * argument, you will interactively set two points to define the parallel offset. If you type in a distance for COPY PARALLEL, a positive value specifies a parallel offset to the left, in the direction of the arc. A negative value specifies a parallel offset to the right, in the directory of the arc. (To see arc directions, use the DRAWENVIRONMENT ARC ARROWS command.)

The DELETE arcs command.

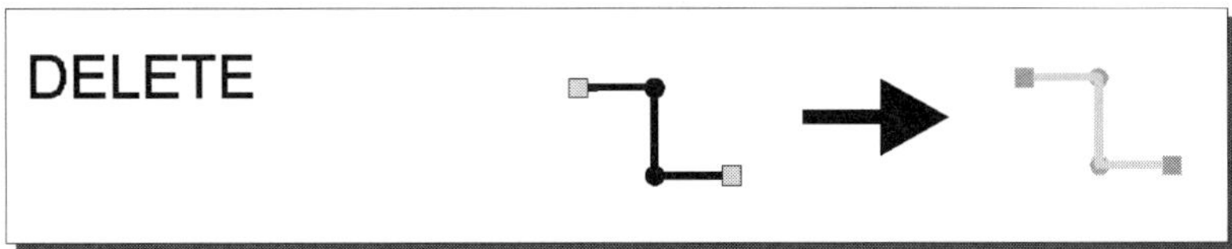

The DELETE command removes the selected arcs from your coverage.

If you accidentally deleted arcs, use the OOPS command to restore them. You cannot restore deleted arcs after a SAVE.

The DENSIFY command.

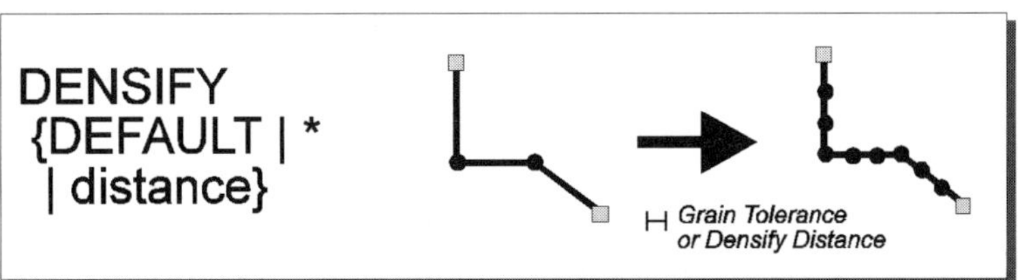

The DENSIFY command splits the selected arcs into many arcs, according to the grain tolerance or the densify tolerance set with this command.

The overall shape of the arcs is not changed. What changes is that several arcs may become many arcs, separated by new pseudo nodes.

This command is sometimes used to make node snapping easier in special circumstances.

The DRAG command.

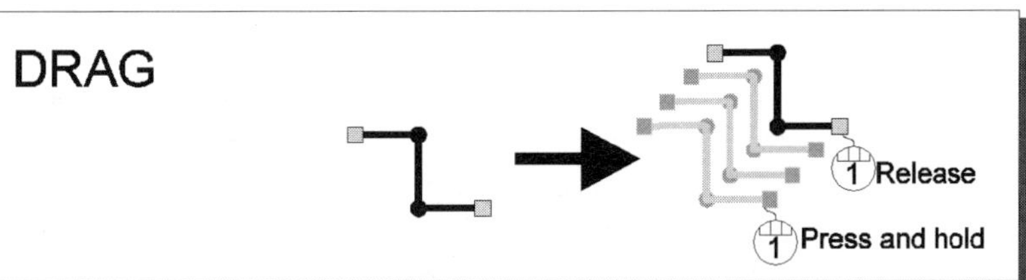

The DRAG command lets you move selected arcs, and graphically display the selected arcs as you move them.

When you use this command, you should first move the cursor to one of the selected features, press and hold down the mouse button or digitizer key, and, while holding down the key, move the cursor until the features are placed exactly where you want.

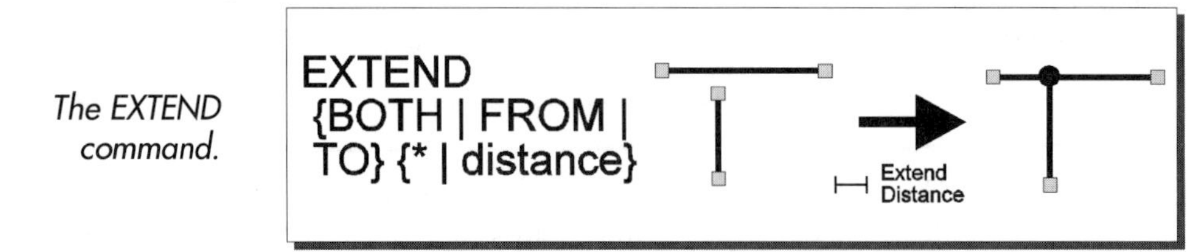

The EXTEND command.

The EXTEND command is used to snap selected arcs together within a specified distance.

This command can be used to selectively snap only the FROM node or the TO node, but usually you will use the BOTH option to snap either node.

You can type in a maximum distance that arc nodes will be snapped to arcs. When you extend arcs with this command, the new arcs are automatically intersected, despite your INTERSECTARCS setting.

This command is very useful and performs a type of snapping that the node snapping, arc snapping, or general snapping does not: it allows you to snap to points on arcs that are in between vertices.

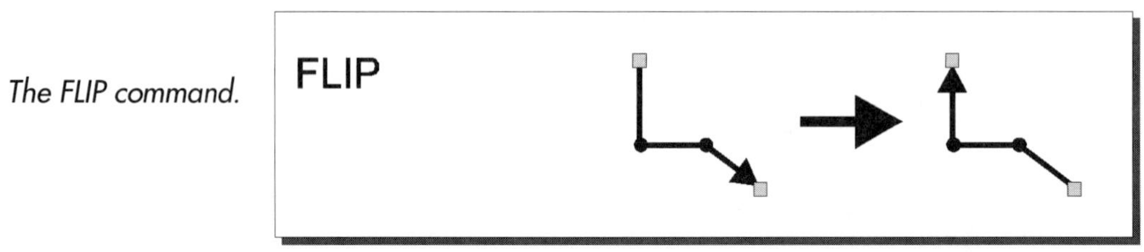

The FLIP command.

The FLIP command reverses the order of the FROM and TO nodes in the selected arcs.

For some applications, the order of FROM and TO nodes is unimportant. For other applications, particularly utility applications, arc direction is very important because network tracing (with the NETWORK module's TRACE command) is simplified if arc directions are in a natural flow direction.

Use the DRAWENVIRONMENT ARC ARROWS command to see arc directions before and after the use of this command.

The GENERALIZE command.

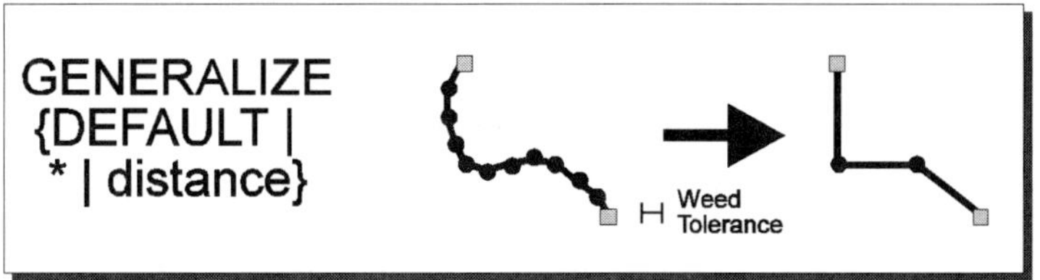

The GENERALIZE command is used to filter out dense vertices in selected arcs. This command either uses your currently set weed tolerance or a distance you can specify as an argument.

This command is sometimes used when you have digitized arcs at a large scale, such as 1:2400, and later want to use them for a map published at a smaller scale, such as 1: 100 000. GENERALIZE can filter out the unnecessary detail in the arcs.

The MOVE command.

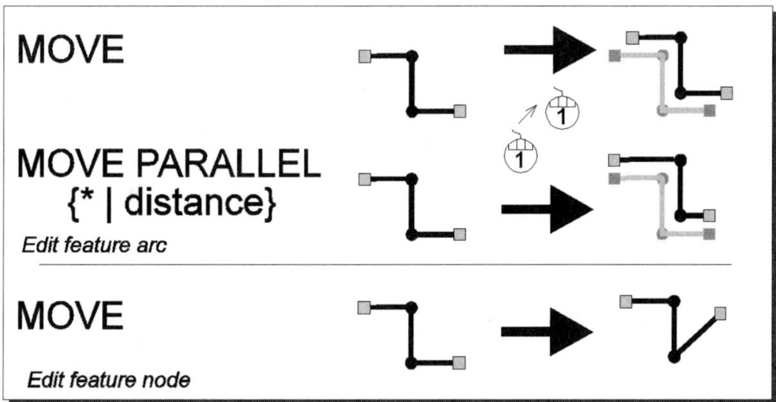

The MOVE command enables you to reposition selected arcs from one point to another. The operation of MOVE is the same as the COPY command, but the original arcs do not remain and are moved. Use the DRAG command if you want to move arcs dynamically with the cursor.

MOVE is one of the few edit commands that is permissible with nodes. If your edit feature is node, when you initiate this command, you will be prompted to select a node. Once you've done so, you will see this submenu appear:

```
1 = Select    2 = Next    3 = Who
4 = Move      9 = Quit
```

Use key 1 to select a different node. Use key 2 to select the next node that is within the edit distance. Use key 3 to flash the currently selected node. Use key 4 to move the node. Key 9 exits the MOVE command.

The RESHAPE command.

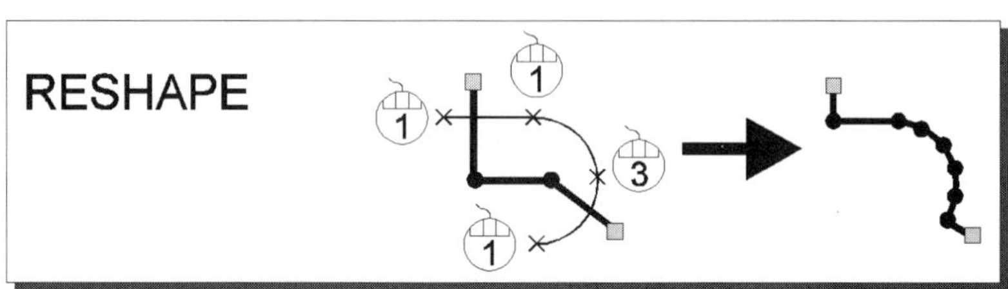

The RESHAPE command gives you an easy way to edit a portion of an arc. Before you use this command, you should select only one arc. After you initiate this command, the cursor is activated. Use input key 1 to digitize a vertex and input key 3 to specify a point along a circular arc. When you are done, use input key 9 to end this command. The new arc portion must intersect the selected arc in at least one location.

The ROTATE command.

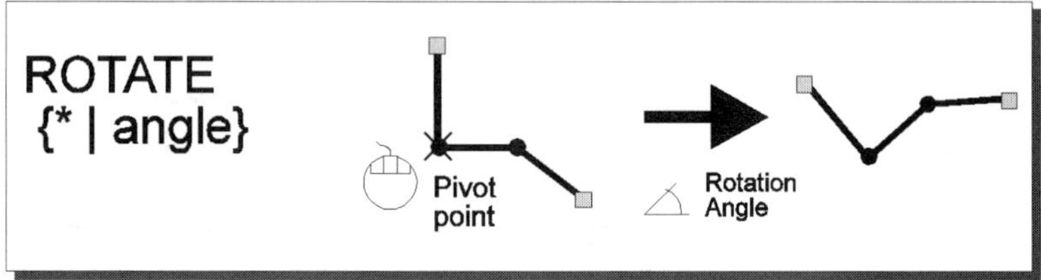

The ROTATE command lets you rotate the selected arcs by any rotation angle you specify. This command needs two pieces of information: a pivot point and a rotation angle.

After initiating this command, specify a pivot with the cursor. The arc(s) will be rotated about this point.

Next, specify the rotation angle by typing in an angle value in degrees. A positive angle rotates to the left, a negative angle rotates to the right. Or, you can interactively define a rotation angle by entering two points with the cursor. Keep in mind that these two points do not set a directional angle, but set an angle by which the arcs will be rotated.

An associated ARCEDIT command, SETANGLE, lets you predefine the rotation angle. If you've set an angle with the SETANGLE command, then you won't need to specify the angle again with the ROTATE command.

The SPLINE command.

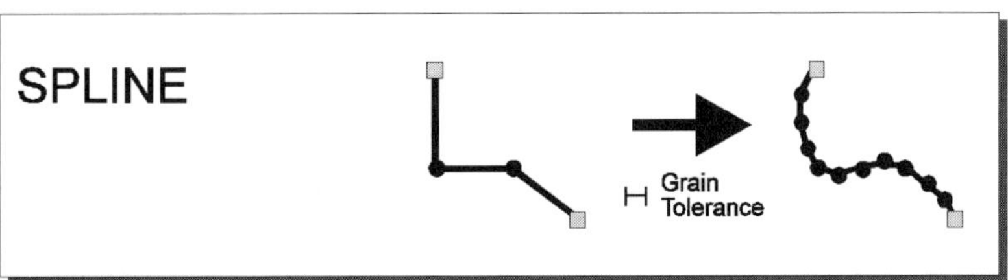

The SPLINE command smoothes the selected arcs using a mathematical calculation.

This command does not "improve" your arcs, but it can enhance their aesthetic appearance if used properly. Splines are frequently applied to natural features, such as streams and contours but are almost never applied to man-made features, such as roads and property lines.

Many vertices may be created with the SPLINE command. You can limit the number of vertices by adjusting your grain tolerance.

The SPLIT command.

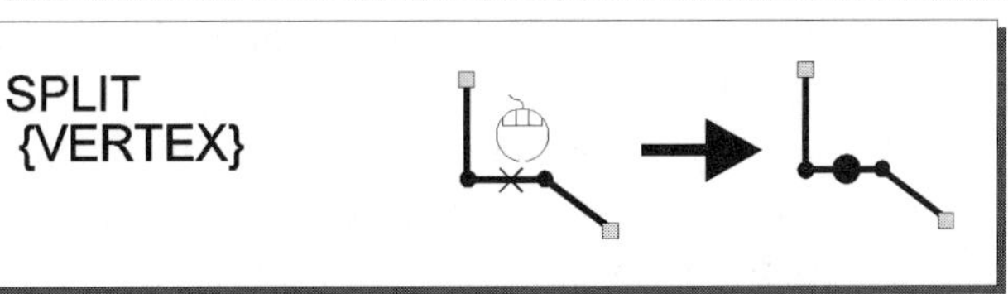

The SPLIT command will split an arc into two arcs. You must have only one arc selected before using this command. When you initiate SPLIT, the cursor is activated to select a point for splitting the arc.

With the VERTEX argument, the arc can only be split at an existing vertex. Without the VERTEX argument, the arc can be split anywhere along its length.

Normally, an arc is split when an attribute is changed along that arc. For instance, the diameter of a water line may decrease from 20 inches to 12 inches. If you are using the Arc Attribute Table to model pipe widths, you can split the arc and update the item values for both arcs.

NOTE: *For some applications, using the Dynamic Segmentation functions and route systems is preferable to splitting arcs. This topic is beyond the scope of this book. Reference the Dynamic Segmentation discussion in ArcDoc or in the ESRI documentation.*

The UNSPLIT command.

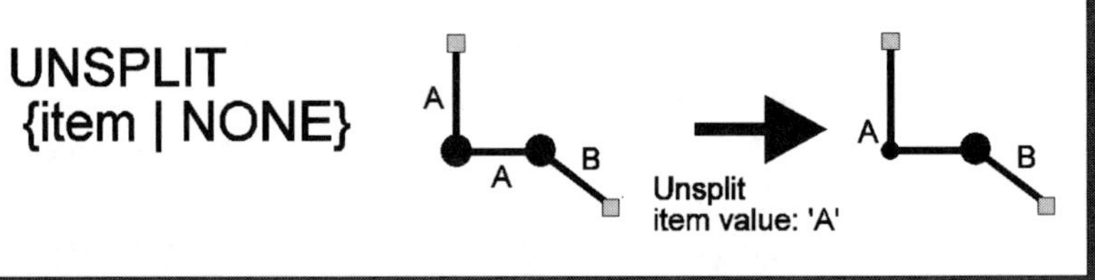

The UNSPLIT command is used to consolidate selected arcs connected by pseudo nodes.

If you use the item argument, only the selected arcs that have the same item value will be consolidated. The NONE option allows any connected arcs to be consolidated; otherwise only arcs with the same User ID will be consolidated.

Sometimes connected arcs cannot be consolidated. The probable reason is that combining adjacent arcs would yield an arc with more than 500 vertices, which is the limit for arcs. If you must consolidate these arcs, consider filtering out vertices with the GENERALIZE command or increase the grain size and run SPLINE.

The VERTEX commands.

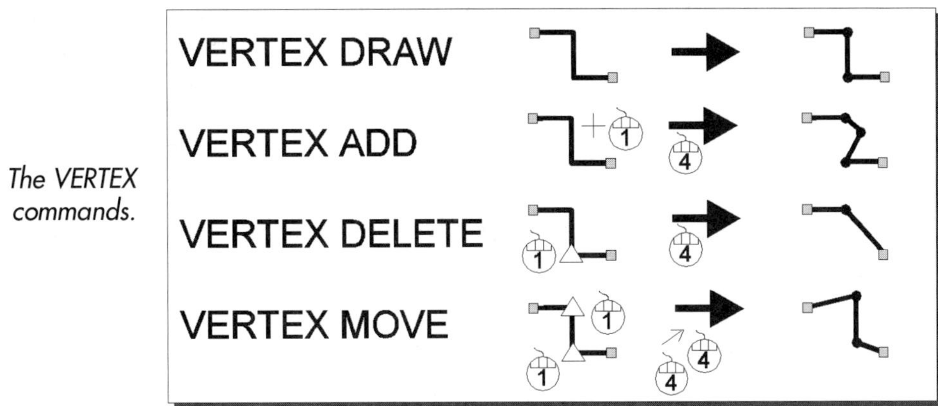

The VERTEX command performs many edit operations on vertices. Before you use this command, you should select only one arc. The full usage for this command is

```
Arcedit: VERTEX DRAW {IDS}
Arcedit: VERTEX <ADD | DELETE | MOVE {MANY}>{DRAW | NODRAW}
```

VERTEX DRAW displays all vertices in the selected node as small, solid-filled circles. When we discussed the DRAWENVIRONMENT command, you may have noticed that there is no draw environment for vertices. The reason is that if you draw all the vertices in a coverage, you may need to take a coffee break before the display is done. Even through there is no vertex DRAW ENVIRONMENT setting, the VERTEX DRAW command allows you to display the vertices for one selected arc.

The ADD, DELETE, or MOVE choice has an optional argument for setting the display of vertices. The default for the optional argument is DRAW, which means vertices will be drawn.

VERTEX ADD lets you add vertices to the selected arc. Initiating this command activates the cursor for you to successively add vertices. When you are done, use input key 9.

VERTEX DELETE lets you remove vertices from the selected arc. When you select a vertex, a triangle will be drawn on it. If that is the vertex you want to remove, use input key 4. To quit this command, use input key 9.

VERTEX MOVE {MANY} lets you remove one or many vertices from the selected arc. As you select vertices, triangles will appear on the selected vertices. When you are done, use input key 4 to specify that you want to move that vertex. Next, use any input key but 9 to position the vertex. Use input key 9 to quit this command.

Label Points

Because label points are one-dimensional features, they are much simpler to edit than arcs. Remember, label points have a dual mission: they represent small geographic features, and they also mark the interior of a polygon and store the polygon's attributes.

Placing Label Points

Labels can be snapped to other features by setting the general snapping environment. (Node snapping and arc snapping apply only to arcs.)

Before we add labels, we can use two commands to affect the rotation and size of labels.

```
Arcedit: LABELANGLE {* | angle}
```

This command sets an angle of rotation for all new labels. That angle will be in effect until changed.

```
Arcedit: LABELSCALE {* | scale}
```

This command sets a scale factor for sizing new labels. If you enter .75 as an argument, labels will be added at 75% of the normal size for the marker symbol.

Before using these commands to edit labels, set your edit feature to labels with the EDITFEATURE command.

The ADD label command.

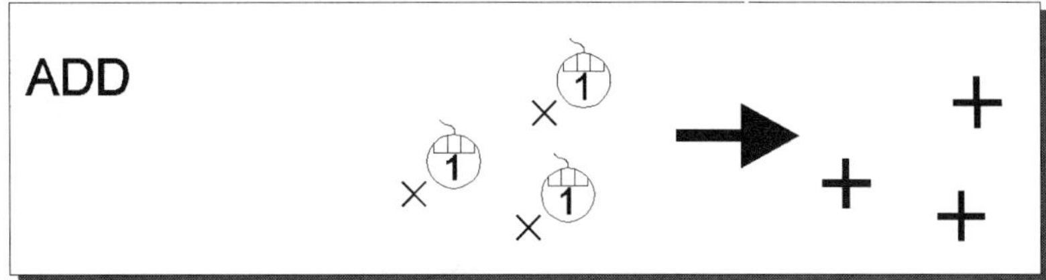

When you begin the ADD command, you will see this submenu:

```
1) Add Label   5) Delete last label  8) Digitizing options
9) Quit
```

Use key 1 to successively add labels. Key 5 removes the most recent label you've added. When you've finished adding labels, they will be your new selected set.

The COPY label command.

COPY {MANY}

The COPY command will activate your screen cursor. Pick two points to set a delta X and Y offset. After the second point, the selected label points will be copied.

The DELETE label command.

DELETE

The DELETE command will remove your selected label points. You can use the OOPS command to undo DELETE.

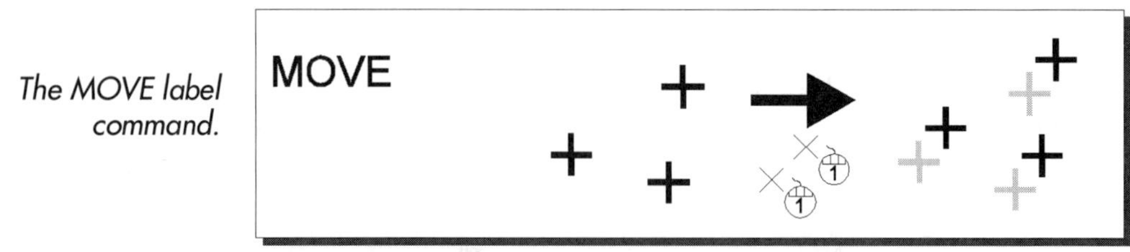

The MOVE label command.

The MOVE command activates your screen cursor to input two points. Your selected label points will be moved by the distance between the two points.

Polygons

Release 7 Improvements

The most profound difference between release 7 of ARC/INFO and previous releases is that you can now interactively edit polygon topology. Users familiar with previous releases are very familiar with the procedure of editing arcs and label points in ARCEDIT, going to the ARC program to run commands to identify potential polygon errors, and finally, running CLEAN or BUILD in the ARC program to build polygon topology. Hopefully, you got it all right. If you had some polygon errors, you were in for another iteration of ARCEDIT and ARC commands.

Building and maintaining polygon topology is now a smooth operation in ARCEDIT, and the need to run ARC commands is greatly decreased. ARC/INFO release 7 also introduces regions, which are compound polygons. Editing regions is an advanced ARCEDIT topic. Refer to ArcDoc for further information on editing regions.

Before editing polygons, set your edit feature to polygon. When you edit polygons, you can most clearly see the polygons with the command DRAWENVIRONMENT POLYGON FILL (or DE POLY FILL). By default, the polygons will be shaded as light gray, and selected polygons will be shaded yellow.

Remember that there are two basic requirements for polygon topology: that the connecting arcs defining a polygon join exactly at nodes, and that each polygon have an interior label point. Some of the polygon editing commands enforce these requirements for you. For instance, if you split a polygon into two, there will be two label points, one for each polygon.

You can create polygons in two basic ways:

1. Use the arc editing and label editing commands to edit components of a polygon, and then use CLEAN and BUILD to update polygon topology.
2. Use the new polygon editing commands such as ADD, MERGE, and SPLIT to directly update polygons.

During an ARCEDIT session, you can use both methods for updating polygons. The advantage of the former is to simplify the updating of a select number of polygons, and the advantage of the latter is more efficient editing of many polygons.

Editing Arcs and Label Points to Update Polygons

You can edit the arcs and label points that make up polygons, and then run the CLEAN and BUILD commands. When you do so, set your edit feature to arc and label when updating those respective features.

CLEAN and BUILD have the same basic purpose that they do in the ARC program, but the arguments are different. The main difference is that CLEAN resolves intersections while BUILD does not.

You can run both the CLEAN and BUILD commands while you have any edit feature set. ARCEDIT indicates when CLEAN or BUILD is necessary. If you set your edit feature to polygon and receive an error message stating the polygon topology is not current, you definitely need to run CLEAN or BUILD.

```
Arcedit: CLEAN {DEFAULT | fuzzy_tolerance | *}
         {DEFAULT | dangle | *} {DUPSOK | NODUPS}
         {DIFFSOK | NODIFFS}
```

CLEAN will resolve polygon topology by calculating arc intersections, removing small dangling arcs, and merging intersection nodes within the fuzzy tolerance.

The first argument specifies the fuzzy tolerance. Your choices are to accept the default, which is 1/10,000 of the coverage width or height, to type in a fuzzy tolerance, or to enter a fuzzy tolerance by picking two points on the graphics window with the screen cursor.

The second argument specifies the maximum allowed distance for dangling arcs. Again, you can specify the default, which is the value stored in your coverage, type in a dangle distance, or use the cursor.

The third argument affects whether duplicate label points within a polygon will be allowed, provided they have the same coverage IDs. Generally, only one label point should occupy each polygon.

The fourth argument specifies whether duplicate label points in polygons are allowed, regardless of the coverage IDs.

```
Arcedit: BUILD {DUPSOK | NODUPS} {DIFFSOK | NODIFFS}
```

BUILD will rebuild polygon topology. It will not resolve intersection errors. If there are arcs that intersect without dividing at a node, you will receive an error message from BUILD.

Like CLEAN, BUILD will add label points to polygons without label points, and optionally remove duplicate label points with the same or different User IDs.

The DUPSOK or NODUPS choice controls whether duplicate label points with the same User IDs are acceptable. The DIFFSOK and NODIFFS choice specifies whether duplicate label points with different User IDs are accepted.

Using the Polygon Editing Commands

The second option for editing polygons is to set your edit feature to polygons and use the following polygon editing commands. When you use these commands, polygon topology is updated for you. Potential polygon errors, such as dangling arcs, missing label points, or multiple label points, are generally prevented by using these commands.

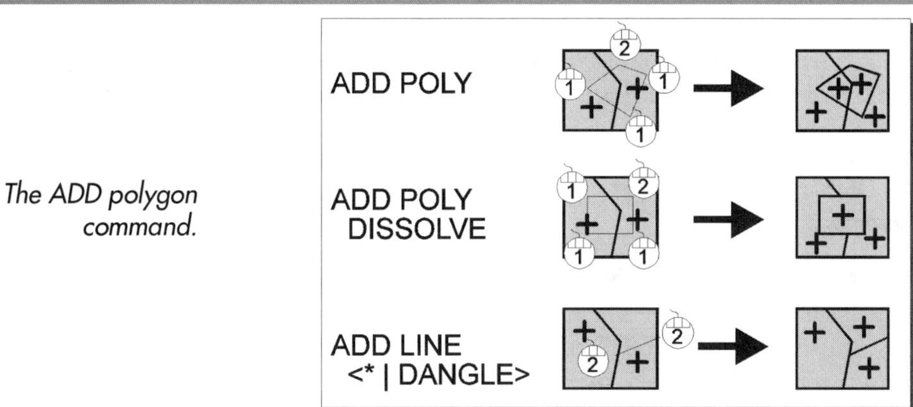

The ADD polygon command.

The ADD command is an important polygon editing command with several usages. When you use ADD with the POLY argument, you will see this submenu:

```
1) Add polygon           2) End polygon          4) Delete last point
5) Delete last polygon  8) Digitizing options 9) Quit
```

Start with input key 1 and successively pick vertices of your new polygon. When you are ready to close the polygon, use input key 2. Input key 9 returns you to the ARCEDIT command prompt.

If you specify the DISSOLVE argument, only a single polygon will be added from your polygon definition. If not, then intersections with existing polygons will be calculated and several or many polygons may result.

When you use ADD with the LINE argument, you will see a different submenu:

```
1) Vertex              2) Node            3) Curve
4) Delete vertex       5) Delete arc      7) Square on/off
8) Digitizing options  9) Quit
```

These input keys work just as if you were digitizing arcs. The difference is that they will be used to split existing polygons into two or more polygons. This command will not create any dangling arcs; instead, intersections are calculated and dangles are removed immediately. Also, when you create new polygons, new label points are automatically created for you.

You cannot use ADD LINE to digitize a new, free-standing polygon because an arc added with this command cannot intersect itself.

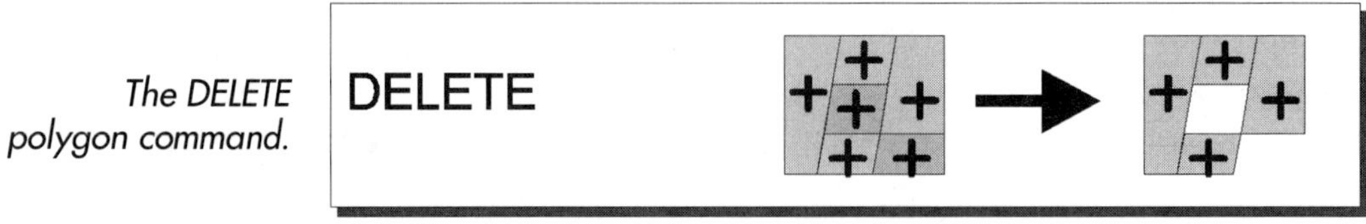

The DELETE polygon command.

You can DELETE polygons in your selected set. The label points for the selected set are removed, and arcs that are not shared by other polygons are deleted too.

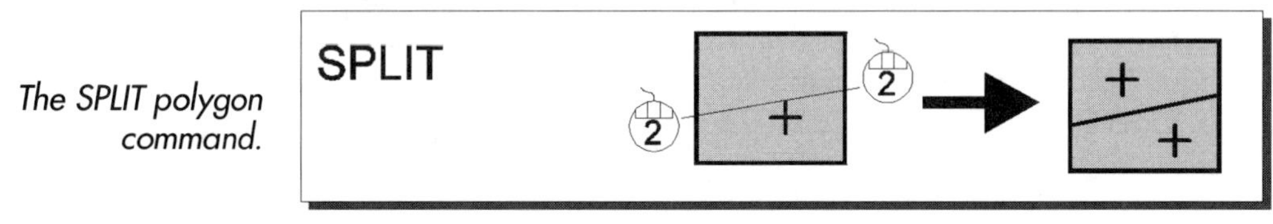

The SPLIT polygon command.

The SPLIT command subdivides polygons in your selected set. When you use SPLIT, you will see this submenu appear:

```
1) Vertex           2) Node          3) Curve
4) Delete vertex    5) Delete arc    7) Square on/off
9) Quit
```

The SPLIT command is similar to the ADD LINE command, but this difference distinguishes the two: SPLIT requires and operates upon a selected set of polygons, whereas ADD LINE operates on any polygons intersected by the lines you define.

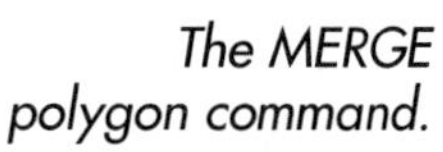

The MERGE polygon command.

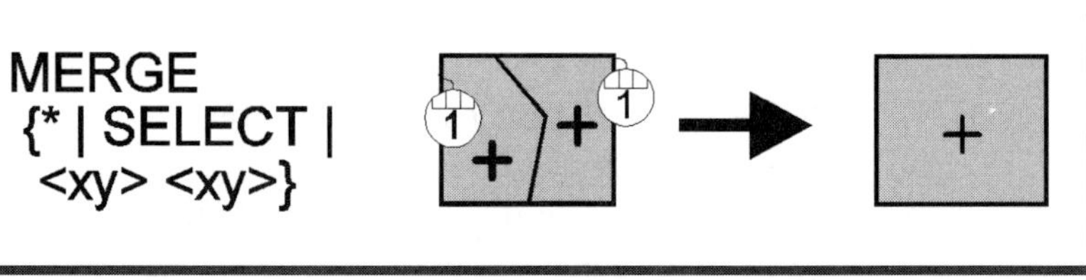

The MERGE command will dissolve two or more polygons into one. Using the * argument activates your screen cursor to select polygons interactively. When you use the SELECT argument, the selected polygons will be merged. The third option lets you specify coordinates for points within polygons. This third option will rarely be used interactively, but might be used within AML programs.

Annotation

Annotation is defined with a number of properties so that you can create attractive maps with clear text placement. Before adding annotation, we need to set some values that control its appearance.

The Annotation Environment

Annotation resides with subclasses in a coverage. The usage of the EDITFEATURE command is bit different from other feature types:

```
Arcedit: EDITFEATURE ANNO.<sub-class>
```

If you want to create a new subclass while you are in ARCEDIT, use this command:

```
Arcedit: CREATEFEATURES ANNO.<sub-class>
```

Following are several important ARCEDIT commands you can use to set default values for annotation.

```
Arcedit: TEXTSET <textset_file>
```

This command specifies the textset to be used during this ARCEDIT session.

```
Arcedit: ANNOFIT <OFF | ON>
```

ANNOFIT specifies whether or not, if you place annotation by two points, the annotation will be stretched or compressed to fit exactly between those two points.

```
Arcedit: ANNOOFFSET {* | y_distance} (* | x_distance}
```

This command is used to specify an offset distance between annotation and the features used to create that annotation.

```
Arcedit: ANNOPOSITION <LU | CU | UU | LL | LC | LR |
         CL | CC | CR | UL | UC | UR>
```

ANNOPOSITION sets the justification for placing annotation. Text justification helps you place annotation precisely along, on top of, below, or to the right or left of other features.

```
Arcedit: ANNOSIZE {* | size}
```

This command sets the size of annotation to be added. The size settings remain in affect until changed.

```
Arcedit: ANNOSYMBOL <symbol>
```

This command assigns a symbol to be stored with the annotation.

```
Arcedit: ANNOFEATURE <NONE | LINE | POINT | POLY |
         NODE | SECTION.subclass | ROUTE.subclass |
         REGION.subclass> {text_item}
```

The ANNOFEATURE command has a dual role. First, it specifies the feature type that annotation will be positioned upon. Second, it can specify an item as a source for creating annotation.

```
Arcedit: ANNOSELECTFEATURE
```

The ANNOSELECTFEATURE command will activate the screen cursor to select features to create annotations from, using a specified item value. Use this command after ANNOFEATURE and before ANNOTEXT. Use key 1 to select and key 9 to finish feature selection.

Editing Annotation

Editing annotation can be quite complex. Here we will illustrate some simple command usages to place new annotation. We'll skip the ANNOADD command because of its complexity. ANNOADD invokes an AML menu for adding annotation. Edit GIScity offers annotation editing menus that are simpler.

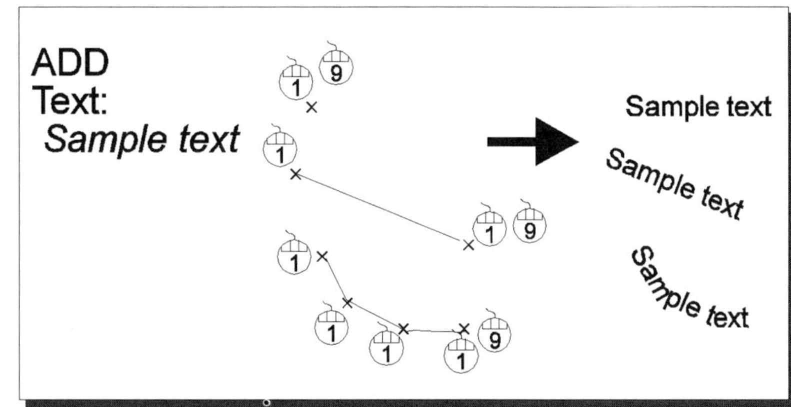

The ADD command for annotation.

The ADD command prompts you to enter a text string. After you've typed in a text string, the screen cursor will be activated to place the annotation. The annotation settings will be used for setting properties of the new annotation. After you've entered an annotation, the text string prompt reappears. You can add a new annotation, or you can just press [Return] on your keyboard to return to the `Arcedit:` prompt.

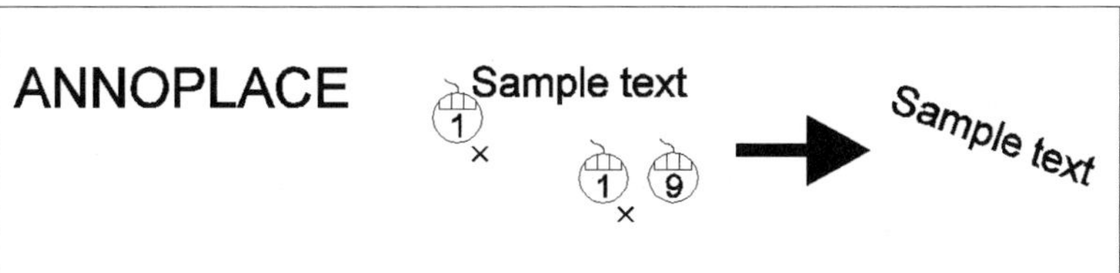

The ANNOPLACE command.

The powerful ANNOPLACE command lets you move or modify the positions of annotation you've placed. The simple usage of this command is 'ANNOPLACE *. When you enter this command, you'll see the following submenu:

```
<1,2 to enter, 4 to remove last point, 5 to remove line, 9
to end>
```

ANNOPLACE lets you change any annotation to one point, two point, or many point annotation. To specify which type of annotation, use key 1 or 2 successively and key 9 to close.

ANNOPLACE will work on many selected annotations, but the result on multiple annotations will be that they are all moved to the same point, which is not desirable. You should normally have only one annotation selected before you use this command.

If you've set a feature type and item with the ANNOFEATURES command and selected some of those features with the ANNOSELECTFEATURES command, the ANNOTEXT command will automatically add annotation using the item values for the text string.

The DELETE command for annotation.

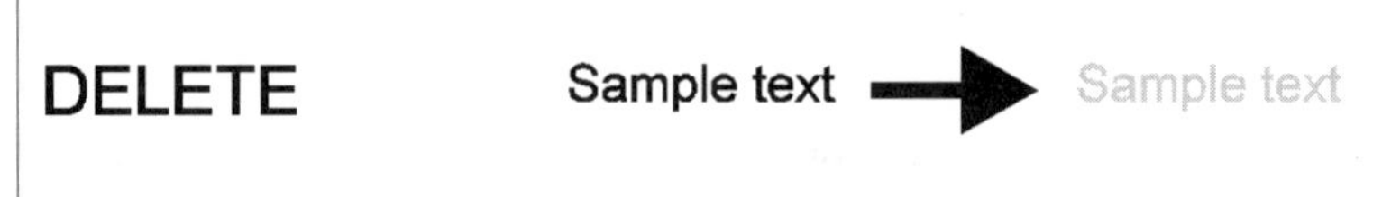

The DELETE command removes the selected annotation from the coverage. This command can be undone with the OOPS command.

The COPY command for annotation.

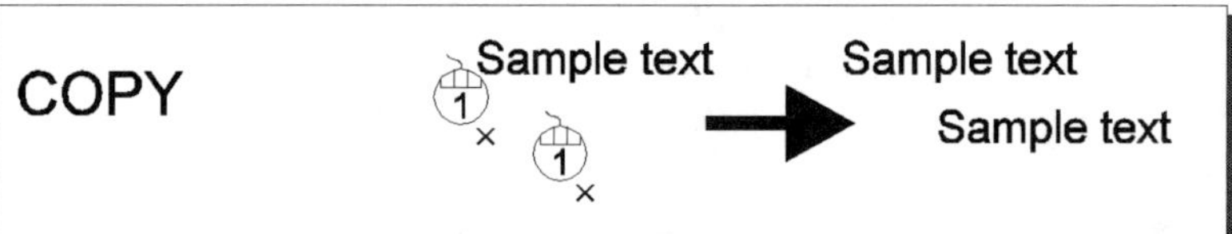

The COPY command copies the selected annotation.

The MOVE command for annotation.

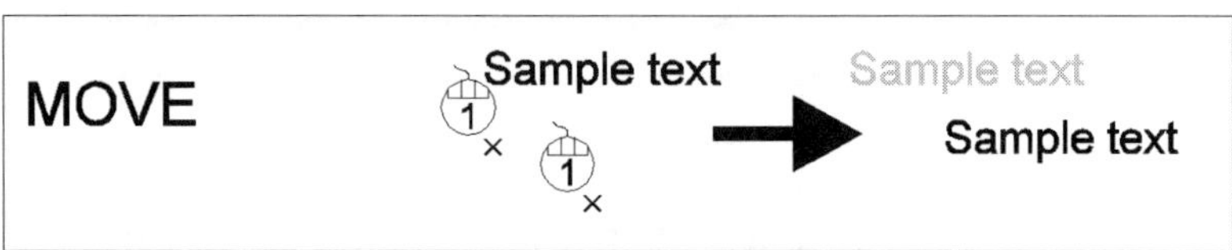

The MOVE command moves the selected annotation.

Tics and Links

We'll finish our discussion of feature editing with two special-purpose feature types.

Editing Tics

Tics are registration points for coverages. They are used for registration on digitizing tables, and for a few ARC commands that perform coordinate adjustments, such as TRANSFORM.

For most feature types in a coverage, the User ID is not important in most applications. However, tics are an exception: the User ID is very important, because it is the basic identifier for specifying common registration points.

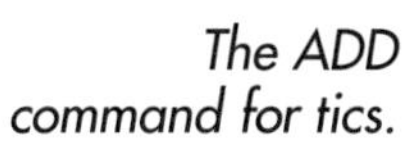

The ADD command for tics.

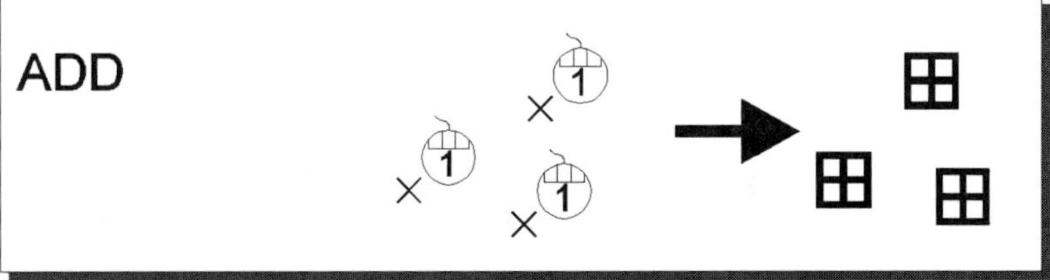

The ADD command for tics has this submenu:

```
1) Add Tic      3) New User ID          5) Delete last tic
6) New symbol   7) Autoincrement OFF    8) Autoincrement RESUME
9) Quit
```

Use key 1 to add a new tic. Key 3 lets you type in a User ID to be applied to the next added tic. Keys 7 and 8 control the incrementing of tic User IDs.

The MOVE tic command.

The DELETE tic command.

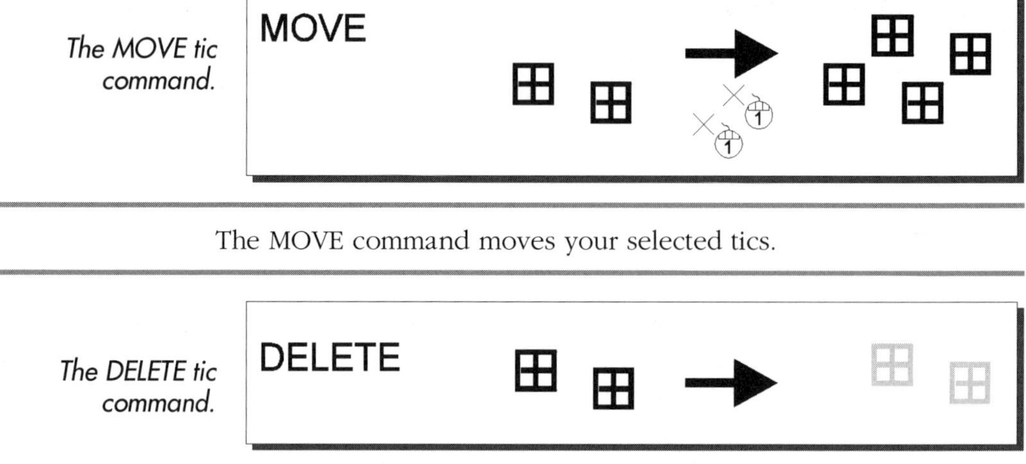

The MOVE command moves your selected tics.

The DELETE command removes your selected tics.

Editing Links

Links are used for only one purpose: to define adjustment vectors for shifting coverage features. They are used by both the ARC and ARCEDIT ADJUST commands.

There are two types of links: normal and identity. Identity links are defined with just one point and are used like a nail to hold down adjustment in that area.

When you set your edit feature to links, if no links exist, eight links will automatically be added around the perimeter of your coverage's map extent.

The ADD command for links.

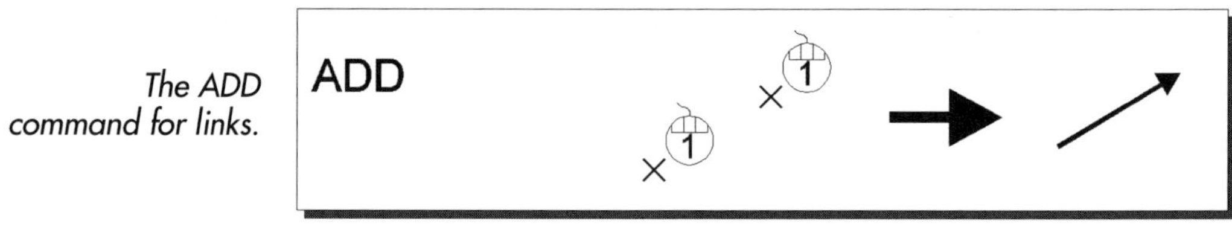

When you use the ADD command, you will see this submenu:

```
1) Free point    2) Snap point   3) Set symbol
5) Delete link   9) Quit
```

Free points are link ends that are placed anywhere. Snap points are link ends that you can snap to existing features using your current snap environment. You should use the LINKFEATURE command before setting snap points to new links.

The DELETE links command.

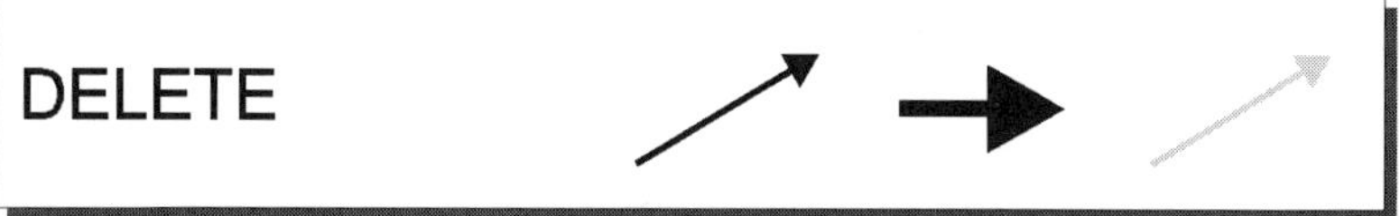

The DELETE command removes the selected links.

The MOVE links command.

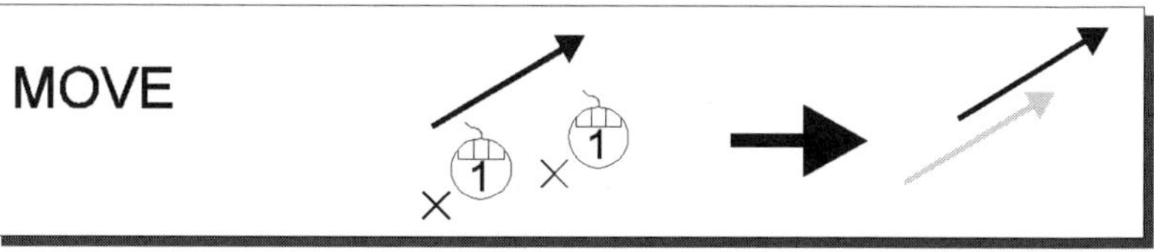

The MOVE command moves the selected links by the offset you interactively define with the cursor.

After you have set links, you can use these ARCEDIT commands to further add or modify links: AUTOLINK, HOLDADJUST, LIMITADJUST, LIMITAUTOLINK, LINKFEATURES. Next, you can perform rubbersheeting through the ADJUST command. Often, you will undo this command and fine-tune your links before running ADJUST again.

Exercise: Stepping Through an Editing Session

This exercise follows up the user exercises from the previous chapter. Refer to those exercises for the commands to start up your ARCEDIT session again.

1. Let's start with arcs. Set your edit coverage to EDGEROAD, edit feature to arc, and enter these draw environments:

```
Arcedit: EDIT EDGEROAD
Arcedit: EF ARC
Arcedit: DE ARC ARROWS
Arcedit: DRAW
Arcedit: DE ARC INTERSECT
Arcedit: DRAW
Arcedit: DE NODE DANGLE
Arcedit: DRAW
Arcedit: DE ARC PSEUDO
```

2. Watch the symbols at nodes and vertices for each of these draw environments.

3. Now, set arc intersections and start adding arcs. If you are using ARC/INFO 7.1 on Windows NT with a two-button mouse, remember to issue the 2BUTTON command before the ADD command.

```
Arcedit: INTERSECTARCS OFF
Arcedit: ADD
```

Add an arc across an existing arc.

```
Arcedit: DRAW
Arcedit: INTERSECTARCS ON
Arcedit: ADD
Arcedit: DRAW
```

Add another arc, also across an existing arc. Notice the difference.

4. Try three different types of arc entry: Line, Circle, and Box. You've already done ARCTYPE LINE because that is the default.

```
Arcedit: ARCTYPE CIRCLE
Arcedit: ADD
```

Pick two points with input key 2. Pick two more points with input key 2. Exit with input key 9.

```
Arcedit: ARCTYPE LINE
Arcedit: ADD
```

Now, try digitizing circular segments within arcs. Start with input key 2, add one or two vertices with input key 1, then mark a point on a circular segment with input key 3. Next, pick the end of the circular segment with input key 1 (or 2). Finish the arc with input key 2. With a little practice, you can become adept at digitizing cul-de-sacs.

```
Arcedit: ARCTYPE BOX
Arcedit: ADD
```

Use input key 2 to mark two corners. A closed arc with four sides will be made for you.

5. Try copying and moving arcs:

```
Arcedit: SELECT MANY
```

Pick several arcs with input key 1, and then close selection with input key 9.

```
Arcedit: COPY
```

Pick two points to set an X and Y shift.

```
Arcedit: SELECT ONE
```

Use input key 1 to pick just one arc.

```
Arcedit: COPY PARALLEL
```

Now pick a point on the right or left side of the selected arc.

6. Try these same steps with the MOVE command. SEL is an abbreviation for SELECT.

```
Arcedit: SEL MANY
Arcedit: MOVE
Arcedit: SEL ONE
Arcedit: MOVE PARALLEL
```

Now try the DRAG command.

```
Arcedit: SEL MANY
Arcedit: DRAG
```

Remember to move the cursor over one of the selected features; keep the mouse button depressed as you move them around.

7. A very useful command is EXTEND. Try this on some dangling arcs:

```
Arcedit: SEL SCREEN
Arcedit: RESELECT DANGLE
Arcedit: EXTEND
```

Set two points with the cursor to specify an extend distance.

8. Now try moving vertices around.

```
Arcedit: SELECT ONE
Arcedit: VERTEX ADD
```

Use input key 1 to add a few vertices, and close with input key 9.

```
Arcedit: VERTEX MOVE
```

Likewise, move some vertices. Close with input key 9.

```
Arcedit: VERTEX DELETE
```

Remove some vertices. Close with input key 9.

```
Arcedit: RESHAPE
```

Start with input key 2, add some vertices with input key 1 and be sure to intersect the selected arc at least once. RESHAPE is like a single command that combines all the VERTEX command usages.

9. Add, delete, and move label points. This is pretty easy.

```
Arcedit: EDITFEATURE LABEL
Arcedit: ADD
```

Use input key 1, close with input key 9.

```
Arcedit: SEL BOX
```

Use input key 1 to mark two corners of a box to collect label points.

```
Arcedit: MOVE
```

Use input key 1 twice to specify the X and Y shift to move label points.

```
Arcedit: COPY
```

Again, use input key 1 twice to specify the X and Y shift to move label points.

10. Editing polygons is fun. First, let's set our environment and try updating polygon topology.

```
Arcedit: EF POLY
Arcedit: DE POLY FILL
Arcedit: CLEAN
Arcedit: EF ARC
Arcedit: ADD
```

Add many arcs that intersect arcs and polygons.

```
Arcedit: CLEAN
```

Watch the new label points appear after the CLEAN command.

11. Now let's use the polygon editing commands.

```
Arcedit: EF POLY
Arcedit: ADD POLY
```

Use input key 1 to mark several vertices and finish polygon with input key 2. Use input key 9 to return to command level.

```
Arcedit: ADD POLY DISSOLVE
```

Try the above steps again. What happens inside the polygon you interactively defined? Now try the third usage of the ADD command for polygons:

```
Arcedit: ADD LINE
```

Use input key 2 to start, add vertices with input 1 while making sure to intersect existing arcs, close the polygon with input key 1, and return with input key 9.

12. Let's try merging and splitting polygons:

```
Arcedit: MERGE
```

Select several adjacent polygons with input key 1. Finish selection with input key 9.

```
Arcedit: SPLIT
```

Now, using input keys 2, 1, and 9, split the polygon you just merged into new polygons.

13. Moving right along to annotation...

```
Arcedit: EDIT EDGEROAD
Arcedit: EDITFEATURE ANNO.NAMES
Arcedit: ANNOSIZE *
```

Pick two points with input key 1 to specify an annotation size.

```
Arcedit: ADD
Text: {type in some text here and press RETURN}
```

Now, place an annotation with just one point using input key 1, then close with input 9.

```
Text: {type in some more text}
```

Place two-point annotation by using input key 1 twice, followed by input key 9.

```
Text: {Press RETURN at this prompt to return to command level}
Arcedit: ANNOFIT ON
Arcedit: ADD
Text: {Add another text string}
```

Place that annotation with two points again. See the difference? Now, try many-point annotation:

```
Text: {Add another text string}
```

Pick about 4 or 5 points with input key 1, and close with input key 9.

Well, what do you know? Now you can place a curved street name along the windiest road you can find.

14. Placing annotation is tricky business. The ANNOPLACE command is your ticket to moving annotation easily to another position.

```
Arcedit: SELECT ONE
Arcedit: ANNOPLACE
```

Use input key 1 once, twice, or several times to reposition the annotation anywhere you want.

15. Let's call it a wrap. Let's NOT save any of the changes we made:

```
Arcedit: QUIT NO
```

You'll be back at the ARC program. To quit ARC, type:

```
Arcedit: QUIT
```

Now you are back at your workstation's operating system prompt. We have now executed many of the basic ARCEDIT commands to edit coverages.

Starting a Session in Edit GIScity

Window, Select, and Session Commands in ARCEDIT

The greatest expense in building a GIS is the automation of geographic data. Over time, this expense far exceeds your investment in GIS software and computer hardware.

We now cover the basics of ARCEDIT through another application that you've installed from the companion CD-ROM, called Edit GIScity.

Entering Geographic Data

ARC/INFO supports a variety of ways to enter geographic data:

- **Table digitizing.** You can tape an existing manuscript map on a digitizing table, making sure that it's registered, and then enter geographic features by tracing the map. This is the most common method for entering geographic data, but it has a significant limitation: what you digitize cannot be more accurate or up-to-date

than your source map. Table digitizing is still popular but is gradually becoming replaced by other methods.

- **Screen digitizing.** If you have access to digital orthophoto maps (aerial or satellite images which have been corrected for scale distortions) or scanned manuscript maps, you can display those images in the background of your graphics window and digitize them with the mouse. This is also called *heads up digitizing*. You can register and display images with the Image Integrator functions within both ARCPLOT and ARCEDIT.
- **Raster to vector conversion.** Raster to vector conversion is the semi-automatic conversion of raster information (such as the pixels on a television set) into vectors (mathematically defined line segments). Until recently, this type of conversion was seldom used because it was usually more trouble to clean converted data than simply to digitize it. However, GIS software has now improved to the point where many users can more efficiently convert scanned maps than digitize them. A new ESRI product called ArcScan guides you through interactive conversion of raster data to vector data. While the ArcScan program is beyond the scope of this book, ARCEDIT is used to clean and correct data converted from raster images.
- **Coordinate geometry input.** The most accurate method to create geographic data is to use information from precise surveys. *COGO* (short for **coordinate geometry**) is an industry term for geographic software that converts raw survey information, usually distances and angles, into geographic coordinates. ESRI offers an optional module called COGO for entering and manipulating survey information. COGO comprises additional commands and data structures within ARCEDIT and is also beyond the scope of this book.
- **Uploading field collected coordinates.** Some new survey instruments, such as *total stations* and *GPS* (Global Positioning System) *receivers*, collect positional information and calculate coordinates internally. COGO commands are not required for converting data from these instruments, and this type of data can be directly downloaded into ARC/INFO. While it is possible to import this data

directly into ARCEDIT, it is more common to use some ARC commands such as GENERATE or DXFARC to import this data. These commands are discussed later in the book.

- **Format conversion.** Finally, geographic information can be converted from other popular CAD and GIS software. The ARC program contains many commands to import and export geographic data from most software systems used for automated mapping. After format conversion, ARCEDIT is used for clean-up and attribute assignment.

ARCEDIT is the ARC/INFO program that edits and corrects *coverages*, the basic unit of storage (see the section of select commands for further elucidation of coverages.). If you are digitizing maps with a digitizing table or mouse, or by entering survey information with the COGO commands, the entire map automation procedure is done within ARCEDIT. If you are importing geographic information from external devices such as GPS equipment, or converting from other CAD and GIS systems, you typically import that data into the ARC program and then use ARCEDIT for refinement and correction.

> **NOTE:** *ARC/INFO contains an optional program called GRID that offers display and editing of the elements of a grid. A grid is a raster geographic dataset, and consists of a two-dimensional array of cells with attributes assigned to them. (A coverage is a vector geographic dataset.) While ARCPLOT and ARCEDIT allow the background display of images, which can be simply created from a grid, these programs do not provide editing of grids. The GRID program is important to many users of aerial photogrammetric and satellite imagery, but is beyond the scope of this book.*

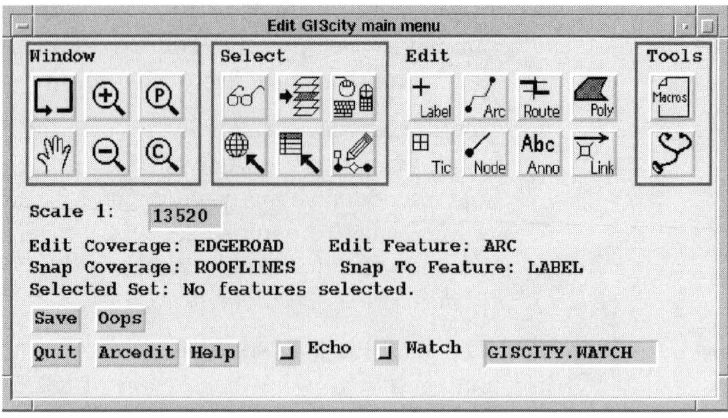

The Window, Select, and Tools commands in the main menu of Edit GIScity are documented in this chapter.

Starting Edit GIScity

To begin Edit GIScity on a UNIX workstation, type the following command:

```
arc &run $GCHOME/edit/start
```

To begin Edit GIScity on a workstation running Open VMS, type:

```
arc &run $GCHOME:[edit]START
```

Once you begin Edit GIScity, there are no differences between UNIX and Open VMS in the user interface of Edit GIScity. As in Display GIScity, the only significant difference you will see is between the two windowing systems in common use: OpenLook on Sun SPARCstations and Motif on all other workstations.

NOTE: *On some workstations, you may have to start the application by typing* `arc` *in the console window first; then in the Arc window type :*

```
&run $GCHOME/edit/start (UNIX) or,
&run $GCHOME:[EDIT]START (VMS)
```

If you are a Windows NT user, start the Arc program from the Start button, then type the following in the Arc console window.

```
&run $CGHOME/edit/start
```

After you type in one of the above commands to start Edit GIScity, you will see a menu in the middle of the workstation display asking you to specify an edit coverage. If this menu does not appear, follow the same troubleshooting steps documented in Chapter 1 for Display GIScity.

After you select an edit coverage (more on this setting later in this chapter), you will see a menu for specifying coordinate input from the digitizer, mouse, or keyboard. After you go through both menus, the display will look like the following illustration.

After setting the edit coverage and coordinate input, you'll see your edit coverage in the graphics window, along with the Edit GIScity main menu.

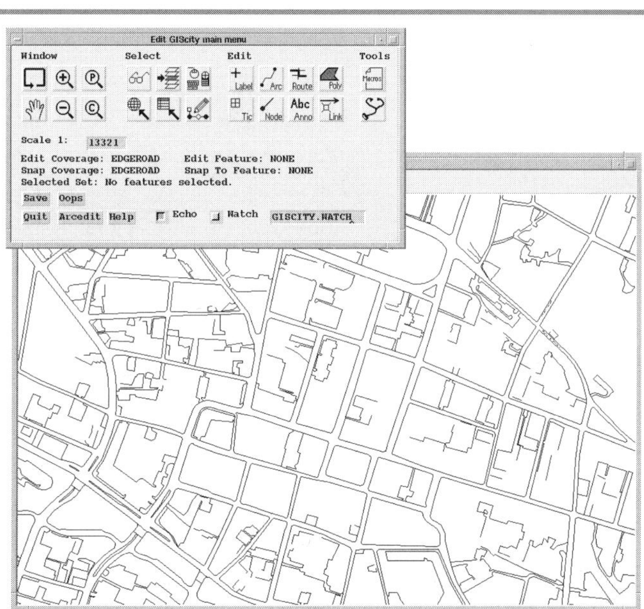

The Window Commands in Edit GIScity

The Edit GIScity Window commands are nearly identical to those in Display GIScity. The primary difference is that Edit GIScity has no Refresh Window command. Another change you'll notice is that just one coverage is initially set for display in Edit GIScity. Later in this chapter, you'll find

out how to display all of the themes as a background to your edit coverage.

The Window commands in the Edit GIScity main menu.

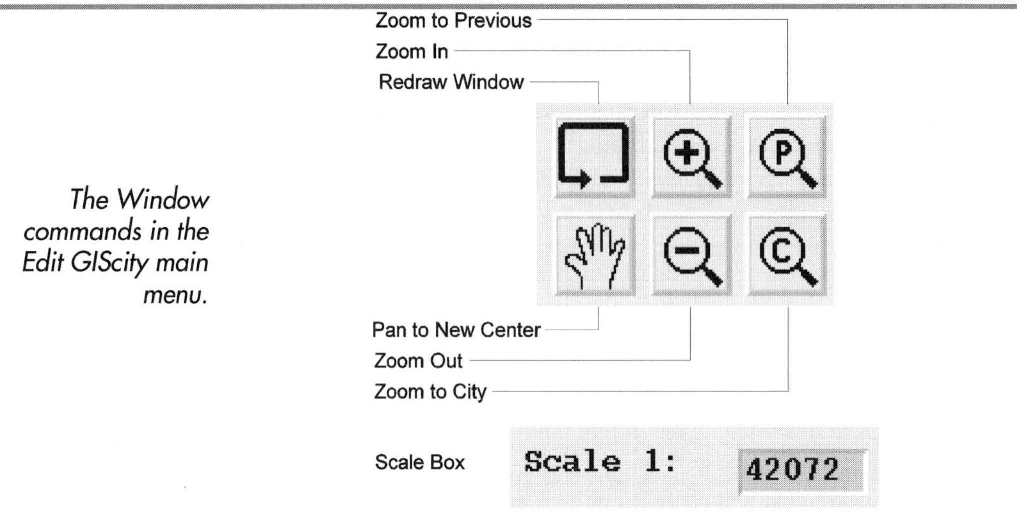

Redraw Window redraws the graphics window at the same scale. This command is used when you change your edit coverage or modify settings for drawing your edit coverage.

Zoom In lets you increase the magnification in your graphics window. After you click on this command, using the left mouse button click twice on the screen to mark two corners for a new map extent.

Zoom to Previous will redraw your graphics window at the previous map scale and extent.

Pan to New Center will redraw your graphics window around the point you click on in the graphics window.

Zoom Out redraws your graphics window by zooming out a factor of two.

Zoom to City will draw the graphics window to span the downtown Santa Fe area spanned by coverages such as EDGEROAD and CONTOUR.

Scale Box operates identically to the same command in Display GIScity. For some GIS applications, such as utility mapping, the relative size and placement of symbols and text at specified scales is critical. This command lets you see the GIS database with exactly the scale proportions you would expect in the map plot.

The Select Commands in Edit GIScity

Set Coordinate Input
Set Edit Coverage
View Manager

Select Features Spatially
Select Features Logically
Set Draw Environment

The Select commands in the Edit GIScity main menu.

Three of the Select commands — View Manager, Select Features Spatially, and Select Features Logically — are very similar to their counterparts in Display GIScity. The other three Select commands — Set Coordinate Input, Set Edit Coverage, and Set Draw Environment — are new.

The Set Edit Coverage Menu

A coverage is the basic unit of storage in ARC/INFO. A coverage can contain:

- Line features (arcs and route systems)
- Point features (label points)
- Area features (polygons and regions)
- Annotation
- Rubbersheeting vectors (links)
- Registration points (tics)

In Display GIScity, you interacted with *themes*. In Edit GIScity, you will be using *coverages* so that you can become familiar with this fundamental storage unit. A coverage is the actual geographic data set or database that includes all information available about a related group of features. A theme is a way of looking at that coverage within a certain context. At this point, you don't have to worry about distinguishing between the two. Display GIScity automatically invokes themes and Edit GIScity automatically invokes coverages.

Most of the themes you've already seen are based on coverages. The exception in GIScity is the Orthophoto theme, which is a photographic image of downtown Santa Fe. Two other exceptions are grids and TINs, which are data sets that are not documented in this book.

NOTE: *In many ARC/INFO applications, you might store data within a map library. A set of tools called Map Librarian organize coverages into layers (like themes) which are subdivided into tiles (map sheets). While this strategy was formerly useful for huge data sets, the new ArcStorm functions released with ARC/INFO 7.0 reduce the need to create explicit map libraries.*

After you begin Edit GIScity, you will see a menu asking you to select an edit coverage. You'll see the actual name of the coverage (like EDGEROAD or CONTOURS).

In ARCEDIT, you edit one coverage at a time. You must select a coverage to begin Edit GIScity, and you can change your edit coverage at any time. Watch the messages on the Edit GIScity main menu to see what your current edit coverage is.

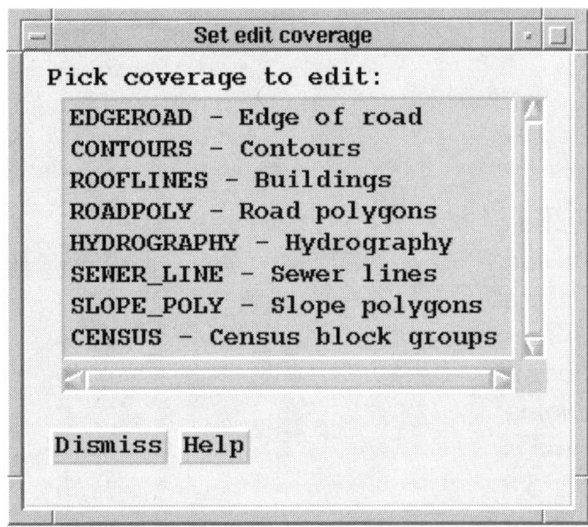

When you begin Edit GIScity, you will see the Set Edit Coverage menu appear in the middle of your workstation screen.

The Set Coordinate Input Menu

There are three ways to enter geographic positions in ARCEDIT: trace map features using a digitizing table; use the mouse to edit maps; or use the keyboard to type in geographic positions (usually survey data). When you begin Edit GIScity, the mouse is the default coordinate input device.

All ARCEDIT edit commands accept input from the mouse, so for now, we'll keep things simple by primarily using the mouse to edit maps in this and succeeding chapters.

Using the Orthophoto theme (the photographic map of downtown Santa Fe) as a background, we'll do some heads up digitizing. In the previous chapters on the command line use of ARCEDIT, we have covered the basic commands for setting up the digitizer. After you've learned how to set up the digitizer for coordinate input, you can return to this menu and use it to edit maps with a digitizing table.

NOTE: *If you are a Windows NT user, the Edit GIScity application automatically applies the 2BUTTON command for you to remap the input keys from a three-button mouse to a two-button mouse. Consult ArcDoc for details on 2BUTTON.*

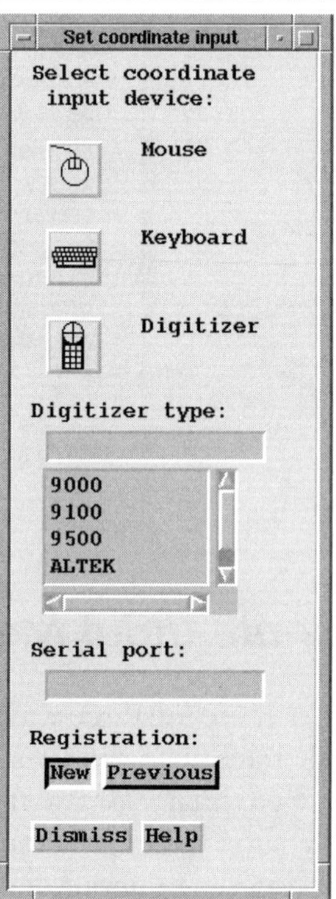

At any time in Edit GIScity, you can activate any of the three input devices to edit maps. Changing coordinate input is simple with the Set Coordinate Input menu.

In Display GIScity, we mainly used input keys 1 and 9. In Edit GIScity, however, many commands use the other nine input keys. As a reminder, here's how the nine input keys are implemented with the mouse in ARC/INFO:

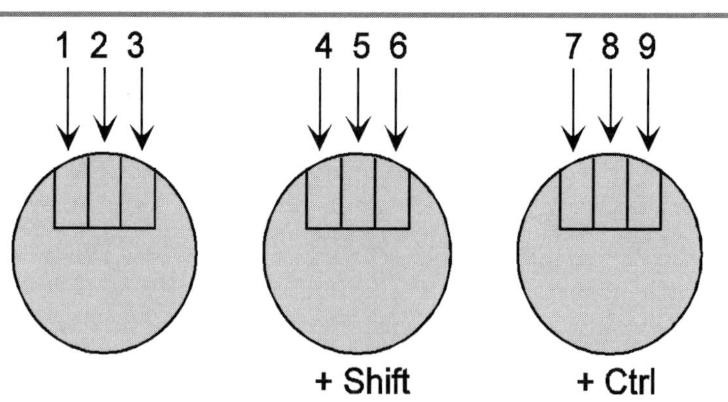

To execute commands with input keys four through six use the Shift key while clicking the mouse button. Use the Control key for input keys seven through nine.

Spatial Feature Selection in Edit GIScity

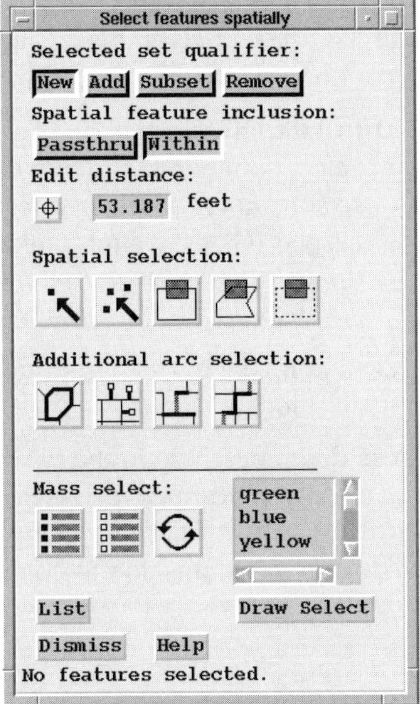

The Select Features Spatially menu.

In Edit GIScity, you select features in the edit coverage. Selection of features is generally the same in Edit GIScity as in Display GIScity. However, there are some differences worth noting:

- There is a new selected set qualifier: **New**. When this qualifier is chosen, a fresh selected set is made every time you perform a selection. (In Display GIScity, this would be the same as clearing your selected set and selecting the Add qualifier before performing each selection.)
- Edit Distance in Edit GIScity is essentially the same as the Search Tolerance in Display GIScity. The difference is semantic.
- You cannot select features within a circle in Edit GIScity.
- You have four additional selection commands that operate on arcs.

- You can have only one selected set in Edit GIScity. (In Display GIScity, you could define any number of selected sets.) You cannot read or write selected sets in Edit GIScity.
- You don't have a Clear Select command in Edit GIScity. Instead, you have both Select All and Select None commands.
- In Edit GIScity, the Highlight command is replaced by the Draw Select command. The two commands are similar, but Draw Select sets the color of selected features after every redraw command. In Display GIScity, after a redraw, only the selected features were drawn. In Edit GIScity, all features are drawn, but selected features are drawn with the specified color.
- In Edit GIScity, the List button will quickly display the attributes of all selected features.

These differences within the two GIScity applications closely reflect the actual differences between selecting features in the ARCEDIT and ARCPLOT programs. These menus are designed to give you insight into making an easy transition to command-level ARCEDIT and ARCPLOT.

These qualifiers at the top of the Select Features Spatially menu set the context for the Selection commands.

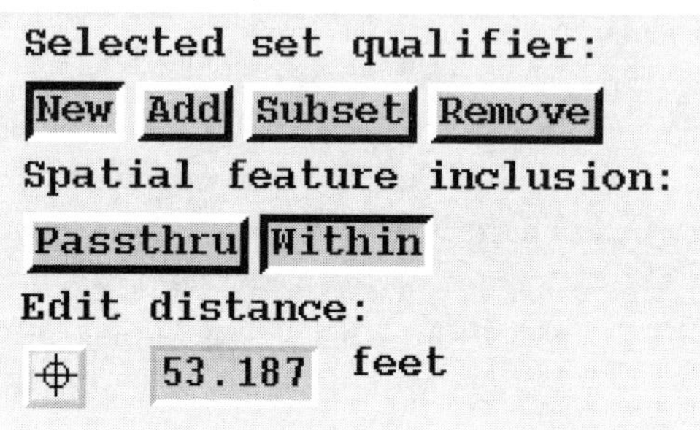

The Spatial Selection commands work the same as those in Display GIScity. The Additional Arc Selection commands are new, and apply only to arcs.

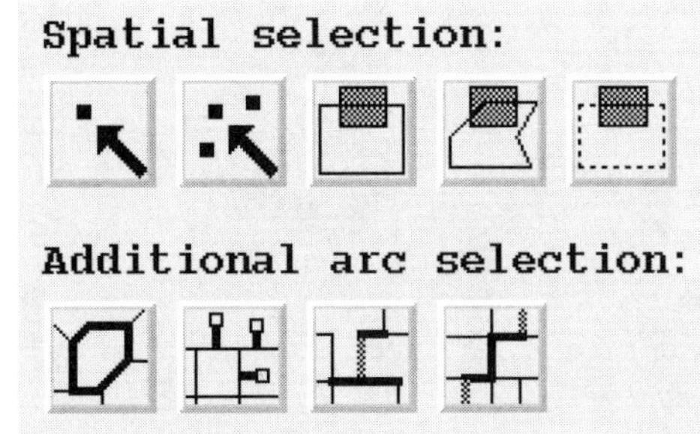

The Mass Select commands are simpler in Edit GIScity than in Display GIScity. You only have one selected set in Edit GIScity, and you cannot read and write selected sets.

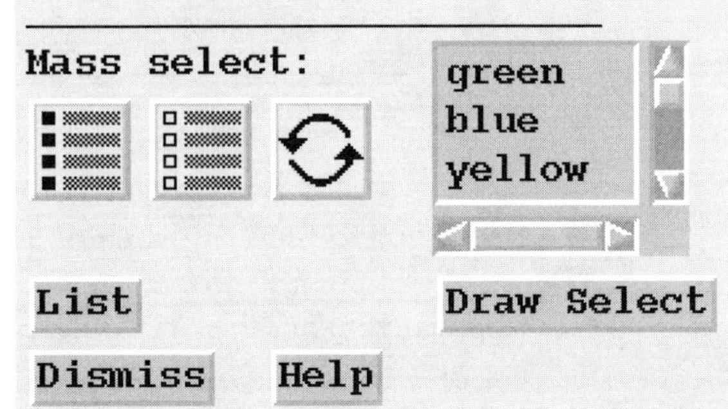

NOTE: *You must invoke a menu for an Edit feature before either of the Select Features menus can be invoked. Both the Select Features Spatially and Select Features Logically menus work on the edit coverage and whichever edit feature you have set with one of the Edit feature commands.*

Logical Feature Selection

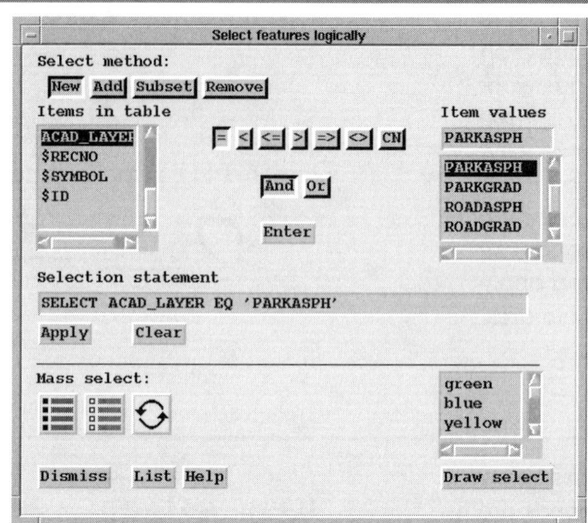

The Select Features Logically menu in Edit GIScity.

Selecting features by matching attribute values is nearly identical to the same process in Display GIScity. As in the Select Features Spatially menu, the only differences are:

- ❑ There is another Select Method qualifier—New.
- ❑ The Mass Select commands are simpler.
- ❑ You can List selected features.
- ❑ Selected features are drawn with the color you chose after every redraw or selection.

The Draw Environment in ARCEDIT

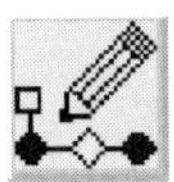

At any time in ARCEDIT, we have only one edit coverage. The draw environment is designed to show clearly how coverage features are connected, and where potential errors may be.

In a GIS, it is important that certain features be properly connected. For example, to draw buildings as shaded areas in ARC/INFO, all the arcs that form the perimeter of a building must connect exactly. If any arcs do not meet precisely, the shaded area cannot be identified and drawn. The draw environment can show that arcs which seem connected on visual inspection do not actually join exactly. Other commands can be used to correct those mismatches.

In addition to the edit coverage, you can set a drawing environment for a backcoverage. In ARCEDIT, there are three types of coverages you can interact with:

- The Edit coverage contains the features that you are editing at the moment.
- You can set and display a Backcoverage to provide a reference to the edit coverage, but the back coverage is not edited.
- A Snap coverage defines which existing features are snapped to when adding new features to the edit coverage.

In Edit GIScity, setting an edit coverage is always mandatory. Setting a back coverage and snap coverage is optional. When you set a snap coverage in Edit GIScity, you are also setting a back coverage. You will find yourself doing this most of the time, so Edit GIScity makes snap and back coverages easy.

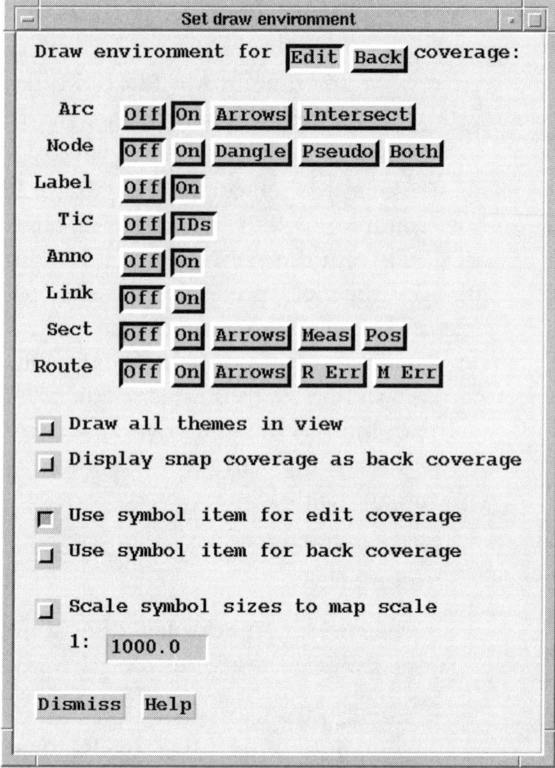

The Set Draw Environment menu in Edit GIScity.

The Edit/Back choice at the top of the menu toggles between displaying current settings for each feature type for either edit coverages or back coverages. When you click this choice, the highlighted choices below will change to the current draw environment settings.

All feature types have an On or Off setting. (Tics are an exception; they have an IDs setting, which will draw the tics together with an ID number used for map registration on a digitizer.) Some of the feature types have additional settings, such as Dangle for nodes or Intersect for arcs, which we'll cover later.

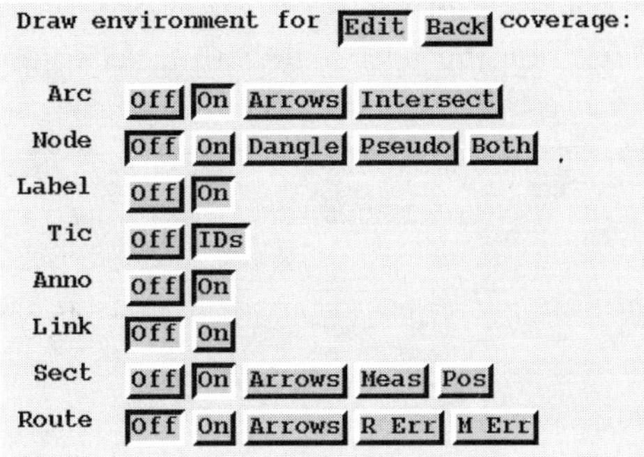

The available choices in setting draw environments.

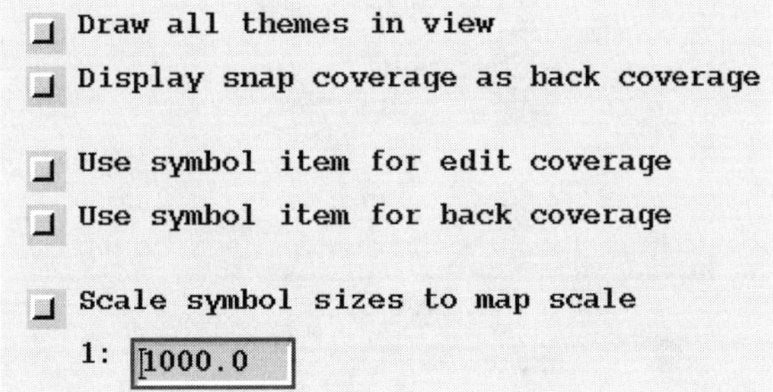

Additional options for setting draw environment in Edit GIScity

There are five additional options for draw environments. The three following illustrations show sample graphics window displays with some of these options.

Clicking on the Draw All Themes in View check box will display all themes as they appear in Display GIScity when you next redraw the graphics window.

After you've set a snap coverage, the Display Snap Coverage as Back Coverage check box will specify that the snap coverage be drawn.

You can either display edit coverage or back coverage as simple points, lines, and areas, or you can set the Use Symbol Item For Edit Coverage and Use Symbol Item For Back Coverage check boxes to specify that point, line or shade symbols be drawn for the features.

Normally, symbols appear at a fixed size on your graphics window. You can set the Scale Symbol Sizes to Map Scale check box, together with typing in a scale value, to control the relative size of symbols on the graphics window.

The Use Symbol Item default settings—off—draws simple features in Edit GIScity.

Clicking on the Use Symbol Item check boxes applies stored symbols to features.

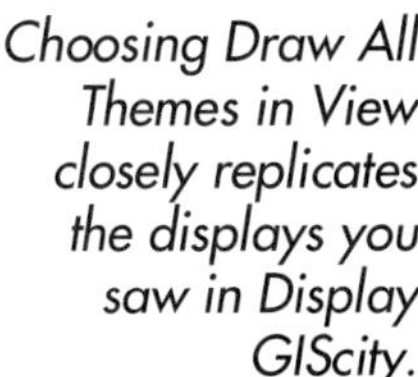

Choosing Draw All Themes in View closely replicates the displays you saw in Display GIScity.

As you use Edit GIScity, you will need all of these options. Sometimes you'll want a clean, simple display that shows only the edit coverage. At other times, you'll want to see how the finished map product will appear. The Set Draw Environment menu gives you control and flexibility in managing the appearance of coverage features in your graphics display.

The View Manager Menu in ARCEDIT

A key difference between Edit GIScity and Display GIScity is that Display GIScity is optimized to deliver geographic information through multiple themes defined in a view, while Edit GIScity is designed to make corrections efficiently to one coverage at a time.

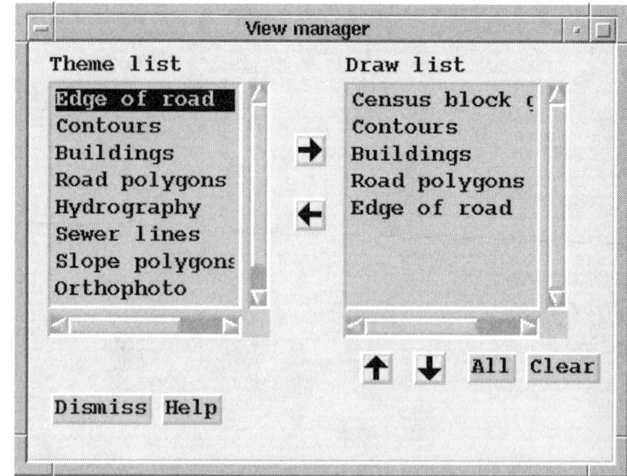

The View Manager in Edit GIScity is identical to the one in Display GIScity.

The draw environment in ARCEDIT controls the display of the edit coverage. Yet, you can display background themes using views in Edit GIScity. To see the themes specified in the current view, be sure to enable the Draw All Themes in View check box in the Set Draw Environment menu.

The Tool Commands in Edit GIScity

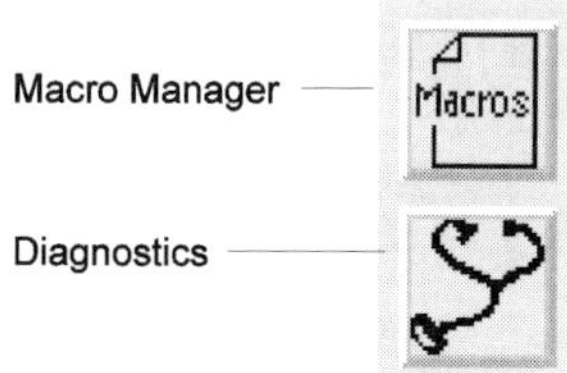

Tools commands in the Edit GIScity main menu.

Running Macros

The Macro Manager in Edit GIScity operates identically to the one in Display GIScity. We'll reserve discussion of this menu until the last part of the book when we cover writing simple AMLs.

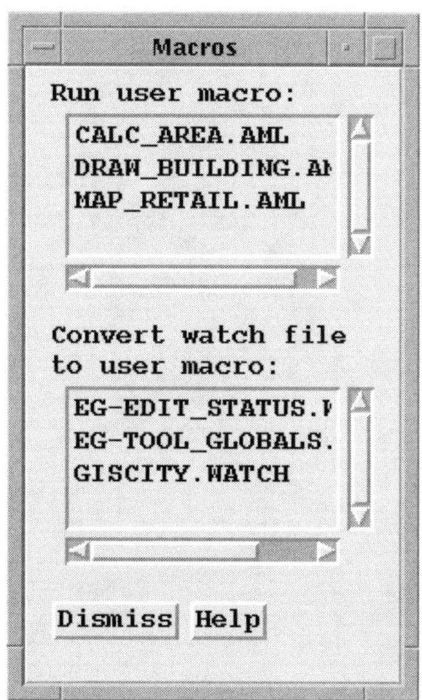

The Macro Manager in Edit GIScity.

Diagnostics

The Diagnostics command offers a quick way to survey all of your substantial environments and settings in ARCEDIT. When you understand the information in each line of this display, you will have mastered the fundamentals of ARCEDIT.

```
arc
Change 1 Editing NONE Total=0 A/D=0,0 Orig=0
Current edit coverage: /GISCITY/DATA//EDGEROAD
Current node angle item: NONE
Edit coverages: /GISCITY/DATA//EDGEROAD
Info files:
Current snap coverage: NO CURRENT SNAP COVERAGE
Snap coverages:
Snap tolerance: 5.851
Snapping off from NONE to first NONE in snap cover
Link features: Edit is FREE  Snap is FREE
Transfer coverage: NO CURRENT TRANSFER COVERAGE
Transfer feature: NONE
Transfer items: $ALL
Snap transfer items: NONE
Tolerances:  Edit=58.506  Weed=5.851  Grain=5.851
Snap nodes within 5.851 to first node
Mapextent 1730067.081,1703446.063,1735845.794,1709296.688 Scale=21594.004
Background coverage(s):
Environments     Drawselect symbol: 0,yellow
                 Drawselect option: ONE
Draw: ALL OFF
Draw: ARC ON
Draw: NODE OFF
Draw: LABEL ON
Draw: ANNO ON
Draw: TICS IDS
Draw: LINK OFF
Back: ALL OFF
Coordinate input: MOUSE
Rotation: Set=0.000  Last=0.000
Text symbolset   : font.txt
Line symbolset   : /arcexe70/symbols/COLOR.lin
Marker symbolset: /arcexe70/symbols/COLOR.mrk
Shade symbolset: /arcexe70/symbols/COLOR.shd
Edit tolerance        : 58.506
Node Snap tolerance : 5.851
Weed tolerance        : 5.851
Grain tolerance       : 5.851
Snapping tolerance  : 5.851
Tracing parameters     arrowhead length(pixel): 32
                       arrowhead width(pixel): 3
                       backtrack: YES
                       branch priority: STRAIGHT
                       dash size: 20.000
                       actions at EOL: NOMANUAL
                       actions at EOS: STOP
                       fan size(degree): 45.000,90.000
                       foreground value: 1
```

Clicking the Diagnostics icon displays a pop-up menu.

Session Control in Edit GIScity

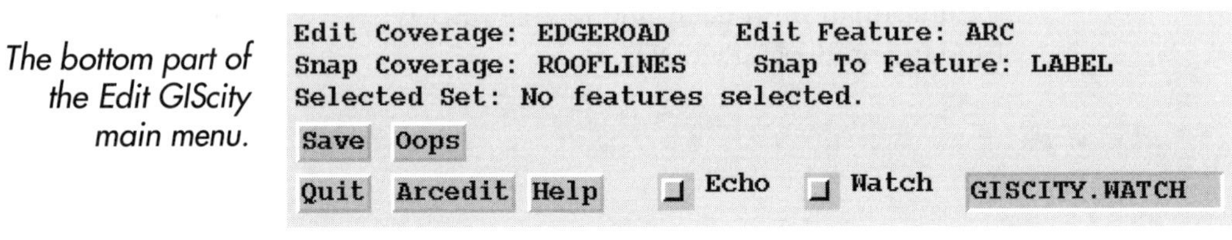

The bottom part of the Edit GIScity main menu.

As in Display GIScity, the bottom section of the Edit GIScity main menu provides message lines that summarize the current Edit GIScity environment and provide some general session controls.

The three message lines tell you the current Edit Coverage, Edit Feature, Snap Coverage, Snap to Feature, and Selected Set. The Echo

and Watch check boxes and Watch text input field operate the same as those in Display GIScity.

Saving and Undoing Changes

Perhaps the most widely used command in ARCEDIT is Oops! Whenever you perform an edit, if the result is not what you wanted, you can undo that change by clicking the Oops command.

At any time in Edit GIScity, you can permanently save edit changes by clicking the Save button. When you do so, you will see a menu asking you if you really want to save the changes made.

NOTE: *You can click Oops successively to undo all the changes you made in the edit session. However, if you have saved your changes, you can only go back to the most recent save. Save has the same effect as exiting ARCEDIT, keeping the changes, and returning right away.*

Exiting ARCEDIT

When you click the Quit button, if you have made any edit changes, a menu will appear asking you if you want to keep the effect of those changes. By answering Yes, you will make those changes permanent. Answering No will quit Edit GIScity without saving changes.

Geographic Features in Edit GIScity

Coverage Features

The Edit commands are the meat of Edit GIScity. The following chapters include discussions and exercises using each of these commands for editing a type of coverage feature.

Each of the Edit commands invokes a menu to edit a specific type of features in a coverage.

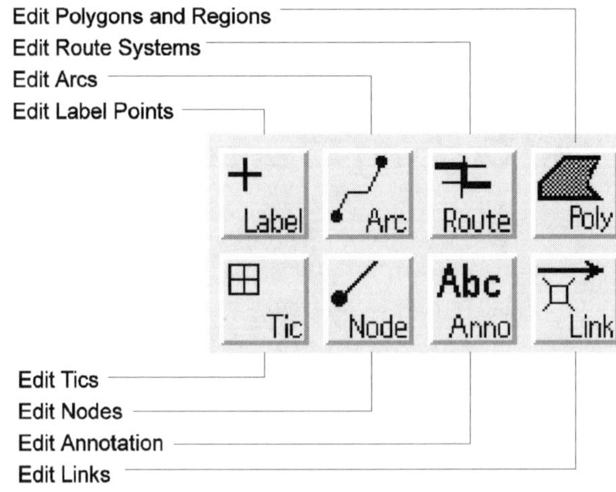

Let's quickly define the types of geographic features you can store in a coverage. We will define these features in greater detail in the following chapters.

- **Arcs** are connected line segments. (Arcs in ARC/INFO are not the arcs you learned in high school geometry because they are generally not segments of a circle.) An arc can have up to 500 connected line segments. Attributes for arcs can be optionally stored in an Arc Attribute Table (AAT).
- **Nodes** are endpoints to arcs. They cannot be defined separately from arcs, but can optionally have attributes stored in a Node Attribute Table (NAT).
- **Label points** have a dual identity. (1) They can be used to locate small geographic features, such as wells. Attributes for points can be optionally stored in a Point Attribute Table (PAT). (2) Or, label points, together with arcs, can define polygons. Attributes for polygons can be optionally stored in a Polygon Attribute Table (PAT).
- **Polygons** are areas bounded by a closed perimeter of arcs and contain one interior label point. Attributes for polygons are stored along with the label point in a Polygon Attribute Table (PAT).

- **Route systems** are linear networks. They comprise sections, which can span partial or complete arcs, and routes, which are collections of sections. Routes have attributes stored in a Route Attribute Table (RAT).
- **Regions** are compound polygons. Attributes for regions are stored in special Polygon Attribute Tables with defined subclasses.
- **Tics** are registration points. They are used for registering maps on a digitizing table and they can also be used to transform map coordinates for coverages. Tics do not have attribute tables.
- **Annotation** is descriptive text stored in a coverage. Annotation is considered descriptive, and not a geographic feature in its own right, but is important in delivering information in a GIS. Optional Text Attribute Tables (TAT) can be defined for a coverage.
- **Links** are rubbersheeting vectors. They are rarely displayed in a GIS, but are created for adjusting maps. Links do not have attribute tables.

The Snapping Environment in ARCEDIT

Snapping is an ARC/INFO technique that ties newly added geographic features to existing features within a distance tolerance. For instance, if you are digitizing segments in a road, you need to ensure that a new arc is exactly connected to an existing arc.

Snapping is a powerful editing technique. It will be explored in the next three chapters in the context of each edit feature.

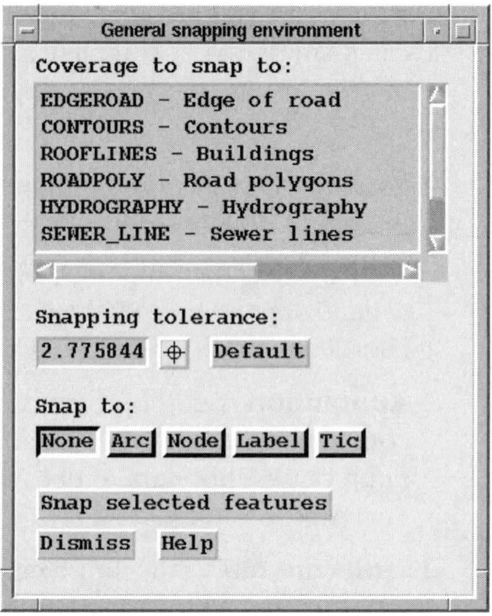

The Snap menu, invoked from the edit feature menus.

Exercise 1: Starting an Edit Session

In this chapter, we've introduced Edit GIScity and covered some preliminary steps for editing coverages. In this exercise, you will start an edit session and get ready for the exercises in the next three chapters. Some of the menus we've shown in this chapter will be employed in the later exercises.

1. Start Edit GIScity on your workstation. For UNIX systems, type `arc` at your console window, then
   `&run $GCHOME/edit/start`
2. For VMS systems, type `arc` at your console window, then
   `&run  $GCHOME:[EDIT]START`
3. For Windows NT systems, start Arc from the Start button and type
   `&run $GCHOME/edit/start`
   at the arc console window.

4. The Set Edit Coverage menu will appear. Select the first edit coverage, EDGEROAD, from the menu. Then click the Dismiss button.

5. The Set Coordinate Input menu will appear. Select the Mouse icon. You have set your coordinate input to the mouse.

6. After the graphics window is drawn and the Edit GIScity main menu appears, try using each of the Window commands in the Main Menu. Although you will see less map information in your graphics window, these commands will work just like those in Display GIScity.

7. Click on the Set Draw Environment command in the Select command group. This is the icon that looks like a pencil pointing to a line with a square, diamonds, and circles. From the menu that appears, click on the Draw All Themes in View check box. Next, click the Redraw Window command on the main menu. Now, your graphics window will look similar to Display GIScity.

That's enough to get started. In the next three chapters, we'll use the other menus in this chapter, together with the Edit feature commands.

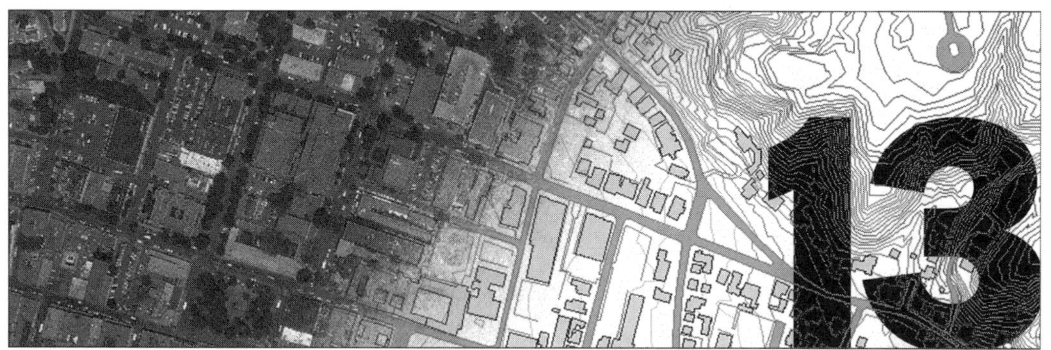

Editing Lines in Edit GIScity

Arcs, Nodes, and Route Systems in ARCEDIT

Arcs are the central type of geographic feature in ARC/INFO. The importance of arcs is reflected in the name of ESRI's flagship GIS product, ARC/INFO.

An **arc** is a series of connected line segments. With arcs, you can model network features such as streets, utility lines, and streams. Arcs are also used to delineate the boundaries of areal features such as lakes, administrative districts, habitats, and environmental zones.

Arc Attribute Tables

Much of ARC/INFO's power and utility derives from the integration of the spatial data that defines an arc with the tabular data you can define for each arc. The tabular data is stored in an *arc attribute table* (AAT).

Because ARC/INFO is a relational database management system, you have the freedom to define additional *items* (the ARC/INFO term for

database rows or fields) in the AAT. Some items you can define are street names, speed limits, flow capacity, classification, and line symbols.

A keystone in ARC/INFO is the integration of spatial features with records in attribute tables. The relation between arcs and AAT records is shown here.

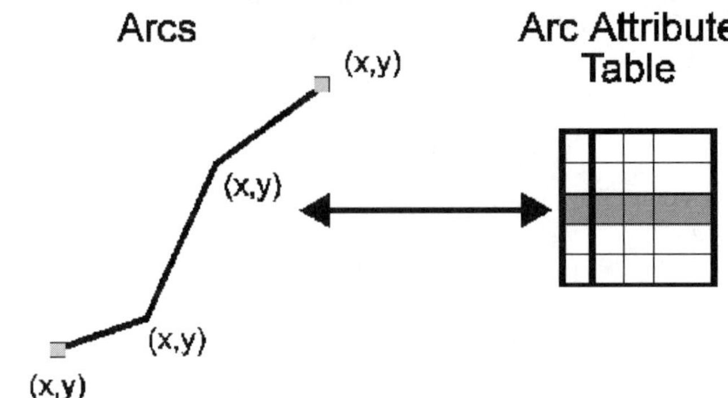

Once you have specified the layout of an AAT, ARCEDIT commands manage the integrity between spatial data and tabular data for you. When you digitize a new arc, ARCEDIT adds a new record to your AAT. And when you delete an arc, the corresponding AAT record is removed as well. There is always a one-to-one correspondence between arcs and AAT records.

NOTE: *A coverage can contain arcs without an associated AAT, but in practice you will usually define an AAT for each coverage that includes arcs. This is because you will nearly always want to define attribute data, such as street names or line symbols, with each arc.*

You can also establish a relate to another table from the arc attribute table, and reference, edit, or visualize information from the relate table.

Node Attribute Tables

The endpoints of an arc are called *nodes*. A node can be shared by two or more arcs if it is located at the precise junction of the multiple arcs.

Establishing this precise junction among adjacent arcs is important, because it allows us to perform the two basic types of GIS analysis—network tracing and polygon processing. You will see a variety of tolerances and commands in ARCEDIT to streamline the creation of these connections.

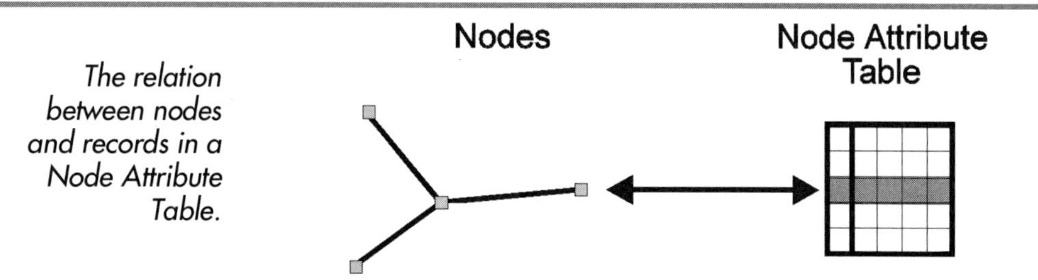

The relation between nodes and records in a Node Attribute Table.

You can optionally define a *Node Attribute Table* (NAT). As with arcs, ARCEDIT fully manages the referential integrity between the spatial and tabular data for nodes. How might you use a NAT? To analyze traffic accident statistics for road intersections, to store flow rates for natural springs feeding streams, or to model electrical substations, the beginning point of electric distribution, are just some examples.

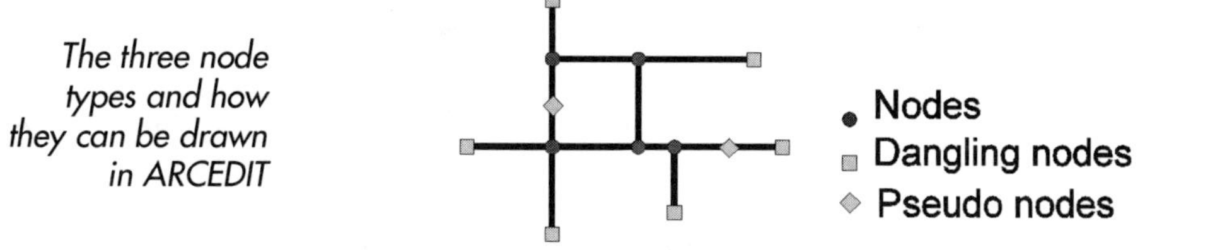

The three node types and how they can be drawn in ARCEDIT

There are three types of nodes: *junction nodes* (simply called *nodes*), *dangling nodes*, and *pseudo nodes*. Nodes are located at the junction of three or more arcs. Pseudo nodes are at the junctions of two arcs. Dangling nodes have only one arc connected to them.

The term *pseudo node* is unfortunate. There is nothing phony about these nodes, and they are quite useful for some purposes, such as when an attribute value changes along connected arcs.

The intermediate points along an arc are *vertices*. Vertices are strictly spatial features. They cannot have tabular data associated with them. However, if you need to record data at a vertex, you can split one arc

into two arcs, and thereby change any vertex into a new node. The attribute data you want to enter can then be added for the new node.

Route Attribute Tables

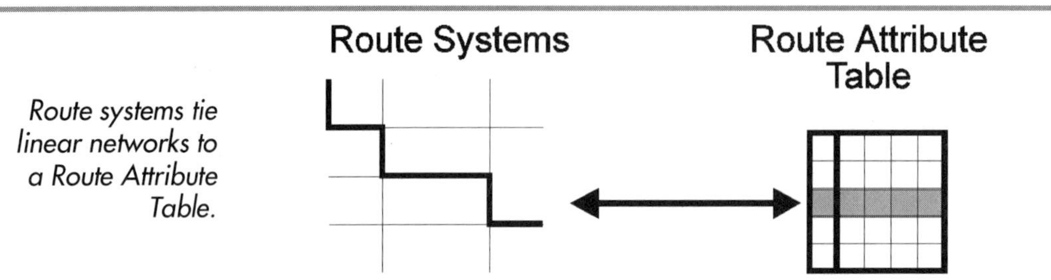

Route systems tie linear networks to a Route Attribute Table.

While arcs and AATs can model many linear features very well, some network features require more sophistication. Transportation networks are an ideal example. You might group together some connected arcs to model a bus route. Or, you might need to reference points by a distance from a starting point, such as a highway mile marker or a station along a centerline for a civil engineering project.

Sets of connected arcs (whole or partial) can be grouped together to define a *route*. Each route can reference a record in a *Route Attribute Table*. In addition, you can define auxiliary tables to reference attribute information at discrete points on a route or segments along a route.

Events at discrete points along a route can be modeled as *Point Events* in a Route Attribute Table. Point events are referenced by a distance measured from the start of a route (not an X,Y coordinate). For example, you could model the occurrence of accidents along a route as point events.

Attributes along segments of a route can be stored as *line events* in a Route Attribute Table. Line events are referenced by a start distance and an end distance. An example of a line event is the speed limit along a segment of a highway route.

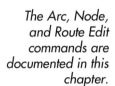

The Arc, Node, and Route Edit commands are documented in this chapter.

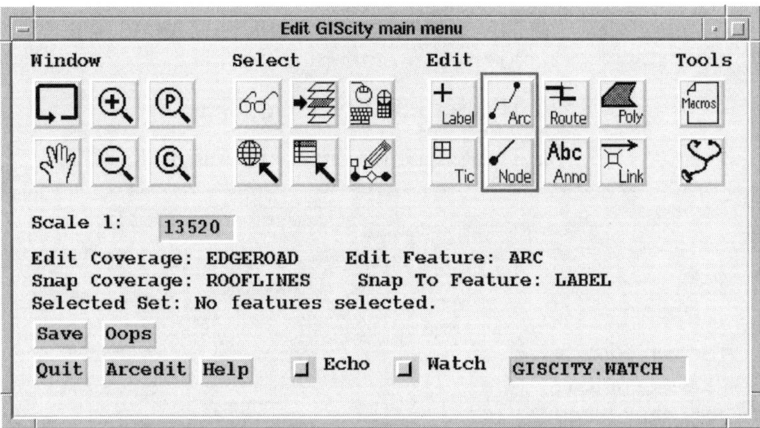

Now that we have degined our terms, we can show you how to use ARCEDIT to edit linear features through two commands and their associated menus: Edit Arcs and Edit Nodes.

Editing Arcs

Edit GIScity contains a rich selection of tools for editing arcs. The Edit Arcs menu is your main "workshop," but you'll make frequent visits to Edit Arcs submenus and to the Edit GIScity Select menus discussed in the previous chapter.

The Edit Arcs menu includes submenus you'll use frequently:

- ❑ Clicking **Arc Environment** displays the Arc Editing Environment menu
- ❑ Clicking **Snap...** displays the General Snapping Environment menu
- ❑ Clicking the **Attribute form** icon displays a form menu for the selected arc.

Edit GIScity Select commands you will often use:

- ❑ The Select Features Spatially menu, which includes four Additional Arc Selection commands
- ❑ The Set Edit Coverage menu
- ❑ The Set Draw Environment menu

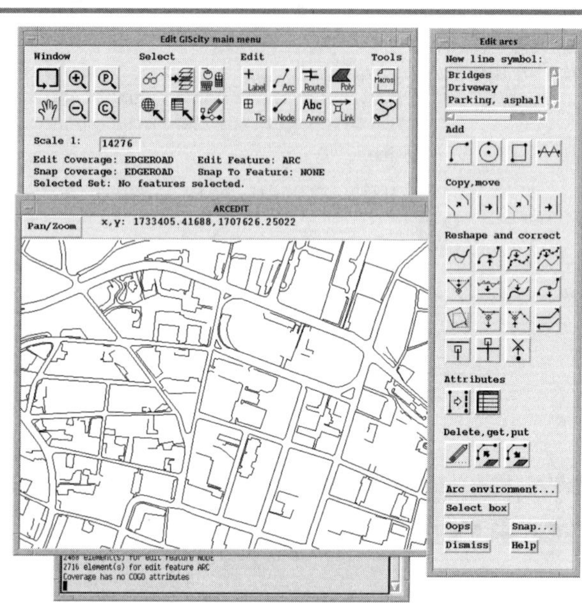

A screen capture showing the Edit GIScity main menu along with the Edit Arc menu, invoked by clicking the Edit Arcs icon in the Edit GIScity main menu.

Adding Arcs

Before you add an arc, you can select a line symbol from the New Line Symbol data list to automatically apply that line symbol to any new arcs you digitize. Whenever you change the edit coverage, the Edit GIScity application will update the line symbol list to another list for that coverage. For example, if you select the CONTOUR coverage, you will see a new set of line symbols appropriate for symbolizing contours.

The Edit Arcs menu has four commands to add new arcs. The arcs that result from each of these commands are the same as any other type of arc. These four commands are simply designed to make it easy to digitize freeform arcs, circles, rectangles, and the centerline.

The Add commands at the top of the Edit Arcs menu.

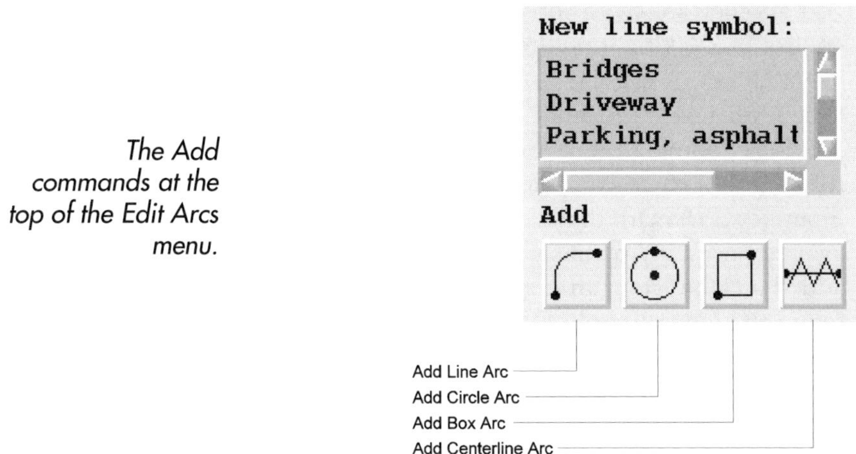

Adding arcs. The numbers in the mouse symbols show which input key to use.

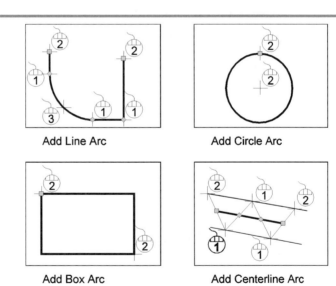

Add Line Arc

Add Circle Arc

Add Box Arc

Add Centerline Arc

Add Line Arc is a general-purpose mode for adding arcs. Of the four edit commands, this is the most frequently used. After you click this command, your input cursor (mouse, digitizer, or keyboard) is activated.

To start an arc, click input key 2. To add vertices, click input key 1 successively. To end an arc, click input key 2 again. Next, you can either

add another arc (with input key 2 again), or you can return to menu control by clicking input key 9.

Input key 3 can be used to digitize circular segments of an arc. Whenever you click input key 3, that point is used, together with the previous and next point, to define a circular segment.

Add Circle Arc easily creates a circle. After you click this command, click input key 2 to mark the center point of a circle. Then, click input key 2 again to mark any point along the circumference of a circle. After you've added a circle, you can either add another circle, or use input key 9 to return to menu control.

Add Box Arc is a quick way to make a rectangle in Edit GIScity. Like Add Circle Arc, it accepts two successive points with input key 2, and creates a rectangle by using those two points as opposite corners.

Add Centerline Arc is a subtle and interesting command that helps you digitize the centerline of something. Though not frequently used, this command is quite helpful in some circumstances. Suppose you have a parcel map and want to add the road centerline between the parallel curbs along a road. By zigzagging from one side of the road to the other, this command will join the midpoints between the curbs and make a nice centerline for you.

Sometimes you will acquire detailed parcel maps. Doing network modeling or address matching using curb lines is awkward, and road centerlines are better for GIS analysis. This is an efficient way to make a centerline from curb lines.

Editing Environment for Arcs

Before we explore the other commands for editing arcs, let's take a look at some of the settings ARCEDIT provides for adding and modifying arcs.

The Edit GIScity Select Features Spatially menu includes a command to set the *edit distance* tolerance. To select a feature, both the screen cursor and the feature must be within the Edit Distance tolerance. For example, if you are selecting arcs and the edit distance is 50 feet, the cursor must be within 50 feet of the selected feature in the graphics window for the selection to be successful.

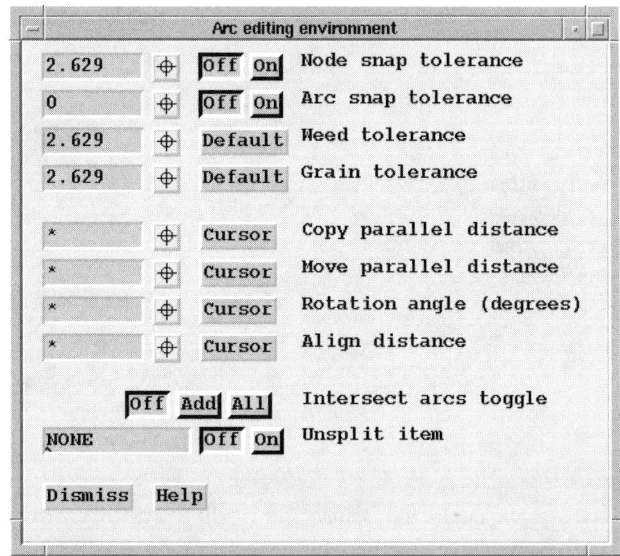

Clicking the Arc Environment... button on the bottom of the Edit Arc menu invokes this menu.

There are three snap environments for arcs: node snapping, arc snapping, and general snapping.

Node snapping allows you to snap newly digitized nodes to existing nodes, within the node snap tolerance. This is an important technique to ensure that adjacent arcs are connected at their common end nodes. Node snapping is on by default when you start ARCEDIT. A default node snap tolerance is set for you, based on the map extent of your edit coverage.

Arc snapping lets newly digitized nodes snap to any part of an arc. With arc snapping, not only can you snap a new node to an arc, you can also split the existing arc into two arcs, divided at the new node point. Arc snapping is on by default. A default arc snap tolerance is also set for you, based on the map extent of your edit coverage.

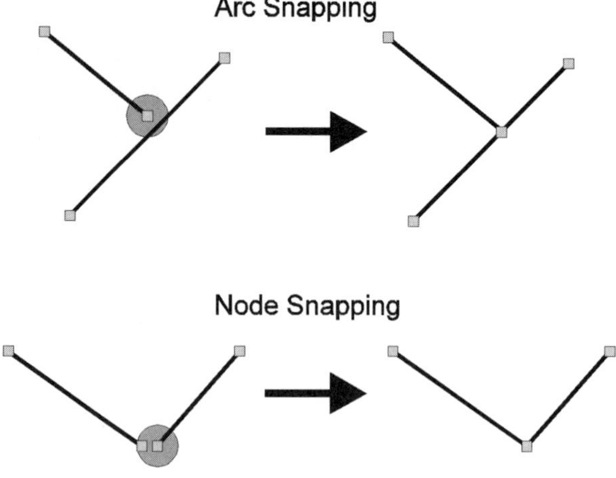

Node snap and arc snap tolerances are shown as shaded circles.

You have three methods to modify node and arc snap tolerances. In the Arc Editing Environment menu (which you invoke by clicking the Arc Environment... button in the Edit Arcs menu), you can:

- ❑ Type a new value directly in the text input field. That value, in map units, becomes your new node or arc snap tolerances.
- ❑ Click the cross-hair icon. The screen cursor is activated to accept two points which mark the arc or node snap distance you want to specify.
- ❑ Click the Off or On choice to enable or disable node or arc snapping.

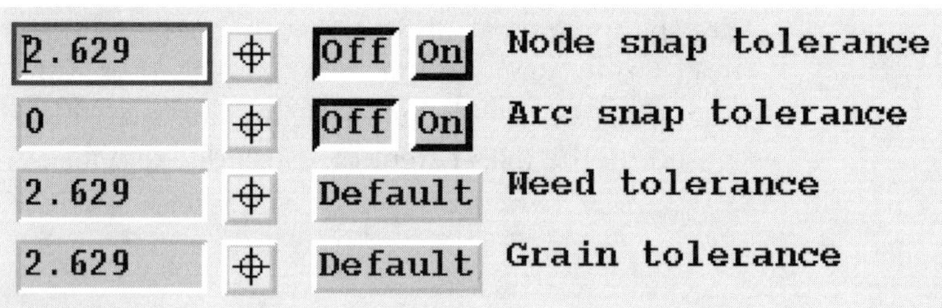

Node and Arc snap tolerances are set at the top of the Arc Editing Environment menu.

General snapping lets you snap new features to any other features. You can snap different features within the same coverage, or between any

features and any coverages. Unlike node snapping and arc snapping, you have to enable general snapping. To set this, click the Snap... button at the bottom of any of the Edit menus.

General snapping lets you snap any feature to any other feature between any two coverages.

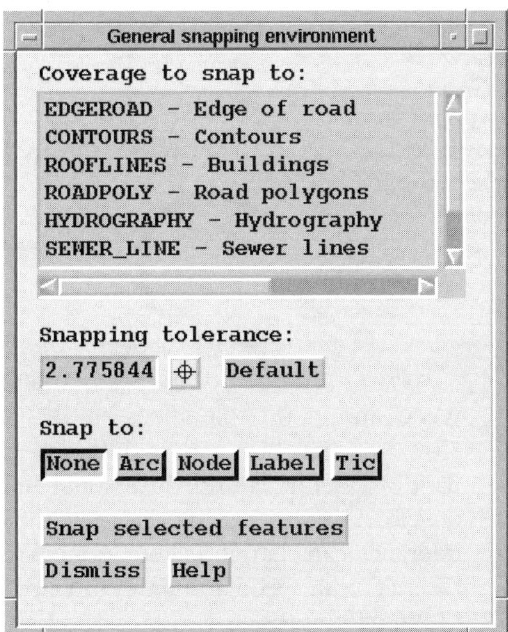

Other terms you will notice on the Edit Arcs menus are *weed tolerance* and *grain tolerance*.

Weed tolerance simply specifies the shortest distance between two vertices that ARCEDIT will accept when you digitize line arcs. If you digitize adjacent vertices within a distance less than the weed tolerance distance, your workstation will beep, and it will ignore the latest input point until it gets another at a greater separation.

Grain tolerance affects circular curves, circles, and splined arcs. If you zoom in and look closely at circular arcs that you create with the Add Circle Arc, you'll see that ARC/INFO approximates circular curves by a tight string of line segments. The finer these segments are, the better your circular curve can look, but the more storage it needs. You can modify the grain tolerance to set the granularity of line segments in circular arcs to your taste.

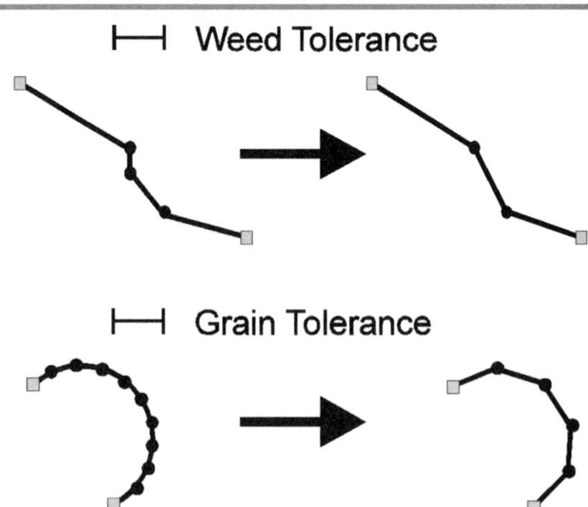

The Weed tolerance is used for filtering detail out of freeform arcs. The Grain tolerance filters out vertices in circular curves, circles, or other smoothed arcs.

As with setting node and snap tolerances, you have three ways to set weed and grain tolerances in the Arc Editing Environment menu. However, the third way is different: Instead of an Off/On choice, you have a Default button. Instead of using the map extent of your edit coverage to set a default distance, the default value for weed and grain tolerance can change as you zoom in and out in Edit GIScity. By clicking Default, your weed and grain tolerances are automatically set to 1/1000 of the width of your graphics window, no matter what the current map scale in your graphics window is.

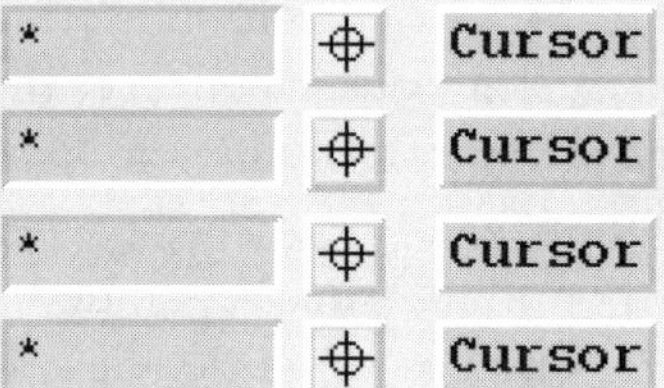

These are distances and angles you can set before using some of the other Edit Arc commands.

In the middle of the Arc Editing Environment dialog box, you will see settings for **Copy Parallel distance** and **Move Parallel distance** are values used by the **Copy Parallel** and **Move Parallel** commands.

This menu allows you to precisely set these distances with the text input field, or to interactively set them with the screen cursor.

Rotation Angle is also a value you can preset before rotating selected arcs. You can type in a precise angle in degrees, or interactively set it.

Align Distance is a value used with the Align command. This distance sets a tolerance for adding a new arc exactly on top of part, or all, of another arc.

Values for these four commands can also be set in three ways. The Cursor option is the default, which will show a star in the dialog box.

- You can type a distance or angle directly in the text input field.
- You can click the cross-hair icon. The screen cursor will be activated to accept two points to set that value. That value will not change, until you change it again.
- You can click the Cursor button, which will display a star in the text input field. Then, at the time you execute the relevant command, you will be prompted to pick two points to set a distance or value.

The cross-hair icon and Cursor button seem similar, but are distinct. The cross-hair icon will set a distance or angle, keep it, and automatically apply it when you execute the relevant command. Clicking the Cursor button says that you always want to be asked to pick two points on the graphics window.

The last two commands on the Arc Editing Environment menu.

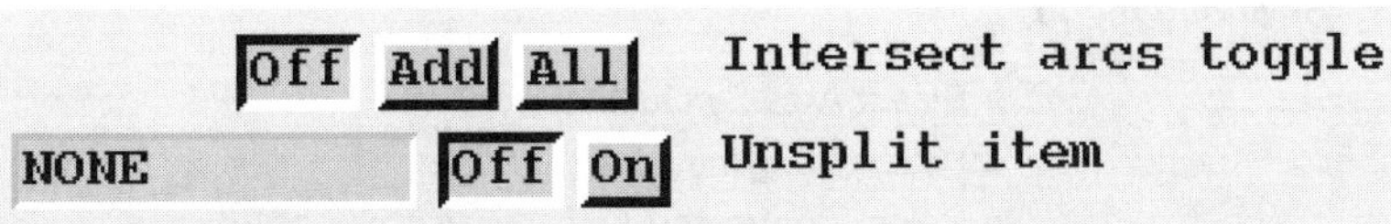

At the bottom of the Arc Editing Environment dialog box are two more setttings. **Intersect Arcs Toggle** sets whether ARCEDIT will automatically split arcs that cross other arcs when you digitize, and calculate their intersection point for you. Most of the time, this toggle should be set to Add or All. You would want it Off if you were digitizing

utility lines that cross but are not physically connected because they cross one another at different heights.

Off means that nodes at intersections will not be calculated and added. Add means that intersections are done only for digitized arcs. All means that intersections are done for any Reshape or Modify command that can change arcs.

Unsplit Item is a criterion for determining whether to merge two or more connected arcs together. The Unsplit command removes pseudo nodes. You control which pseudo items get eliminated or not.

Arc Selection Commands

Many of the commands in the Edit Arc menu operate on whichever arcs you have selected.

In the last chapter, we reviewed the Select Features Spatially menu and noted the differences between the similar menus in Edit GIScity and Display GIScity. We'll not cover that menu again, but the Select Features Spatially menu includes some useful commands that work only on arcs:

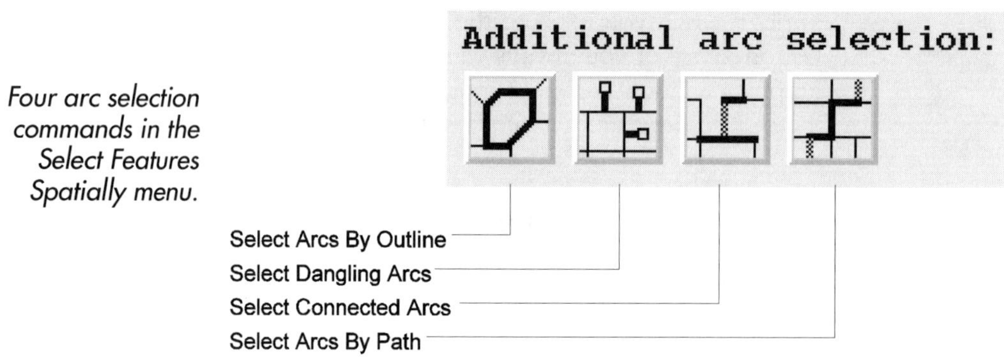

Four arc selection commands in the Select Features Spatially menu.

Select Arcs by Outline lets you select a group of arcs that form a perimeter. When you click this command, the screen cursor is activated to select one arc. This command then determines whether this arc is part of a closed set of arcs, and if so, returns those arcs to you in a selected set.

Select Dangling Arcs lets you automatically select arcs that have at least one node to which no other arc is connected. Sometimes you may digitize dangling arcs in error. This command can locate and delete dangling arcs.

Select Connected Arcs adds to your selected set any arcs which are connected to the arcs already in your selected set. You should usually set the "New" qualifier before using this command; otherwise the previous arcs are deselected.

Select Arcs by Path is a very powerful command for coverages that model networks. For example, if your coverage is street centerlines, you can easily find the shortest path between two points. Click this command to activate the screen cursor, and pick two arcs. This command will then determine the shortest distance on the arc network and return that path to you as a selected set of arcs.

Copying and Moving Arcs

Now that you've gotten an overview of commands to control your edit environment and to select arcs, we'll get back to the Edit Arcs menu.

The Copy and Move commands operate on your selected arcs. Before you use these commands, use the Edit GIScity Select Features Spatially or Select Features Logically menus to select some arcs. You can see the status of your selected set in the message area of the Edit GIScity main menu.

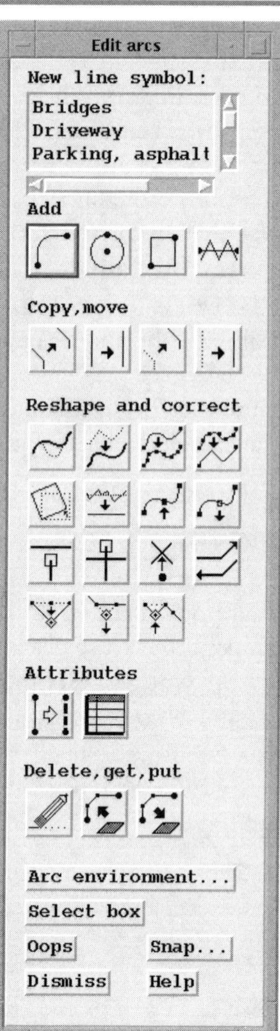

The Edit Arcs menu.

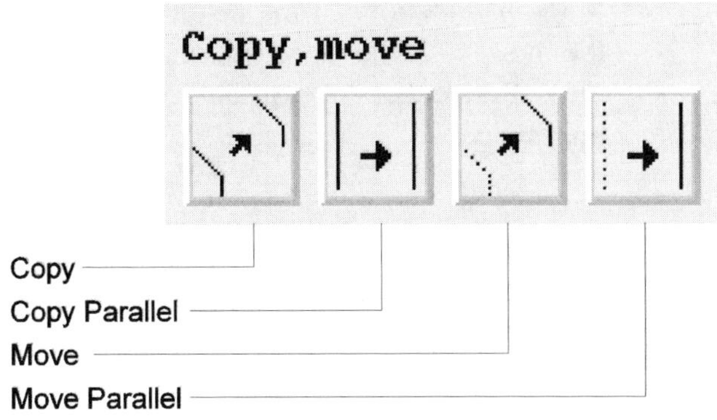

The Copy and Move commands in the Edit Arcs menu.

Copy will copy your selected arcs to another position on your graphics window. The original arcs are unchanged, and the copied arcs inherit the same attribute values as the original arcs. When you click this command, your input cursor is activated to accept two points. These two points are used to set the displacement in both the X and Y directions. The selected arcs will be copied by this displacement.

Copy Parallel copies your selected arc so that each part of the new arc is a parallel offset of the original arc. You can only have one arc in your selected set; otherwise this command will return an error message. Unless you have already set a Copy Parallel distance in the Arc Editing Environment menu, clicking this command will activate the input cursor for you to click two points. After you do so, a new arc will appear on your graphics window.

This command is very useful for adding rights-of-way or easement lines based on existing property lines or curb lines.

NOTE: *By default, you select two points with your input cursor to set the Copy Parallel distance. However, you can Copy Parallel by a precise distance; type in a value in the Arc Editing Environment menu (invoked by clicking the Arc Environment... button on the bottom of the Edit Arcs menu.)*

Move will move all your selected arcs to a new location. This command does not change the number of arcs, or modify their attribute values; it moves them to a new location. Clicking this command activates your input cursor to place two points that mark a displacement vector in the

X and Y directions. If the result is not what you wanted, you can undo the move with the Oops button.

Move Parallel operates just like Copy Parallel, but it doesn't leave behind the original arc.

Reshaping and Correcting Arcs

Generalize
Densify
Spline
Reshape
Rotate
Align
Split
Unsplit

These fifteen commands give you a rich tool set for refining the position and shape of arcs.

Undershoots
Overshoots
Intersect
Flip
Move Vertex
Delete Vertex
Add Vertex

The Reshape and Correct commands will modify the shapes and vertices of your arcs, but won't modify any values in your Arc Attribute Table. Some of these commands require that only one arc be selected. Other commands will accept any number of selected arcs.

The last three commands in this part of the Edit Arcs menu, Move, Delete, and Add Vertex will prompt you to interactively select arcs and will only modify vertex positions.

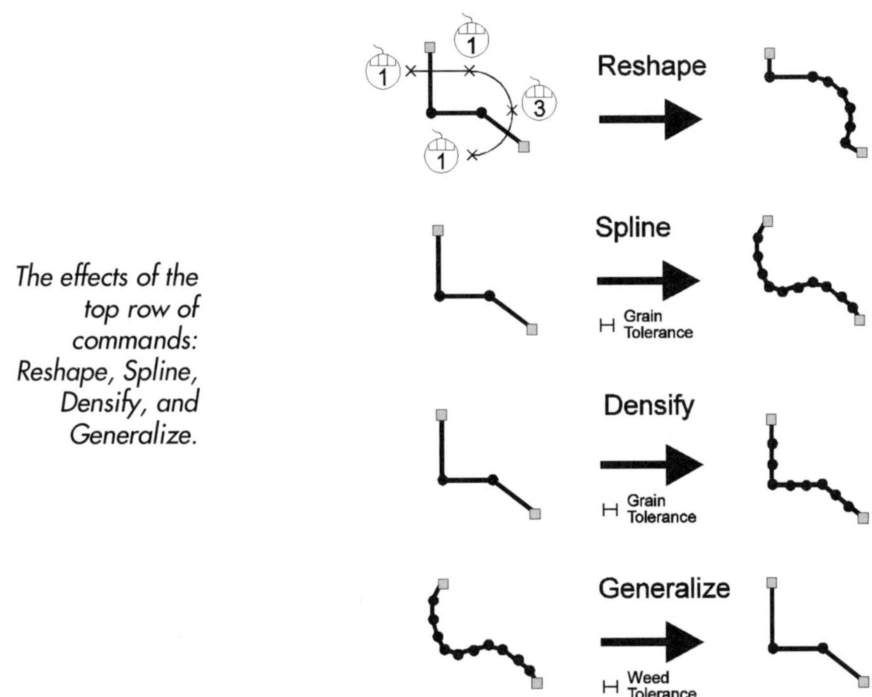

The effects of the top row of commands: Reshape, Spline, Densify, and Generalize.

Reshape lets you change the shape of a part of the selected arc. You can only have one arc selected when you issue this command. Click Reshape and then digitize a new portion of the arc. Reshape automatically finds the intersections of the new shape with the original arc. The command modifies the arc only between the continuous segments you add.

Spline lets you smooth arcs. Spline uses your current grain tolerance and works on your current selected arcs to improve the cartographic appear-

ance of geographic features such as contours or streams. You would rarely use the Spline command on man-made features, which are usually angular rather than smooth like natural features.

Densify adds new vertices to your selected arcs, using the grain tolerance. The primary use of this command is to assist arc snapping by adding more vertices. The Densify command is usually followed by the Snap command.

Generalize is the opposite of the Densify command. This command filters out closely spaced vertices, using the weed tolerance you have set. Generalize is often used on arcs imported from other systems, and for thinning out detail.

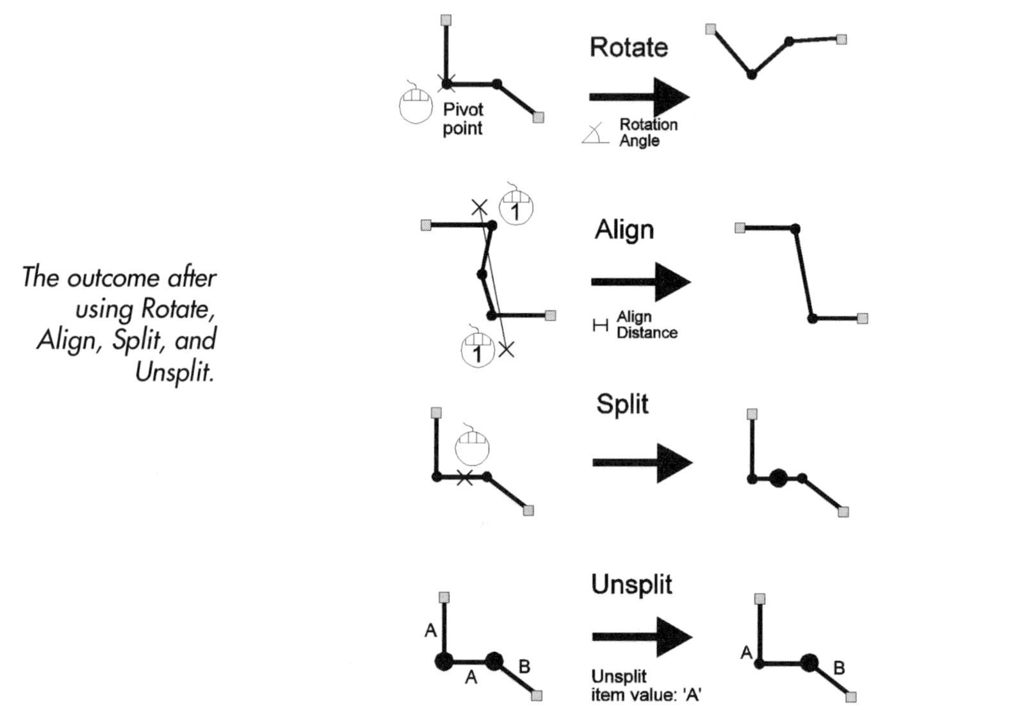

The outcome after using Rotate, Align, Split, and Unsplit.

Rotate turns all arcs in your selected set to any new angle. You can set an exact rotation angle in the Arc Editing Environment menu by typing in an angle. Alternately, you can set this command for interactive cursor input (the default), which activates your cursor to input two points when

you click this command. The first point serves as a pivot point. The second point defines the angle to rotate, relative to the first point. If you've typed in an angle, your screen cursor will be activated to accept one point for a pivot point.

Align lets you snap the nodes and vertices of an existing arc to a line segment you digitize. You can preset an Align distance in the Arc Editing Environment menu, or choose interactive cursor input, which activates your cursor to input two points to mark the alignment distance. Next, you will see a message asking you to enter two points to mark a line segment. All nodes and vertices within the Align distance will be snapped to this line segment. This command works repeatedly until you enter input key 9. Align is commonly used for correcting arcs that are created by converting raster images into vector data sets; it can efficiently remove jaggies from these arcs.

Split lets you divide one arc into two connected arcs with a common pseudonode. Before you use this command, select one (and only one) arc. Clicking this command activates your screen cursor to accept a point along the selected arc. The arc will be split right there. Splitting arcs is useful for when you want to specify different arc attributes for the two new arcs.

Unsplit merges sets of arcs joined by pseudo nodes. Any number of arcs can be in your selected set. By default, all pseudo nodes will be removed from your selected set of arcs. However, you may want arcs that have different item values to remain unsplit. In that case, set an Unsplit Item in the Arc Editing Environment menu.

Undershoots will take your selected set of arcs and attempt to join them within a distance you define. Before you use this command, select some arcs. Clicking Undershoots activates your screen cursor to accept two points to mark an Extend distance. Undershoots then uses that distance to extend all selected sets to try to join them.

Overshoots takes your selected set of arcs, reselects any dangling arcs, and deletes any dangling arcs that are smaller than a Dangle distance which you specify. Before using this command, select arcs from which you want to remove short, dangling arcs. Click Overshoots and pick two points with your screen cursor. All dangling arcs in the selected set that are smaller than the Dangle distance will be automatically deleted.

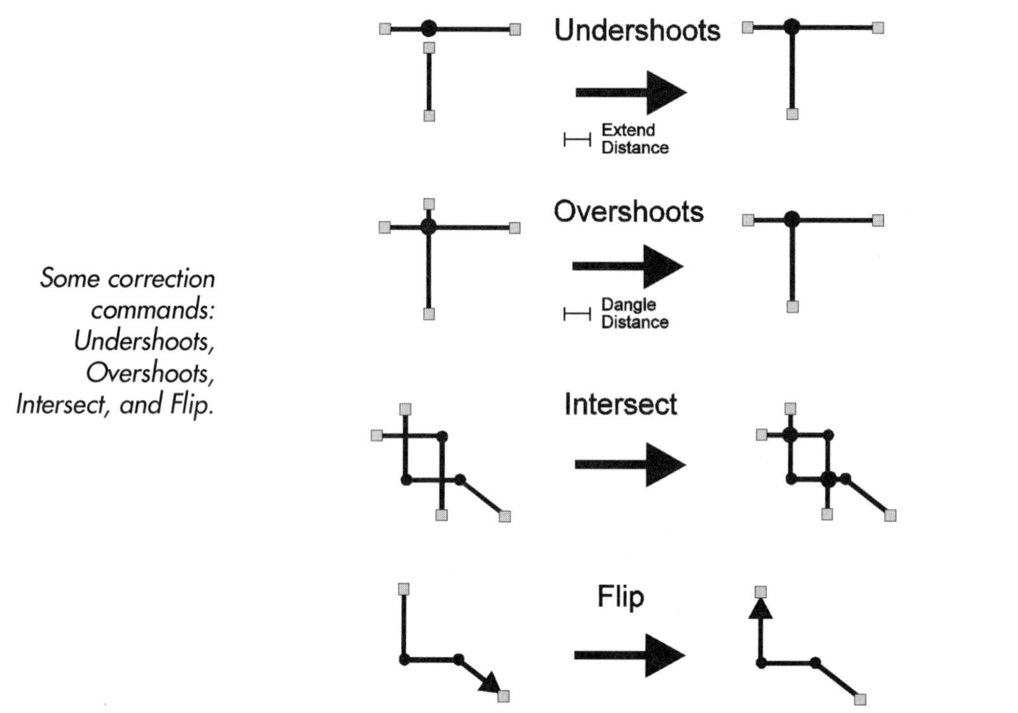

Some correction commands: Undershoots, Overshoots, Intersect, and Flip.

Intersect calculates intersections and makes new nodes where non-intersecting arcs lie over each other. Before you use this command, you can set the Draw Environment for arcs to Intersect, which will show non-intersecting arcs that cross each other. Then select some arcs. When you click this command, intersections will be calculated and new nodes created.

Flip reverses the direction of your selected arcs. In the Set Draw Environment menu, set Arc to the Arrows choice to show arc directions. Then, select arcs whose direction you want to reverse. The Flip command will quickly change their direction. This command is often used to prepare coverages for network tracing.

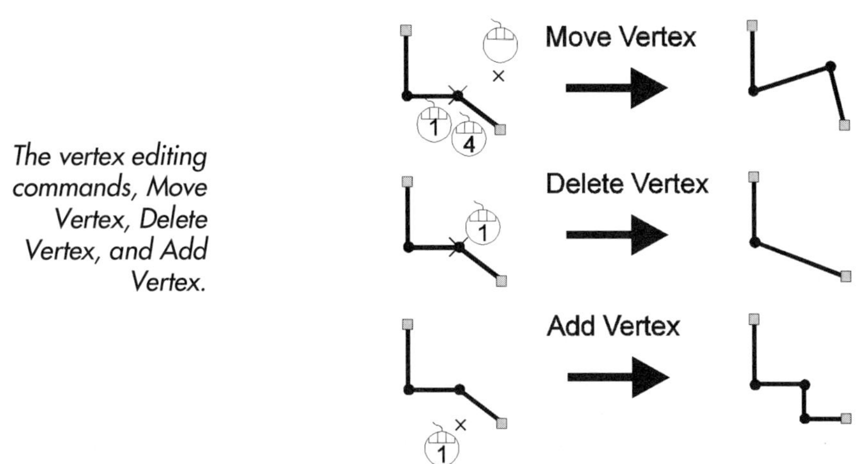

The vertex editing commands, Move Vertex, Delete Vertex, and Add Vertex.

These three commands edit vertices exclusively. They cannot be used to reposition or modify nodes. Before using any of these commands, select one and only one arc. Selecting one of the vertex commands then automatically highlights the vertices in your selected arc. These commands use a combination of input keys. If you get stuck, or when you are ready to leave any of these commands, use input key 9.

Move Vertex lets you select a vertex and move it. When you click this command, use input key 1 to select a vertex. A triangle will appear on the vertex. Move the cursor over that vertex and press input key 4 to indicate that you want to move that vertex. Then move the cursor to any new position on your graphics window and press any input key. You will see the vertex moved.

Delete Vertex lets you remove any vertex from the selected arc. Click the command, move the cursor over the vertex you want to remove, and press input key 1. If the desired vertex has a triangle drawn over it, then you can press input 4 to remove it. Otherwise, use input key 2 to select another vertex or input key 9 to exit.

Add Vertex lets you add new vertices to an existing arc. When you click this command, move the cursor into any position to add a new vertex and press any input key. You can do this successively. When you are done, press input key 9.

Editing Arc Attributes

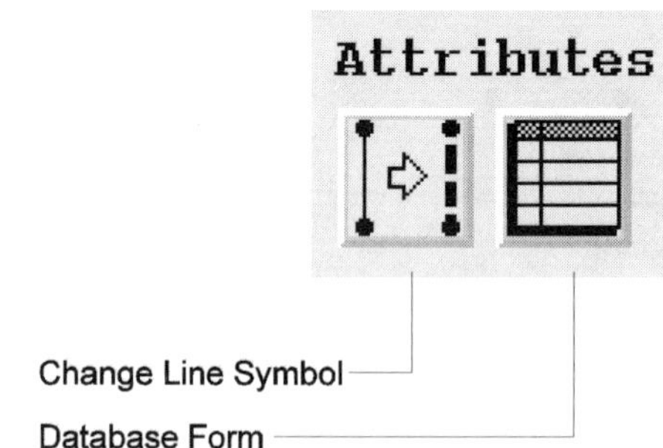

The Change Line Symbol and Database Form commands.

Change Line Symbol allows you to modify the line symbol for existing arcs. To do so, first select arcs with either the Select Box command in the Edit Arcs menu, or select arcs with the Select Features Spatially menu. Next, choose a new line symbol for the selected arc. Then, click the Change Line Symbol command.

If you have set your draw environment to use a symbol item for the edit coverage (see the Set Draw Environment menu), then you will see the line symbols for selected arcs change right away.

Database Form lets you edit any Arc Attribute Table items for selected arcs. Clicking this command creates a menu which has exactly the items in your Arc Attribute Table. There will be a pause of several seconds while the menu is being made for you.

When you first invoke a database form, it points to the first Arc Attribute Table record in your selected set.

The first half dozen items listed on the form are items that ARC/INFO manages for you. It is not recommended that you attempt to modify them. The database form displays these values: the from node and to node numbers of the arc, the left polygon and right polygon bounding the arc, the calculated length of the arc, and the internal identifier for the arc.

NOTE: *You can move your screen cursor to any of the input fields in the database form and type in a new value. The item value is updated as soon as you press [Enter] on the keyboard.*

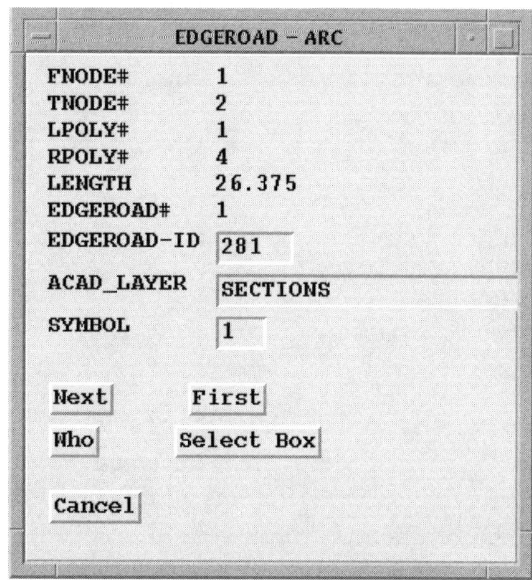

A sample database form created for arcs in Edit GIScity.

The **Next** button takes you to the next record in the Arc Attribute Table. When you reach the last record, you will see a message in the message area of the menu telling you've reached the end of the selected set.

The **First** button returns you to the beginning of the selected set.

Who flashes the current arc on the graphics window, showing you which arcs you are viewing or editing attributes for.

Select Box lets you select new arcs without leaving this menu. After you select new arcs, you will see the first record of that selected set appear in the database form.

Cancel dismisses the database form.

NOTE: *This menu establishes a database cursor, which is a logical pointer through selected database records, similar to cursors used in commercial database management systems. When you open a cursor, it points to the first record in your selected set. You can move to the next record, return to the first selected record, or exit. Although we often use the word "cursor" to indicate the pointer on your workstation screen, "cursor" here means a logical pointer through your selected set.*

Delete, Get, and Put

The Delete, Get, and Put commands in the Edit Arcs menu.

Delete,get,put

Delete

Get from Coverage

Put to Coverage

Delete simply removes all selected arcs from your edit coverage. Be careful with this command! If you've made a mistake, use the Oops command to undo deletion of arcs.

Get Arcs from Coverage invokes the Get Features from Coverage menu. With this menu, you can import arcs from another coverage. Be aware that all arcs from the other coverage are imported. If you want to import only some of the arcs, change your edit coverage to the other coverage and use the Put Arcs to Coverage command instead.

Put Arcs to Coverage invokes the Put Features to Coverage menu. With this menu, you can export selected arcs to an existing or new coverage.

Exercise 1: Editing Arcs

The commands for editing arcs in the Edit Arc menu are quite comprehensive and we'll not try to cover them all. Instead, we'll cover some of the main commands through some "heads-up digitizing" in this exercise.

Remember, if you are a Windows NT user, the 2BUTTON command is automatically run for you. Reference ArcDoc on 2BUTTON for details on the mouse mapping you need to apply in this exercise.

1. Click on the Set Edit Coverage icon (an arrow pointing at a stack of papers) in the Edit GIScity main menu. Set your edit coverage to EDGEROAD.
2. Click the Edit Arcs command to invoke the Edit Arcs menu.
3. Click the Select Features Spatially command (the globe and arrow icon on the main menu) to invoke that menu.
4. Click the Set Draw Environment icon (the pencil pointing at a chain of nodes). When the Set Draw Environment menu appears, turn the choice for Arc to Intersect and turn the choice for Node to Both. The Intersect setting for arcs shows all of the arcs and where they cross without intersecting. The crossovers are shown as small boxes with an "X." The Both setting for nodes will show pseudo nodes as diamonds and dangling nodes as small squares.
5. Use the Zoom In command (the magnifying glass with a plus sign in the Window commands in the Edit GIScity main menu) to zoom to a part of Santa Fe without too many streets.
6. Click the Arc Environment... button in the bottom of the Edit Arcs menu. When that Arc Environment menu appears, click the Node Snap Tolerance cross-hair icon. Pick two points in your graphics window spaced about 1/4 inch apart. Watch the menu to see what tolerance was calculated for you. This is your node snap tolerance. Do the same for arc snap tolerance. Then go down to the Intersect Arcs Toggle and click on the All choice.

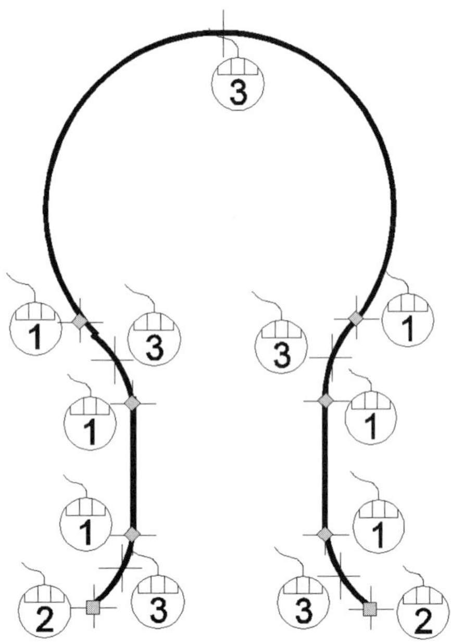

Illustration of how to digitize a cul-de-sac.

7. Now you're all set up to edit some arcs. On the Edit Arcs menu, click the first of the Add commands, Add Line Arc. Using the illustration above as a guide, try adding a cul-de-sac along one of the road curbs. It may take several tries and Oops commands, but practice using the input keys illustrated until you get a nice cul-de-sac.

8. Your cul-de-sac should be highlighted to indicate that it is the selected arc(s). If it is not, reselect it with the Select Box command on the Edit Arcs menu. Make a copy of the cul-de-sac you've digitized at a point up the road. Click the Copy Arcs command and pick two points to mark a displacement vector. You will see an identical copy of the cul-de-sac appear.

9. Click the Arc Environment... button on the Edit Arcs menu. On the Arc Editing Environment menu, set a new Weed Tolerance. Type in a new value about five times greater than the present value. Now click on the Add Line Arc icon and digitize a new cul-de-sac elsewhere. How does the curved part of the cul-de-sac look now,

compared with your first cul-de-sac drawn with a smaller weed tolerance?

10. Use the main menu icon to invoke the Set Draw Environment menu again. Click the check box for Use Symbol Item for Edit Coverage. Then click the Redraw Window command on the Edit GIScity main menu. You'll see line symbols appear for all arcs in the EDGEROAD coverage.

11. In the Set Draw Environment menu, click the check box for Draw All Themes in View. Again, click the Redraw Window command. You'll see a display similar to the ones in Display GIScity. Whichever option you use to display the GIS database largely depends on your preference. Sometimes you'll want to see the finished map appearance. At other times, you'll need a simple and clear display of only your edit coverage.

12. On the top of the Edit Arcs menu, pick a New Line Symbol. Select one of your cul-de-sacs. Then click the Change Line Symbol icon (the left icon in the Attributes group.) This procedure sets the cartographic appearance of your arcs.

13. Click the Database Forms icon in the Attributes group. Go through the database cursor using the Next button. This is a key way of editing associated attributes to all geographic features in ARC/INFO.

14. Experiment with the Reshape and Correct commands. Always be aware of what your selected set is, and what the selected set is after these commands. Some commands will modify the selected set and some will not.

15. When you are done, don't use the Save command or when you Quit, don't elect to save the changes. You've probably made a mess and would like a clean slate for the next time!

Editing Nodes

Nodes are endpoints of arcs. They cannot be defined separately from arcs.

Node Attribute Tables (NAT) can be optionally defined for nodes in a coverage. As with the other attribute tables, you can define any additional items appropriate for your application.

Attributes for nodes are assigned independently of arcs. For example, you can model attributes of features that occur at junctions of arcs, such as intersections, switches, or valves. Nodes can also have point symbols assigned to them.

When two or more arcs join at a common node, all the arcs share that node. There is only one record in the NAT for that node. It is the relationship between arcs and nodes that allows a GIS to perform much of its analysis.

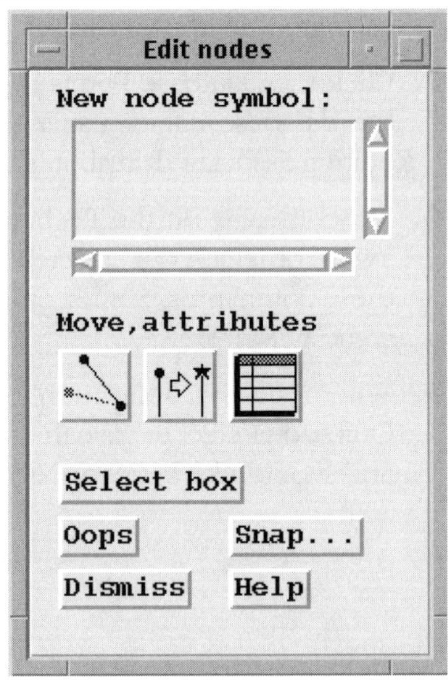

The Edit Nodes menu.

While arcs usually have an associated Arc Attribute Table, NATs are less frequent. If the edit coverage has an associated NAT, all of the node edit commands are functional. If there is no NAT, only the Move Node command is operational, and it works differently depending on whether a NAT is present.

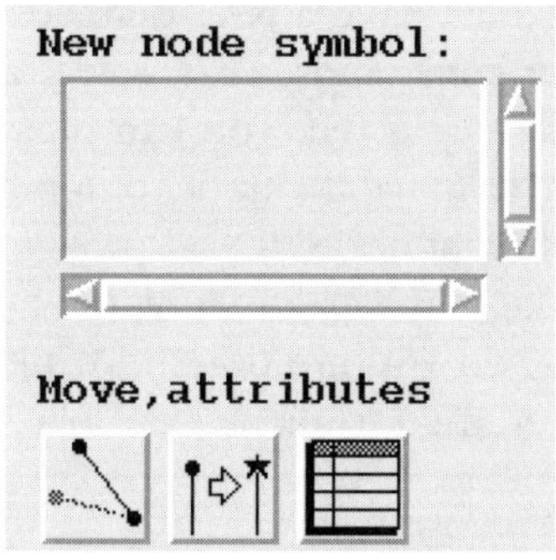

The top of the Edit Nodes menu.

Editing Nodes with a NAT

Move Node works two different ways, depending on whether you have a NAT defined. If your edit coverage has a NAT, then you can select nodes with the Select Box command or the Select Features Spatially menu. After you click the Move Node command, your screen cursor is activated to accept a "from" point and a "to" point. The selected nodes will be moved by the displacement between the from and to points.

If your coverage has a NAT, **Change Node Symbol** lets you quickly change node symbols. First, go to the Set Draw Environment menu, turn the draw environment for nodes to any setting but Off, and click on the check box for Use Symbol Item for Edit Coverage.

Next, choose a New Node Symbol from the data list on the top of the Edit Node menu. When you click the Change Node Symbol icon, your node symbols will be immediately changed to the new node symbol.

Database Forms works on selected nodes just as the same command works for selected arcs. Reference the ESRI documentation for Database Forms for arcs.

Editing Nodes without a NAT

The Move command is the only node editing command you can use if you don't have a NAT defined, and it's a bit more complicated than the other commands. First, select a node with the screen cursor. After the node is selected, you can press either input key 4 to indicate that you want to move the node, or input key 1 to indicate that you want to select a different node. If you press input key 4, then pick a destination point on the graphics window; the node will move to that selected point.

When you are done moving nodes, press input key 9, and you will be returned to menu control.

Control commands for the Edit Nodes menu.

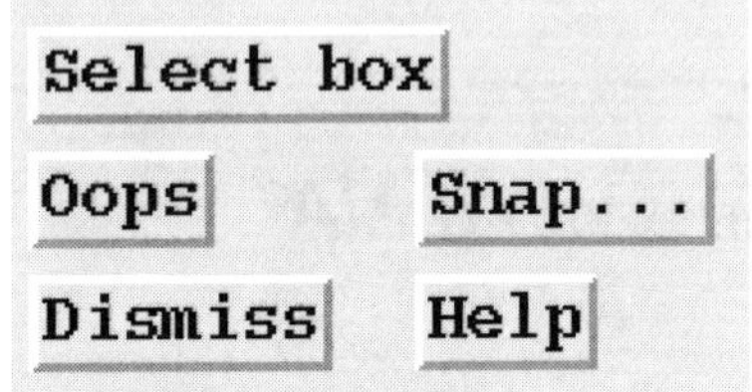

The control commands for nodes work just the same as the control commands for arcs. Refer to the section on arcs (this chapter) for instructions on these commands.

Exercise 2: Editing Nodes

Thankfully, editing nodes is much simpler than editing arcs. We'll quickly try it for two situations: a coverage with a NAT and a coverage without a NAT.

1. Use the Set Edit Coverage menu to select the EDGEROAD coverage.
2. Click the Edit Node icon in the Edit GIScity main menu to invoke the Edit Nodes menu.
3. The EDGEROAD coverage does NOT have a NAT, so we can only move nodes. Moving nodes without NATs is not immediately intuitive, so remember that input key 9 is always your way out.

Zoom to any part of Santa Fe and click the Move Nodes command. Use input key 1 to select a node. If the selection is successful, use input key 4 (positioned anywhere) to indicate that you want to move that node. Then use any input key to position where the node should go. Either repeat the steps to move more nodes, or use input key 9 to exit.

4. Now change your edit coverage to SEWER, which has a NAT.
5. Use the Select Box command on the bottom of the Edit Nodes menu to select one node. Click the Move Node command and pick two points for a displacement vector. Now, isn't that simpler?
6. Pick an item from the New Node Symbol data list and click the Change Node Symbol command. The appearance of the selected node will change.
7. Use the Select Box command to select a number of nodes. Click the Database Forms command, and use the Next button in the forms menu to "walk" through the NAT.

Editing Points and Areas in Edit GIScity

Label Points and Polygons in ARCEDIT

In ARC/INFO, label points and polygons are closely related.

A label point is a one-dimensional feature, with only a single X and Y coordinate. It has a dual role in ARC/INFO: it can describe a small feature as a point, or, when placed inside a polygon, it can be used to store the attributes for the polygon.

A polygon in ARC/INFO is defined by a set of arcs that connect exactly at their nodes to form a closed shape and by an interior label point. The arcs define the shape of the polygon. The label point stores the attributes for the polygon. Polygons are used to represent any geographic features that span an area, such as lakes, forest stands, or, at a suitable map scale, buildings.

Any coverage can have label points to represent point features or to represent polygon features, but not both. A coverage with points has a Point Attribute Table. A coverage with polygons has a Polygon Attribute Table. It turns out that the items for both tables are identical, and they are both referred to as PATs.

Polygon Attribute Table and Point Attribute Table.

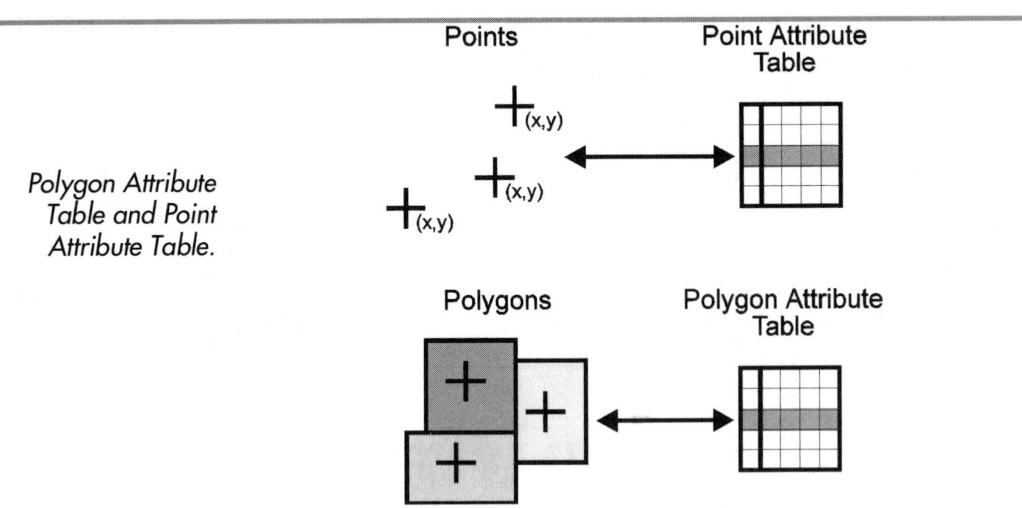

This chapter documents how you can use label points for both point and polygon features, how to update polygon topology, and some special polygon editing commands.

The Edit Label Point and Edit Polygon commands are documented in this chapter.

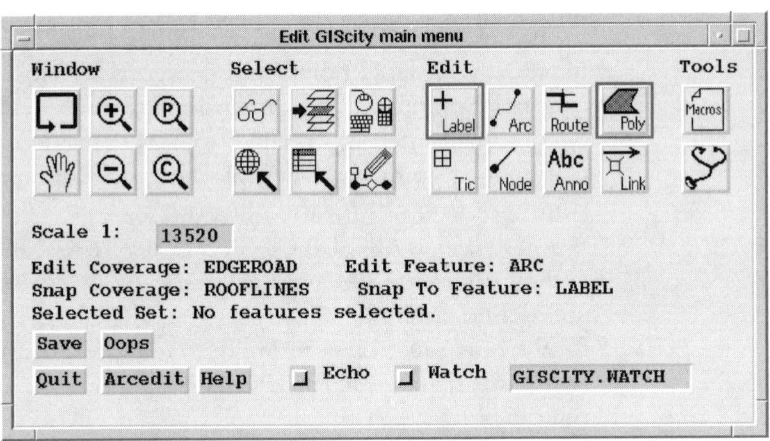

Editing Label Points

Label points are simple to add in ARC/INFO. The procedure is the same whether your coverage contains points or polygons. (Later in this chapter, we learn some additional polygon commands that can automatically place or remove label points that mark polygons.)

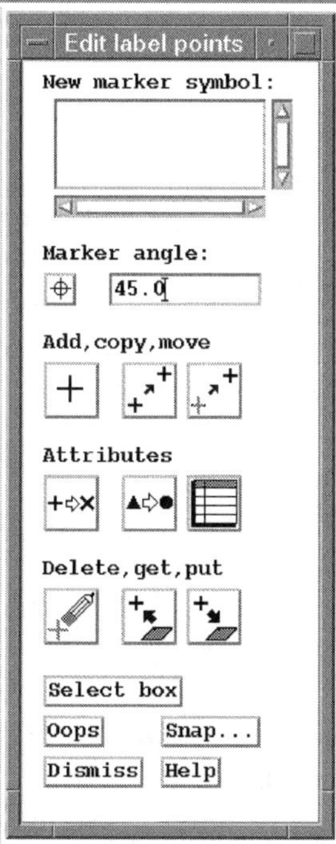

The Edit Label Point menu.

Click the Edit Label Point command on the Edit GIScity main menu to invoke the menu to edit label points. You may want to change the edit coverage first if the current edit coverage comprises line features. Select a coverage you know has point or polygon features in it.

Many of these commands operate on your selected set of label points. To update your selected set, you can use the Spatial Selection menu, the Logical Selection menu, or the Select Box button in this menu.

Adding, Copying, and Moving Label Points

Commands to add, copy, and move label points and affect their properties.

Before adding a label point, choose a marker symbol from the scrolling list. Symbols applied to label points in ARC/INFO are called *marker symbols*. These symbols are preset for you in the Edit GIScity application. ARC/INFO references marker symbols by number, but Edit GIScity gives you names so that it is easier to assign the symbol you want.

Another option you have is to set a rotation angle for marker symbols you are about to add. Now, if your marker symbol has a circular shape, then setting a marker angle will not make a difference visually. However, if your marker symbol has another shape, such as a triangle or square, then you can optionally set a rotation angle for placing new markers. If you click the cross-hair cursor, then the screen cursor is activated to accept two points. The angle is calculated between these two points. Or, you can simply type in a rotation angle in the input field.

NOTE: *It is not required that you use marker symbols when adding label points. Generally, label points used for marking polygons do not have marker symbols, and label points for point features usually do. For coverages in Edit GIScity with polygons, the scrolling list for marker symbols will be blank. If you add a label point without a marker symbol, then it will appear as a small cross, the default marker symbol.*

For the **Add Label Point** command, the screen cursor will be activated. Move the cursor to points inside the graphics window. Click button 1 to add a label point, and click button 9 to return to menu control. When you are done adding label points, the new ones are now your selected set.

Copy Label Points adds a new set of label points copied by a distance and direction. First, select one, several, or many label points. Next, click this command. The screen cursor will be activated to accept two positions on the graphics window. Click button 1 to mark a begin point and button 1 again to mark an endpoint. ARCEDIT will calculate the distance and direction between those two points and copy the selected label points.

Move Label Points works just like the copy command, but the previous label points will not remain as they would when copying.

Rotating, Changing Symbols, and Editing Attributes

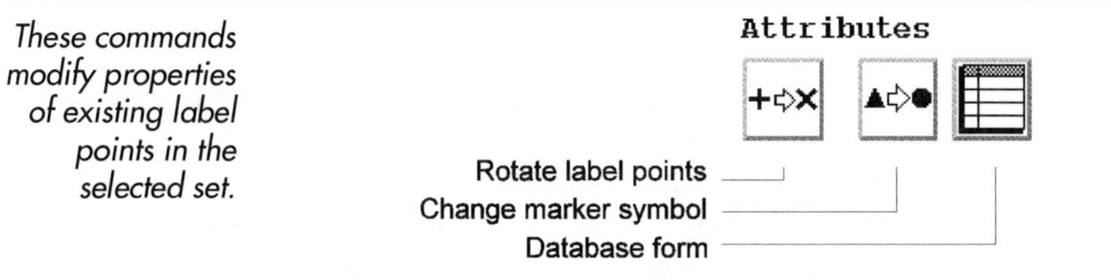

Rotate Marker Symbol rotates all the label points in your selected set by the angle you have set for the marker angle.

Change Marker Symbol modifies the marker symbol of all label points in your selected set to the new marker symbol you've set in the scrolling list.

Database Form invokes a menu, letting you update all the attributes for the label points in your selected set. This works just like the Database Form command for editing arcs.

Delete, Get, and Put

More commands to edit label points.

Delete,get,put

Delete

Get from Coverage

Put to Coverage

Delete removes all the label points in your selected set.

Get from Coverage extracts label points from another coverage, the one you choose from a menu of coverages that appears when you invoke this command, into your edit coverage.

Put to Coverage writes out label points in your selected set into the coverage you choose from a menu that pops up when you click this command. You cannot undo this command.

Editing Polygons

There are two ways to edit polygons: you can edit their constituent arcs and label points, or you can use some of the commands in the Edit Polygons menu.

Each polygon ideally has one and exactly one label point inside. These commands are designed to automatically update label points for edited polygons. This means that when you split a polygon, a new label point for that polygon is added for you. When you merge two polygons into one, then one of the label points is removed for you. You can see how these commands make polygon editing easier for you.

Let's start with the Add commands:

Adding Polygons

Three variations on the Add command.

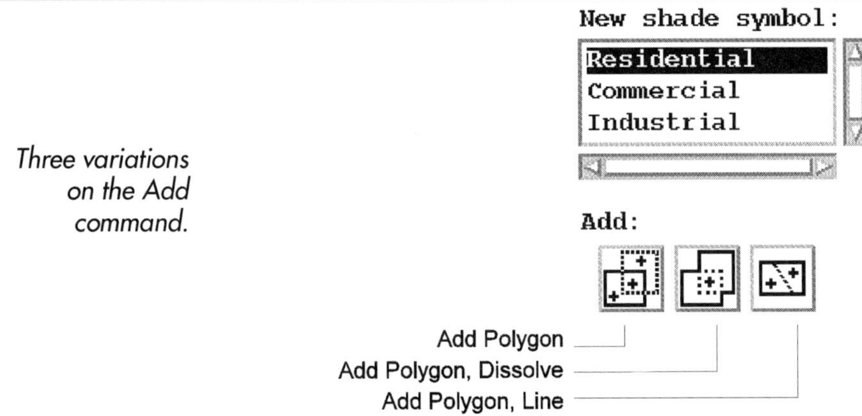

First, you can optionally set a New Shade Symbol for polygons. Any new polygons you add will inherit this shade symbol. Just click one of the choices in the scrolling list.

The first command, Add Polygon, adds new polygons, and if there are intersections with existing polygons, makes additional polygons at the intersections.

The icon for this command depicts an existing polygon being intersected by a new polygon. The overlap area becomes an additional polygon in its own right. So, if you digitize polygons that intersect other polygons, you may have many polygons resulting.

When you click this command, your screen cursor is activated to accept new lines. As with arcs, you control the polygon definition using the input key buttons. Button 1 starts a polygon, button 2 closes a polygon, and button 9 terminates this command.

A variation is Add Polygon, Dissolve. This works like Add Polygon, but intersecting areas do not become new and separate polygons. Rather, they are merged to span the previous and newly defined polygons. As the icon shows, this command can join several polygons into one. Any excess label points are automatically removed.

When you use this command, the screen cursor is active. Use buttons 1, 2, and 9 to add a polygon. When you are done, any polygons that intersect your new polygon are merged with the existing polygon. The

arcs along the intersection and any additional label points are removed for you.

Another version of this command is Add Polygon, Line. Unlike the first two commands to add polygons, you don't have to add a closed shape. Rather, you can draw a line across a polygon and quickly split it in two. This is useful for something like splitting a parcel lot into two.

When you use this command, note that the new arc must fully cross an existing polygon to split it. If not, the arc will be removed. Also, if you add an arc that crosses polygons, yet leaves dangle arcs, then those dangle arcs are removed.

Editing Polygons and Modifying Attributes

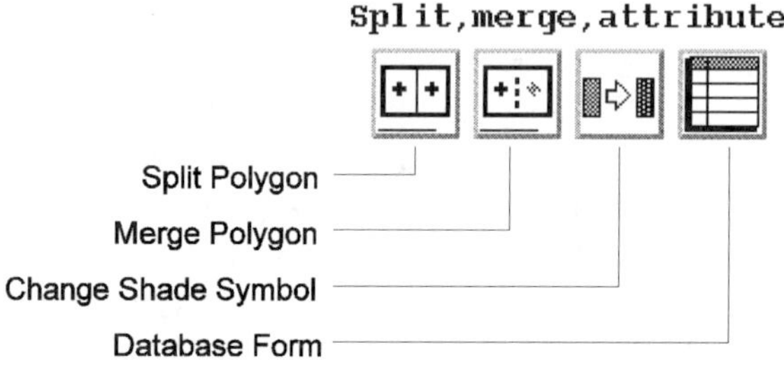

Split Polygon works on the selected set of polygons. First, you must have at least one polygon selected. Note that when you select polygons, you can pick any position in the interior of the polygon, and it will become selected. After you click the Split Polygon command, the screen cursor is activated. Use buttons 2, 1, and 9 to add lines and return to menu control.

Merge Polygon requires that you have at least two polygons selected. If they are adjacent, then this command will join them into one polygon. It does this by removing arcs that are the boundaries between adjacent and selected polygons, and any redundant label points.

Change Shade Symbol modifies the fill area of the selected polygons. First, select a new shade symbol in the scrolling list on top of the menu. Then, select as many polygons as you like. When you click this command, the shade symbol for all of these polygons is changed to the new shade symbol.

Delete, Get, and Put

Delete,get,put

Delete
Get from Coverage
Put to Coverage

When you do a Get from Coverage or a Put from Coverage command on polygons, then both the arcs and label points for the selected arcs are written out or imported in.

Delete removes the selected polygons. In some circumstances, this command may remove a label point, but no arcs. Consider a polygon that is fully surrounded by other polygons. Delete removes that label point, but because those surrounding arcs are shared by other polygons, no arcs will be removed. If a polygon is freestanding, then all the arcs and the label point for that coverage will be removed.

Get from Coverage imports all the polygons from another coverage into the edit coverage. Both arcs and label points are imported. A menu with coverage names will appear when you click this command. After you've selected a coverage, you'll see new arcs and label points appear. If they intersect existing polygons in your coverage, then new intersection polygons, as result with the Add Polygon command, will appear.

Put to Coverage outputs selected polygons to another coverage. First, select one or many polygons. Then when you click this command, a menu with coverages appears. When you've selected one, the label points and arcs for those polygons are output to that coverage.

Updating Polygon Topology

Earlier, we mentioned that there are two ways to edit polygons: you can either edit the components of polygons, edit label points and arcs, separately, or you can use the commands in the Edit Polygon menu we just covered.

If you use the polygon editing commands in the Edit Polygon menu, then the relationships between label points and arcs are updated

automatically for you. This includes an internal description in ARC/INFO of which arcs bound a given polygon. This is called *polygon topology.* If you use the Edit Label Points menu and Edit Arcs menu to edit polygons, then ARC/INFO needs a way to keep the relationships up to date between label points and arcs that define polygons. The two commands to do this are Build and Clean.

As you update polygon topology in the Edit Polygons menu, these two conditions should be met:

- ❑ There should be no dangling arcs that bound polygons. In the Draw Environment menu, use the Node Dangle or Node Error setting when editing arcs, so that you can ensure closed arcs. Also, put the snap tolerances to work for you to close arcs while adding them.
- ❑ There should be only one label point inside each polygon. It doesn't matter where this label point is, as long as it is inside.

You can tell whether polygon topology is current by whether polygons are drawn with shaded areas. Before using these commands, you should go to the Draw Environment menu and click the Fill choice for polygons. This will ensure that polygons are shaded.

Build and Clean. The fuzzy distance and dangle distance is used by the Clean command.

```
Update topology:
[Build]
[Clean]
[⌖]  [        ]  fuzzy
[⌖]  [        ]  dangle
```

Build updates polygon topology without modifying arcs. When you run this command, you may see polygons that were not shaded, now appear shaded.

Clean is similar to Build, but this command makes some minor edits on your polygon coverages. Clean uses the specified fuzzy distance to resolve node points that are close to each other into clean intersections, and uses the specified dangle distance to remove small dangling arcs.

You can set both the fuzzy distance and the dangle distance with either the cursor or by typing in values. These distances are in coverage units: usually feet or meters.

The fuzzy distance merges nodes that are within the fuzzy distance into a clean junction. The dangle distance is a tolerance to automatically remove small dangling arcs.

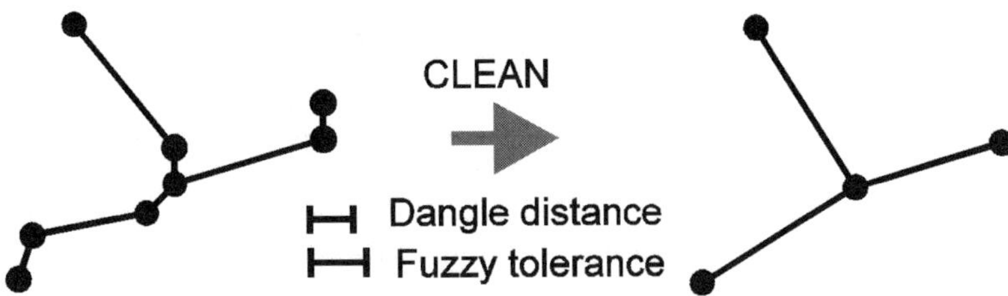
CLEAN
Dangle distance
Fuzzy tolerance
Dangling arcs smaller than dangle distance removed,
arcs at junctions smaller than fuzzy distance removed.

NOTE: *Be very conservative with the fuzzy and dangle values for the Clean command. If in doubt, start with small values and work to larger values.*

NOTE: *If your coverage is large, then both the Build and Clean commands may take several moments or longer to complete.*

Don't worry about the details of how ARC/INFO keeps track of arcs and label points that define polygons: Build and Clean update these relationships automatically for you.

Exercise: Editing Polygons

Let's edit some building polygons in Edit GIScity.

1. Make the Roof Lines coverage your edit coverage. Click the Coverage Manager command in the Edit GIScity main menu and pick the ROOFLINES coverage.

Edit
GIScity

2. Zoom to an area where you only see about a dozen buildings.
3. Click the Draw Environment command in the Edit GIScity main menu. From the Draw Environment menu, select the On choice for arcs, the On choice for labels, the Dangle choice for nodes, and the Fill choice for polygons.
4. Click the Edit Polygon command in the Edit GIScity main menu. The Edit Polygon menu will be invoked.
5. Let's use the Add Polygon command to add a new, imaginary building. Click the Add Polygon command. Your screen cursor is active. Use button 1 to mark all the corners of a new building. When done with the corners, use button 2 to close the polygon. Then, use button 9 to return to the menu. After adding one polygon, try this command again, but this time, add a building that intersects other buildings. Watch what happens, and count the label points.

You will see free-standing polygons with one arc, one label point, and a shaded area. A new polygon that intersects other polygons will be split, with a label point added for each split polygon.

6. Let's try the Add Polygon, Dissolve command to make new polygons that intersect with existing polygons. Click the Add Polygon, Dissolve command. Use buttons 1, 2 and 9 as before, but this time, make some of the building lines intersect an existing building.

When you add a polygon with the Dissolve option, you can effectively enlarge polygons. This is like adding a room to a building; you don't need to add the room as a separate building.

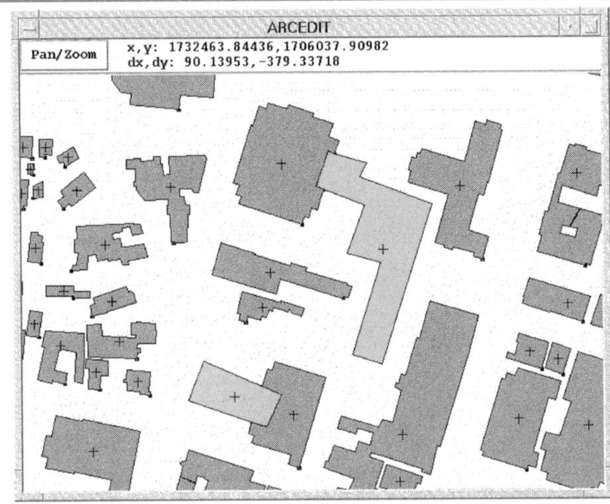

7. Two commands in the Edit Polygon menu are similar: Add Polygon, Line, and Split Polygon. We will try both of these.

First, click the Add Polygon, Line command. The screen cursor is activated. Use button 2 to start a line. Then, use button 1 to add some vertices. Make sure the line completely crosses an existing polygon. Use button 2 to close the line and button 9 to return to the menu.

Next, select a polygon. Either use the Select Spatial menu or the Select Box button at the bottom of the Edit Polygons menu. You'll see selected polygons highlighted in yellow. Click the Split Polygon command. Just as with the other command, start with button 2, use button 1 for intermediate points, and close with button 2. Use button 9 to return to the menu.

You should see something like the following figure.

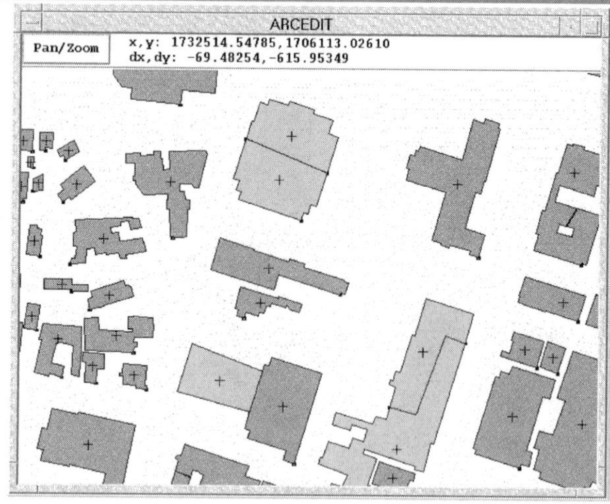

Adding polygons by line and splitting polygons.

8. Finally, let's use the other way of editing polygons: adding arcs and label points separately and then updating the polygon topology with Clean or Build.

Open the Edit Arcs menu. Digitize a set of arcs that form an area. Then, open the Edit Label Points menu. Add label points, one for the interior of each area. Open the Edit Polygons menu. Type in a value of 5 for the fuzzy distance. This means that any node points within 5 feet of each other will be automatically closed. Type in a value of 2.5 for the dangle distance. Any small dangling arcs less than 2.5 feet will be removed. Click the Clean command.

The end result is the same as using the polygon edit commands. When you edit coverages, you'll find yourself using both methods.

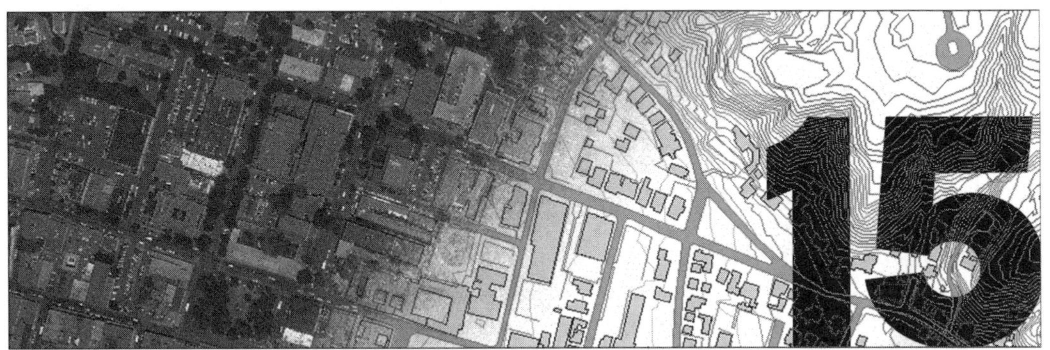

Editing Secondary Features in Edit GIScity

Annotation, Tics, and Links in ARCEDIT

The secondary features in a coverage—annotation, tics, and links—do not directly model geographic features that you can find on the earth, but they are important in a GIS for description, registration, and correction.

Annotations are text features stored in coverages. They are primarily used for place names such as towns, parks, buildings, and roads. A map without place names is a poor map.

Tics are registration marks. They are mainly used in the process of registering map manuscripts upon a digitizing table. Tics are also used for transforming the coordinates of features from one coverage to another. Tics usually correspond to map sheet corners or survey control points.

Links are rubbersheeting vectors, used for making corrections to digitized data. Sometimes, you have to digitize from less than perfect

maps; links can be used to adjust digitized data so that it better conforms to an accurate land base.

This chapter documents the three Edit commands outlined in this illustration of the Edit GIScity main menu.

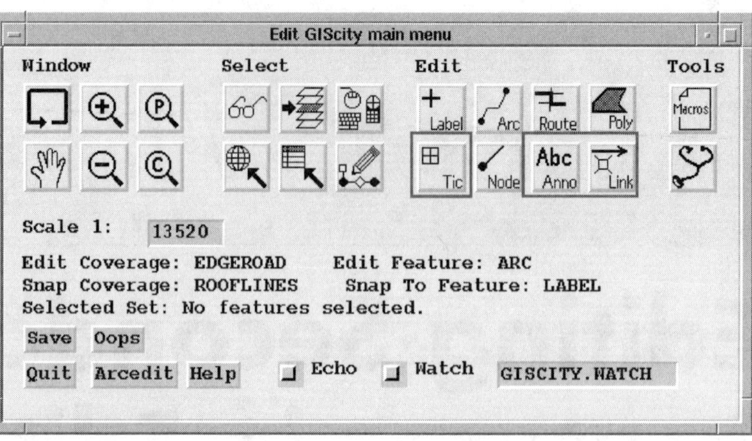

Editing Annotation

Abc Anno

Imagine a map of a city you are visiting for the first time. All the streets, parks, rivers, and buildings are drawn with arcs and polygons. However, no descriptive text is on your map. That map would be almost useless. Devoid of descriptions, a map has little reference and meaning.

I once had an interesting experience while visiting a city in eastern Europe. At a tourist agency in the railroad station, I got a map of the city with street names annotated using the Roman alphabet. However, on the streets, every street sign used Cyrillic characters. Instead of being helpful, the tourist agency map gave me the real sense of being lost.

Annotation is an important component in a GIS database. While it is not directly used in geographic analysis (you can't do a network trace or polygon overlay with annotation), annotation is essential in providing a reference and context to users of the GIS database.

In Display GIScity, you've seen information displayed to you as text. A powerful capability of ARCPLOT is to dynamically display attribute

information as text. However, you also need to add and edit permanent annotation to your coverages. Some information in a GIS, such as street names, is relatively permanent. While street names can be dynamically displayed, adding annotation as features in a coverage gives you greater control over cartographic placement, and the annotation will be consistently displayed whenever you visit that area in your GIS display.

NOTE: *"Annotation" and "Text" in a GIS are easily confused. Annotation is a permanent feature that you can edit and store in a coverage in ARCEDIT. Text is an ephemeral feature that appears when you issue text display commands in ARCPLOT. Text is not stored in a coverage. Display GIScity displays both annotation and text. Edit GIScity displays only annotation. Properties for text and annotation are similar, but distinct.*

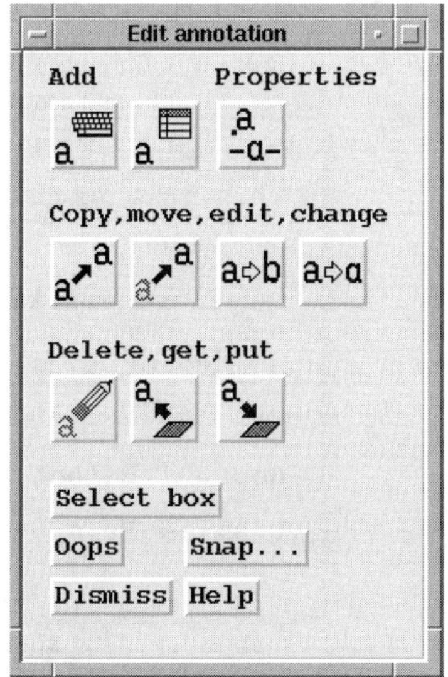

The Edit Annotation menu.

All of the commands for adding annotation to your coverages are in this menu. Some of the commands invoke other menus; others perform their action based on your selected set and edit environment.

Adding Annotation and Setting Properties

Add and Properties.

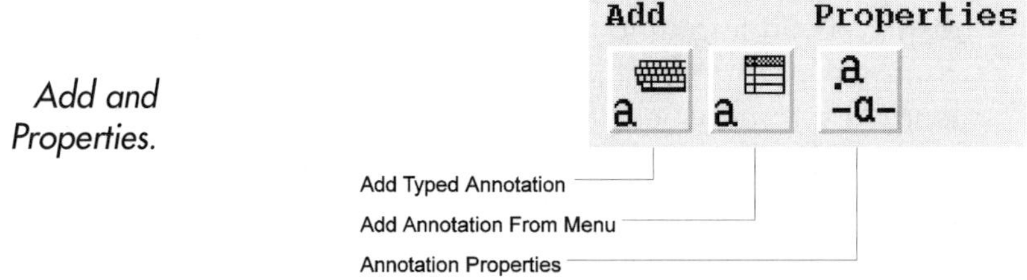

All three commands in the first row of the Edit Annotation menu—Add Typed Annotation, Add Annotation From Item, and Annotation Properties—invoke other menus. The first two menus present the two basic ways to add annotation to a coverage, and the third menu sets default properties to be applied when you add annotation.

Adding Typed Annotation

As well as a place to type in annotation, the Add Annotation from Text String menu gives you several options for positioning annotation.

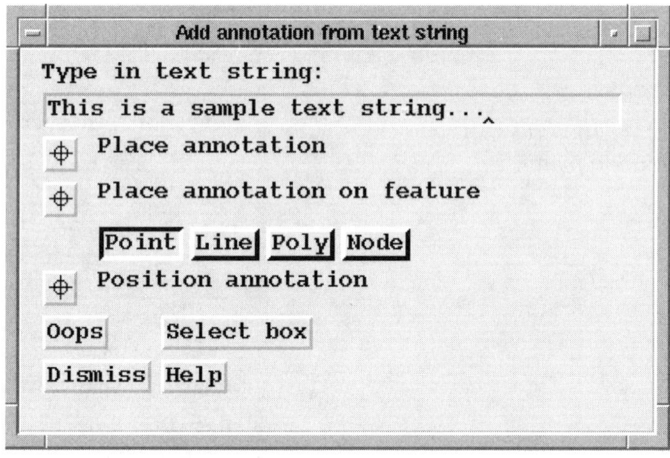

In the text input field in the Add Annotation from Text String menu, you can type in any text string you want to place on the coverage.

After you type in a text string, you have two choices for placing the annotation. If you click the cross-hair icon for the Place Annotation command, you can add annotation horizontally, at an angle, or along a

user-defined set of points. Or, you can use the Place Annotation on Feature cross-hair to place annotation along other features in your edit coverage.

NOTE: *Anytime you add annotation, that new annotation is your selected set. It will appear on your canvas with the current Draw Select color. To change the color that new annotations are drawn as, invoke the Select Features Spatially menu and change the Draw Select color.*

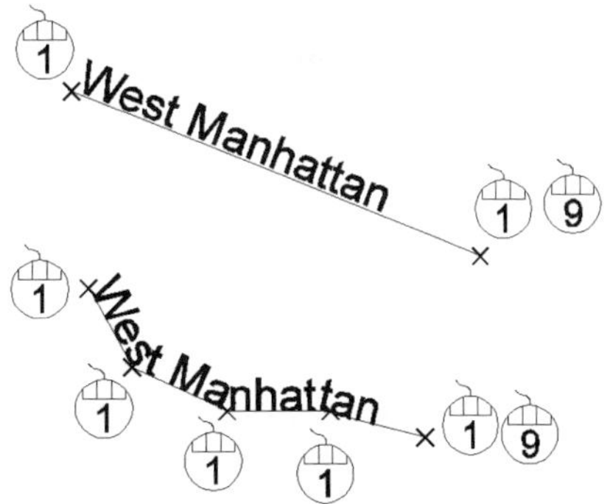

Options for free form placement of annotation.

To add a text string as horizontal annotation, click the cross-hair icon for the Place Annotation command, move the screen cursor to the point where you want to anchor that annotation, and press input key 1. Next, press input key 9. The position of the screen cursor when you press input key 9 doesn't matter.

To add the text string at an angle, click the cross-hair icon for the Place Annotation command, move the screen cursor to a start point, and press input key 1. Move the cursor to another point to set the desired angle and press input key 1. Press input key 9. You will see the text string placed at an angle along the line between the two points.

To add the text string along a user-defined set of points, click the cross-hair icon for the Place Annotation command and move the screen

cursor to the first point. Then move the screen cursor and successively press input key 1. (You can define up to 500 points to place curved annotation!) When you press input key 9, you will see the annotation neatly splined.

Annotation placed along a coverage feature.

To add annotation adjacent to another coverage feature, first select the point, line, polygon, or node feature type under the Place Annotation on Feature command. Then click the cross-hair icon.

Your screen cursor will become active to select a feature of the type you chose. If you picked Line, select an arc on your display. If the arc is not the one you meant, just select another one. When you have picked the arc you want, press input key 9, and your text string will be placed exactly by that feature.

After you have placed annotation with the freeform method or along a coverage feature, you can easily reposition it. Press the cross-hair icon next to Position Annotation. Then use the same steps for freeform placement of annotation. You can do this several times to get the placement of annotation exactly where you want it.

NOTE: *The Position Annotation command works on your selected set. Whatever changes you make instantly apply to your selection annotations. If you make a mistake, you can click Oops to undo that change. Or, you can use the Select Box command to change your selected annotation before you reposition it. (The Select Box command is identical to the command in the Select Features Spatially and is in this menu for your convenience.)*

Adding Annotation from Item Values

The Add Annotation from Item Values menu makes it easy to turn item values into annotation.

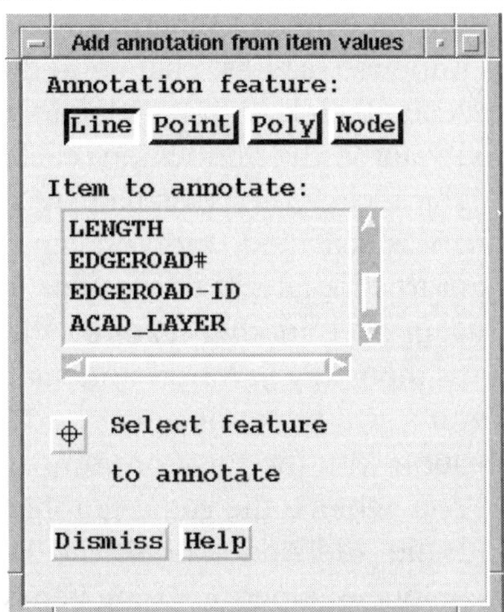

If you must add lots of annotation, there might be a shortcut for you. For example, you may have the job of adding street names as annotation. If you have an item for street names in your Arc Attribute Table in your edit coverage, the Add Annotation from Item Values command can save you time. This menu lets you automatically add annotation along selected features with item values for selected features. You can save a lot of typing this way!

Annotation can be item values from attribute tables.

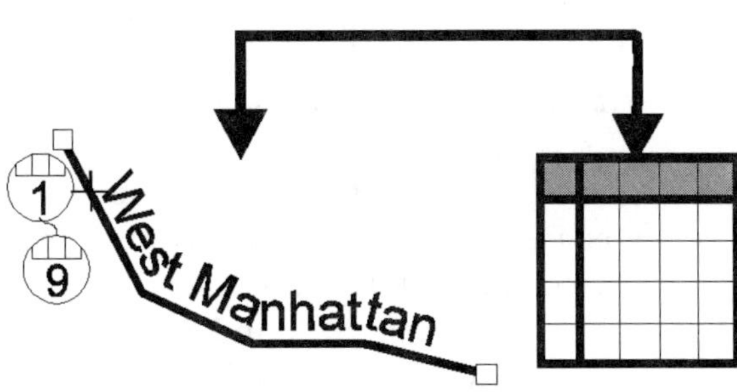

Use the following steps to add annotation from item values.

First, select a feature type to annotate: Line, Point, Polygon, or Node. (For this command to work, the feature type must have an attribute table associated with it.) When you pick a feature type, you will see the Item to Annotate data list change to the items for that feature attribute table.

Second, pick an item value to annotate. This item contains the values you want in the annotation; each feature can have a different value.

Third, click the cross-hair by Select Feature to Annotate. The screen cursor is activated, and you can press input key 1 over the feature you want to annotate. If the feature isn't the right one, just press input key 1 again over another feature. When you have chosen the right feature, press input key 9. A new annotation, with the chosen item value for that feature, will be automatically added for you. (You can later change its location with the Position Annotation command.)

You achieve the same graphical result as when adding annotations with the Add Annotation from Text String, but wherever possible, this technique is superior. Using item values reduces the possibility for error and streamlines your job of adding annotation.

Annotation Properties

On the Annotation Properties menu, the Set Text Symbol data list gives you a variety of text fonts to choose from. Text symbol 1 is the default when you begin Edit GIScity. Whenever you choose a new font, that choice remains active and outlined in the data list until you change it. Any annotation you now add will assume this text symbol.

In Edit GIScity, you can specify the size of text by clicking the cross-hair icon and picking two points on your graphics window, or you can directly type in a text size in map units.

You can specify that annotation be offset from a feature, or from points you pick with the screen cursor. To set an offset for the annotation you are about to add, you can click the cross-hair icon under the word Offset, and pick two points on your graphics display. Edit GIScity will calculate the shift in X and Y between those two points and set the X and Y offsets to those values. Alternately, you can move your screen cursor inside the input fields for X and Y offsets and directly type in those values.

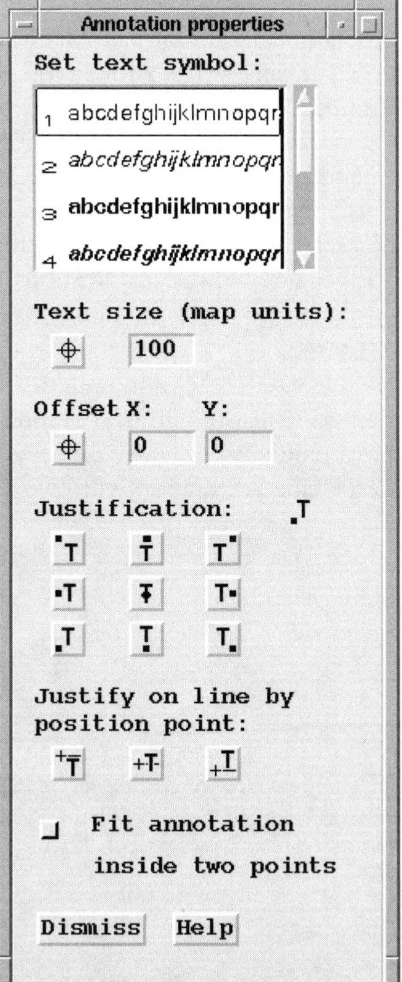

The Annotation Properties menu.

NOTE: *If you want to offset an annotation placed along a line, you only need to set a distance for the Y offset. Edit GIScity will use the Y value to offset from the line, no matter the orientation of the line. The value for X offset is ignored.*

As with text in Display GIScity, you can set justification points for annotation. Most often, you'll choose the lower left icon, which is lower left justification. The Annotation Properties menu indicates the selected justification by displaying an icon to the right of the nine input icons. When you select a new justification, the display icon is immediately updated.

There are three additional options for justification. They are to place annotation below, on, or above lines. When you place annotation along existing lines, these justifications let you pick a point along the line to automatically place the annotation above, on, or below.

There is one more option in this menu for placing annotation. You can choose to fit annotation automatically inside two points. This method will stretch or compress your annotation to exactly span between the two points you set. This option takes effect only when you are placing annotation with two points.

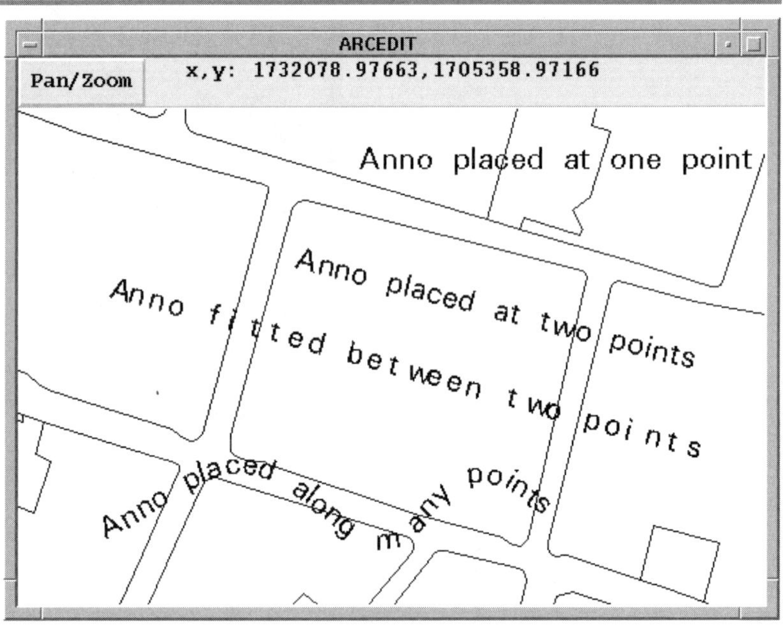

A sampling of annotation placed with the Edit Annotation commands.

Copying, Moving, and Changing Annotation

The second row of commands in the Edit Annotation Menu.

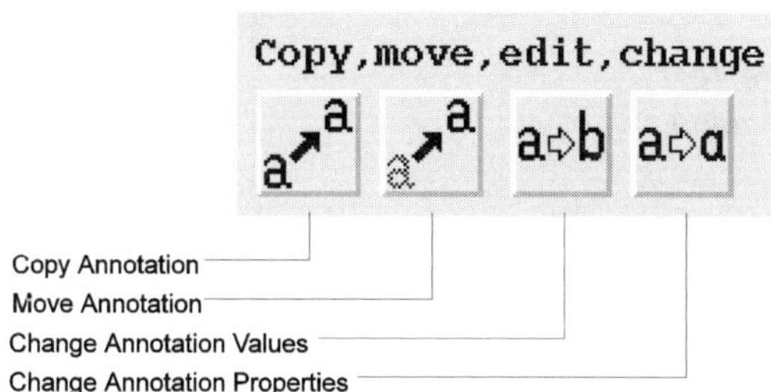

You can modify all of the characteristics of annotations you have added. All of these commands operate on your current selected set of arcs. Either use the Select Box command on the bottom of the Edit Annotation menu or invoke the Select Features Spatially menu. The first two commands move or copy annotation. The second two change the wording in your annotation, or change the annotation properties.

Use the Copy Selected Annotations command to make additional copies of selected annotations. After you click this command, move your screen cursor and use any input key to mark a begin point and an endpoint. These two points will mark a displacement by which your duplicate annotation will be moved.

The Move Selected Annotations command works just like Copy Selected Annotations, but the annotation at the previous position will disappear. When you copy or move annotation, you can undo these commands either by using the Oops command or performing the copy or move again.

Clicking the Edit icon on the Edit Annotation menu invokes the Substitute Text Strings menu. The menu lets you replace the text within selected annotations. To change text, follow these steps.

Step 1. Select the annotation with the Select Box command or the Select Features Spatially menu.

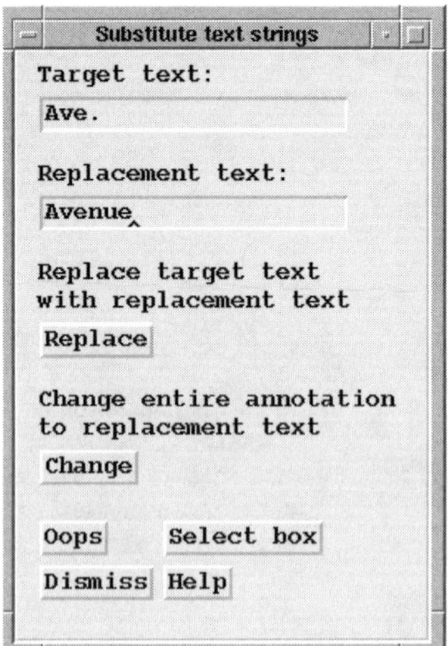

The Substitute Text Strings menu.

Step 2. Type in the text you want to change in the Target Text input field. It can be part of the text or the complete text.

Step 3. Type the new text that you want in the Replacement Text input field.

Step 4. Click either the Replace or Change button. The Replace button will allow partial replacement of text in the selected annotations. The Change button will replace only the text which fully matches the target text with replacement text.

Clicking the Change Annotation Values command invokes the preceding menu, which lets you easily change the properties of existing annotation. These are the steps:

Step 1. Select the annotation or annotations whose properties you want to modify properties for. You can use the Select Features Spatially menu or the Select Box on the bottom of the Change Annotation Properties menu.

Step 2. Use the Annotation Properties menu to set new properties. For example, you can set a new text size in that menu.

Step 3. Choose any of the buttons in this menu to change the properties of the selected annotations to the new properties you've set with the Annotation Properties menu.

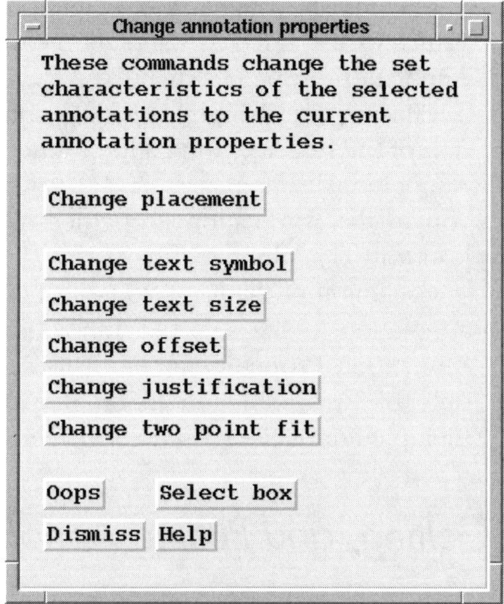

The Change Annotation Properties menu.

Change Placement will change the placement of any selection annotations. It actually encompasses the function of the Move Annotation command, and also lets you change how many points define the placement of the annotation. For example, you can change one-point annotation to two-point annotation. Or change one-point annotation to many-point annotation. Or you can redefine the points for many-point annotation.

NOTE: *Don't use this command when you have selected multiple annotations. This command will place all selected annotation on top of one another. You must select only one annotation when you issue the Change Placement command.*

Clicking the Change Text Symbol button will change the font of the selected annotation to the symbol you've set in the Annotation Properties menu.

Clicking the Change Text Size button will change the text size of the selected annotation to the size you've set in the Annotation Properties menu.

Clicking the Change Offset button will change the X and Y offset values of the selected annotation to the new offsets you've set in the Annotation Properties menu.

Clicking the Change Justification button will change the text justification of the selected annotation to the new justification you've set in the Annotation Properties menu. (The new justification will be shown to you in the Annotation Properties menu as an icon with one of the justification settings.)

Clicking the Change Two Point Fit button will change the two point annotation in the selected annotation to the value you've set in the Annotation Properties menu. This command will change the spacing only of two-point annotation in your selected set. (One-point and multiple annotation will not change.)

Deleting, Getting, and Putting Annotation

The third row of commands in the Edit Annotation menu.

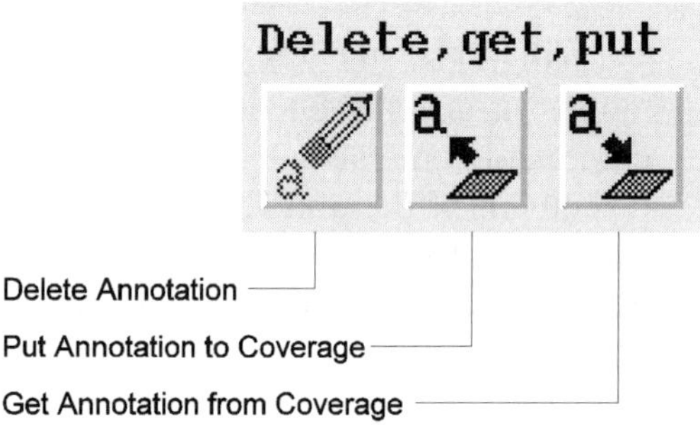

The Delete command will remove any annotation you have in your current selected set. If you've made a mistake, you can click the Oops button to restore the deleted text.

As with some of the other edit features, you can copy selected annotation from the current edit coverage out to another coverage, or you can import annotation from another coverage into the current edit coverage.

The Put Annotation to Coverage command will invoke the Put Coverage Features menu that you've seen with some of the other edit

features. When you select an existing coverage, or type in the name of a new coverage, the selected annotation will be written out to it.

The Get Annotation from Coverage command will invoke the standard Get Features from Coverage menu. To import all annotation from another coverage, just select that coverage name from the data list in the Get Features from Coverage menu.

Selecting, Snapping, and Undoing Annotation Edits

The Control commands from the Edit Annotation menu.

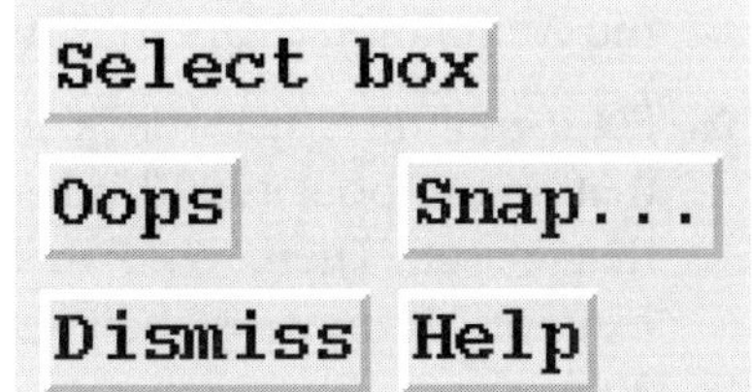

The Select Box command is a short-cut to let you quickly select annotation. You can use it for simple selections in a box, without invoking the Select Features Spatially menu. When you use this command, your previous selected set will be replaced.

The Oops command will undo any annotation edit command you perform. You can successively Oops your way back to the beginning of your edit session.

The Snap command will invoke the standard General Snapping Environment menu. You can snap selected annotation to label points, arcs, nodes, or polygons. The tolerance for snapping is set in the General Snapping Environment menu.

Exercise 1: Editing Annotation

Probably the most common annotations on maps are street names. Let's practice adding and editing street names.

1. From the Set Edit Coverage menu, select EDGEROAD for your coverage.
2. Zoom into an area of Santa Fe that includes a winding road. Zoom to a scale where you can see the curb lines clearly separated, which would roughly be a scale of 1:6 000.
3. Click the Edit Annotation command on the main menu. The Edit Annotations menu will appear.
4. Click the Annotation Properties command. This is the third command on the first row of the Edit Annotation menu.
5. Pick a text symbol from the Set Text Symbol data list at the top of the Annotation Properties menu. Symbol 1 is the default.
6. Set a new text size. Click the cross-hair icon and pick two points that span about two-thirds of a road width. Look at the text size that is set. Then, type in a rounded number that is close. For example, if the cross-hair selection yielded 17.82, type in 15, 17, or 20.
7. Click the Add Annotation from Text String command (the icon with a keyboard). You will see a menu for typing in and placing annotation.
8. Type a street name in the Type in Text String input field. Some street names in Santa Fe are Otero Road, Paseo de Peralta, Washington Street, and Water Street. Press the cross-hair button for Place Annotation.
9. Press input key 1 twice to place that street name along a defined line. Press input key 9 to exit.
10. Whoops! You put it in the wrong place. Click the cross-hair icon for Position Annotation. Press input key 1 twice again, followed by input key 9.
11. Whoops! You typed in the wrong street name. Click the Substitute Text Strings command (third command in the second row in the Edit Annotation menu). In the menu, type in the old street name in the Target Text input field. In the Replacement Text input field, type in a new name. Then click the Change command. You'll see the entire street name changed.

12. Now go to that winding road. In the Add Annotation from Text String menu, type a long street name. Pick the Line choice and then click the cross-hair icon for Place Annotation On Feature. Move the screen cursor to the right side of a winding curb line.

13. Type in that long street name again in the Add Annotation from Text String menu. This time, click Place Annotation and pick about 5 or 6 points with input key 1 along the curb line, and exit with input key 9. Which method do you like better?

14. Try changing any (or all) of the annotation properties using Select Box. Make changes in the Annotation Properties menu (last icon on the first row, Edit Annotation menu) and then click the corresponding buttons in the Change Annotation Properties menu (the fourth command on the second line in the Edit Annotation menu.)

15. Now, change your edit coverage to TGR_ROAD. This is a less detailed road database, but it has centerlines with more attributes assigned. This data is from the United States Census Bureau and is called TIGER data. Redraw your graphics window. It's not as pretty, is it? But just wait.

16. Click the Add Annotation from Item Values command (second command on the first line in the Edit Annotation menu). Select Line for the Annotation Feature. Select FNAME as the item to annotate. Then click the cross-hair icon for the Select Feature to Annotate command. Place the cursor on a TIGER road centerline. Press input key 1. You should see a street name automatically annotated from the TIGER item for street names, FNAME. Press input key 1 successively. When you are done, press input key 9.

17. Now select the annotation you've just added in the TGR_ROAD coverage, and click the Put Annotations to Coverages command (third command on the third row of the Edit Annotation menu). Choose EDGEROAD as the coverage to put to.

18. Change your edit coverage to EDGEROAD and redraw your graphics window. Now you've transferred annotation from one coverage to another. You can use the Move command on the Edit Annotation menu to move the street names from TIGER around to line up better with the more detailed curb lines in EDGEROAD.

Editing Tics

These are all the commands for adding and editing Tics.

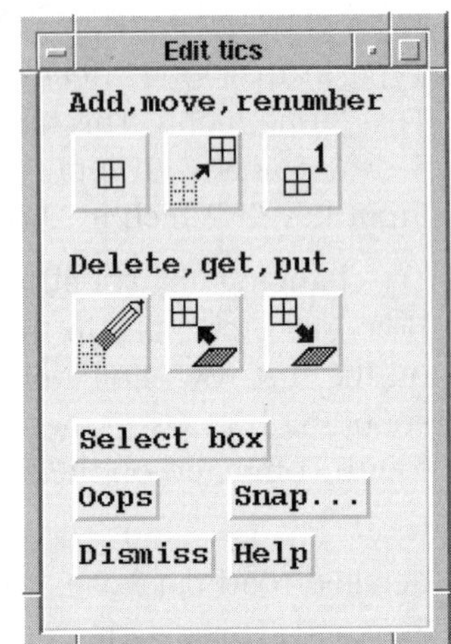

Tic

Tics are probably the simplest feature in a coverage. As registration marks, only two things count about a tic: its location and its number. Normally, you only display tics when you are registering a map manuscript or preparing a coverage for digitizing from another map manuscript.

Adding, Moving, and Renumbering Tics

The top row of the Edit Tics menu.

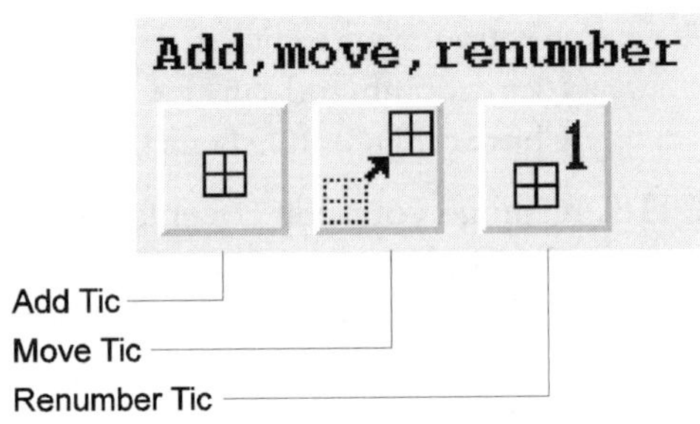

To add tics, click on the Add Tic command. When you start this command, the screen cursor will be activated for you to add tics.

In your dialog window, you will see a line that contains text that might look like this: "(Tic) User-ID: 5." This means that the new tic you are about to add will receive the number 5. If you want to assign a different number, at this point you can press input key 3. Within the dialog window, you will be asked to type in a new tic number. After you press [Enter], you can move your screen cursor inside the graphics window.

You can add multiple tics by repeatedly placing the screen cursor over points and clicking input key 1. To terminate the addition of tics, press input key 9.

Once you have placed tics, if they are not exactly where you want them, you can move them. Click the Move Tics command. Then your screen cursor is activated to accept a begin point and endpoint. These two points define a displacement vector. The selected tic(s) will be immediately moved. (You can Oops this command.)

Renumbering tics.

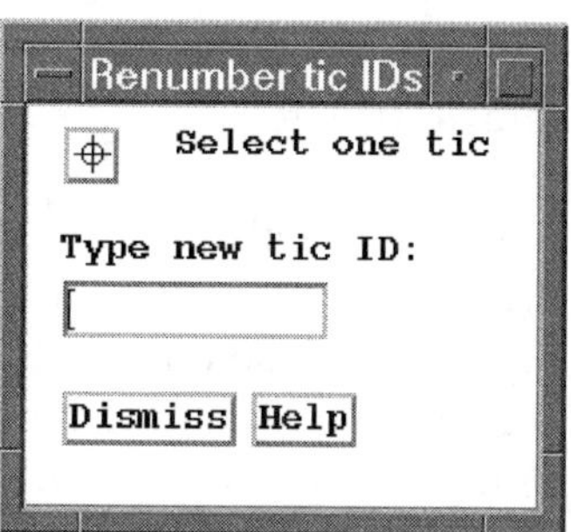

Clicking the Renumber Tic IDs command invokes this menu. To renumber existing tics, first select a tic. Click the cross-hair icon and move the screen cursor to select a tic. If your selection is within the edit distance (set in the Select Features Spatially menu), the tic will be selected and you will see the current tic ID in the input field. To renumber the selected tic, type in a new tic ID. Be sure to press [Enter] after you type in the value; the tic is renumbered at that point.

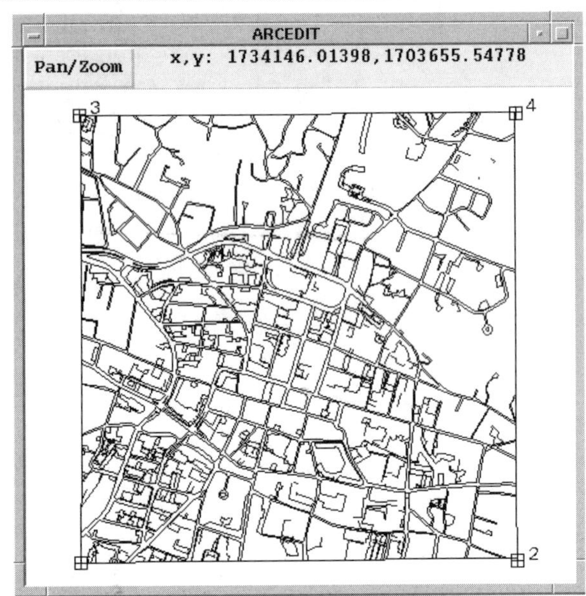

The EDGEROAD coverage in GIScity has tics defined at the four corners of the map coverage.

Tics are not required for coverages. If coverages are imported from other digital sources and are corrected with "heads-up" digitizing, you may not need any tics.

When you digitize on a coverage using a digitizing table, you will usually have the tics at map sheet corners or survey control points. In GIScity, some of the coverages have tics at the sheet corners, as shown.

To get strong registration results, you should have at least four tics in a coverage. Also, tics should be well-spaced from each other in both distance and angle.

Deleting, Getting, and Putting Tics

The Delete command will remove any tics you have in your current selected set.

The Put Tics to Coverage command will bring up the Put Coverage Features menu and let you copy tics to another coverage.

The second row of commands in the Edit Tics menu

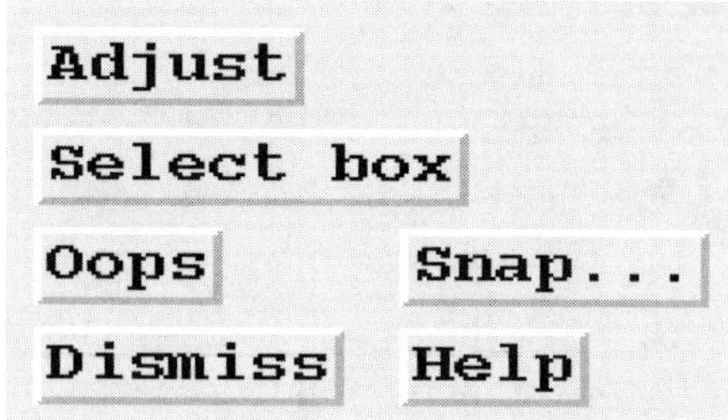

The Get Tics from Coverage command will invoke the Get Features from Coverage menu. If you select a coverage from this menu, all the tics from that coverage will be added to the edit coverage. You will probably want to inspect the tic IDs of the new tics and perhaps renumber the new IDs.

The Put and Get commands are used more frequently for tics than other feature types. These commands are a handy way of ensuring that different coverages with the same geographic extent share the same registration.

The Control commands in the Edit Tics menu.

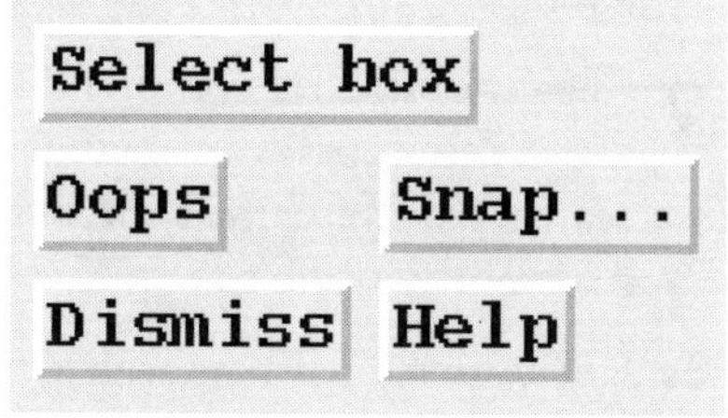

The Control commands work the same as those in other edit feature menus.

Select Box is a shortcut to selecting tics quickly. Oops undoes your last edit command, whether it was on a tic or other feature type. Snap invokes the General Snapping Environment menu.

Snap is particularly useful for tics. If you have label points marking survey control points, you will probably want to snap some tics to those points.

Exercise 2: Editing Tics

1. Select the EDGEROAD coverage.
2. Click the Edit Tics command from the main menu to invoke the Edit Tics menu.
3. Click the Snap... button to invoke the General Snapping Environment menu. On that menu, choose SURVEY from the Coverage to Snap To data list, and Label as the Snap To feature. Type in a value of 10 for the Snapping Tolerance.
4. Click the Set Draw Environment command in the Edit GIScity main menu (the pencil pointing at squares and diamonds). Click the Back coverage choice on the top of that menu, and turn the choice for Label to On. Click on the check box for Display Snap Coverage As Back Coverage.
5. Click the Redraw Window command in the Edit GIScity main menu.
6. Now zoom to a location where you can see a survey control symbol. You should see several of these in the middle of some roads in Santa Fe.
7. Click the Add Tic command in the Edit Tics menu. Add one tic while the screen cursor is very close to a survey control symbol. If you watch carefully, you might see the tic snap to the symbol at the moment you digitize it.
8. Say you want to reference that tic by the number 106. Click the Renumber Tic IDs command. Type in 106. You'll see the tic ID renumbered for you.

Editing Links

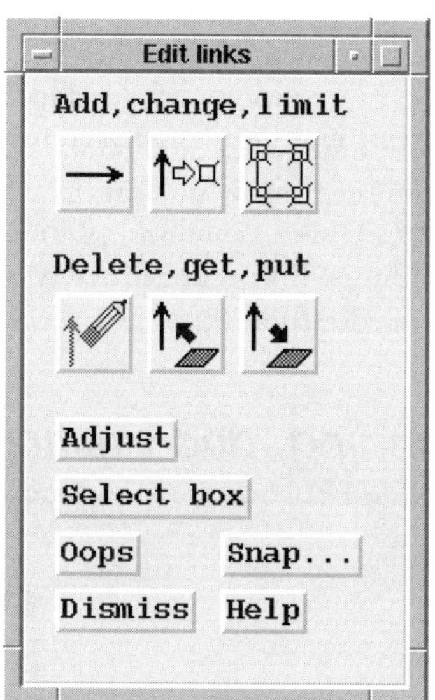

The menu for adding and editing links.

Links are rubbersheeting vectors. A link simply comprises a start point and an endpoint. Links are not geographic features that are normally displayed or published on a map. Links are created and used when you are making corrections to a map. Although links can be stored indefinitely in a coverage, they are usually used and removed soon after you make them.

A possible scenario for the use of links unfolds like this: You have some old insurance maps from the turn of the century. They contain valuable historic property records but are based on an inaccurate map base. Also, these old map manuscripts have shrunk differentially (shrunk in one direction more than the other) due to the passage of time. You also have some modern GIS land base data for the same area. You would like to digitize and register the historic land records upon the modern GIS data.

These are the basic steps: First, establish tics to common points on the old maps and the new GIS data. Register the old maps as best you can on a digitizing table. After registration, because of the condition of the maps, your RMS (Root Mean Square) error may not be close to the tolerance you would like. Then digitize the old map's parcel lines and code the property ownership data in attribute or relate tables.

After you have digitized the old insurance map, create links between common points to both the modern GIS data and the newly digitized lines. These common points will usually be street intersections or buildings. If you are lucky, you might find survey control symbols, but if you do, then the maps probably won't need rubbersheeting.

Adding, Changing, and Limiting Links

The first row of Edit Link commands.

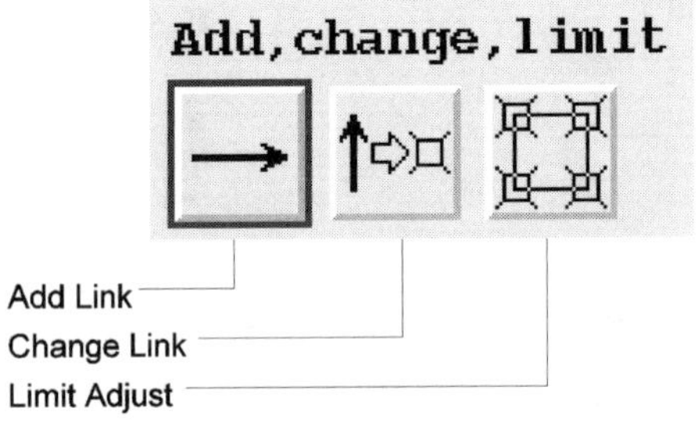

Click the Add Link icon to start digitizing links. You probably want the other themes displayed, as a visual guide for the start point of each link. If you want very precise place of links, you can also set your snapping environment (using the General Snapping Environment menu), so that link start points snap exactly.

When you start adding links, you will see several options in your dialog window. We'll cover all of these later, but for now, we'll cover what input keys 1, 2, and 9 do.

Each link has two points. If you use input key 1 for both the start and endpoints, a new link will be added at exactly those points. If you use input 2 for either the start or endpoints, and features in your snap

coverage are within the snap distance, the end of the link will snap to that snap feature.

Typically, the first point of a link will correspond to a point on your edit coverage, and the second point will correspond to a point on your snap coverage. Therefore, you will frequently use either a combination of input key 1 and input key 2 for each link, or input key 1 twice. Rarely would you use input key 2 for the first point of a link.

Adding links is a tricky business. Getting the results you desire might take several tries. You might go through a cycle like this: Add some normal links, set identity links, click the Adjust command, click the Oops command if the result is not quite right, and then start over. With some experience, you will gain judgment about how to make links maximize the improvement in positioning in your edit coverage.

There are two types of links—*normal* and *identity*. Normal links look like a vector with an arrowhead. Identity links look like a small square with four radiating diagonal lines. (See the icon in the Edit GIScity main menu, or the following illustration.)

Identity links prevent rubbersheeting from taking place in the immediate vicinity of that link. This has the same effect as making a link with the same start and endpoints.

The Change Link command will turn any selected links into identity links. You will use Change Link for any parts of your map that are already accurate.

The Limit Adjust command will automatically make a large number of links for you. When you perform rubbersheeting, you often want to restrict rubbersheeting to a certain part of a map. When you click Limit Adjust, your screen cursor is activated to accept two points. After you pick the second point in your graphics window, a number of links, maybe a hundred or so, will be created in the shape of a box. This will prevent rubbersheeting from occurring outside this area.

Deleting, Getting, and Putting Links

The second row of commands in the Edit Links menu.

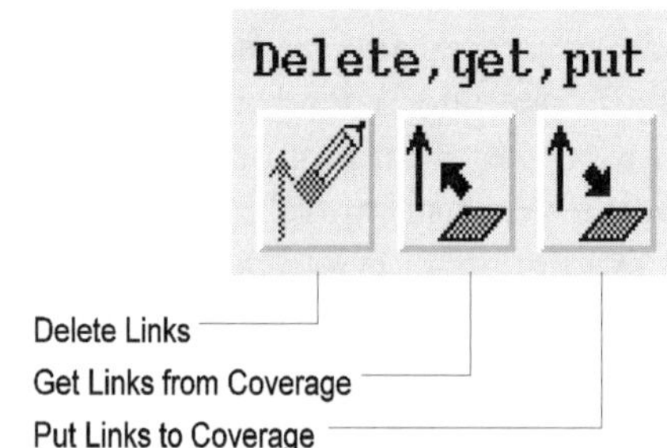

As with the other feature edit commands, you can delete links, put them to another coverage, and get them from a coverage.

The Put and Get commands allow you to perform rubbersheeting on one coverage, and then perform the same adjustment on any other coverage. This is important if you have digitized two or more coverages from the same map manuscript.

The control commands in the Edit Links menu.

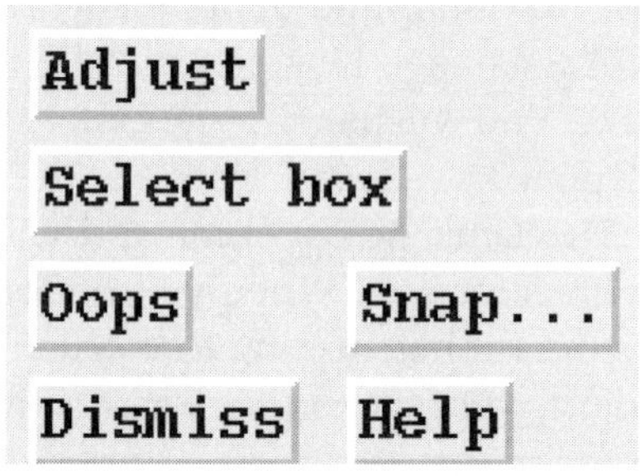

Most of the control commands in this part of the Edit Links menu work identically to those in other edit feature commands, but there is one more command that is interesting—Adjust.

Adjust performs rubbersheeting on all features in your edit coverage, using the links you've made. When you click this command, ARCEDIT might take several minutes to calculate the result of the rubbersheeting operation. The Adjust command may take several tries to get right, so you can Oops this command and modify your links.

Besides modifying the position of all coverage features within the identity links, Adjust has an interesting effect: all normal links are converted into identity links, positioned at the second point of each link. This makes sense because you may want to refine the map with successive rubbersheet operations, and presumably you want to fix the same areas you've already rubbersheeted. If this is not true, then you can delete these links and make some more, or Oops your way back.

This is what the result of rubbersheeting looks like:

This is a sample of the GIScity database with normal and identity links defined.

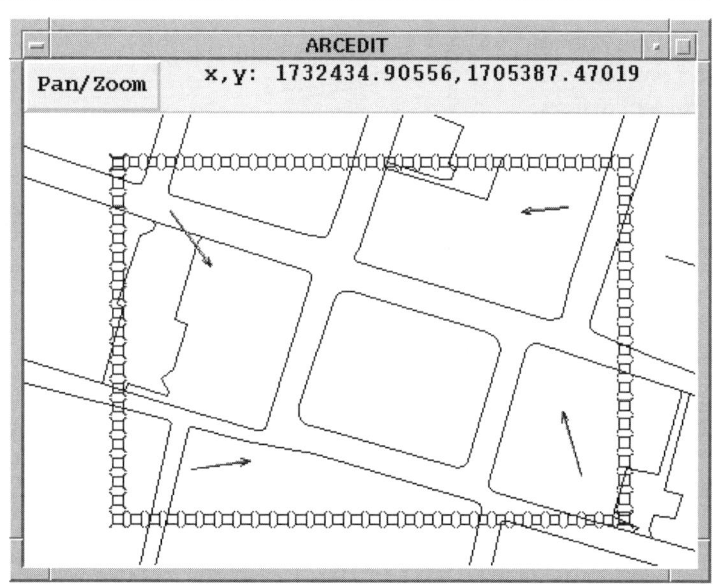

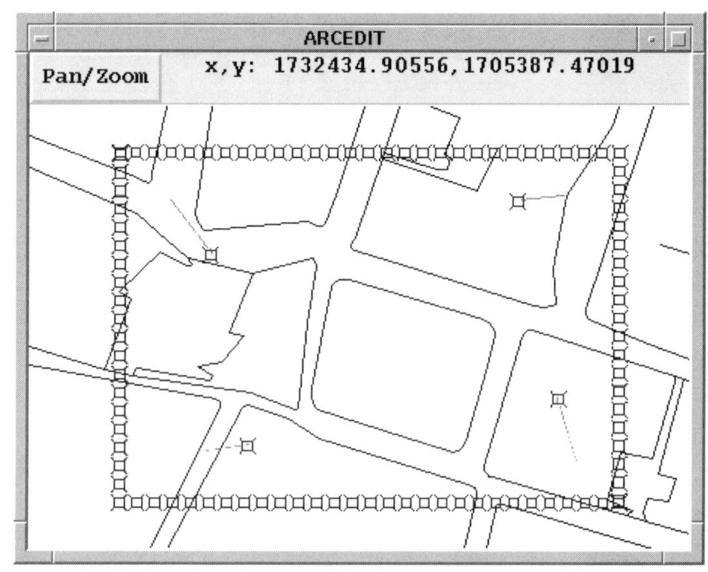

This is the result after clicking the Adjust command. Note how features are warped in the direction of the normal links, and that the normal links are changed to identity links.

Exercise 3: Adding Links

Many GIS professionals frown on rubbersheeting. Indeed, this technique should be used only as a last resort, and only for correcting digitized features from poor map manuscripts. Rubbersheeting is not a substitute for locating better sources or more carefully registering and digitizing maps.

Keep in mind that links and rubbersheeting is one of several ways to correct digitized map information. You don't want to casually rubbersheet anything that doesn't line up right. However, as you follow the steps below, you will see that rubbersheeting is one of the most interesting and fun tasks in ARCEDIT!

1. Make EDGEROAD your edit coverage. Zoom to the center of Santa Fe.
2. Click the Edit Links command on the main menu.

3. On the Edit Links menu, click the Add command. Add four normal links, like those in the preceding illustration that shows normal and identity links.
4. Click the Limit Adjust command (third command on the first row). When the screen cursor is activated, pick two points to define a rectangle outside the four links you've added.
5. Click the Adjust button. After a few seconds, or several minutes, you'll see all the features in your coverage rubbersheeted.
6. Click the Oops command, and try the above steps again. It takes some practice to predict the results of rubbersheeting. Link placement is important.

Part IV

ARC

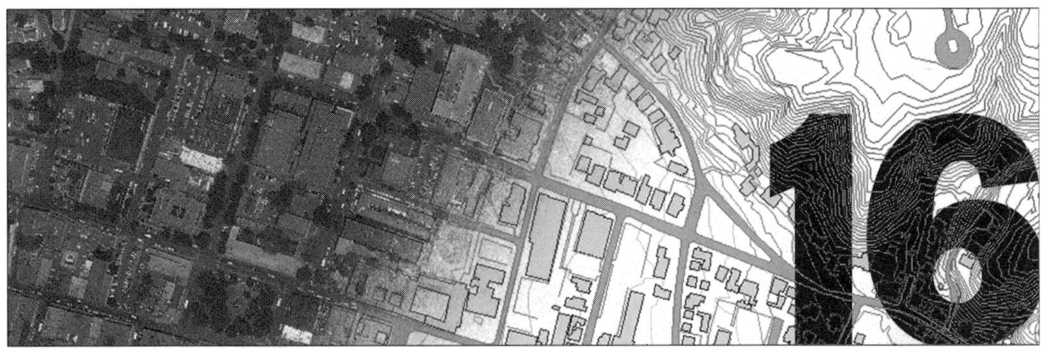

Managing Workspaces and Coverages in ARC

Creating Geographic Data Sets with Topology

The ARC program is the gateway to the rest of ARC/INFO. It is a non-graphic program for managing workspaces, and for creating, tuning, troubleshooting, converting, and analyzing coverages.

Earlier, we explored coverages in some detail. This chapter describes how coverages are organized into workspaces. A workspace is a logical area on your computer's disk that ARC/INFO uses to store coverages. It is implemented as a series of directories, subdirectories, and files in your computer's operating system. In this chapter, we'll learn how to use ARC commands to manage workspaces.

We'll also learn techniques in ARC to create coverages and modify their definitions. The ARC program is where you create coverages and specify exactly their organization: which topological relations, feature attribute tables, and items exist for a given coverage.

Because of the space limitations of this book, this chapter (and succeeding chapters) will not attempt to provide you with a comprehensive listing of all the ARC/INFO commands. Rather, we'll concentrate on the most important commands and the situations in which you might use them.

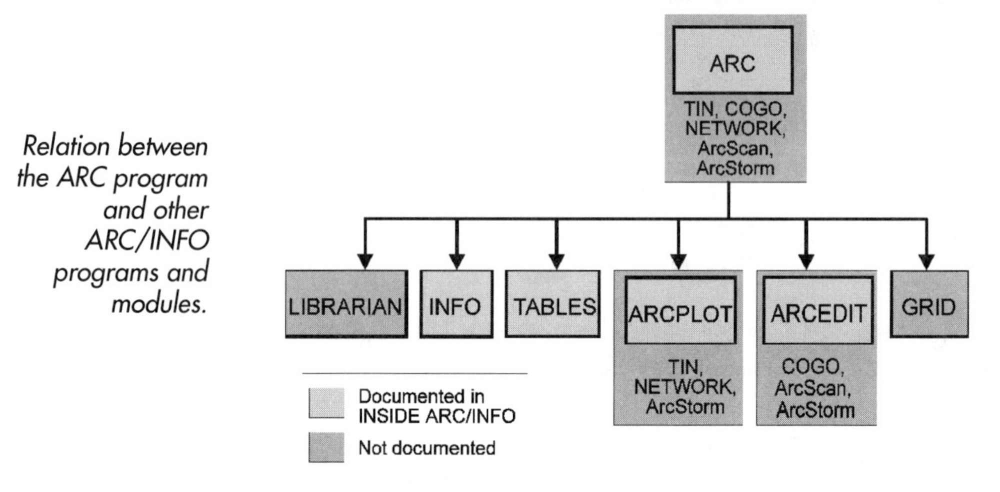

Relation between the ARC program and other ARC/INFO programs and modules.

Managing Coverages and Workspaces

ARC/INFO stores coverages and other geographic data sets in workspaces

A Workspace is a directory that ARC/INFO creates on your computer's disk. The following illustration shows you what the files and directories look like when you use operating system commands ("ls" in UNIX or "DIRECTORY" in VMS) to look inside workspaces.

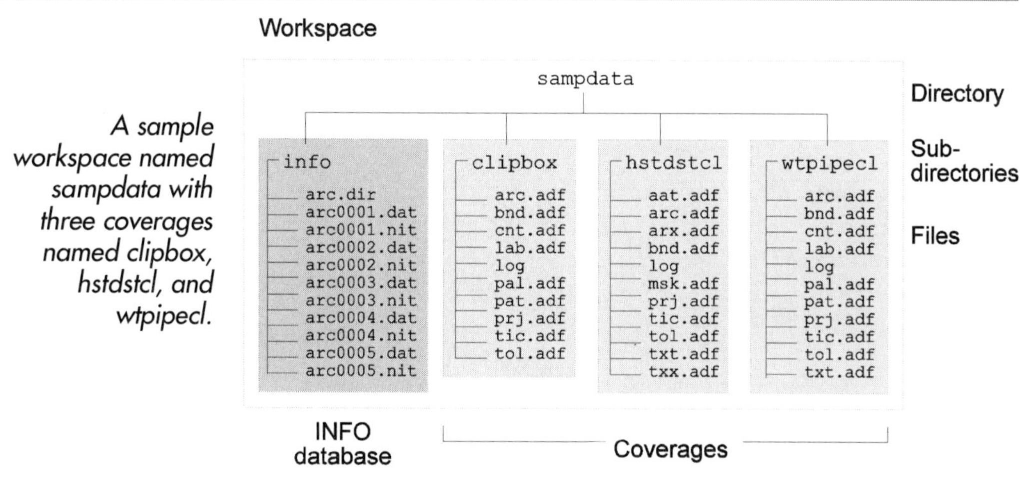

A sample workspace named sampdata with three coverages named clipbox, hstdstcl, and wtpipecl.

There are two types of directories within a workspace: the INFO directory and coverage directories. A workspace always has exactly one INFO directory. A workspace can have one or many coverages.

Under the coverage directories, you'll recognize files with some abbreviations from the last chapter, like aat, pat, and so on. They share the common suffix .adf, which stands for "ARC data file." It is unnecessary for most users to know what all these files are. ARC/INFO does a good job of hiding the details of files and coverages from you, and you never need to enter any of these file names to execute an ARC/INFO command.

NOTE: *If you have used versions of ARC/INFO prior to release 7.0, you may notice that the INFO and coverage file names are different from before. What used to be the arcdr9 file in UNIX, or ARC.DR9 in VMS, is now the arc.dir file. Release 7.0 changed the naming standard of these files for two reasons: to increase the limit of coverages within a workspace from 1,000 to 10,000, and to facilitate network access of ARC/INFO coverages from the ArcView program on PCs by using only file names recognizable by the DOS operating system.*

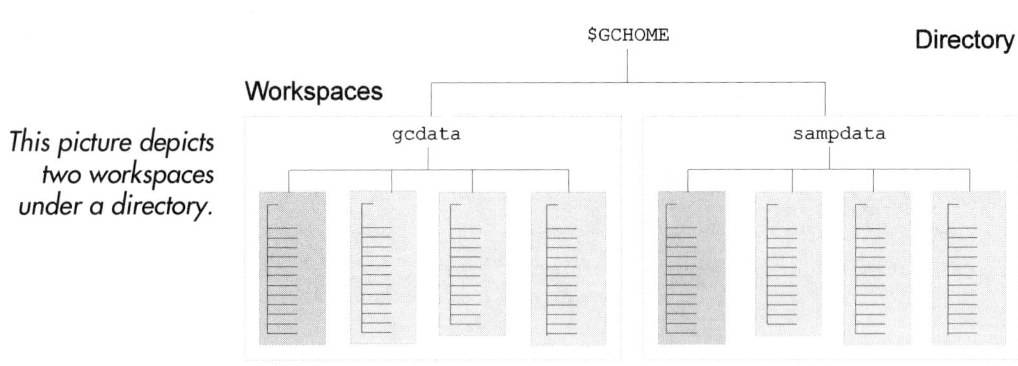

This picture depicts two workspaces under a directory.

When you loaded this book's companion CD-ROM on your workstation, a workspace was created for you. They are called $GCHOME/data on UNIX systems and $GCHOME:[DATA] on VMS systems.

WARNING: *ARC/INFO contains all the commands you need to manage the files and directories in your workspaces and coverages. Unless you are expert in both your operating system and ARC/INFO, you should never use UNIX or VMS commands to modify any of the files or directories in a workspace. There is a substantial risk of irrevocably corrupting coverages by using operating system commands on coverages and workspaces.*

Workspace Management Commands

When you begin working in the ARC program, you'll go first to a workspace so that ARC can locate the coverages you want to interact with. You will need some basic knowledge of the directories on your system to execute these commands. All command arguments that specify a workspace can be either relative or full paths. For example, if you type `gisdata`, the WORKSPACE command assumes there is (or will be) a workspace named gisdata under the computer directory you are presently set to. Or, you can type in `/disk1/gisdata/test` in UNIX or `DISK1:[GISDATA.TEST]` in VMS as a full path to completely specify the workspace location on your computer's disk.

ARC provides the following commands to create and maintain workspaces.

WORKSPACE Abbreviation: W

```
Arc: WORKSPACE {workspace}
```

This command moves you to a different workspace. {workspace} is the name of the workspace to move to. If you don't enter a workspace name, this command will inform you of the workspace you are now in. This command is usually one of the first commands you type when you begin the ARC program.

WORKSPACE is like the cd command in UNIX and DOS or the SET DEFAULT command in VMS. You can alternately use cd or SET DEFAULT to move to a workspace *before* you enter ARC.

COPYWORKSPACE Abbreviation: CPW

```
Arc: COPYWORKSPACE <from_workspace> <to_workspace>
```

This command copies a workspace and all its contents to a new workspace. <from_workspace> and <to_workspace> are the names of the existing and new workspaces. Both arguments are mandatory. This command is useful if you need to make a spare copy of all the coverages in a workspace before making changes.

CREATEWORKSPACE Abbreviation: CW

```
Arc: CREATEWORKSPACE <workspace>
```

This command creates a new, blank workspace. <workspace> is the name of the new workspace. After this command is done, you will find the INFO directory, but no coverages yet in this new workspace.

This command can also convert an existing directory into a workspace.

DELETEWORKSPACE Abbreviation: DW

```
Arc: DELETEWORKSPACE <workspace>
```

This command removes a workspace and all the data sets within it. <workspace> is the name of the workspace that you want to completely remove. After you execute this command, you will be asked if you really want to delete the workspace. Be careful; you cannot undo this command.

LISTWORKSPACES Abbreviation: LW

```
Arc: LISTWORKSPACES {directory}
```

This command lists all workspaces under a directory. {directory} is the name of a directory and can include a pathname. You should use the UNIX or VMS syntax for directory specification.

RENAMEWORKSPACE Abbreviation: RW

```
Arc: RENAMEWORKSPACE <old_workspace> <new_workspace>
```

This command assigns a new name to the specified workspace. <old_workspace> is the original workspace name. <new_workspace> is the new workspace name. This command updates some internal values in the INFO directory, but the coverages and data sets below the workspace are otherwise unaffected. When you return to your operating system prompt, you will notice that the directory name corresponding to the workspace has changed as well.

Coverage Commands

ARC includes general coverage maintenance commands. Some of these commands are specific to coverages; others work on any geographic data sets within ARC/INFO, such as grids, tins, and lattices. We'll concentrate on coverage uses, which is the geographic data set we are focusing on in this book.

PRECISION

```
Arc: PRECISION <SINGLE | DOUBLE> {HIGHEST | LOWEST |
     SINGLE | DOUBLE}
```

This command specifies the precision level of any new, blank coverages you create. <SINGLE | DOUBLE> is a mandatory argument. Single precision means that coordinate values are stored with about seven digits of accuracy before mathematical round-off errors take place. Double precision means that there are about 14 digits of accuracy before round-off error. SINGLE is the default value.

{HIGHEST | LOWEST | SINGLE | DOUBLE} is an optional argument that sets the precision level of coverages created by processing other coverages. HIGHEST and LOWEST apply when you have two or more coverages that you process to create a new coverage. They specify that the highest or lowest precision among the coverages will apply. SINGLE or DOUBLE fixes the precision level. HIGHEST is the default value.

Note that this command cannot modify the precision level of any coverage. If you want to change the precision level of a coverage, then you should copy that coverage to another coverage, while setting a new precision level.

CREATE

```
Arc: CREATE <out_cover> {tic_bnd_cover}
```

This command makes a new, blank coverage. <out_cover> is the name of the new coverage. {tic_bnd_cover} is an optional argument indicating a coverage from which to copy tics and boundary. You should first be in the workspace you want to add the coverage to, and set the precision level for the coverage.

The output coverage name can be up to thirteen characters long. In addition to the letters of the alphabet and numerals, you can use a hyphen (-) and underscore (_) for the name. Your keyboard has other characters that ARC may accept in a coverage name, but these are not advisable because of the potential for confusing your operating system, or different operating systems you might transfer coverages to.

The second, optional command argument lets you specify that the new coverage will adopt the tics and boundary values from another coverage. This is handy; many times coverages in a workspace share common tics, and this option will save you the trouble of making new tics.

This command does not establish feature attribute tables for you. Later, we'll document the ARCEDIT command, CREATEFEATURES, which does that task.

COPY

```
Arc: COPY <from_geo_dataset> {to_geo_dataset} {DEFAULT
     | SINGLE | DOUBLE}
```

This command copies coverages and other geographic data sets. <from_geo_dataset> is the name of the existing data set to copy. {to_geo_dataset} is the name of the new data set to create. If you do not specify the {to_geo_dataset} argument, the name of the new data set will be the same as that of the old data set, but only if the old data set is in a different workspace. {DEFAULT | SINGLE | DOUBLE} specifies the precision level and overrides the default precision level from the PRECISION command.

All the components of a coverage, including all feature attribute tables and INFO relate tables with the same name prefix as the coverage, will be copied with this command.

RENAME

```
Arc: RENAME <old_geo_dataset> <new_geo_dataset>
```

This command renames coverages and other geographic data sets. <old_geo_dataset> is the name of the existing data set. <new_geo_dataset> is the new name of the data set. All the components of a coverage, including all feature attribute tables and INFO relate tables with the same name prefix as the coverage, will be renamed with this command.

KILL

```
Arc: KILL <geo_dataset> {ARC | INFO | ALL}
```

This command removes a coverage or geographic dataset from a workspace and your computer's disk. <geo_dataset> is the name of the data set to remove. {ARC | INFO | ALL} is an optional argument to specify which parts of a coverage to remove. The default, ARC, means that all coverage files and feature attribute tables are removed. INFO means that only INFO tables with the same prefix as a data set name will be removed. ALL means that both coverage files, feature attribute tables, and all INFO tables with the coverage prefix will be removed.

WARNING: *Be very, very careful with the kill command. If you accidentally remove coverages with this command, they are gone from your computer's disk. Consult with your system administrator to restore the coverage from backup tapes that are kept (you hope) for your computer system (and bring a suitable bribe).*

DESCRIBE

```
Arc: DESCRIBE <geo_dataset>
```

This command summarizes the contents of a coverage or geographic data set, designated by the <geo_dataset> argument. The summary appears in your dialog window with information such as the number of arcs, tics, and labels in your coverage. Also listed are which feature attribute tables are defined, and the map extent of the coverage. This command is used for verifying and troubleshooting coverages.

LISTCOVERAGES Abbreviation: LC

```
Arc: LISTCOVERAGES {workspace} {NOSTATUS | STATUS |
     PRECISION}
```

This command lists all coverages in a workspace. If you don't specify a {workspace} value, the coverages for the workspace you are in are listed. {NOSTATUS | STATUS | PRECISION} is an optional argument to specify whether you also want to see the editing status for a coverage and its precision level.

If you specify STATUS, you will see either CLEAN or EDITED next to each coverage, indicating whether the coverage has clean topology (which we will soon discuss.) If you specify PRECISION, you will see the letter "S" or "D" next to each coverage name, denoting single or double precision.

REBOX

```
Arc: REBOX <cover>
```

This command adjusts the boundary limits of a designated coverage (<cover>) to span all arcs and label points in the coverage. It also removes excess tics outside boundary limits. The BND file in your coverage is updated with this command. Remember that the CREATE command lets you copy tics from another coverage in a newly created coverage. REBOX can be used as a short-cut to eliminate unnecessary tics outside the map extent of your features.

This command is not frequently used but is useful if you've accidentally placed geographic features way outside your coverage's normal map extent and you want to locate those stray features.

EXTERNAL

```
Arc: EXTERNAL <geo_dataset>
```

This command corrects INFO external file pathnames for the feature attribute tables stored for a coverage. <geo_dataset> is the name of the geographic data set to correct. This command can be safely used for troubleshooting problems with coverages.

The INFO database for a workspace keeps track of the pathnames for all attribute tables for coverages. For example, someone else on your computer system might change the name of a disk or directory above your workspace. If you start getting error messages in ARC/INFO about missing attribute tables, using EXTERNAL might fix the problem for you.

EXTERNALALL

```
Arc: EXTERNALALL {directory}
```

This command corrects INFO external file pathnames for all coverages in a workspace. If you don't specify a workspace directory with {directory}, this command will work on the workspace you are in.

This command performs the same function as EXTERNAL, but on an entire workspace. You can safely use it to troubleshoot problems with both coverages and workspaces.

INDEX

```
Arc: INDEX <cover> {ALL | POINT | LINE | POLY | ANNOTATION}
```

This command indexes a coverage to optimize graphics performance in ARCPLOT. <cover> is the name of the coverage to be tuned. The optional choice of ALL, POINT, LINE, POLY, or ANNOTATION lets you choose whether all or just some features are indexed.

This command creates a spatial index for a coverage. Depending on which features are selected, several files are added to your coverage. There is a trade-off when using INDEX; in return for improved performance for many ARCPLOT commands, the disk size of your coverages increases by about 8% to 10%.

Whenever you modify a coverage through an ARC command or in an ARCEDIT session, you must run the INDEX command again. Many AML applications incorporate this command when exiting ARCEDIT.

The use of INDEX is highly recommended for large coverages. The increase in performance for drawing and spatial selection in ARCPLOT can be dramatic.

RESTOREARCEDIT

```
Arc: RESTOREARCEDIT <edit_cover> {out_cover}
```

This command recovers edits to a coverage which was abnormally interrupted in ARCEDIT. <edit_cover> is the name of the interrupted coverage. {out_cover} is the name of an optional new coverage to restore the edits to. The use of {out_cover} is recommended, because you can first verify that the restoration is successful before committing changes to your edit coverage.

ARCEDIT creates temporary files during the course of an edit session. If you experience an interruption from a power outage or a fatal error in ARCEDIT, these temporary files remain, but the edits are not added to the coverage. This command incorporates all the edits in the temporary files and posts them to the coverage. This command can restore many hours of work that would otherwise be lost.

Exercise 2: Working with Coverages

In this exercise, we'll make a copy of some GIS data supplied with GIScity, and look at some of its coverage information. There is one workspace supplied: $GCHOME/data (in VMS, $GCHOME:[DATA]).

1. First, let's copy the complete workspace in $GCHOME/sampdata to a new workspace. We'll call the new workspace "testdata." Type in the following:

   ```
   UNIX or NT - Arc: COPYWORKSPACE $GCHOME/data $GCHOME/testdata
   VMS - Arc: COPYWORKSPACE $GCHOME:[DATA] $GCHOME:[TESTDATA]
   ```

2. To get to our new workspace, type the following:

   ```
   UNIX or NT - Arc: WORKSPACE $GCHOME/testdata
   VMS - Arc: WORKSPACE $GCHOME:[TESTDATA]
   ```

3. Now try getting a listing of the workspace coverages by using the abbreviation for LISTCOVERAGES. (From here on, the commands are the same for UNIX and VMS.)

   ```
   Arc: LC
   ```

4. You will see a listing of several sample coverages we'll use in later exercises. To get more information about the coverages, pick any one of the coverages and type in the following:

   ```
   Arc: DESCRIBE <coverage name>
   ```

5. You'll get a summary listing of the contents of the coverage.
6. After you have finished these and some following exercises, you can use the DELETEWORKSPACE command to remove the workspace you've just copied.

Building Topology and Detecting Errors

Topology Maintenance

The ARC program contains two important commands for building and maintaining topology in a coverage: CLEAN and BUILD. We'll allot some extra space for these two commands because of their importance. Two additional commands which are sometimes useful are RENODE and CREATELABELS.

CLEAN

```
Arc: CLEAN <in_cover> {out_cover} {dangle_length}
     {fuzzy_tolerance} {POLY | LINE}
```

This command creates topology and corrects geometric relations in a coverage. <in_cover> is the input coverage to clean. {out_cover} is an optional argument for an output coverage. If you leave this value blank, the input coverage is replaced by the cleaned output coverage.

{dangle_length} is the length of the smallest dangling arc to survive this command; any smaller arc will be assumed a digitizing error and deleted. {fuzzy_tolerance} is the distance tolerance for joining close intersections. {POLY | LINE} specifies whether you want to create and update polygon topology or arc-node topology.

If you do not specify values for dangle length and fuzzy tolerance, CLEAN will set values for the tolerances defined for the coverage TOL file. If that file does not exist, it will use a value of 0.002 if the bad file has a boundary of less than 100 units, or one ten-thousandth of the width or height of your coverage extent.

CLEAN is an important command in ARC/INFO. CLEAN and BUILD (covered next) are the essential commands for creating and maintaining polygon topology. CLEAN is also available in the ARCEDIT program. The command has the same purpose in ARCEDIT as in ARC, but the command arguments are different.

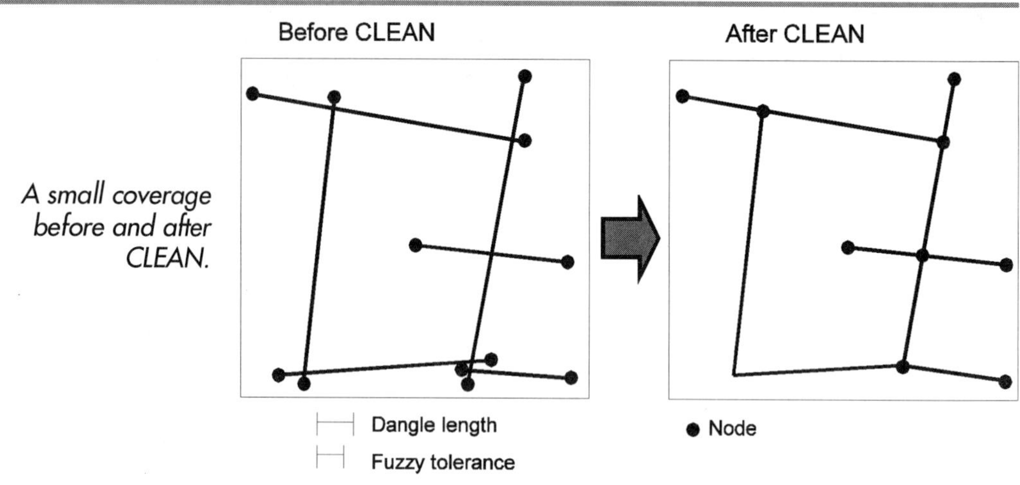

A small coverage before and after CLEAN.

Here are three situations where you might use CLEAN:

- ❑ You have just created a coverage and want to define polygon topology for that coverage.
- ❑ You have converted a digital map from another automated mapping system into a coverage. CLEAN will perform the geometric corrections that are usually necessary and build polygon or arc-node topology.
- ❑ You might find it efficient to digitize coverages without careful correction of intersection errors. This is called *spaghetti digitizing*. CLEAN can perform mass corrections on that coverage for you.

These are some of the steps that CLEAN performs on a coverage:

- ❑ Locates arcs that intersect without a node point. CLEAN calculates the point of intersection and splits those arcs to make an intersection node.
- ❑ Uses the fuzzy tolerance to merge nodes that are near each other into clean intersections.
- ❑ Uses the fuzzy tolerance to merge vertices that are close to each other.

- ❑ If you set the LINE option, creates an Arc Attribute Table if none existed. If an AAT existed, CLEAN updates the AAT by removing or adding arcs modified by resolving intersections or other changes.
- ❑ If you set the POLY option, CLEAN creates a Polygon Attribute Table if none existed. If a PAT exists, CLEAN updates the PAT (and AAT if it exists) to match the geometric changes.

You cannot use the CLEAN command with the POLY option on coverages with existing Point Attribute Tables. You will get an error message if you try. (To create a Point Attribute Table, use the BUILD command.)

Some of the rules for how CLEAN works on coverages with route systems and regions are complicated. Refer to the ArcDoc help listing for CLEAN for a detailed summary of the implications of using CLEAN on complex features.

CLEAN is a double-edged sword; it can swiftly make mass corrections on your coverage, but if improperly used, CLEAN can delete far more features and vertices than you intended.

WARNING: *Be very careful with the tolerances you set with CLEAN. If you are liberal when setting the size of dangle lengths and fuzzy tolerances, then you will notice smaller features collapsing or condensing. Start with small values at first, and increase until many corrections are made without generalizing too much.*

BUILD

```
Arc: BUILD <cover> {POLY | LINE | POINT | NODE | ANNO.<subclass>}
```

This command defines and updates feature attribute tables for a coverage. <cover> is the coverage name. {POLY | LINE | POINT | NODE | ANNO.<subclass>} specifies the type of feature for which you are trying to create, or update, an attribute table.

The primary use of BUILD is to update polygon topology. When you edit arcs, arc-node topology is updated within ARCEDIT if you have an AAT defined. However, when you edit arcs to define polygons in ARCEDIT, you must run BUILD to update a coverage's polygon topology.

BUILD used with the POLY or LINE option is similar to the effects of CLEAN, but it won't resolve geometric intersections or dangling arcs. If BUILD detects unresolved intersections, it will abort and give you a message advising you to execute CLEAN instead. The advantage of BUILD is that it executes in a shorter time than CLEAN and will not change any geometry in a coverage.

Occasionally, you will get an error message in ARC/INFO stating that the feature IDs do not match the coverage internal identifier. This may result from a system error. When you get this message, the BUILD command will usually resolve this for you.

Another frequent use of BUILD is to create feature attribute tables for a coverage. POLY creates or updates a PAT, LINE creates or updates an AAT, POINT creates or updates a PAT, NODE creates or updates a NAT, and ANNO.<subclass> creates or updates a TAT. Remember, Point Attribute Tables and Polygon Attribute Tables are mutually exclusive. If you have one defined, BUILD will disallow the creation of the other.

BUILD and CLEAN are among the most frequently used ARC commands.

CREATELABELS

```
Arc: CREATELABELS <cover> {id_base}
```

This command creates label points for polygons in a coverage. <cover> is the coverage name. {id_base} is an optional argument to specify a starting number for User IDs for label points.

Polygons in ARC/INFO are defined by a list of arcs that form the perimeter of an area, and a label point to associate polygon attributes. When you add arcs that form polygons in ARCEDIT, you can either add label points while you are within ARCEDIT or you can have the label points added for you afterward with this command.

Adding label points while in ARCEDIT is usually the preferred method because you can assign attribute values at that time. However, some situations, such as processing coverages converted from other automated mapping systems, prove the value of the CREATELABELS command by painlessly adding polygon label points for you. Existing label points are unaffected. New label points are placed in a calculated centroid position, one for each polygon.

RENODE

```
Arc: RENODE <cover>
```

This command renumbers internal IDs for nodes. <cover> specifies the coverage whose node numbers you want to update.

Some ARC commands process coverages but do not update node numbers. Node numbers are unique, but arcs share a common node where they intersect. It is desirable to distinguish between normal, pseudo, and dangling nodes.

RENODE quickly renumbers all the nodes in a coverage and updates the FNODE# and TNODE# items in the Arc Attribute Table, if one exists. This command is usually used after data conversion.

DROPFEATURES

```
Arc: DROPFEATURES <cover> <feature_class> {ATTRIBUTES |
     GEOMETRY}
```

This command removes a feature attribute table or actual features from a coverage. <cover> is the name of the coverage from which to drop features. <feature_class> specifies the type of feature to drop. The possible values are LINE, LINK, NODE, POINT, POLY, ANNO.<subclass>, ROUTE.<subclass>, SECTION.<subclass>, REGION.<subclass>. {ATTRIBUTES | GEOMETRY} specifies whether to drop only the attribute tables, or delete the geographic features from a coverage as well.

The primary use of DROPFEATURES is if you have accidentally defined a feature attribute table with the BUILD or CLEAN command. For example, you may have intended a coverage to have a Point Attribute Table, but an accidental CLEAN with the POLY option on the coverage would preclude this.

With the GEOMETRY option, DROPFEATURES can also be used to delete all geographic features from a coverage.

Error Detection

Generally, ARCEDIT is used to identify and correct errors in coverages. Here is a trio of ARC commands that complement ARCEDIT in identifying any errors that may have escaped your attention in ARCEDIT.

INTERSECTERR

```
Arc: INTERSECTERR <cover>
```

This command displays a list of arcs that intersect without a node point. <cover> is the coverage to inspect.

When you run INTERSECTERR on a coverage, all arcs that intersect without a common node point at the intersection are identified for you. The list will display the internal and User IDs of the arcs, together with the X and Y coordinate positions of the intersections. To correct intersecting arcs, you can use the CLEAN command, or make the corrections in ARCEDIT.

NOTE: *Sometimes intersecting arcs without nodes at their intersections are desirable. An example is two electrical lines which cross but are not connected.*

LABELERRORS

```
Arc: LABELERRORS <cover>
```

This command displays a list of polygons which have either no label points or multiple label points. <cover> is the coverage to inspect. The polygon internal number and label User-IDs are listed. CREATELABELS can be used to correct the polygons without label points. To correct multiple label point errors in ARCEDIT, use the values from this listing to locate and remove the extra label points.

NODEERRORS

```
NODEERRORS <cover> {ALL | DANGLES | PSEUDOS}
```

This command generates a listing of all nodes according to the node type option. The node type, node number, and X, Y coordinates are listed. <cover> is the coverage to inspect. ALL specifies that both dangling and pseudo nodes will be reported. DANGLE and PSEUDO specify that that type of node is reported.

Processing and Converting Coverages in ARC

Coordinate Adjustment, Polygon Overlay, and Data Conversion

The ARC program supplies you with a wealth of commands for processing geographic data sets. These commands modify coordinates, calculate geometric intersections, make buffer polygons, calculate the proximity between features, and convert coverages to and from ARC/INFO and many other GIS, CAD, and graphics formats.

The ARC program is the gateway to the rest of ARC/INFO. Just by typing "arc" at your workstation's command level, you enter the ARC program.

The key difference between the ARC program and the ARCPLOT and ARCEDIT programs is that commands in ARC operate on entire coverages. Commands in ARCPLOT and ARCEDIT act primarily on selected sets of features within a coverage.

Another difference is that ARC is a non-graphic program. The reason for this is that some ARC commands can demand most of your computer's processing power and a graphic display is not essential.

You can think of many ARC commands as "batch commands." You initiate an ARC command, wait a few seconds, a few minutes, or longer, and then inspect the results in ARCPLOT or ARCEDIT. You can also run these commands as a batch job in your computer's operating system. Chapter 18 will document simple AML scripts which you can use for this and other purposes.

We've already covered several ways to get information into a GIS. The ARCEDIT program (which we used in Edit GIScity) is an important way to enter and correct geographic data. However, often the best and most economic way to input GIS information is to convert existing geographic data from another automated mapping system. Of course, this option is contingent on the availability and quality of the digital mapping data. When using this option, you'll find that the ARC program contains an impressive suite of commands for importing and exporting GIS data between the ARC/INFO format and most common GIS, CAD, and graphics file formats.

Adjusting Coordinates

Changing Coverage Coordinate Systems

There are three basic commands that adjust coordinate values in the ARC program: PROJECT, ADJUST, and TRANSFORM.

When you need to transform coordinate systems, the PROJECT command is the preferred method because it is the most mathematically accurate. However, the use of this command is not always possible. The prerequisites for PROJECT are (1) that you know the original and intended map projection parameters, and (2) that the original coverage is accurately digitized in the original map projection.

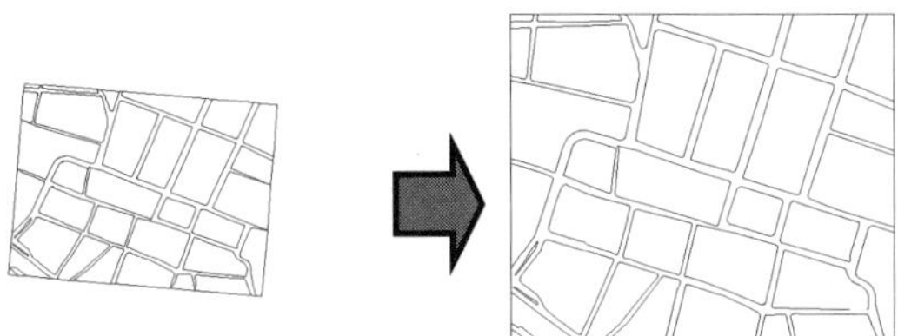

PROJECT uses map projection parameters

Which of these three methods for transforming coordinate systems is best depends on what information you have about the original coordinate system.

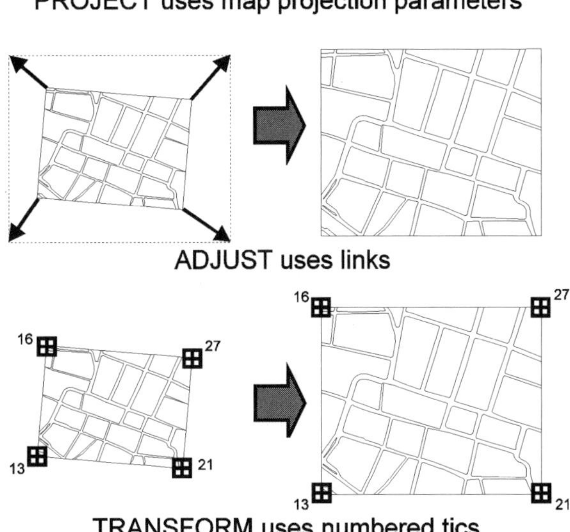

ADJUST uses links

TRANSFORM uses numbered tics

ADJUST and TRANSFORM use corrections that you've specified through either links or tics. On the surface, the results from ADJUST and TRANSFORM can be similar, but they use different mathematical methods. Another difference is that you can localize coordinate changes in ADJUST through the use of normal and identity links, whereas TRANSFORM globally changes coordinates in a coverage.

PROJECT

```
Arc: PROJECT <COVER | GRID | FILE> <input> <output>
     {projection_file} {NEAREST | BILINEAR | CUBIC}
     {out_cellsize}
```

This command changes the coordinates for all features in a coverage from one map projection to another. <COVER | GRID | FILE> specifies whether you are projecting a coverage, grid, or ASCII file. <input> is the input coverage, grid, or file. <output> is the output coverage, grid, or file. {projection_file} is an optional map projection file. If you don't enter this value, PROJECT engages you in a dialog to set the map projection parameters. {NEAREST | BILINEAR | CUBIC} and {out_cellsize} are optional arguments that apply only to grids.

Three coordinate systems frequently used in the United States are geographic (latitude and longitude), Universal Transverse Mercator, and the State Plane Coordinate System. Most countries in the world have other map projection systems specific to that country or region.

PROJECT is used when you need to share data with other agencies or companies. You may receive a DLG file from the USGS which has UTM coordinates. You want to merge that data into your GIS database, which is defined in State Plane coordinates. The PROJECT command precisely converts all coordinates in a coverage from one map projection system to another.

Sometimes, people extend map projection zones into another area, so the same territory could be described by two map zones. PROJECT can also be used to convert coordinates from one zone to another.

ADJUST

```
Arc: ADJUST <in_geo_dataset> <link_cover | link_file>
     <out_geo_dataset> {FORWARD | BACKWARD} {NOTICS |
     TICS} {NEAREST | BILINEAR | CUBIC} {BIVARIATE |
     LINEAR}
```

This command uses links to rubbersheet the coordinates in a coverage. <in_geo_dataset> is the input coverage or grid. <link_cover | link_file> is a coverage containing links. It may be the same as the input coverage. <out_geo_dataset> is the output coverage or grid. {FORWARD | BACKWARD} specifies whether adjustment uses the forward or backward direction of links. {NOTICS | TICS} specifies whether tics are rubbersheeted along with other features. {NEAREST | BILINEAR | CUBIC} and {BIVARIATE | LINEAR} are options that apply only to grids.

This command works the same way as the ADJUST command in ARCEDIT. After ADJUST is run, all links become identity links.

TRANSFORM

```
Arc: TRANSFORM <in_cover> <out_cover> {AFFINE | PROJECTIVE}
```

This command uses tics to change the coordinates in a coverage. <in_cover> is the input coverage. <out_cover> is the output coverage after transformation. {AFFINE | PROJECTIVE} is an option for the mathematical method employed. AFFINE is the default; PROJECTIVE is often used for maps digitized from high-altitude aerial photographs.

To use TRANSFORM, you must have tics common to both coverages. They must mark the same locations on both coverages, and must have exactly the same tic numbers.

TRANSFORM is sometimesused for coverages digitized from sources with unknown or inaccurate coordinates. Although not as desirable as the use of the PROJECT command, TRANSFORM can successfully transform coverage coordinates to another coordinate system.

Exercise 1: Adjusting Coordinate Systems

For this exercise, we'll use a coverage called "sf-latlong" from the workspace that you copied in the preceding chapter. We'll also use an AML command to see the results of our work. The AML command, called "testdraw.aml," is already in your testdata workspace. This AML will briefly enter the ARCPLOT program and show you a simple view of what the coverages look like.

1. Start ARC and set your workspace to the one you made in the preceding chapter by typing:

   ```
   UNIX or NT - Arc: WORKSPACE $GCHOME/testdata
   VMS - Arc: WORKSPACE $GCHOME:[TESTDATA]
   ```

2. Next, list your coverages and describe each one. You should find two coverages—sf-latlong and sf-transf. Use DESCRIBE and note the number of arc segments and what the coverage boundary looks like. Type in:

   ```
   Arc: LC (abbreviation for LISTCOVERAGES)
   Arc: DESCRIBE sf-latlong
   Arc: DESCRIBE sf-transf
   ```

3. Run the PROJECT command by typing in the following:

```
Arc: PROJECT COVER sf-latlong sf-proj sf-ll2sp.prj
```

4. This command will take a few moments to run (depending on the speed of your computer). When you are done, you've created a new coverage called "sf-proj." Run DESCRIBE on that coverage, and note in particular the coverage boundary compared to sf-latlong.

5. The file called sf-ll2sp.prj is an ASCII file defining the coordinate systems for both the input and output. If you want, use your operating system's text editor to look at the contents of this file.

6. To look at the original coverage, run the TESTDRAW.AML by typing:

```
Arc: TESTDRAW sf-latlong
```

7. After you have looked at the coverage, press [Enter] on your keyboard to return to the ARC program. Then look at the new coverage by typing:

```
Arc: TESTDRAW sf-proj
```

8. Try running the TRANSFORM command by typing:

```
Arc: TRANSFORM sf-latlong sf-transf
```

9. The sf-transf coverage already existed before you ran TRANSFORM, but all it contained was four tics. Use DESCRIBE on this command. While the coverage boundary should look the same, you'll see the numbers of features, such as arcs, go from zero to several hundred. Features from sf-latlong have just been moved into sf-transf using the tics for coordinate transformation. If you want, run the TESTDRAW.AML on both coverages.

10. Run the ADJUST command by typing the following:

```
Arc: ADJUST sf-latlong sf-latlong sf-adjust
```

11. The coverage sf-latlong had links in it. In the command above, you've specified that both the original features and the links are in sf-latlong, and the output is directed to a new coverage, sf-adjust. Use DESCRIBE on sf-adjust and compare the coverage boundary range. If you want, run the TESTDRAW.AML on both coverages.

Filtering Vertices

If the arcs in your coverage have too many or too few vertices, you can use the two following commands.

DENSIFYARC

```
Arc: DENSIFYARC <in_cover> {out_cover} <interval> {VERTEX | ARC}
```

This command adds new vertices along arcs at the specified interval. <in_cover> is the input coverage. {out_cover} is the output coverage; if left blank, the output from DENSIFYARC is sent to the input coverage. <interval> is the distance at which new vertices or arcs are created. {VERTEX | ARC} specifies whether this command adds new vertices, or whether new arcs are created by adding new nodes at interval lengths.

This command sometimes assists feature snapping, because it is easier to snap to nodes or vertices than along the length of a line segment.

This command performs the same function as the DENSIFY command in ARCEDIT.

Densifying arcs is sometimes done before using the PROJECT command. If your arcs have long distances between vertices, DENSIFY ARC will give you a better result.

GENERALIZE

```
Arc: GENERALIZE <in_cover> <out_cover> <weed_tolerance>
```

This command filters out vertices for all arcs in a coverage. <in_cover> is the input coverage. <out_cover> is the output coverage. <weed_tolerance> is the value used to filter out vertices.

This command uses a mathematical algorithm that considers the distances and angles between vertices along an arc and determines which vertices can be dropped without losing significant cartographic detail.

Some of the uses for this command are for coverages which are converted from other mapping systems; for thinning continuous, digitized features such as streams and contours; for decreasing the disk size of very large coverages; and for increasing the performance of other ARC commands.

This command performs the same function as the GENERALIZE command in ARCEDIT. Again, the difference is that GENERALIZE works on all the arcs in a coverage.

Processing Coverages

Joining Coverages

Some time in your work, you may find that you have data organized in separate map sheets, and you wish to join them into a larger, or seamless, coverage.

The ARC program offers two similar commands to merge separate coverages into one. The MAPJOIN command is recommended for coverages with a Polygon Attribute Table or coverages with both a Polygon Attribute Table and Arc Attribute Table because it automates the assignment of polygon topology for the newly joined polygons. The APPEND command is used for coverages with Point Attribute Tables and coverages with Arc Attribute Tables.

APPEND

```
Arc: APPEND <out_cover> {NOTEST | template_cover |
     feature_class...feature_class} {NONE | FEATURES |
     TICS | ALL}
```

This command appends multiple coverages into one coverage. <out_cover> is the output coverage. You will enter the input coverages through a dialog initiated by this command. {NOTEST | template_cover | feature_class...feature_class} specifies whether any and all coverages are appended, or whether they must adhere to the feature attribute tables in a template coverage, or a list of acceptable feature classes. {NONE | FEATURES | TICS | ALL} specifies whether and which features have internal IDs renumbered when they are appended. The default is NONE, but TICS is useful if you want to join tics and maintain unique TIC IDs. The FEATURES choice guarantees that all coverage IDs are unique within the output coverage.

The BUILD command must be run after APPEND to create feature topology.

After you enter the command, you will be asked to type in a list of all the input coverages. This command accepts up to 500 input coverages and may take quite some time to complete.

MAPJOIN

```
Arc: MAPJOIN <out_cover> {feature_class...feature_class
     | template_cover} {NONE | FEATURES | TICS | ALL}
     {clip_cover}
```

This command appends multiple coverages into one coverage and builds polygon topology. <out_cover> is the output coverage. The input coverages are specified through a dialog after you begin this command. {feature_class...feature_class | template_cover} specifies the feature classes acceptable for input. {NONE | FEATURES | TICS | ALL} specifies which, if any, features have internal IDs renumbered. {clip_cover} optionally defines a coverage that can be used to clip the output coverage. Clipping a coverage means that all features outside the area you are defining will be clipped or removed.

MAPJOIN is like the combination of the APPEND and CLEAN commands.

Polygon Processing

Here's a trio of ARC commands that process polygons in a coverage.

BUFFER

```
Arc: BUFFER <in_cover> <out_cover> {buffer_item}
     {buffer_table} {buffer_distance}
     {fuzzy_tolerance} {LINE | POLY | POINT | NODE}
     {ROUND | FLAT} {FULL | LEFT | RIGHT}
```

This command creates polygons that are buffers around features in a coverage. <in_cover> is the input coverage. <out_cover> is the output coverage. {buffer_item} and {buffer_table} are optional items that can be used to make variable width buffer polygons. {buffer_distance} specifies a fixed buffer width. {fuzzy_tolerance} sets the minimum distance between vertices in the created polygons. {LINE | POLY | POINT | NODE} specifies which feature type buffers are created about. {ROUND | FLAT} is an optional argument that controls the appearance of the ends of polygons. {FULL | LEFT | RIGHT} specifies whether half buffers are made and on which side.

The BUFFER command creates a buffer polygon around features in a coverage. You can create buffers around points, lines, and polygons.

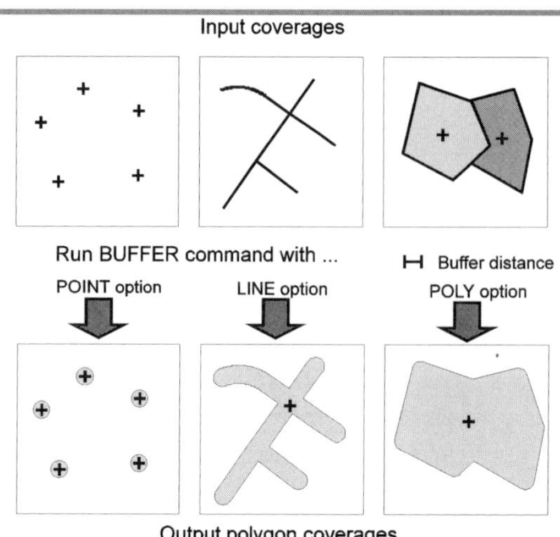

Buffer analysis is an important analytic capability in ARC/INFO. You can use the coverage created by this command to find features that are within a buffer distance of another feature.

DISSOLVE

```
Arc: DISSOLVE <in_cover> <out_cover> <dissolve_item |
     #ALL> {POLY | LINE | NET | REGION.subclass}
```

This command merges polygons with similar attributes. <in_cover> is the input coverage. <out_cover> is the output coverage. <dissolve_item | #ALL> specifies whether one or all items in a Polygon Attribute Table are used to check for like attributes. {POLY | LINE | NET | REGION.subclass} specifies the feature class that remains after the use of this command. POLY is the default. DISSOLVE is used to simplify coverages for geographic analysis.

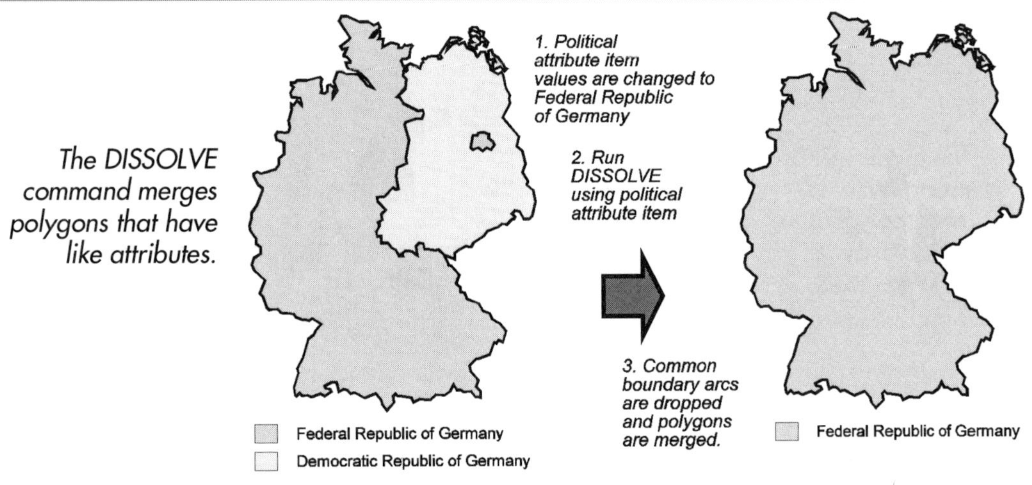

1. Political attribute item values are changed to Federal Republic of Germany
2. Run DISSOLVE using political attribute item
3. Common boundary arcs are dropped and polygons are merged.
Federal Republic of Germany
Democratic Republic of Germany
Federal Republic of Germany

The DISSOLVE command merges polygons that have like attributes.

ELIMINATE

```
Arc: ELIMINATE <in_cover> <out_cover> {NOKEEPEDGE |
     KEEPEDGE} {POLY | LINE} {selection_file} {BORDER
     | AREA}
```

This command removes small or narrow polygons by removing the longest shared arc or connected arcs. It also removes pseudo nodes from a coverage. <in_cover> is the input coverage. <out_cover> is the output coverage. {NOKEEPEDGE | KEEPEDGE} specifies whether polygons on the fringe are protected. {POLY | LINE} sets whether polygons or lines are merged. {selection_file} and {BORDER | AREA} are additional options for merging polygons. ELIMINATE is normally used to remove polygons that are very small or very narrow, which are likely digitizing errors.

The ELIMINATE command removes small polygons that may be digitizing errors.

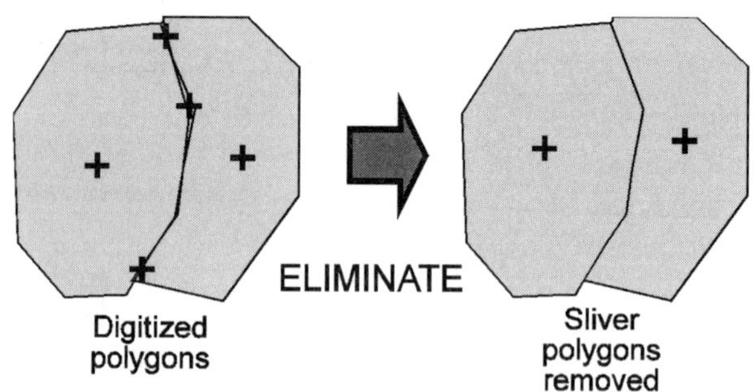

Exercise 2: Manipulating Polygons

In this exercise, we'll use the BUFFER and DISSOLVE commands.

1. Let's try displaying and buffering three coverages. They have point, arc, and polygon feature classes. First, we'll use the point coverage and create a 500-foot buffer around each well in the original coverage. Type in:

```
Arc: &run TESTDRAW SF-WELLS
Arc: BUFFER SF-WELLS SF-WELLSBUF # # 500.0 # POINT
Arc: &run TESTDRAW SF-WELLSBUF
```

2. Now try using the line coverage to create a 100-foot buffer around a sample of roads. Type in:

```
Arc: &run TESTDRAW SF-ROAD
Arc: BUFFER SF-ROAD SF-ROADBUF # # 100.0
Arc: &run TESTDRAW SF-ROADBUF
```

3. For the polygon coverage, we'll create a buffer of 1000 feet around two historic districts in Santa Fe. Type in:

```
Arc: &run TESTDRAW SF-DIST
Arc: BUFFER SF-DIST SF-DISTBUF # # 1000.0
Arc: &run TESTDRAW SF-DISTBUF
```

4. Now, let's try a couple of DISSOLVE operations. We'll use a coverage with census districts in it. Type in the following:

```
Arc: &run TESTDRAW CENSUSTRACT
Arc: DISSOLVE CENSUSTRACT CENSUSBLOCK BLOCK
Arc: DISSOLVE CENSUSTRACT CENSUSZIP ZIPCODE
Arc: &run TESTDRAW CENSUSBLOCK
Arc: &run TESTDRAW CENSUSZIP
```

5. We've just made two dissolved polygon coverages, a census block coverage from census tracts, and a simple zip code map from the census tracts.

Polygon Overlay

There are several commands that span a class of geographic analysis called *polygon overlay.*

This set of commands involves three coverages: an input coverage, an overlay coverage, and an output coverage. All of these commands require that the overlay coverage contain polygon topology. The input coverage can usually have any topology. The output coverages usually have topology specified by feature class arguments (POLY, LINE, POINT, NET, LINK). For valid results, the input coverage topology and feature class argument must be compatible. For example, you can't do a polygon buffer on an input coverage that doesn't have polygon topology.

CLIP

```
Arc: CLIP <in_cover> <clip_cover> <out_cover> {POLY |
     LINE | POINT | NET | LINK | RAW} {fuzzy_tolerance}
```

CLIP is the most frequently used polygon overlay command. It is the means to extract a portion of a large GIS database for analysis or for producing a special-purpose map.

The CLIP command creates a new coverage by removing all features outside the outermost polygon of a clip coverage.

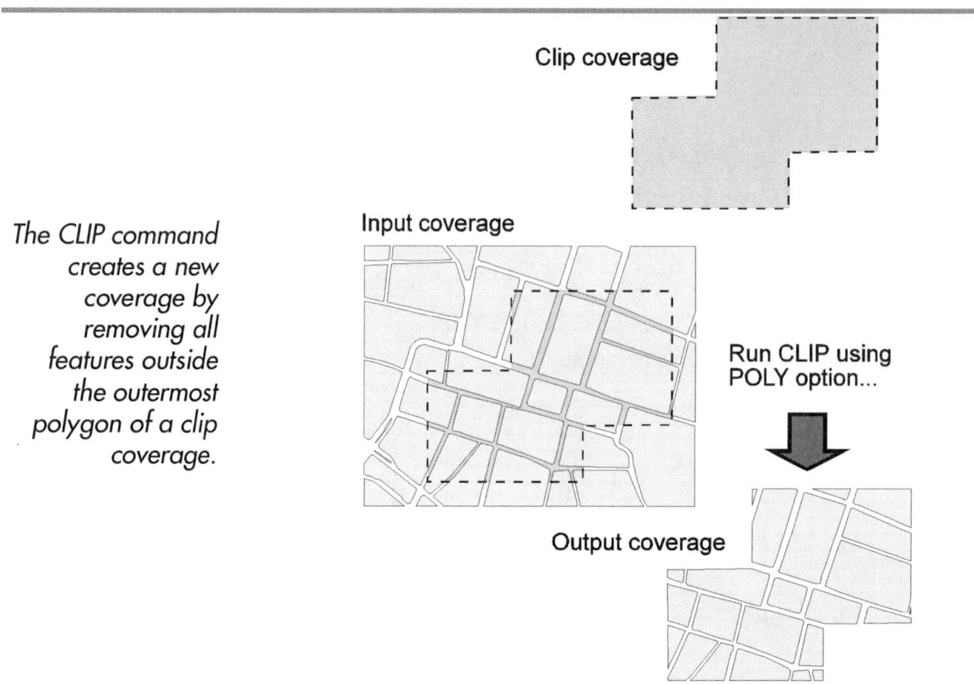

This command clips a coverage using the outermost polygon in the clip coverage. <in_cover> is the input coverage. <clip_cover> contains a clip polygon. If more than one polygon exists, the outermost polygon is used. <out_cover> is the resulting coverage. {POLY | LINE | POINT | NET | LINK | RAW} are the feature classes that are clipped. {fuzzy_tolerance} is the smallest distance between two adjacent coordinates in the output coverage.

Usually, the clip coverage has only one polygon defined, and its only purpose is to enable this command. However, any polygon coverage can be used, and CLIP will determine the outermost polygon (by dissolving the interior polygons) and use that polygon for clipping.

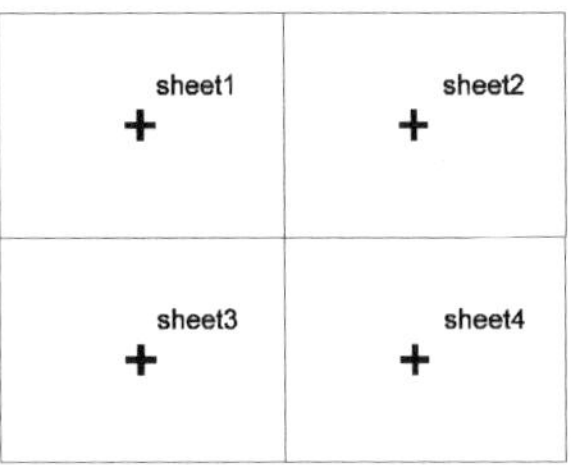

The SPLIT command creates many coverages from an input coverage and a split coverage.

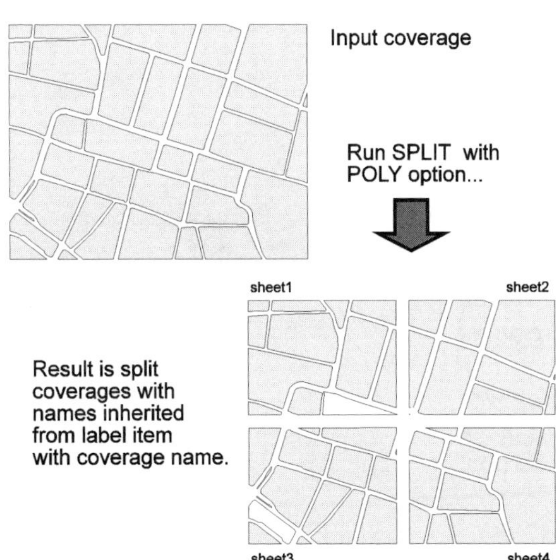

SPLIT

```
Arc: SPLIT <in_cover> <split_cover> <split_item> {POLY
     | LINE | POINT | NET | LINK | RAW} {fuzzy_tolerance}
```

This command splits one coverage into many coverages. <in_cover> is the input coverage. <split_cover> is a coverage with many polygons. <split_item> is an item with values corresponding to map sheet names. {POLY | LINE | POINT | NET | LINK | RAW} specifies the coverage class you want to output. {fuzzy_tolerance} is the smallest distance by which two vertices or nodes can be separated.

SPLIT can create many output coverages. You are prompted to type in these coverage names.

Sometimes, it is useful to split one large seamless coverage into many map sheets. For example, you may want to share your GIS data with another agency using a CAD system that is not equipped for large data sets. You can create a master polygon coverage defining exactly the boundaries of a map graticule, such as a 7 1/2 minute or 15 minute quadrangle map series. By entering sheet names as an item in the PAT for the polygon coverage, you are ready to routinely split your seamless databases for use by others.

The ERASE command removes features that overlay with polygons in the erase coverage.

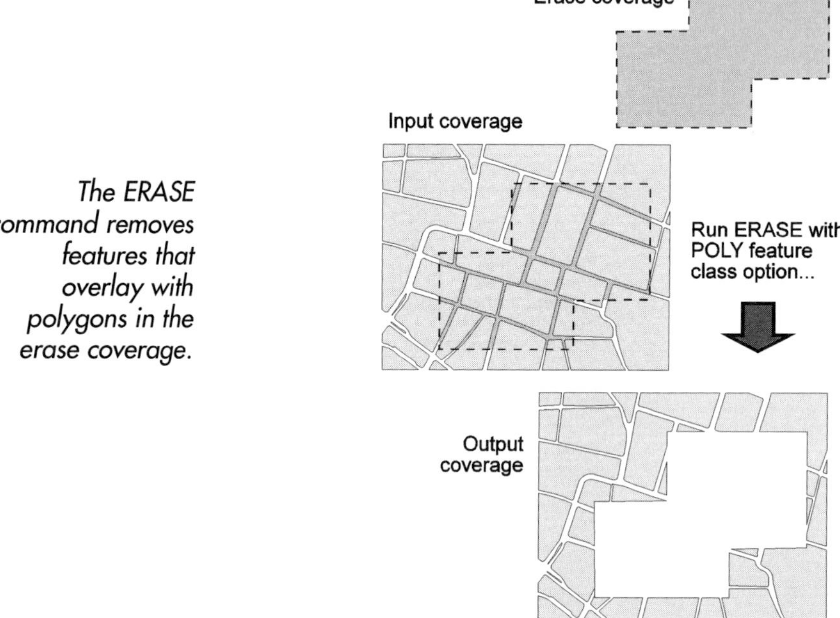

ERASE

```
Arc: ERASE <in_cover> <erase_cover> <out_cover> {POLY
     | LINE | POINT | NET | LINK | RAW} {fuzzy_tolerance}
```

This command removes features by overlaying an erase coverage on an input coverage. <in_cover> is the input coverage. <erase_cover> is the overlay coverage. <out_cover> is the output coverage. {POLY | LINE | POINT | NET | LINK | RAW} is the desired feature class for the output coverage. {fuzzy_tolerance} is the smallest distance allowed between two vertices or nodes.

The UPDATE command replaces features in the overlay coverage into the input coverage.

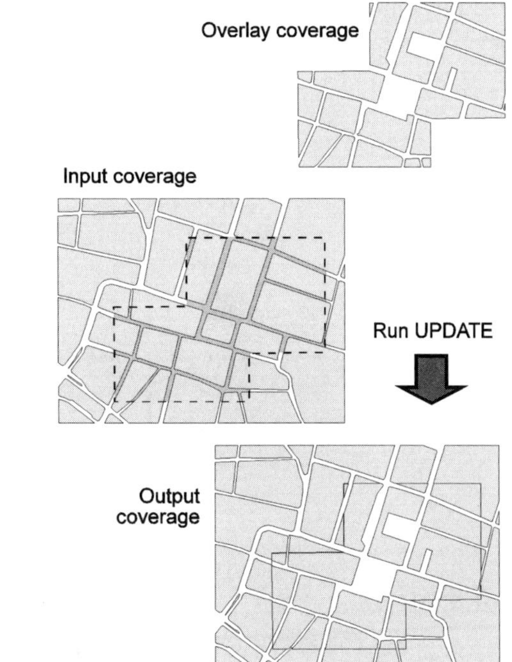

UPDATE

```
Arc: UPDATE <in_cover> <update_cover> <out_cover>
     {POLY | NET} {fuzzy_tolerance} {KEEPBORDER |
     DROPBORDER}
```

This command replaces features in the input coverage with features from the overlap coverage. <in_cover> is the original coverage. <update_cover> is the overlay coverage. <out_cover> is the output coverage with {POLY | NET} topology. {fuzzy_tolerance} is the smallest distance between two nodes or vertices. {KEEPBORDER | DROPBORDER} preserves or deletes the outer border of the overlay coverage.

UPDATE is a curious polygon overlay command and demonstrates the imagination of the ARC/INFO GIS programmers. It allows you to make "cookie cutter" updates to your coverages. This means you can remove a part of your map and replace it with other map data. While this type of update is not your normal map editing procedure, it will surely find innovative uses.

IDENTITY, UNION, and INTERSECT are three similar polygon overlay commands.

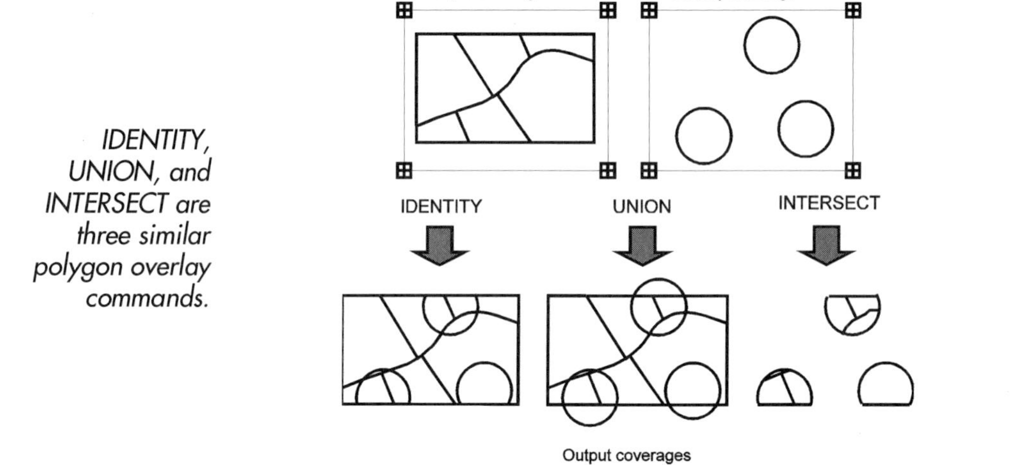

IDENTITY

```
Arc: IDENTITY <in_cover> <identity_cover> <out_cover>
     {POLY | LINE | POINT} {fuzzy_tolerance} {JOIN |
     NOJOIN}
```

This command intersects the features in two coverages and keeps the features within the overlay coverage. <in_cover> is the input coverage. <identity_cover> is the overlay coverage and must have polygon topology. <out_cover> is the output coverage. {POLY | LINE | POINT} are options for the type of overlay: polygon-on-polygon, line-in-polygon, or point-in-polygon. {fuzzy_tolerance} is the smallest distance between adjacent nodes or vertices. {JOIN | NOJOIN} specifies whether a superset of the feature attribute items in the input and overlay coverages are kept, or whether most feature attribute items are dropped.

IDENTITY is an important overlay command that is invaluable for analysis. With the results from IDENTITY, you can perform analysis on selected areas and preserve all the attribute information within the intersection.

UNION

```
Arc: UNION <in_cover> <union_cover> <out_cover>
     {fuzzy_tolerance} {JOIN | NOJOIN}
```

This command intersects the features in two coverages and keeps the union of both coverages. <in_cover> is the input coverage and must have polygon topology. <union_cover> is the overlay coverage, and also must have polygon topology. <out_cover> is the output coverage. {fuzzy_tolerance} is the closest distance between two nodes or vertices. {JOIN | NOJOIN} specifies whether items from the two polygon attribute tables are joined or dropped.

UNION is closely related to IDENTITY. The difference is that features outside the overlay coverage are kept.

INTERSECT

```
Arc: INTERSECT <in_cover> <intersect_cover> <out_cover>
     (POLY | LINE | POINT} {fuzzy_tolerance} {JOIN |
   NOJOIN}
```

This command intersects features in the input coverage and overlay coverage and keeps only intersected features common to both. <in_cover> is the input coverage. It can have point, line, or polygon features. <intersect_cover> is the overlay coverage and must contain polygons. <out_cover> is the coverage to be created. (POLY | LINE | POINT} specifies polygon-on-polygon overlay, line-in-polygon overlay, or point-in-polygon overlay. {fuzzy_tolerance} is the minimum distance between two nodes or vertices. {JOIN | NOJOIN} specifies whether the two feature attribute tables are joined or dropped.

INTERSECT is similar to IDENTITY and UNION but only keeps features that are common to both the input and overlay coverage.

Exercise 3: Using Polygon Overlay Commands

In this exercise we'll use a selection of polygon overlay commands. Our input coverage is "sf-input," and the overlay coverage is "sf-overlay." For the split command, we have a split coverage called "sf-split."

1. Type in your choice of the following commands and arguments.

```
Arc: CLIP sf-input sf-overlay sf-clippoly POLY
Arc: CLIP sf-input sf-overlay sf-clipline LINE
Arc: SPLIT sf-input sf-split sheetname POLY
Arc: SPLIT sf-input sf-split sheetname LINE
Arc: ERASE sf-input sf-overlay sf-erase
Arc: UPDATE sf-input sf-overlay sf-update
Arc: IDENTITY sf-input sf-overlay sf-idenpoly POLY
Arc: IDENTITY sf-input sf-overlay sf-idenline LINE
Arc: UNION sf-input sf-overlay sf-union
Arc: INTERSECT sf-input sf-overlay sf-intersect
```

2. If you try all these commands, you will have these new coverages in your workspace: sf-clippoly, sf-clipline, sf-erase, sf-update, sf-idenpoly, sf-idenline, sf-union, sf-intersect, and from the SPLIT commands, sf-sheet1, sf-sheet2, sf-sheet3, and sf-sheet4.
3. Use the DESCRIBE and TESTDRAW.AML on any of these coverages to see any of the input, overlay, and output coverages.

Proximity Commands

Much geographic analysis asks, What is my closest neighbor, and how far away is this neighbor? ARC provides the commands NEAR and POINTDISTANCE to answer these questions. These two related commands find the closest neighbors between the specified feature types, and stores those values in a feature attribute table.

One use of these commands is to automate the generation of database keys between two distinct sets of databases. For example, suppose you work for a utility company. The utility keeps extensive records on both customers and equipment, but these two databases often don't have a direct database key linking one to the other. Knowing how consumers are connected to equipment is important for keeping the quality of service high. When there is an outage of electrical, gas, or water delivery, your consumers let you know right away. It becomes your job to efficiently locate the equipment serving those users and repair it.

With the ARC/INFO proximity commands, you can determine closest neighbors between consumers and equipment and use this information to build database keys among a utility's corporate databases. This is an example of GIS integrating diverse information.

NEAR

```
Arc: NEAR <in_cover> <near_cover> {LINE | POINT |
     NODE} {search_radius} {out_cover} {NOLOCATION |
     LOCATION}
```

For each point in the input coverage, this command calculates the distance to the closest neighboring feature in the near coverage. <in_cover> is the input coverage. <near_cover> is the coverage to search for neighboring features (must be different from the input coverage). {LINE | POINT | NODE} is the feature class to search for in the near coverage. {search_radius} is the maximum distance for which the search will be executed. {out_cover} is the output coverage. If not specified, the input coverage is replaced by the output coverage. {NOLOCATION | LOCATION} specifies whether locational information is also calculated and stored.

When you run NEAR on two coverages, the Point Attribute Table of the input coverage (or the output coverage if you specify one) has several items added and calculated. If you specify the NOLOCATION option, the first two items are added. If you specify LOCATION, four items are added:

- **<near_cover>#.** internal ID of closest neighbor in near coverage.
- **DISTANCE.** length from point in input coverage to closest neighbor in near coverage.
- **X-COORD.** x location of neighbor if LOCATION option is set.
- **Y-COORD.** y location of neighbor if LOCATION option is set.

After the NEAR command is finished, you can use this information in the PAT with other commands or ARC/INFO programs to set the associations you want.

POINTDISTANCE

```
Arc: POINTDISTANCE <from_cover> <to_cover> <out_info_file>
     {search_radius}
```

For each point in the input coverage, this command calculates the distance to many or all of the neighboring features in the to coverage. <from_cover> is the input coverage. <to_cover> is the coverage to which you are making the proximity search, and can be the same as the input coverage. <out_info_file> is the name of an INFO table to which the results will be sent. {search_radius} is a tolerance that limits the distance to neighboring features. Any features exceeding this distance will not be added to the INFO table.

POINTDISTANCE is useful when you need to find not only the closest neighbor, but several, many, or all neighbors.

The items stored in the INFO table are:

- **<from_cover>#.** The internal ID of the feature in the from coverage.
- **<to_cover>#.** The internal ID of the found feature in the to coverage.
- **DISTANCE.** The calculated distance.

When you look through the resulting INFO table, you'll see that for each <from_cover># value, there can be several or very many features indicated by the item value of <to_cover>#.

WARNING: *Indiscriminate use of POINTDISTANCE can cause very long execution times and huge output INFO tables. The potential number of features in the output INFO table can be calculated as the product of the number of features in the two coverages. For example, if you are searching from a coverage with 5,000 points to another coverage of 10,000 points, the number of possibilities is 50,000,000. Use the search radius to restrict the numbers of features found and stored in the INFO table.*

Data Conversion

Among the many strengths of ARC/INFO is the ability to convert an impressive list of graphics data formats from other automated mapping systems or standard data exchange formats.

Some of the data conversion commands can convert between the different types of geographic data sets: raster to vector, tin to vector, lattices to tins, and so forth. Because we are most interested in ARC/INFO's vector data set—the coverage—we'll only consider the vector formats that ARC/INFO is compatible with.

For the data formats that are not directly compatible with ARC/INFO, data conversion can still be performed through intermediary data formats. For example, nearly every automated mapping system supports the AutoCAD.DXF format, which can be used as the intermediary.

A salient point about data converters is that there are no fully automated data conversions. With every conversion, there is "something lost in the translation" and work to be done to extract the best conversion. This is particularly true when converting ARC/INFO to other formats, because most other data formats do not support two important features of the ARC/INFO model: arc-node and polygon topology, and the flexible definition of feature attribute tables.

In this section, we'll introduce the ARC/INFO interchange format, list the compatible data exchange formats, and discuss the conversion of the AutoCAD.DXF format, a CAD vector format which has become a de facto standard.

ARC/INFO Interchange Data Format

ARC/INFO has an interchange data format designed to let you easily exchange coverages and other data sets with any other ARC/INFO system. ARC/INFO users sometimes call this interchange format the *EXPORT format* after the command that creates these files.

The purpose of the interchange data format is to let two users of ARC/INFO on any hardware platform or operating system exchange any data set or file created with ARC/INFO. For example, there are some fundamental differences between how coverages are stored in UNIX and VMS systems. The interchange data format is the bridge between any distinct systems.

The interchange data format contains all the information in a coverage or other data set in an ASCII format. It stores the positional information, topology, and feature attribute tables.

When you do directory listings on your computer system, you can recognize interchange files because they have a standard file extension of .E00.

All of the geographic data delivered on this book's companion CD-ROM is in the ARC/INFO interchange format. The AMLs you (or your system administrator) ran to unpack the data sets unpacked coverages and other data sets onto your computer hard drive.

There are two ARC commands that read in and write out interchange files: IMPORT and EXPORT. They are easy to use and frequently used.

IMPORT

```
Arc: IMPORT <option> <interchange_file> <output>
```

This command converts an ARC/INFO interchange file into a data set or file. <option> is the type of data or coverage to be imported. COVER is the most frequent choice. Other option choices allow you to convert tins, grids, symbol tables, and any file created with ARC/INFO. <interchange_file> is the name of the interchange file, which usually has a file extension of .E00. It is not necessary to type in the file extension. <output> is the name of the coverage or other ARC/INFO data set or file you want to import.

EXPORT

```
Arc: EXPORT <option> <input> <interchange_file> {FULL
     | PARTIAL | NONE} {max_lines}
```

This command converts an ARC/INFO data set or file into an interchange file. <option> is the type of data set or file to be exported, usually COVER. <input> is the name of the coverage, data set, or file. <interchange_file> is the file name of the export format. You don't need to enter a file extension; an extension of .E00 will be automatically appended for you.

{FULL | PARTIAL | NONE} is the compression option. {max_lines} is the maximum number of lines in the interchange file and is rarely used.

TIP: *The FULL option creates an ASCII file which contains information in a compressed format. If you look at this file in a text editor, it is not readable. Nearly always, the FULL option is preferred because it takes the least space. However, if you want to get at the information within a coverage at a basic level, you can use the NONE compression option. This creates an ASCII file that is readable and somewhat understandable. You can process this file in a text editor and extract information for use by programs outside of ARC/INFO. This technique requires expertise with your text editor, and particularly in edit macros. Most users will never use this technique, but it is a "back door" for accessing ARC/INFO data if you have good operating system and programming skills.*

Exercise 4: Importing and Exporting Coverages

Unless you work in complete isolation, you will surely use IMPORT and EXPORT to share ARC/INFO coverages.

1. An export file has been prepared for you, so type in the following:

```
Arc: IMPORT COVER sf-import sf-import
```

2. Now, try the DESCRIBE command and/or the TESTDRAW.AML and take a look at this coverage.

3. Pick any of the coverages you made with the polygon overlay commands. Type

```
Arc: EXPORT COVER <overlay coverage> <file name
     for export file>
```

4. You can check for the existence of the export file in your operating system by typing in the following:

```
UNIX - Arc: ls *.e00
VMS - Arc: dir *.e00
Windows NT - Arc: dir/w *.e00
```

5. You've just used the command interpretation of ARC/INFO to issue a system command to get a listing of the export files. You've now successfully imported and exported interchange files in ARC/INFO.

Vector Data Formats

ARC/INFO supports a variety of external data exchange formats for data conversion. Some of these formats are specific to commercial mapping or graphics software, and some are government or industry standard formats.

In most cases, ARC/INFO supports two-way translation of data formats—conversion from the exchange data format into a coverage and vice versa. Most ARC commands that are data converters share a naming convention. Commands that convert a coverage to another format have a name in the form ARC<format> Commands that convert a foreign data format to a coverage have a name in the form <format>ARC.

These are the vector formats supported, together with the ARC commands for converting them:

- **AMS, Automated Mapping System.** This is a digitizing format for MOSS, a U.S. Government GIS system. The data converter is AMSARC.
- **DIME, Dual Independent Map Encoding.** This format was used for the U.S. 1980 Census but is rarely used anymore. Data converters are ARCDIME and DIMEARC.
- **DLG, Digital Line Graph.** These are data files provided by the United States Geological Survey. Data converters are ARCDLG and DLGARC.
- **DXF, Drawing Exchange File.** This is the exchange for AutoCAD and is the most commonly used exchange format. Virtually every mapping system can use DXF as an intermediary format. Data converters are ARCDXF and DXFARC. An auxiliary utility command for reporting DXF contents is DXFINFO.
- **ETAK, MapBase file.** This is an ASCII format for a commercially available United States street network database. The data converter is ETAKARC.

- **GIRAS, Land Use and Land Cover Data.** This is a United States format for digital data that includes political units, land use, and federal and state land ownership. The data converter is GIRASARC.
- **IDGS, Interactive Graphics Design Software.** This is the binary format for the Intergraph MicroStation system. These files are called *design* files. Data converters are ARCIGDS and IGDSARC. An auxiliary utility command for reporting on the IGDS file is IGDSINFO.
- **IGES, Initial Graphics Exchange Standard.** This is a government standard data format, which is being superseded by a newer standard. Data converters are ARCIGES and IGESARC.
- **MOSS, Moss Export File.** This is an exchange format compatible with the MOSS GIS system. Data converters are ARCMOSS and MOSSARC.
- **TIGER, Topologically Integrated Geographic Encoding and Referencing System.** This is the format of the complete national map series published on CD-ROM by the United States Bureau of the Census. Data converters are ARCTIGER and TIGERARC. An AML supplied with ARC/INFO which may be useful is TIGERTOOL.

Each of these data converters requires careful consideration. Usually, after you import a file from another system, you will need to follow through with other ARC commands. Some of these commands include: CLEAN and BUILD (to build topology), GENERALIZE (to filter coordinates), PROJECT (to change map projections and to process and manipulate attribute and symbol information in the TABLES or INFO programs.

Part V

AML

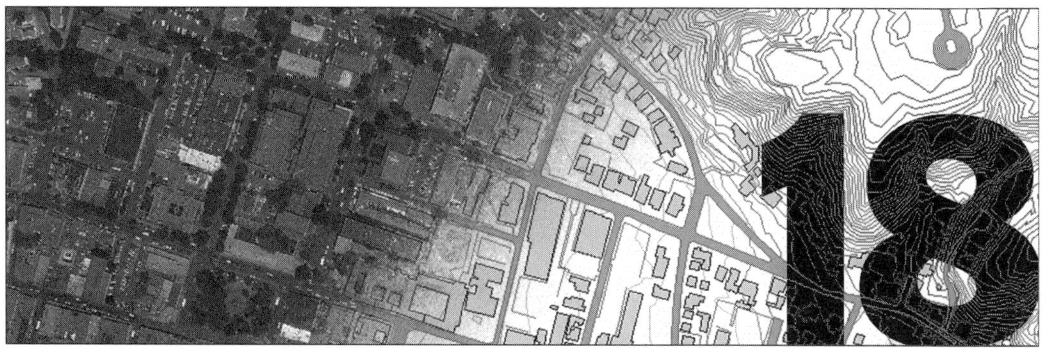

Putting ARC/INFO to Work with AML

Macros and Menus in the ARC Macro Language

Using AML applications such as GIScity is vital for putting ARC/INFO to work. ARC/INFO comprises an exceptionally rich set of tools for managing geographic and attribute data, but it is cumbersome if you use the command level exclusively. You must acquire or build applications written in the ARC Macro Language.

So there is a tradeoff. ARC/INFO is unquestionably the richest set of geographic tools that exists. It is a general-purpose software toolkit, one that can be adapted to any application. But, straight out of the box, you'll be doing a lot of typing at the command level.

The designers of ARC/INFO made a conscious decision: they did not build GIS software oriented toward specific applications, such as AM/FM, forestry, or municipal management. They built GIS software that is general, flexible, and extendible. Industry- or agency-specific applications are layered on ARC/INFO with the ARC Macro Language.

So an important strength of ARC/INFO is the ease of building powerful GIS applications. You'll find that if you have programming skills, you can build an application to do anything you want. You might say ARC/INFO is "programmer friendly." This attribute is important to nonprogrammers as well as programmers, because application customers can then expect superior software.

This chapter is for both programmers and nonprogrammers. You don't have to be a programmer to write a macro or make a menu. If you know how to use your text editor and the ARC/INFO commands you want to use, you can write a macro. If you have programming skills, you'll find that you can build powerful applications and it's not too difficult. At the end of the chapter, we'll explore how you can use the GIScity applications to study AML or jump-start your application design.

Getting Started with Script Macros in AML

What Is AML?

The ARC Macro Language lets you customize ARC/INFO.

Typically, you will use ARC/INFO's command level when you simply want to enter ARC, ARCEDIT, or ARCPLOT and to perform just a few commands. For example, you can make a simple graphics file in ARCPLOT and perform some basic edits on a coverage. But as soon as you find yourself repeating a series of steps, then you are wise to write a macro to capture these steps to be easily repeated.

You can use AML in two ways: you can build a simple script AML to run repetitive command sequences, or you can build comprehensive applications that completely shield your end users from the ARC/INFO command level.

An AML file is an ASCII text file that contains four kinds of text lines: ARC/INFO commands, AML directives, AML functions, and comment lines. An AML directive is a statement that performs variable assignment, flow of control, or feedback on current settings. An AML function performs calculations and returns information. We will further define directives and functions later in this chapter.

Comment lines are for keeping notes for the programmer about the AML. Comment lines always begin with two characters: /*. Any characters after the forward slash and star are ignored by ARC/INFO when you run the AML.

If you find that commands you need to type are longer than the width of your text window, you can use a tilde (~) at the end of the line to continue the command on the next line.

AML is a subject worthy of a book of its own. This chapter only touches on some major points of the ARC Macro Language with several examples. From this chapter, you should get an overview of the main capabilities of AML. Consult ArcDoc and other reference materials for more complete documentation on AML.

Making a Script File in the Text Editor

You don't have to be a programmer to use AML. All you need to know to is how to use the text editor on your system and which ARC/INFO commands you want to use. To start running script macros, you just need to know two AML directives, &RUN and &RETURN.

```
&RUN <aml_file> {argument...argument}
```

runs an AML, with optional arguments.

When you use this command, ARC/INFO will read in the AML file and execute all the commands. You can optionally specify arguments, which are values you want to pass into an AML, such as coverage name.

NOTE: *AML is recognized in all ARC/INFO programs. However, keep in mind that an AML written for ARCEDIT will not work if you try to run it from the ARC or ARCPLOT programs. GIScity uses this convention for naming AMLs: AMLs that work in ARC, or in ARC, ARCEDIT, and ARCPLOT, have a name that starts with "tool-", such as "tool-macro.aml." AMLs designed for ARCPLOT begin with "disp-", such as "disp-legend.aml." AMLs designed for ARCEDIT start with "edit-," such as "edit-arc.aml."*

NOTE: *Different operating systems have different text editors. All UNIX workstations have text editors such as vi or emacs. VMS workstations typically use the TPU editor, sometimes with EDT emulation. Whatever you use, it is important to produce ASCII files, without special control characters.*

```
&RETURN
```

returns control to the calling AML, to the command prompt to menu input, or from a routine to the calling statement.

You will usually use &RETURN in macros at the bottom of an AML. You can also have several &RETURN statements in case your AML detects a need to bail out before the AML is finished.

Exercise: Writing a Simple AML Script

Let's write a very simple AML script, so that you can see that it's easy. This AML will be run from the ARC program, will enter ARCPLOT, draw some coverage features, pause for you to press [Return] on your keyboard, and then exit back to ARC.

1. In your text editor, open a new text file call DRAWTEST.AML. Type this into your text file:

```
arcplot
display 9999 1
&workspace $GCHOME/data
mapextent $GCHOME/data/tgr_road
linecolor red
arcs tgr_road
linecolor green
arcs tgr_cult
linecolor blue
arcs tgr_hydr
```

```
&pause
quit
&return
```

If you are using the VMS operating system, substitute this command for setting the map extent:

```
mapextent $GCHOME:[DATA]tgr_road
```

2. That's it. Now, open a dialog window on your workstation with your operating system's prompt. First, check that you are in the same directory as the AML. You can use `ls drawtest.aml` in UNIX (be careful with upper- and lowercase in UNIX) or `directory drawtest.aml` in VMS.

   If the AML file does not appear on the directory listing, then use your operating system's commands to change your current directory. In UNIX, use `cd <path>` and in VMS, use `SET DEFAULT <directory>`.

3. When you have verified the location of the AML, then enter the ARC program (type `arc` at your operating system prompt) and then run the AML by typing `&run drawtest.aml` at the `Arc:` prompt.

4. You will immediately see a small graphics window appear with arcs from three small scale coverages in GIScity, derived from U.S. Census TIGER data sets. After the three coverages are drawn, then go to your dialog window and press return to finish the AML. (The &PAUSE directive in the previously shown AML does this.)

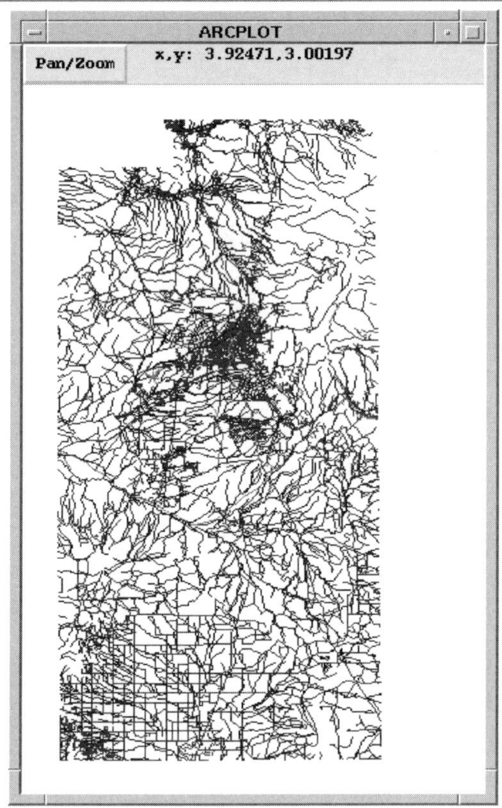

Graphics window that appears from the script AML listed previously.

Making a Script AML from Watch Files

Another way to make a watch file is to capture a series of ARC/INFO commands executed by an AML application like GIScity. You can use a couple of AML directives to record and run macros.

```
&WATCH <watch_file> {&APPEND} {&COMMANDS} {&COORDINATES}
&WATCH <&OFF | &DISPLAY>
```

This opens a file that records everything that appears in your dialog window.

&WATCH has two functions: it is used for troubleshooting problems, and it is also used as another way to prepare a script AML.

The first usage of &WATCH opens a watch file. You can give the file any valid file name and optional directory path. The other, optional arguments have this effect: &APPEND adds new lines to a preexisting watch file (normally preexisting watch files are overwritten), &COMMANDS causes all commands to be sent to the watch file regardless of the input mode, and &COORDINATES displays the X and Y coordinates along with the commands.

The second usage of &WATCH closes the watch file (&WATCH &OFF) and can tell you the name of the current watch file (&WATCH &DISPLAY).

```
&CONV_WATCH_TO_AML <watch_file> <aml_file> {&COORDINATES}
```

Once you've made a watch file, you have an option to convert it into a script macro. &CONV_WATCH_TO_AML can be abbreviated as &CWTA. Immediately after running this command to make a script AML, you can execute that AML with the &RUN directive. You can think of &WATCH and &CONV_WATCH_TO_AML as a macro recorder. Here are the basic steps you would use on a watch file called SESSION.WATCH and an AML file REDRAW.AML. This session could be in either ARC, ARCPLOT, or ARCEDIT.

```
Arc: &watch session.watch
... Execute a number of ARC/INFO commands
Arc: &watch &off
Arc: &cwta session.watch redraw.aml &coordinates
Arc: &run redraw.aml
```

When you run the macro, you will see the exact sequence of ARC/INFO commands replicated. Also, you can edit this AML file in your text editor and make any modifications you want.

Making Menus with AML

Menus are the key to making ARC/INFO easy to use.

A *menu file* is a special kind of AML file: it contains instructions for the placement and actions of *widgets*, which are the elements of a menu:

buttons, icons, slider bars, scrolling lists, and so on. A menu file usually has a file extension of .menu.

Nonprogrammers and programmers can both produce AML menus. This section introduces a powerful menu editing tool, FORMEDIT, and walks you through the creation of a simple menu.

Once you've created a menu file, you can invoke it from any ARC/INFO prompt with the `&menu` directive. We'll do this in the last step of the next exercise.

Editing Menus with FORMEDIT

You can make a menu file by using a text editor, but there is an easier way: the FORMEDIT program in ARC/INFO lets you visually build a menu.

Type `formedit` at the `Arc:` prompt, and you will see two windows appear like this:

The FORMEDIT program lets you easily make changes to menus.

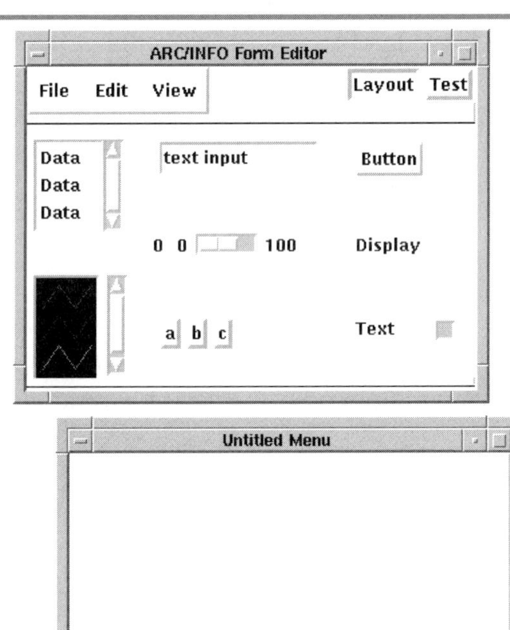

NOTE to Windows NT Users: *FORMEDIT has been extensively revised for ARC/INFO 7.1 on Windows NT. Its function is the same, but it now sports a Windows-like interface.*

This is the basic idea of FORMEDIT: You have two windows, one for selecting widgets, and one for the menu you are making. You can select a widget by moving the cursor over a widget, pressing mouse button 1, and while keeping the mouse button pressed, moving the cursor to a location in the menu window. When you let go of the mouse button, then the widget is dropped in the menu for you.

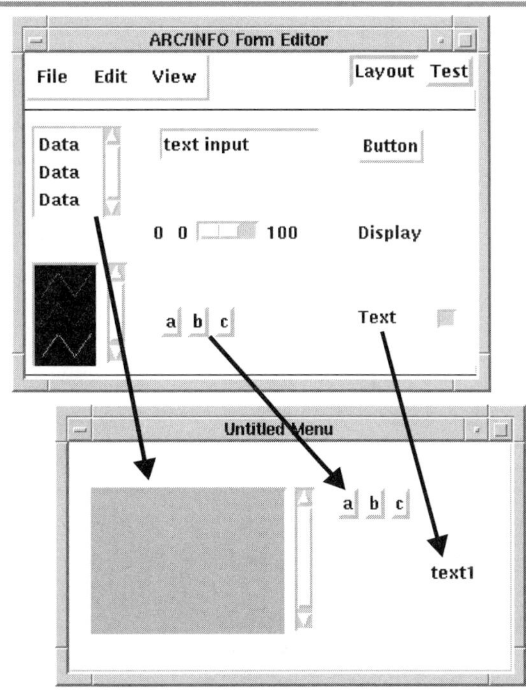

Placing a widget on the menu.

Now you have a widget. To make the widget work the way you want it to, move the cursor over the new widget in the menu and double-click mouse button 1. Each type of widget has a different menu to define its properties. For a button widget, you will see something like the following illustration.

Setting the properties for a button widget.

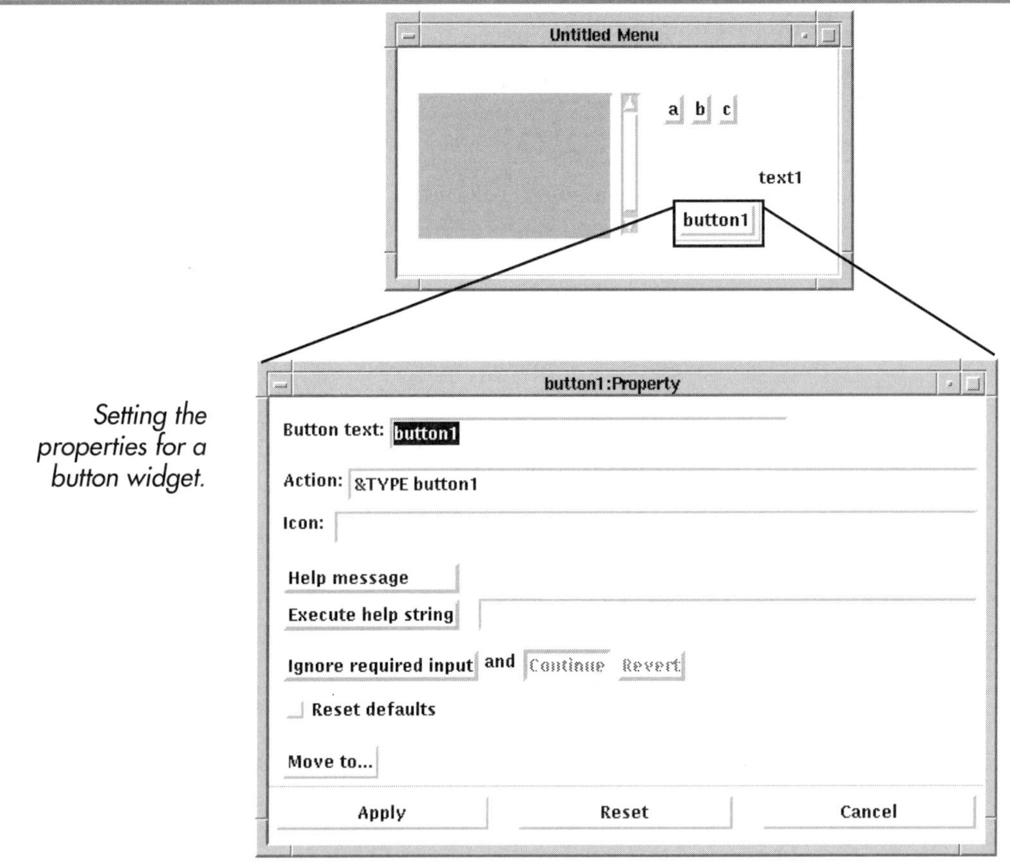

Refer to the FORMEDIT documentation in ArcDoc for a full reference of all the properties for each type of widget.

Exercise: Building a Menu

Now, let's put menus to work for us. This example is simple and doesn't require programming skills.

We've going to build a menu that works in the ARC program and gives you some basic information about coverages.

1. Start the ARC program by typing `arc`.

2. Type `formedit` at the `Arc:` prompt.
3. From the window titled ARC/INFO Form Editor, grab a button widget and drag it to the window titled Untitled Menu.
4. The button widget will be called button1 after you drop it. Double-click on that widget and another window called button1:Property will appear. Set the first two properties like this:

   Button text: **Describe**
   Action: **describe $GCHOME/data/%cover% [UNIX]**
   or...Action: **describe $GCHOME:[DATA]%cover% [VMS]**

Then press the Apply button in the property window, followed by the Cancel button. You'll see the button name changed to Describe.

5. Do the same with three more button widgets. Set these properties:

   Button text: **List point items**
   Action: **listitem $GCHOME/data/%cover%.pat**

   Button text: **List arc items**
   Action: **listitem $GCHOME/data/%cover%.aat**

   Button text: **List arc records**
   Action: **list $GCHOME/data/%cover%.aat**

VMS users should substitute &GCHOME:[DATA]%cover%.aat or $GCHOME:[DATA]%cover%.pat for the path.

6. Add one more button, the button to exit the menu. Add a button with these properties:

   Button text: **Quit**
   Action: **&return**

7. Add a data list. It's the widget that has a scroll bar with the choices Data, Data and Data in the Form Editor window. Drop it onto the side of your other buttons. You can change the size of your data

list by clicking on the widget in your menu window and then resizing with your cursor.

After you've placed, sized, and moved the data list, double-click on that widget and enter only these two properties:

Variable name: **cover**
Workspace: **$GCHOME/data** (UNIX)
$GCHOME:[DATA] (VMS)

You'll see many other property choices for a data list, but for this example, you can ignore them. This widget will give you a listing of all the coverages in the GIScity data workspace.

8. After you have entered these widgets, five buttons, and a data list, then save your menu by clicking on File in the Form Editor menu, and click the Save as... choice. From the menu it invokes, enter `$GCHOME/test/list_cover.menu` (UNIX) or `$GCHOME:[TEST]LISTCOVER.MENU` (VMS).

What your menu might look like after you add five buttons and a scrolling list.

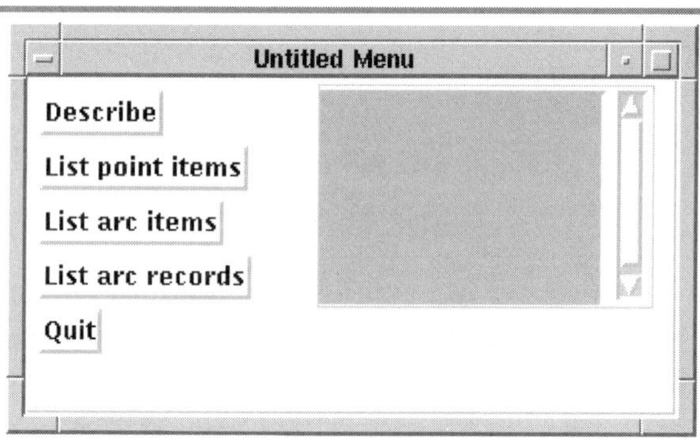

9. Run the menu. At the `Arc:` prompt, type `&menu list_cover.menu`. First, select a coverage from the scrolling list. Then press any of the four command buttons, and you will see the result in the dialog window. When you are finished with the menu, click the Quit button.

When you look at the menu file in a text editor after leaving FORMEDIT, your menu file will look like this:

```
7 list_cover.menu
  %b1
  %button2       %datalist
  %button3
  %button4


  %button5
%b1 BUTTON KEEP 'Describe' describe
%datalist1 INPUT cover 14 TYPEIN NO SCROLL YES ROWS 6 COVER
  * -ALL -SORT
%button2 BUTTON KEEP 'List point items' listitem
  $GCHOME/data/%cover%.pat


%button3 BUTTON KEEP 'List arc items' listitem
  $GCHOME/data/%cover%.aat
%button4 BUTTON KEEP 'List arc records' list
  $GCHOME/data/%cover%.aat
%button5 BUTTON KEEP 'Quit' quit
```

The menu file list invoked from the ARC program.

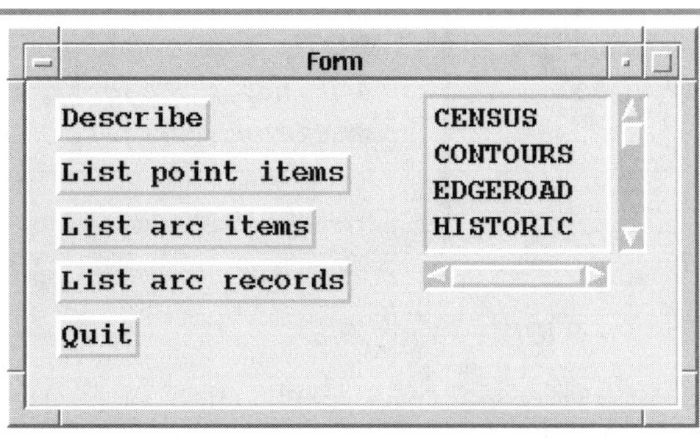

We've now created a menu that lets you select a coverage and then perform four commands that report item definitions and item values. When you think about it, this is powerful stuff. Whatever ease of use that ARC/INFO lacks in its native command level mode, it more than makes up in the flexibility, power, and easy construction of AML menus.

In the menu file shown previously, we have also introduced an AML variable called *cover*. Variables are used in macros to hold values, in this case, the selected coverage from the data list. The other commands in the menu use this variable to perform their actions, like listing item definitions for a coverage.

This menu illustrates how easy it can be to make an AML menu in FORMEDIT. No special programming experience is required to make simple AML menus. But this menu would not be complete for production use. One thing we have not done with this menu is to check the validity of some choices. For example, you could select a coverage without a Point Attribute Table and you would get an error when you tried to list the point items.

You can add validation of choices to be added to a menu, but this usually requires that you run an AML from inside the menu. In the menu example, all the actions in the menu are ARC commands. But you can also put &run statements as a widget action, which will run that AML whenever that widget is clicked in ARC/INFO.

NOTE: *FORMEDIT edits only one of the available types of AML menus, the form menu. It cannot update pulldown, sidebar, or other types of menus. This is not really a restriction, though, because the form menu is nearly always the preferred type of AML menu.*

Menu Widgets

Here's a quick explanation of each type of widget you'll find in FORMEDIT. You can add any of these to an AML menu.

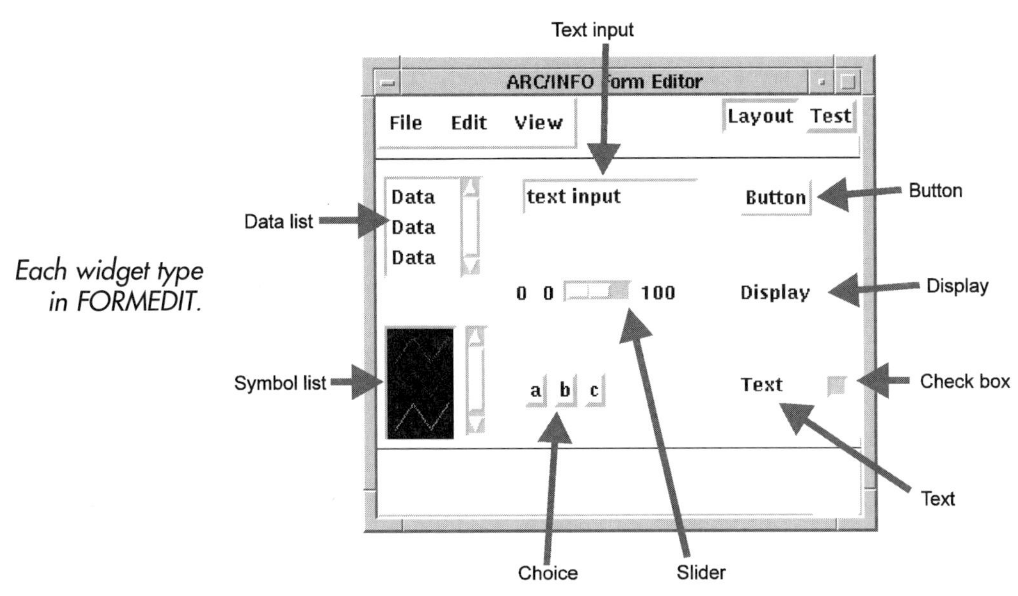

Each widget type in FORMEDIT.

You can surmise how each of these widgets works by inspecting their respective property sheets in FORMEDIT. We can't fully discuss the properties for each widget, but here is a description of how each widget can be used:

- **Button** is used to perform a direct action.
- **Check box** lets you make a quick true/false setting.
- **Choice** lets you choose one value from several choices.
- **Data list** lets you choose coverages, tables, items, and other objects from a scrolling list.
- **Display** displays the value of any AML variable.
- **Slider** lets you choose a numeric value from a sliding bar on a scale.
- **Symbol list** lets you visually choose a marker, line, shade, or text symbol from a scrolling list.
- **Text input** lets you type in a character string.

See ArcDoc for a complete discussion of widgets and their properties. The GIScity applications contain examples of each widget type.

Editing Icons

In the GIScity applications, you also saw icons used for many ARCPLOT or ARCEDIT commands. Here's how you can add icons to menus:

1. Create an icon with the icon editor that comes with your workstation. Reference your ARC/INFO System Dependencies manual for instructions on running the icon editor for your workstation.
2. Add a button in a menu with FORMEDIT. Double-click on the button and type in the name of the icon you created for the Icon: property.

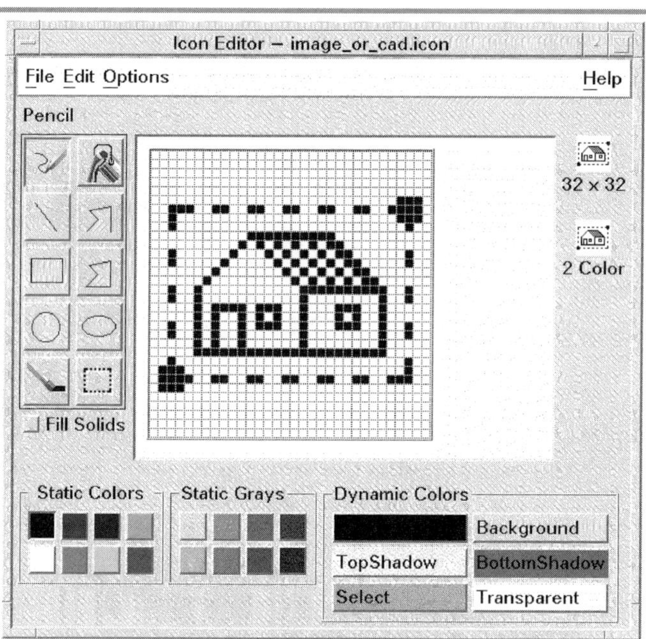

The icon editor for the HP-UX operating system (UNIX for Hewlett-Packard workstations).

Each operating system has its own icon editor. Reference your workstation documentation on instructions for running the icon editor on your system.

Basics of the ARC Macro Language

Programming in the ARC Macro Language is a substantial topic that deserves a book of its own. We can't cover all the details here, but we'll outline the basic commands of AML.

Every ARC/INFO program recognizes AML. The commands in AML mostly work the same whether you are in ARC, ARCPLOT, or ARCEDIT.

Also, AML is identical on all workstations. This means that you can write an AML on one type of workstation and expect it to run on another workstation. You can also craft AML applications that will run on all workstations supported by ARC/INFO.

TIP: *If you are developing AMLs to run on both UNIX and VMS, you're bound to run into differences between the two operating systems. Although the syntax of AML commands is identical for both VMS and UNIX, directory paths have a different syntax. This will be 90% of the system dependency in a substantial AML application. Setting directory paths as global variables in AML can solve this problem. For an example of setting paths as variables, look at this AML in GIScity: $ARCHOME/tools/tool-init.aml (UNIX) or $ARCHOME:[TOOLS]TOOL-INIT.AML (VMS).*

These are some of the characteristics of AML that make it a procedural language:

- Local and global variables
- Program modules and program routines
- Conditional branching and looping
- Mathematical functions
- String functions
- System reporting
- File management

Functions and Directives

There are two basic types of commands in AML: *directives* and *functions.* Functions are designed to return information to you or to ask you for information. Directives execute the assignment of variables, the flow of control in an AML, and many control and diagnostic purposes.

Getting Help

There are a number of ways of getting help in AML.

```
HELP
```

invokes ArcDoc, ARC/INFO's on-line documentation system. Within ArcDoc, you can find alphabetic and functional listings of all AML directives and functions.

```
&COMMANDS {directive | function}
```

lists all or some of the available AML directives and functions.

To specify a directive, use the ampersand (&) to precede the directive name. To specify a function, just type the key word for the function, without the brackets. As with the USAGE command in ARC, ARCPLOT, and ARCEDIT, you can use a wildcard to list AML directives and functions. When you use &COMMAND without any arguments, all the directives and functions are listed.

```
&USAGE <directive | function}
```

lists the mandatory and optional arguments and keywords for the directive or function.

Again, specify a directive with an ampersand, and a function by the keyword without the brackets.

```
&SHOW <parameter> {argument..argument}
```

The &SHOW directive can be used to display many AML settings, as well as certain ARC, ARCPLOT, and ARCEDIT settings. There are too many possible parameters for &SHOW to document here, but the following are a few useful parameters: &AMLPATH, &ECHO, &FORMAT, &GETPOINT, &MENU, &MENUPATH, &MESSAGES, PRECISION, PROGRAM, SCREEN-

SIZE, &SEVERITY, &TERMINAL, &THREAD, VERSION, &WATCH, and WINDOWS.

A similar command to the &SHOW directive is the [SHOW] function.

The &SHOW directive and the [SHOW] function are similar and invaluable for getting information in ARC, ARCEDIT, or ARCPLOT. The difference is that &SHOW displays the result in the dialog windows and the [SHOW] function can be used to set a variable.

AML Directives

Environment

```
&WORKSPACE <workspace>
```

moves you from your current workspace to another.

This is a commonly used command. &WORKSPACE reduces typing in ARC/INFO, because with it, you don't have to type in the full path of a coverage, just the coverage name. You can avoid the use of &WORKSPACE by always using full path specifications when using ARC/INFO commands, but you won't. It's too tedious to constantly type in path names with ARC/INFO commands.

Be careful when using &WORKSPACE in ARCEDIT. One rule for ARCEDIT sessions is that before you quit you should always return to the workspace where you began your ARCEDIT session. ARCEDIT keeps temporary files in that first workspace and requires those files to successively end an ARCEDIT session.

```
&AMLPATH <path...path>
```

sets a search list of directories for ARC/INFO to find an AML.

This command is used by programmers as a shortcut. If you have AMLs in several directories, then you set a list of paths for ARC/INFO to automatically scan and locate AMLs.

WARNING: *If you use multiple directories, be careful that all the names of the AMLs are unique. If you have two AMLs with the same name but in different directories, then which AML file will run is unpredictable.*

TIP: *If you already have several paths specified with &AMLPATH, but want to add one more, the quickest way is to combine the &AMLPATH directive and the [SHOW AMLPATH] function like this:*

```
Arc: &amlpath [show amlpath] /path/to/add
```

```
&MENUPATH <path...path>
```

sets a search list of directories for ARC/INFO to locate a menu file.

This command works just like &AMLPATH, but for menus. The same instructions, warning, and tip apply.

```
&TERMINAL <device> {&CURSOR | &TABLET | &MOUSE | &KEYPAD}
```

specifies which type of workstation or display you are using and which type of input is active for menus.

The device you specify for &TERMINAL will almost always be `9999` for an X-windows compatible workstation.

```
&SYSTEM {operating system commands}
```

lets you execute operating system commands from within ARC/INFO.

If you use &SYSTEM without arguments, then you will be at the operating system level until you leave. To leave the operating system level in UNIX, type `bye`. In VMS, type `logout`.

The &SYSTEM directive is a powerful command. Not only can you issue routine system commands, such as checking the files in a directory, but you can also use &SYSTEM to run other programs from inside ARC/INFO.

Assigning and Using Variables

There are two types of variables: *local* and *global*. A local variable is defined and can be evaluated within just one AML file. When you exit an AML with the &RETURN directive, the local variables for that AML file disappear. Global variables, once set, remain for the duration of your ARC/INFO session unless they are removed.

Given the properties of local and global variables, you may decide to use mainly global variables. Overuse of global variables, however, may cause problems with an AML application's reliability. Good programming techniques call for the maximum use of local variables and passing arguments to other AMLs as arguments and minimizing the use of global variables.

Here's how ARC/INFO distinguishes between the local and global variables: if the variable name begins with a period, it is a global variable. If it does not, then it is a local variable.

The basic command to assign values to variables is &setvar. Here's its basic usage:

```
&setvar <variable> {value}
```

A variable can have almost any name you can devise for it. The following characters are safe to use with variable names: A through Z, 0 through 9, $, -, and _.

An important property of variables in AML is that they are polymorphic. This means they assume the type of value you set them to. The four types of values for AML variables are text strings, real valued numbers, integers, and logicals.

The polymorphism of AML variables is both good and bad: it makes using variables in AML easier, but you sacrifice some control over what type of values can be assigned to a variable. You can fully enforce type validation on variables, but this requires some additional AML programming.

When you assign a variable with the &setvar directive (which can be abbreviated as &s), then you would just type in the variable name for the first argument. If you want to set one variable to a value in another variable, you must surround the variable in the second argument to &setvar with a % symbol.

Here's an example:

```
&setvar pi 3.14159
&setvar pi-squared %pi% * %pi%
```

The percent sign is important, because AML uses it to tell the difference between a text string and a variable.

```
&DELVAR <variable...variable>
```

removes the specified variables.

It's a good programming practice to remove variables when they are no longer needed in an AML application. You can use the wildcard * for the variable name. To remove global variables, you must begin the variable name with a period.

```
&ARGS <variable...variable> {:REST}
```

passes variables into an AML.

The &ARGS directive is an important way to pass variables. Although you could bypass the use of &ARGS by using global variables, it is preferable to accept variables as arguments.

```
&LISTGLOBAL {global_variable...global_variable}
```

lists the specified global variables.

When you are writing an AML application, and are troubleshooting problems, you often look at which global variables are set.

```
&LISTLOCAL {local_variable...local_variable}
```

lists the specified local variables. This is just like &LISTGLOBAL, but for local variables.

Flow of Control

Within an AML, you can define routines. A routine is a part of an AML that is called from within the AML. Routines are used extensively in GIScity.

```
&CALL <routine>
```

jumps the AML to the line in the AML, starting with &ROUTINE followed by the routine name.

```
&ROUTINE <routine>
```

marks the line in the AML where a routine begins.

Here's an example of an AML calling a routine named test-method:

```
{ARC/INFO commands and AML statements}
&call test-method
{More ARC/INFO commands and AML statements}
&routine test-method
{commands and statements for the routine}
&return
```

The return statements return control to the AML line right after the &CALL statement.

Jumping Program Control in an AML File

You can jump from one line in an AML to another line with the &GOTO and &LABEL directives.

```
&GOTO <label>
```

Transfers program control to the AML line with the &LABEL directive and matching label value.

```
&LABEL <label>
```

receives program control from a &GOTO directive.

Here is an example of using &GOTO in an AML file:

```
<ARC/INFO commands and AML statements>
&goto next
<More ARC/INFO commands and AML statements>
&label next
```

As soon as the &GOTO is reached, control jumps to the label statement with matching label value. All lines in between are skipped.

NOTE: *It is poor programming practice to use &GOTO frequently because this type of branching can lead to AML files that are hard to follow. Use &GOTO sparingly, and analyze whether the branching can be better performed with the &IF or &DO directives.*

Looping in AML

The &DO directive is a general command for looping in AML. Loops have a beginning and end, and a condition by which the loop is ended. The usage types for this command are the &DO &LIST, &DO &REPEAT, &DO &TO &BY, &DO &UNTIL, and &DO &WHILE.

The &DO directive is always matched by the &END directive. Whenever you add an &DO directive in an AML file, you must always have the &END directive denoting the end of the loop.

The very simplest usage of the &DO directive is

```
&DO
    <ARC/INFO commands and AML statements>
&END
```

This usage would execute the enclosed commands once and proceed. This may seem a trivial usage, but it is actually quite important in AML. If you are using conditional branching in AML, you will frequently use the simple &DO .. &END combination.

Here's an example:

```
&if %feat% eq ARC &then
  &do
    <commands related to arcs>
  &end
&else &if %feat% eq LABEL &then
  &do
    <commands related to label points>
  &end
```

This example uses a variable named feat and checks whether it matches the text string ARC or LABEL. If you want to execute multiple commands in one of the conditional branches, you must enclose the statements with the &DO .. &END combination.

Also, note that the commands within the loop and condition branch are indented. This is not required in AML, but is a good programming practice to visually keep track of where loops begin and end.

Here's a summary of the other usages.

```
&DO <index_variable> &LIST <list> {&UNTIL
<expression> | &WHILE <expression>}
```

If you have a variable that contains a list of values, for example, red green yellow orange blue, then you can loop through this list. This loop stops when you've gone through the list. You can use the index variable for other commands. In this example, you can use each color for a graphics command in ARCPLOT or ARCEDIT.

```
&DO <index_variable> <start-value> {&REPEAT
<repeat_expression> {UNTIL <expression> |
&WHILE <expression>}}
```

This usage of &DO lets you increment an index variable (by a string or mathematic expression) and continue the loop until or while another expression remains true.

```
&DO <index_variable> <start_value> {&TO
<end-value> {&BY <increment>}} {&UNTIL
<expression> | &WHILE <expression>}
```

loops from one numeric value, by a user-defined increment, until an upper range is reached. This variant of the &DO directive is frequently used.

```
&DO &UNTIL <expression>
```

loops continuously until the expression turns from false to true. Sometimes you want to set up a loop whose endpoint can't be predetermined. With this usage of &DO, you can loop until some condition is satisfied, such as reaching the end of a text file.

```
&DO &WHILE <expression>
```

This is like &DO &UNTIL, but loops until the expression turns from true to false.

Conditional Branching

Two commands for conditional branching in AML are the &IF and &SELECT directives. The &IF directive has a syntax similar to many procedural languages such as FORTRAN.

```
&IF <logical_expression> &THEN <statement | directive>
   {&ELSE <statement | directive}
```

lets you conditionally branch control on an AML, depending on the value of the logical expression.

Here is a code fragment that uses the &IF directives with a simple &DO loop:

```
&if %cur-status% eq .TRUE. &then

&do

[statements based on the variable called cur-status
     having a true value]

 &end
```

The &SELECT directive is very much like the CASE command in the Pascal programming language. It lets you branch to a block of code where the values in the &WHEN statement matches the value of the expression.

```
&SELECT <expression>

 {&WHEN <value..value>...

    <statement>

  {&OTHERWISE <statement>}

&END
```

branches program code to the statement following the &WHEN directive, with the value matching the expression in the &SELECT directive.

Here's an example of using the &SELECT directive:

```
&setvar color green

&select %cover%

&when red

  <statement for red color>
```

```
&when blue
  <statement for blue color>
&when green
  <statement for green color>
&otherwise
  <statement for any other color>
&end
```

In this example, the AML control jumps from the &SELECT directive to the &WHEN directive with a value of green. The statement(s) for this choice will be executed. When done, program control will go down to the &END directive and proceed through the remainder of the AML.

Capturing Coordinate Values

Sometimes, you want to capture a coordinate position in AML for later use. For example, if you want to ask the user of your AML to select a point on the graphic window, you can use the &GETPOINT directive to set variables with mapping coordinates as their value.

```
&GETPOINT {&CURRENT | &MAP | &PAGE} {&PUSH}
&GETPOINT {&CURRENT | &CURSOR | &DIGITIZER | &TABLET |
    &MOUSE}
```

lets you select a point with your coordinate input device and stores the coordinates in program variables.

The &GETPOINT directive is used for AMLs that want to capture coordinate values and then use those coordinates for other operations. When you use &GETPOINT, three AML reserved variables are set for you: pntkey, pntx, and pnt$y. pnt$key stores the input key you used, usually between 1 and 9. pnt$x and pnt$y store the X and Y coordinates of the point you picked. Although the names of these variables look like local variables, they are special variables that the &GETPOINT function sets for you.

Here is an example of using the &GETPOINT directive:

```
&getpoint &map
&setvar key %pnt$key%
```

```
&setvar xy %pnt$x%, %pnt$y%
```

Now we've captured an input and coordinate pair as new, local variables.

The Point Buffer in AML

With AML, you have access to a point buffer. The point buffer in ARC/INFO is what ARC/INFO remembers about previously selected locations selected with the current input device. In programming terms, it is a "first-in, first-out stack." This means you can store multiple points in the point buffer, and when you retrieve them, the point you get out is the first point you placed.

```
&PUSHPOINT {key x y}
```

adds points to the point buffer.

If you are editing features in ARCEDIT, you can preset the point buffer with coordinate values. For example, if you want to add an arc with two endpoints with coordinate pair values of (130000, 240000) and (140000, 260000), you can use this sequence of steps:

```
editfeature arc
&pushpoint 2,130000,240000
&pushpoint 2,140000,260000
&pushpoint 9,0,0
add
```

After the ADD command, the arc connecting the two points is immediately digitized. Normally the ADD command for arcs would prompt you to digitize nodes and vertices with input keys 2, 1, and 9. But commands like ADD first check the point buffer. If enough points are in the point buffer, then AML will use up those points and not ask you for further coordinate input.

&PUSHPOINTS has a companion directive,

```
&FLUSHPOINTS
```

which clears out the point buffer.

Digitizing the example arc using the point buffer.

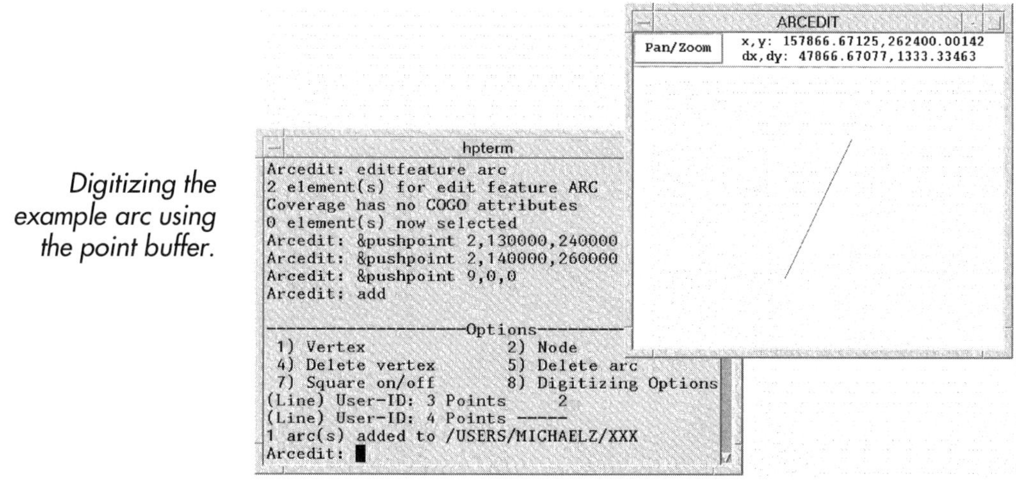

The &FLUSHPOINTS directive is mainly used as a check before using &PUSHPOINT. It guarantees that the point buffer is empty before adding any other points.

Program Testing

These are some commands used to verify program status and troubleshoot problems.

```
&ECHO <&OFF | &ON>
```

enables or disables the display of every ARC/INFO and AML command in the dialog window.

You usually use &ECHO &ON when you are testing a new AML and want to see exactly the actions done in the AML. Using &ECHO &ON does not modify the function of your AML application, but it may slow down your application because displaying all the ARC/INFO and AML commands may take a few moments.

```
&WATCH <watch_file> {&APPEND} {&COMMANDS} {&COORDINATES}
&WATCH <&OFF | &DISPLAY>
```

records everything in the dialog window into a text file. Also, it closes the text file.

The &WATCH directive is used for troubleshooting problems. Often, &ECHO is set to &ON when a watch file is recording. After you close a watch file, you can use the &POPUP directive to display the watch file.

You can also use &WATCH with &CWTA (convert watch to AML) to record a sequence of commands into an AML.

```
&SEVERITY <&WARNING | &ERROR> <&IGNORE | &ROUTINE
   <routine> | &FAIL>
```

specifies the response in an AML file when an error is encountered.

There are two severity conditions in AML, warnings and errors. There are three actions that can be applied to the severity conditions. &IGNORE specifies that the AML should resume after an error. &ROUTINE specifies that the program control be sent to a &ROUTINE statement with the input routine name. &FAIL says that an AML should be terminated when an error is encountered.

This directive is usually at or near the top of an AML.

AML Functions

AML functions are designed to take settings or values from the ARC/INFO session and define those settings as variables.

These functions are usually used with the &SETVAR or &TYPE directives. Here's an example setting a list of items from a coverage:

```
Arc: &setvar item-list [items $GCHOME/data/edgeroad.aat]

Arc: &type %item-list%

FNODE# TNODE# LPOLY# RPOLY# LENGTH EDGEROAD# EDGEROAD-ID
    ACAD_LAYER SYMBOL

Arc: &type [items $GCHOME/data/edgeroad.aat]

    FNODE# TNODE# LPOLY# RPOLY# LENGTH EDGEROAD#
      EDGEROAD-ID ACAD_LAYER SYMBOL
```

You'll notice that &type [function] and &type %item-list% commands give the same result.

In this example, the function returned a list of text strings. Functions can also return integers, real numbers, and logical values. We'll outline some of the common AML functions and describe the type of value returned.

System Reporting

Here is a suite of commands that provide you with information about your system. Most of these functions return character strings.

```
[DATE {format}]
```

returns the current date in one of several specific date formats.

```
[SHOW <parameter> {argument...argument}]
```

returns many ARC/INFO settings.

There are far too many parameters to cover here. The parameters are the same as for the &SHOW directive. See ArcDoc for a complete listing of parameters and usages.

```
[USERNAME]
```

returns the name of the user, as it is known to the operating system.

```
[VALUE <variable>]
```

returns the value of the variable. This is a useful command if you are using variables that contain an index value, such as an integer, within its name.

```
[VARIABLE <possible_variable>]
```

returns true or false, depending on whether a variable with the specified name exists.

Mathematical Functions

All of these functions perform calculations and return a real value.

```
[ANGRAD <formatted_angle>]
```

returns the angle in radians. The formatted angle can be in several formats, such as a survey bearing.

```
[SIN <angle_in_radians>]
```

returns the sine of the input angle.

```
[COS <angle_in_radians>]
```

returns the cosine of the input angle.

```
[TAN <angle_in_radians>]
```

returns the tangent of the input angle.

```
[SQRT <number>]
```

returns the square root of the specified number.

```
[INVANGLE <x1 y1 x2 y2>]
```

returns the angle between the two specified points.

```
[INVDISTANCE <x1 y1 x2 y2>]
```

returns the distance between the two specified points.

String Manipulation

All of these string manipulation functions return a character value. Remember to put single quotes around strings.

```
[AFTER <string> <search_string>]
```

returns the remainder of the string that is after the search string.

```
[BEFORE <string> <search_string>]
```

returns the part of the string that is before the search string.

```
[EXTRACT <element_number> <element...element>]
```

extracts one element from a text string.

```
[QUOTE <string1...string20>]
```

places quotes around the specified strings.

```
[UNQUOTE {string1...string20}]
```

removes the outermost quotes around the specified string.

```
[SUBST <string> <search_string> {replace_string}]
```

substitutes a search string in the specified string with a replacement string.

```
[SUBSTR <string> <position> {number_of_characters}]
```

extracts a portion of a string according to the start position and number of characters.

File Management

In AML, you can open, read from and write to, and close text files. These are the commands for reading and writing text files:

```
[OPEN <file> <status_variable> <-READ | -WRITE | -APPEND>
[CLOSE <aml_file_unit> | -ALL]
[READ <aml_file_unit> <status_variable>]
[WRITE <aml_file_unit> <record>]
```

Here are some more functions that give you information about your files and system:

```
[DIR <file>]
```

sets a list of files in a directory.

```
[EXISTS <object> {-type}]
```

checks for the existence of the specified object and type.

For object, type in the name of the data set, which will usually be a coverage name. The types of objects are -FILE, -ADDRESS, -ANNOTATIONS{.subclass} -ARC, -CLEAN, -COVER, -DIRECTORY, -GRID, -INFO, -LAYER, -LIBRARY, -LINE, -LINK, -NETWORK, -NODE, -POINT, -POLYGON, -ROUTE{.subclass}, -SECTION{.subclass}, TAT{.subclass}, -TIN, -VAT, and -WORKSPACE.

```
[ENTRYNAME <file>]
```

returns just the file (or object) name, without the full directory path.

```
[PATHNAME <file>]
```

takes a file name and expands it to include the full path name.

Specifying Objects

Here are a set of AML functions that invoke a menu from which you can select a value. These functions are designed to run with the &SETVAR directive, to store your selected value and use it elsewhere in the AML files.

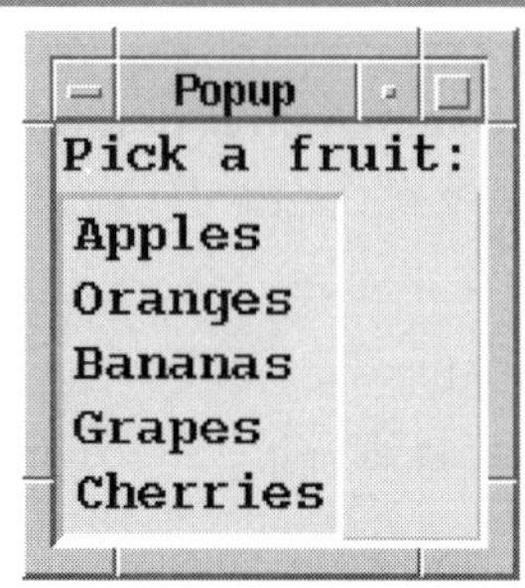

Using the [GETCHOICE] function.

```
[GETCHOICE <choice...choice> {-PROMPT  <prompt_string>}
   {{-NOSORT } -SORT} {-NONE} {-OTHER}]
```

lets you make one choice from a list of choices specified as a list to the function.

Using the [GETCOVER] function.

```
[GETCOVER <workspace | wild_card> {<-coverage_type>
    {prompt_string}} {-SORT | -NOSORT} {-NONE} {-OTHER}]
```

lets you select a coverage.

You can qualify the coverages for consideration by setting the coverage type. The valid <coverage_types> are -ALL, -ADDRESS, -ANNOTATIONS{.subclass}, -ARC, -LINE, -LINK, -NETWORK, -NODE,

-POINT, -POLY, -ROUTE{.subclass}, -SECTION{.subclass}, -TAT{.subclass}, and -TIC.

Using the [GETITEM] function.

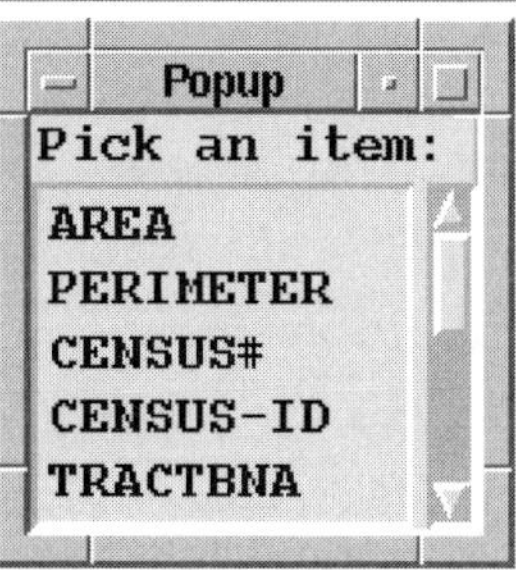


```
[GETITEM <specifier> <-specifier_type> {-item_class}
   {-item_type} {prompt_string} {-NOSORT | -SORT} {-NONE}
   {-OTHER}]
```

gets the name of the selected item from a scrolling list of item values for the specified object.

Using the [GETSYMBOL] function.

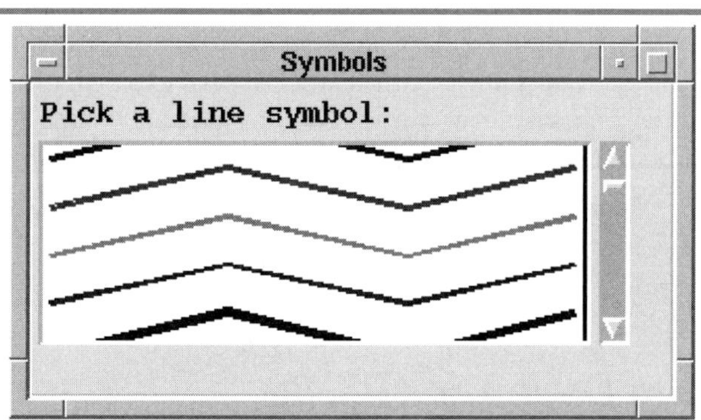


```
[GETSYMBOL -LINE | -MARKER | -SHADE | -TEXT> {-RANGE
   <range_specification>} {-NONUMBER | -NUMBER}
   {prompt_string} {-NOSORT | -SORT} {-NONE} {-OTHER}]
```

gets the symbol number for the selected line, marker, shade, or text.

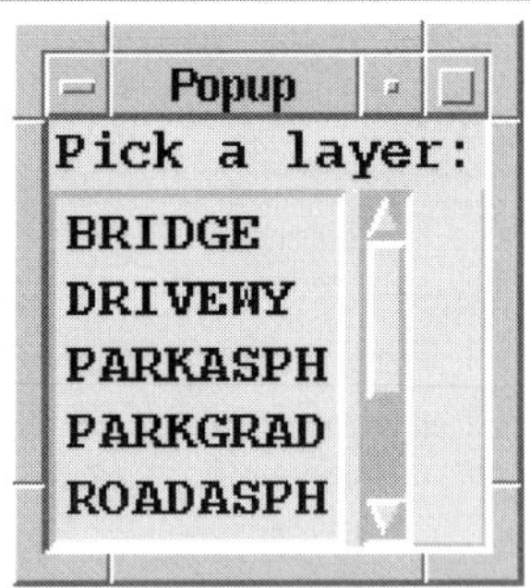

Using the [GETUNIQUE] function.

```
[GETUNIQUE <specifier> <-specifier_type> <item_name>
   {prompt_string} {-NONE} {-OTHER}]
```

gets a unique value from values in feature attribute tables or other tables.

How To Tailor GIScity for Your Applications

The GIScity applications provided on the companion CD-ROM to this book serve three basic purposes. First, GIScity illustrates the most basic ARC/INFO functions through an easy-to-use graphical user interface. Second, you can study the GIScity AML code to learn AML programming by example (if you are so inclined). Third, GIScity is a substantial set of macros and menus that you can use and modify for your own purposes.

Program Design of GIScity

GIScity was designed to instruct you in using a GIS and to make the most commonly used ARC/INFO commands easy to use. Most of GIScity is original AML code, but GIScity borrows its programming methods from ArcTools, which is a set of AMLs supplied with ARC/INFO to get you started.

GIScity, like ArcTools, employs some object-oriented programming techniques. Although AML is not an object-oriented language, some basic object properties such as encapsulation can be programmed in

AML. To understand the programming techniques in GIScity, study the programmer documentation for ArcTools. The methods are the same.

GIScity and ArcTools have different purposes. GIScity only includes the basic commands of ARC/INFO. ArcTools is more complex, because it is a complete graphical user interface to all the ARC/INFO commands. ArcTools is better suited for full access to ARC/INFO, and GIScity is designed as a learning GIS application.

Tailoring GIScity's Themes and Views

GIScity can be modified to run on your data sets instead of the ones supplied with this book. You would take these steps:

1. Go to the $GCHOME/data directory.
2. List the files that begin with THEME- and have an .AML file extension. Make copies of these files to correspond to themes you would like to model for your data sets.
3. Edit each of these AML files. You'll find they consist of just &SETVAR directives with a &RETURN directive at the end. Modify the values as you see appropriate and update the workstation value that is set.
4. Edit the file called DETAIL_VIEW.AML. Check the theme list in this AML for the themes you've defined for your data sets.

After you've completed these four steps, try starting GIScity again. If you have made no mistakes, your data set will appear in the GIScity applications.

Modifying GIScity

You are welcome to take the AMLs supplied with GIScity and modify them for your own use.

Here is a broad outline of steps you could take to add a menu in the GIScity applications:

1. Decide on a name for your menu and AML.

2. Move to the $GCHOME/display or $GCHOME/edit directories, depending on whether you want the menu in Display GIScity or Edit GIScity.
3. Make a copy of TEMPLATE.AML and TEMPLATE.MENU to the new object name. This AML and menu contain the basic parts common to all menus in the design of GIScity.
4. Go in your text editor and edit the new AML file. Change all occurrences of TEMPLATE to your new object name. Edit the new MENU file and change the occurrences of TEMPLATE to the new object name.

Troubleshooting GIScity

I would be happy to assist you with advice on installing, troubleshooting, and modifying GIScity. I would especially appreciate reports on software errors in GIScity so that they can be corrected, and your comments for improvement of the content and organization of this book. To contact me, send electronic mail care of readers@ipg@hmp.com.

Index

D

E

Q

R

S

More OnWord Press Titles

Computing/Business

Lotus Notes for Web Workgroups
$34.95

Mapping with Microsoft Office
$29.95

Geographic Information Systems (GIS)

INSIDE ARC/INFO, Rev. Ed.
Book $59.95 Includes CD-ROM

ARC/INFO Quick Reference
Book $24.95

INSIDE ArcView
Book $39.95 Includes CD-ROM

ArcView Developer's Guide
Book $49.95

ArcView/Avenue Programmer's Reference
Book $49.95

101 ArcView/Avenue Scripts: The Disk
Disk $101.00

ArcView Exercise Book
Book $49.95 Includes CD-ROM

INSIDE ArcCAD
Book $39.95 Includes Disk

The GIS Book, 3E
Book $34.95

INSIDE MapInfo Professional
Book $49.95

Raster Imagery in Geographic Information Systems
Book $59.95

GIS: A Visual Approach
Book $39.95

Pro/ENGINEER and Pro/JR. Books

INSIDE Pro/ENGINEER, 2E
Book $49.95 Includes Disk

Pro/ENGINEER Quick Reference, 2E
Book $24.95

Pro/ENGINEER Exercise Book, 2E
Book $39.95 Includes Disk

Thinking Pro/ENGINEER
Book $49.95

Pro/ENGINEER Tips and Techniques
Book $59.95

INSIDE Pro/JR.
Book $49.95

Automating Design in Pro/ENGINEER with Pro/PROGRAM
Book $59.95 Includes CD-ROM

MicroStation Books

INSIDE MicroStation 5X, 3E
Book $34.95 Includes Disk

INSIDE MicroStation 95, 4E
Book $39.95 Includes Disk

MicroStation Reference Guide 5.X
Book $18.95

MicroStation 95 Quick Reference
Book $24.95

MicroStation 95 Productivity Book
Book $49.95

MicroStation Exercise Book 5.X
Book $34.95 Includes Disk
Optional Instructor's Guide $14.95

MicroStation 95 Exercise Book
Book $39.95 Includes Disk
Optional Instructor's Guide $14.95

MicroStation for AutoCAD Users, 2E
Book $34.95

Adventures in MicroStation 3D
Book $49.95 Includes CD-ROM

Build Cell for 5.X
Software $69.95

101 MDL Commands (5.X and 95)
Optional Executable Disk $101.00
Optional Source Disks (6) $259.95

Softdesk Books

Softdesk Architecture 1 Certified Courseware
Book $34.95 Includes CD-ROM

Softdesk Architecture 2 Certified Courseware
Book $34.95 Includes CD-ROM

Softdesk Civil 1 Certified Courseware
Book $34.95 Includes CD-ROM

Softdesk Civil 2 Certified Courseware
Book $34.95 Includes CD-ROM

INSIDE Softdesk Architectural
Book $49.95 Includes Disk

INSIDE Softdesk Civil
Book $49.95 Includes Disk

Other CAD

Manager's Guide to Computer-Aided Engineering
Book $49.95

Fallingwater in 3D Studio: A Case Study and Tutorial
Book $39.95 Includes Disk

Windows NT

Windows NT for the Technical Professional
Book $39.95

SunSoft Solaris Series

SunSoft Solaris 2. User's Guide*
Book $29.95 Includes Disk

SunSoft Solaris 2. for Managers and Administrators*
Book $34.95

SunSoft Solaris 2. Quick Reference*
Book $18.95

*Five Steps to SunSoft Solaris 2.**
Book $24.95 Includes Disk

SunSoft Solaris 2. for Windows Users*
Book $24.95

The Hewlett Packard HP-UX Series

HP-UX User's Guide
Book $29.95 Includes Disk

HP-UX Quick Reference
Book $18.95

Five Steps to HP-UX
Book $24.95 Includes Disk

Interleaf Books

INSIDE Interleaf (v. 6)
Book $49.95 Includes Disk

Adventurer's Guide to Interleaf Lisp
Book $49.95 Includes Disk

Interleaf Exercise Book
Book $39.95 Includes Disk

Interleaf Quick Reference (v. 6)
Book $24.95

Interleaf Tips and Tricks
Book $49.95 Includes Disk

OnWord Press Distribution

End Users/User Groups/Corporate Sales

OnWord Press books are available worldwide to end users, user groups, and corporate accounts from your local bookseller or computer/software dealer, or from Softstore/CADNEWS Bookstore: call 1-800-CADNEWS (1-800-223-6397) or 505-474-5120; fax 505-474-5020; write to CADNEWS Bookstore, 2530 Camino Entrada, Santa Fe, NM 87505-4835, or e-mail ORDERS@HMP.COM. CADNEWS Bookstore is a division of SoftStore, Inc., a High Mountain Press Company.

Wholesale, Including Overseas Distribution

High Mountain Press distributes OnWord Press books internationally. For terms call 1-800-4-ONWORD (1-800-466-9673) or 505-474-5130; fax to 505-474-5030; e-mail ORDERS@HMP.COM, or write to High Mountain Press, 2530 Camino Entrada, Santa Fe, NM 87505-4835, USA. Outside North America, call 505-474-5130.

On the Internet: http://www.hmp.com

OnWord Press 2530 Camino Entrada, Santa Fe, NM 87505-4835 USA

Please Register Your *INSIDE ARC/INFO* CD for Updates and News

Do you use:

- ❑ ARC/INFO
- ❑ ArcView
- ❑ ArcCAD
- ❑ Other__________

Number of seats in your organization ____________________

Primary hardware platform ____________________

Name ____________________

Company ____________________

Address ____________________

City, State, Zip ____________________

Phone ____________ FAX ____________

E-Mail ____________________

ONWORD® PRESS

Complete this registration and mail today!

RSES9412-1

First-Class
Postage
Required

Post Office will
not deliver
without proper
postage.

HIGH MOUNTAIN PRESS
2530 CAMINO ENTRADA
SANTA FE NM 87505-4835
USA